Springer Undergraduate Mathematics Series

More information about this series at http://www.springer.com/series/3423

Michael Field

Essential Real Analysis

 Springer

Michael Field
Engineering Mathematics Department
Merchant Venturers School of Engineering
Bristol University
UK

Department of Mathematics
Rice University
Houston, Texas
USA

ISSN 1615-2085   ISSN 2197-4144 (electronic)
Springer Undergraduate Mathematics Series
ISBN 978-3-319-67545-9   ISBN 978-3-319-67546-6 (eBook)
https://doi.org/10.1007/978-3-319-67546-6

Library of Congress Control Number: 2017955015

Mathematics Subject Classification: 26-01, 40-01, 26Axx, 26Bxx (especially 26B05, 26B10), 26Exx
(especially 26E05, 26E10), 33Bxx (especially 33B15), 34A12, 40Axx, 42Axx (especially 42A10),
54Exx (especially 54-01, 54E35)

Printed on acid-free paper

This Springer imprint is published by Springer Nature
The registered company is Springer International Publishing AG
The registered company address is: Gewerbestrasse 11, 6330 Cham, Switzerland

Preface

This book is an introduction to *real analysis*: the foundations, the nature of the subject, and some of the big results. It is based on long personal experience of teaching undergraduate and graduate level analysis to a diverse range of classes in England (Warwick), Australia (Sydney) and the United States (Houston) and is intended to appeal to mathematicians with either a pure or applied focus.

Topics in the book are drawn from seventeenth century to late twentieth century analysis. While the techniques of analysis naturally form an important part of the book, the emphasis is on presenting a broad spectrum of some of the powerful and beautiful results that can be proved using analytic methods. Many of the results are important in applications—for example, Fourier series and asymptotics—and should help provide a sound foundation for work in Applied Mathematics.

At the conclusion of the preface, we give a detailed description of the contents of the book, together with advice for the reader and comments about what is not included and why. For now, we mention a few of the highlights. We include a wide range of results on Fourier series. Applications include the infinite product formula for $\sin(x)$, used in the analysis of the Gamma function, and results on Bernoulli polynomials leading to explicit formulas for the sum of a number of well-known infinite series. We give an introduction to the theory of smooth (infinitely differentiable, non-analytic) functions with an emphasis on how to *construct* a smooth function with specific properties. There is an extensive development of the theory of metric spaces (including applications of both the Arzelà–Ascoli and contraction mapping theorems to differential equations). This is followed by a short chapter on the Hausdorff metric, which includes a pretty application of the contraction mapping lemma to fractal geometry and iterated function systems. The final chapter contains a systematic development of the differential calculus on finite-dimensional vector spaces and includes results ranging from the equivalence of norms on a finite-dimensional vector space, through the implicit function and rank theorems to a proof of the strong version of the existence and regularity theorem for differential equations. The guiding principle throughout is that the generality of concepts and language introduced in the text has to be justified by the application. We have more to say about this below but, as an example, we emphasize sequential compactness rather than compactness (open cover definition) simply because it is difficult to find good applications at this level that need the more abstract definition of compactness. In this book, abstraction and generality has to be justified by application and/or transparency. We abstract as much as we can but no more than we need to.

In the next few paragraphs, I sketch some of the reasons and concerns that led to the writing of the book. I hope these comments will help both student and instructor to make the best use of the text.

Analysis is a beautiful, powerful and central part of mathematics. Yet I feel that the way the subject is sometimes presented in undergraduate courses can be deceptive, misleading and uninspiring.

Many of the insights, ideas and practices of mathematics are coded into the language and notation of mathematics. As Alexandre Borovik remarks in his illuminating book: *Mathematics under the Microscope. Notes on Cognitive Aspects of Mathematical Practice*

> 'Mathematical languages unstoppably develop towards an ever increasing degree of compression of information' [4, page 68].

How does one learn a foreign language? It depends on the context. If conversational skills are what is required, then the best way to proceed is to be immersed in the culture; to be surrounded by people who are native speakers. On the other hand, if the goal is to read classic Roman or Greek literature, then it is a painstaking process of learning the language and grammar step-by-step with the eventual aim of reading Virgil or Sophocles in the original. Immersion in the culture can be effective and pleasurable whereas learning from the book can be a slow and painful process (my own experience translating Caesar's *Gallic War III* from the Latin).

Spoken languages evolve relatively slowly; even though the seventeenth century plays of Shakespeare can be difficult to *read*, a production of a Shakespeare is usually easy to understand—though some of the puns and word plays can be missed. Learning and reading mathematics is a trickier proposition.

In contrast to spoken language, mathematical languages can evolve and change rapidly and, as Borovik notes, contain in compressed form much information about the ideas and practice. The modern abstract language of analysis would be hard for a seventeenth or eighteenth century mathematician to grasp. They might reasonably well ask, 'why on earth are you going to so much trouble?' It is not just the compressed way definitions are handled; there is all the infrastructure of logic and set theory that we now take for granted but which was originally regarded as controversial, even at the beginning of the twentieth century. How then does language mesh with mathematical content in contemporary undergraduate classes in calculus and analysis? In substance, as opposed to language, much contemporary undergraduate analysis barely goes beyond eighteenth century mathematics. For example, the one-variable part of a calculus sequence is largely stuck in the seventeenth century and multivariable calculus rarely gets far into the nineteenth century. The mathematical language used in these courses can often be a fractured mix of the old and new which does not resonate well with the content. It may be helpful to discuss one example in detail.

The ε, δ-definition of continuity. This is invariably given in calculus texts, usually as part of a discussion on limits. Subsequently, the topic is largely ignored—with proofs or arguments deferred as being too difficult. Versions of the ε, δ-definition of continuity were originally proposed by Bernard Bolzano (1817) and Augustin-Louis Cauchy (about 1820) and then put on a more sound and, to our eyes, familiar footing by Karl Weierstrass about 20 years later. Bolzano and Cauchy were among the first mathematicians to stress the importance of rigour in analysis. Even so, Cauchy's original use of the continuity definition was incorrect: he argued that a pointwise limit of a sequence of continuous functions was continuous.[1] Although questions about the meaning and nature of limits, especially associated to tangent lines, have played a role in analysis since the time of Leibniz and Newton, the idea of a *continuous function* is relatively recent, spurred in part by the early nineteenth century development of Fourier series. Continuity plays a peripheral role in most contemporary calculus classes and even in introductory analysis classes it is often poorly motivated—it is quite possible to complete a mathematics degree and have little, if any, contact with Fourier series or functions outside of combinations of polynomials, trigonometric and exponential functions.

The ε, δ-definition of continuity *is* difficult: a continuous function is not quite what you think it is—a typical continuous function is nowhere differentiable. Like in much of analysis, the definition is framed in terms of inequality (tricky) rather than the easier equality seen in elementary algebra. Frankly, the continuity definition is ugly and uses too many quantifiers (for all x, for all $\varepsilon > 0$, there exists...). In the twentieth century, a more natural topological definition of continuity appeared which is based on preservation of structure (open sets). The definition is far reaching and applies to many areas of mathematics, including algebra, but unfortunately the definition is even more remote from the content of a standard calculus sequence.

Can continuity be motivated in a calculus or introductory analysis course? In particular, are there significant results or examples? One possible approach is to prove, or at least discuss, the uniqueness and existence of the integral of a continuous function.

In a calculus class, the word 'integral' is, *in practice*, synonymous with 'anti-derivative'. It is easy to show rigorously that if the integral of a continuous bounded function exists, then it must be given by the anti-derivative (this is uniqueness of the integral). Conversely, the proof of existence of the anti-derivative for a continuous bounded function is not hard (the proof does *not* require uniform continuity). While existence does depend on completeness properties of the real numbers (as is so for most results in real analysis), it does not depend on general properties of continuous functions (for example, that a continuous function on a closed interval attains its bounds). In spite of the simplicity of proving the existence of the integral of a

---

[1]The correct result was obtained later by Weierstrass using the idea of uniform convergence. This, and all the other results and definitions we mention, can be found in the main body of the text.

continuous bounded function, calculus texts persist in giving a development based on upper and lower Riemann sums (so presenting the theory as an extension of the method of exhaustion developed by Eudoxus and later Archimedes) and almost universally claim the proof of existence is too difficult to include. It is not. (We refer to Chap. 2, Sect. 2.8 for a simple presentation of the existence and uniqueness of the integral of a continuous bounded function.) The whole point about the integral of a continuous bounded function is that one does not need to do approximation by Riemann sums if the function has an anti-derivative. This is part of the magic of the integral and differential calculus.[2]

In practice, calculus and introductory analysis courses only consider functions constructed from polynomials, trigonometric, exponential and logarithmic functions. For these functions, it is usually easy to prove boundedness on a closed and bounded interval. The differential calculus can then be used to find the bounds and where they are attained. There is no real connection between the abstract theory of continuous functions on a closed and bounded interval and the functions considered in a calculus class which are invariably analytic (in particular, infinitely differentiable). What is interesting and remarkable is the existence theorem that every continuous function, such as $\cos(x^2)$, does have an anti-derivative even though it cannot always be given in closed form (in terms of combinations of known functions). These issues are often not discussed in modern textbooks. In summary, highly sophisticated definitions and language are introduced for the solution of problems that are never mentioned. It is not surprising that students can be baffled by calculus classes: there is a disconnect between the material and the language.

These problems can be reinforced when students take an introductory class on analysis. These courses are often presented[3] as 'an introduction to proof' or 'calculus done properly'. Aside from the poor psychology ('what you did before was wrong and so a waste of time'), the premise is wrong: this abstraction is not needed to do calculus or much of classical analysis (as is testified to by the work of Euler and other eighteenth century mathematicians).

Amongst mathematicians there are strongly held views about teaching students how to write proofs. My personal view is that it is inadvisable to overemphasize the nuances of logic, truth tables, existential quantifiers etc. Often this leads to overuse of symbols combined with a lack of understanding of the underlying mathematics. Symbols seem powerful. Even though (perhaps because) I was brought up on Axiomatic Euclidean Geometry (the mathematical equivalent of learning Latin), I am sceptical about doing serious mathematics in the axiomatic-didactic style. It confuses the process of *doing* mathematics with the activity of *writing* mathematics. Figuring out a proof is an intuitive process that results from increasing understanding of the mathematical structure; formalism comes in when

---

[2]The interest of the Riemann sum approach to the integral is that the construction works for bounded functions which have countably many discontinuities. In that context, upper and lower sums *are* needed. This was highly relevant for nineteenth century mathematics.

[3]In the United States.

one writes the proof down (in a terse coded form) so that others can understand and use the result. Introductory analysis presents rich problems that can naturally lead to the logical style one needs for writing proofs (for example, the proof that a sequentially continuous function is continuous). An essential part of this process is developing the skill to construct good examples and counterexamples. In my view, the way to learn logic and mathematical expression is through application not through a formal course. Emphasis solely on correctness is antithetical to developing the intuitive understanding that one needs to do mathematics.

There can also be disconnects between language and application in more advanced undergraduate classes on analysis. Metric spaces are a terrific subject to learn at any level. There is geometric intuition combined with many powerful results (notably the contraction mapping lemma) and good applications. The subject also provides a beautiful and easily accessible abstraction, generalization and clarification of much foundational one variable real analysis. Definitions and results can be given using elementary sequence-based definitions such as sequential continuity or sequential compactness. These definitions lead to simple and transparent proofs. In contrast, if one emphasizes the topological approach, for example the open cover definition of compactness, proofs are not so transparent and often harder.[4] Most function spaces encountered in elementary analysis are separable metric spaces where sequence-based methods work very well. At undergraduate level, it is not so easy to give interesting examples of non-separable or non-metrizable spaces.[5] Quoting Borovik again

> 'Always test a mathematical theory on the simplest possible example—and explore the example to the utmost limits' [4, Page 3].

As the late Christopher Zeeman said, 'a good example is worth 10 theorems'. Counterexamples are important too—as a way of testing limits of the theory (as well as the need for the theory).[6] In summary, and paraphrasing Albert Einstein, 'Everything should be made as simple as possible, but no simpler.' If there is a good reason to consider the topology of non-separable or non-metrizable metric spaces in the course, then give the topological definition of compactness. Else, keep it simple.

One reason given to stress form over application in a final year analysis course is that the course should be preparation for graduate classes in analysis. I feel this approach is mistaken. As a professional mathematician, I should be inspiring undergraduate students about the nature and power of mathematics and not covering the preliminaries for a possible future graduate class. *A final year undergraduate course should be complete in itself and contain interesting and exciting applications.* If this is not done, it is like learning Latin to read Virgil but never actually reading

---

[4] A simple example is given by the proof of uniform continuity of a continuous function on a closed and bounded interval.

[5] Spaces of functions of bounded variation are metrizable but not separable. Spaces of smooth functions on $\mathbb{R}$ with the Whitney C^∞ topology are neither separable nor metrizable.

[6] Examples showing that results on boundedness of continuous functions on closed intervals fail if we work over the rational numbers.

any Virgil ('we never had the time to reach that part of the course'). Pure form, no content. This can be a problem with introducing the Lebesgue integral at the end of a mathematics degree without giving any applications in probability, ergodic theory or Fourier analysis.

On occasions I advise students in my analysis classes not to spend too much time reading mathematics texts. That view is based on my own experience—an effective way to learn mathematics is to do it, play with it but generally avoid spending too much time reading books about it. Reading a mathematics book can give a veneer of superficial understanding that dissolves the moment one tries to use the theory described in the book. An analogy might be learning carpentry, plumbing or a foreign language—knowing the theory is important but not that helpful; knowing how to use the tools is crucial. That takes time, practice and serious effort. As an example, think about hiring a personal trainer at the gym. You pay him or her rather a lot of money and sit back two or three times a week and watch them exercise, lift weights and generally work out and suffer. As a result you lose weight and gain a svelte figure.... It is the same with mathematics and learning mathematics. Much more is required than finding the ultimate book (or teacher).

So how does one approach a book on mathematics? Certainly not like a novel, to be read breathlessly from cover to cover. Although there are classics of mathematics literature, ranging from Euclid's *Elements* and Newton's *Principia* to the collected works of Euler or Poincaré, rather few mathematicians have read these works cover to cover. Dipping into these books is another matter. So perhaps one should regard a mathematics text as similar to a computer or software manual? Not quite. A good software manual should explain how to do standard tasks and have lots of good examples (they often do not). Although all this is required of a serious mathematics text, more is needed: why do we need this hypothesis, can we relax this condition, why do we have to go to all this trouble to prove this result? Not just operational skill but understanding and insight is required. The language and theory also need to be motivated throughout by good and significant applications.

It is time to say a few words about the book, the contents and how to proceed.

Chapters 1 and 2 play multiple roles. There is a review of basic set theory (Chap. 1) and the introduction of terminology and notation used throughout the text. The main item in Chap. 1 is an elementary rigorous construction of the real numbers using decimal expansions. This material should make for good classroom or small group discussion. The approach is old and originally due to the Flemish (Dutch) mathematician Simon Stevin in the sixteenth century. It predates the more abstract nineteenth century approach to real numbers developed by Weierstrass, Dedekind and others. It has the merit of a direct practical construction, done in the familiar context of decimal expansions, and the approach fits naturally with the methods used in Chap. 2 (for example, in the proof of existence of least upper bounds). Although Stevin's approach is currently unfashionable (or unknown), it does in my opinion have one *outstanding* merit over the more abstract approach—every irrational has a natural sequence of rational approximations given by decimal truncations.[7]

Chapter 2 reviews completeness properties of the real line and basic analysis of continuous functions. Key results, such as the Bolzano–Weierstrass theorem, are proved using natural constructions based on the representation of real numbers as decimal expansions. We include an appendix giving a simple presentation of the existence and uniqueness of the integral for continuous functions. Discussion topics for extension and exploration of this approach to multiple integrals are given in the exercises. There is also an appendix on the more abstract approach to the construction of the real numbers based on Cauchy sequences. There are also review sections in Chap. 2 on complex numbers, a little calculus, and the log and exponential functions. Most readers will know this material already—indeed much of what is in Chap. 2—but my guess is that everyone at some point will get a queasy feeling that there is a detail they need to check and so the details are provided. Note that in both Chaps. 1 and 2, the definition and elementary properties of limits are assumed known—we do not replicate uninteresting proofs about sums, products and quotients of limits often given in elementary calculus texts. One exception is that we do indicate a careful proof of convergence of geometric series over the *rational* numbers (Lemma 1.5.9).

Chapters 3 and 4 are about infinite series, infinite products and uniform convergence. In Chap. 3, we consider infinite series and infinite products of real and complex numbers (mostly real rather than complex). With a view to later applications to Fourier series and power series, we include Dirichlet and Abel's tests. We also give the statement and proof of Tannery's theorem—used in several applications, notably the first proof of the infinite product formula for $\sin(x)$. In Chap. 4 we investigate characteristic problems in analysis involving interchange of limit operations in infinite sums (and products) of functions. A highlight of Chap. 4 is the construction of a continuous nowhere differentiable function.

In Chap. 5 we get to the heart of our subject: functions. Using Bernstein polynomials, we give a constructive proof of the Weierstrass approximation theorem: every continuous function on a closed and bounded interval can be uniformly approximated by polynomials. After giving applications of the Weierstrass theorem we turn to smooth (infinitely differentiable) and real analytic functions. We start by emphasizing examples and give elementary methods for the construction of smooth (non-analytic) functions with specified properties. In so doing, we make our first brief contact with twentieth century mathematics. We also develop a little of the theory of real analytic functions, including results on analytic differential equations (the methods we give apply equally to complex analytic functions). In the remainder of Chap. 5 we develop the foundational theory of Fourier series. The main result we prove is that the Fourier series of a continuous piecewise C^1 periodic function converges uniformly to the function. Using Fourier series we give a second proof of

---

[7]That merit leads to the natural question of 'best possible' rational approximations—Diophantine approximation.

the infinite product formula for $\sin(x)$. We use Fourier series methods to compute the sums of several infinite series.

In Chap. 6 we discuss two topics from eighteenth century analysis. We start with the Gamma-function and verify most of the standard properties. Along the way we introduce important techniques from analysis such as differentiation under the integral sign. In the remainder of the chapter we discuss Bernoulli polynomials and the Euler–Maclaurin formula. We use quite elementary mathematics to obtain remarkably powerful results. For example, we use the Euler–Maclaurin formula to give sharp estimates on the sums of several standard infinite series and also prove versions of Stirling's formula estimating $n!$.

In Chap. 7 we give an introduction to metric spaces. This is a chapter about constructing the infrastructure needed for doing analysis on spaces more general than domains in Euclidean space. We emphasize the metric structure and geometric intuition. For example, a proper subset U of the metric space (X, d) is defined to be open if $d(x, X \smallsetminus U) > 0$ for all $x \in U$ (the alternative is to use an ε, δ definition in terms of balls or disks). Major results proved in this chapter include the Arzelà–Ascoli theorem (the Bolzano–Weierstrass theorem for spaces of continuous functions, uniform metric) and the contraction mapping lemma. We conclude the chapter with some simple yet powerful applications of the contraction mapping lemma to differential equations and the inverse function theorem (this is developed further in Chap. 9).

In Chap. 8 we give a non-trivial application of the contraction mapping lemma to the theory of iterated function systems. The results in this chapter give a beautiful illustration of the power of the abstract methods developed in the chapter on metric spaces. We show how to construct a complete metric on the (non-linear) space $\mathcal{H}(\mathbb{R}^n)$ of compact subsets of $\mathbb{R}^n$. We show that an iterated function system on $\mathbb{R}^n$ naturally defines a contraction operator on $\mathcal{H}(\mathbb{R}^n)$ and thereby deduce that there is a unique fractal defined by the iterated function system. The result is not difficult: the problem lies in organizing the concepts and this is dealt with elegantly and efficiently when we use the language of metric spaces.

Finally, in Chap. 9, we give a systematic account of the modern theory of differential calculus on normed vector spaces. Apart from providing proofs and statements of many standard results, such as the mean value theorem and Taylor's theorem, there are versions of Leibniz's rule and Faà di Bruno's formula for the higher derivatives of a composite of vector-valued maps. We include applications of the contraction mapping lemma to several versions of the implicit function theorem, including the rank theorem. Also proved is the C^r existence theorem for ordinary differential equations—the proof, based on the equation of variations, uses the contraction mapping lemma and uniform approximation by smooth functions. This result is fundamental in the development of the modern theory of dynamical systems.

Although the book makes some use of complex numbers, we have not developed the techniques and results of complex analysis based on Cauchy's theorem and the Cauchy–Riemann equations. The main reason for this omission is the current practice of offering a first self-contained course on complex analysis, including Cauchy's theorem and applications, followed perhaps by a more advanced course including topics such as the Riemann mapping theorem or the Weierstrass and Mittag-Leffler theorems. On integration, we have included a simple exposition of the integral and indicated extensions to functions with countably many discontinuities in the exercises. We have also included, mainly in exercises, results on monotone functions and functions of bounded variation. We have not, however, developed the Riemann–Stieltjes integral—it seemed difficult to give good applications appropriate to the general style and content of the book (for example, applications in probability or Riesz's theorem on the dual space of $C^0([a, b])$, uniform norm). We do not develop the general theory of multiple integrals. Our feeling here is that the key result—the change of variables formula for multiple integrals—is hard to prove (correctly) using Riemann sums and is better done in the context of Lebesgue integration, a topic that lies outside the scope of this text. At a few points in the text we make use of elementary results on multiple integrals (with one exception, always on rectangular domains).

In Sydney, Australia, we gave a year long second year course in analysis approximately based on chapters two though six. In Houston, I have given senior level two semester courses that cover most of the topics from the first eight chapters and sometimes a little from Chap. 9, depending on the background and knowledge of the class (for an undergraduate class, one needs to be fairly selective in the choice of material from Chap. 9).

The exercises: there are approximately 570, which range in difficulty from routine practice to *serious* challenges. Some of the exercises are suitable for class or group discussion and projects.

Acknowledgements are due to Don Cartwright and John McMullen who collaborated with me in the design of a second year honours analysis course given at Sydney University from 1977 and which is the foundation for substantial parts of Chap. 3 through 6. Senior undergraduate and graduate students at the University of Houston have taken analysis courses based on material from all chapters of the book and I would like to record my appreciation for all the many helpful comments and good questions I received from those classes. Last, but not least, many thanks to Springer—most especially Anne-Kathrin Birchley-Brun, Remi Lodh and Angela Schulze-Thomim—the anonymous copy-editor, who did great work on the manuscript, and the production team at Spi for their fine work.

Houston, TX, USA   Michael Field

Contents

1 Sets, Functions and the Real Numbers 1
  1.1   Introduction 1
  1.2   Sets 2
  1.3   Functions 6
  1.4   Countable Sets 8
  1.5   The Real Numbers 13
  1.6   The Structure of the Real Numbers 19

**2 Basic Properties of Real Numbers, Sequences and Continuous
Functions** 31
  2.1   Introduction 31
  2.2   Sequences 32
  2.3   Bounded Subsets of $\mathbb{R}$ and the Supremum and Infimum 37
  2.4   The Bolzano–Weierstrass Theorem 48
  2.5   **lim sup** and **lim inf** 58
  2.6   Complex Numbers 63
  2.7   Appendix: Results from the Differential Calculus 67
  2.8   Appendix: The Riemann Integral 71
  2.9   Appendix: The Log and Exponential Functions 82
  2.10 Appendix: Construction of $\mathbb{R}$ Revisited 86

3 Infinite Series 91
  3.1   Introduction 91
  3.2   Generalities 91
  3.3   Series of Eventually Positive Terms 92
  3.4   General Principle of Convergence 100
  3.5   Absolute Convergence 100
  3.6   Conditionally Convergent Series 106
  3.7   Abel's and Dirichlet's Tests 109
  3.8   Double Series 112
  3.9   Infinite Products 116
  3.10 Appendix: Trigonometric Identities 125

4 Uniform Convergence 129
   4.1   Introduction 129
   4.2   Pointwise Convergence 130
   4.3   Uniform Convergence of Sequences 131
   4.4   Uniform Convergence of Infinite Series 138
   4.5   Power Series 142
   4.6   Abel and Dirichlet's Test for Uniform Convergence 147
   4.7   Integrating and Differentiating Term-by-Term 150
   4.8   A Continuous Nowhere Differentiable Function 155

5 Functions 161
   5.1   Introduction 161
   5.2   Smooth Functions 162
   5.3   The Weierstrass Approximation Theorem 170
   5.4   Analytic Functions 178
   5.5   Trigonometric and Fourier Series 186
   5.6   Mean Square Convergence 204
   5.7   Appendix: Second Weierstrass Approximation Theorem 209

6 Topics from Classical Analysis: The Gamma-Function and the
**   Euler–Maclaurin Formula** 211
   6.1   The Gamma-Function 211
   6.2   Bernoulli Numbers and Bernoulli Polynomials 223
   6.3   The Euler–Maclaurin Formula 231

7 Metric Spaces 245
   7.1   Basic Definitions and Examples 245
   7.2   Distance from a Subset 250
   7.3   Open and Closed Subsets of a Metric Space: Intuition 251
   7.4   Open and Closed Sets 252
   7.5   Interior and Closure 258
   7.6   Open and Closed Subsets of a Subspace 260
   7.7   Dense Subsets and the Boundary of a Set 261
   7.8   Neighbourhoods 263
   7.9   Summary and Discussion 264
   7.10   Sequences and Limit Points 266
   7.11   Continuous Functions 273
   7.12   Construction and Extension of Continuous Functions 276
   7.13   Sequential Compactness 281
   7.14   Compact Subsets of $\mathbb{R}$: The Middle Thirds Cantor Set 289
   7.15   Complete Metric Spaces 297
   7.16   Equicontinuity and the Arzelà–Ascoli Theorem 305
   7.17   The Contraction Mapping Lemma 312
   7.18   Connectedness 321

8 Fractals and Iterated Function Systems 329
   8.1   The Space $\mathcal{H}(\mathbb{R}^n)$ 330
   8.2   Iterated Function Systems 337
   8.3   Examples of Iterated Function Systems 339
   8.4   Concluding Remarks 344

9 Differential Calculus on $\mathbb{R}^m$ 349
   9.1   Normed Vector Spaces 349
   9.2   Linear Maps 353
   9.3   The Derivative 358
   9.4   Properties of the Derivative 362
   9.5   Maps to and from Products 372
   9.6   Inverse and Implicit Function Theorems 376
   9.7   Local Existence and Uniqueness Theorem for Ordinary
            Differential Equations 391
   9.8   Higher Derivatives as Approximations 395
   9.9   Multi-Linear Maps and Polynomials 396
   9.10 Higher-Order Derivatives 407
   9.11 Extension of Results from C^1 to C^r-Maps 412
   9.12 Taylor's Theorem 414
   9.13 The Leibniz Rule and Faà di Bruno's Formula 417
   9.14 Smooth Functions and Uniform Approximation 423
   9.15 The Local C^r Existence Theorem for ODEs 430
   9.16 Diffeomorphisms and Flows 435
   9.17 Concluding Comments 438
   9.18 Appendix: Finite-Dimensional Normed Vector Spaces 439

References 443

Index 445

Chapter 1
Sets, Functions and the Real Numbers

1.1 Introduction

We start by reviewing some of the basic definitions, notations and properties of sets and functions. Although much of this material should be familiar, notations vary and readers should at least skim through the sections on sets and functions so as to familiarize themselves with the notational conventions used throughout the book. The remainder of the chapter is devoted to a leisurely but careful discussion of one approach to defining the real number system. Roughly speaking, we think of a real number as *defined* by its decimal approximations. This will prove useful in Chap. 2 when we prove general results on convergence (when we do not know the limit). Overall, the section on real numbers is intended to motivate group discussion and investigation. (What are the problems? How might we solve them?) At the conclusion of Chap. 2, we return to the problem of the construction of the real numbers and give an elegant, though more abstract, construction.

We assume some familiarity with *proof by induction* and *recursive* or *inductive definitions*. We briefly recall the ideas; first, proof by induction. If for each natural number n, we are given a statement $S(n)$, then $S(n)$ will be true for all n if $S(1)$ is true and the truth of $S(n)$ implies the truth of $S(n+1)$ for all $n \geq 1$. For a recursive or inductive definition, we aim to give definitions or mathematical statements $S(n)$ for $n \geq 1$. We can do this if $S(1)$ is given, and $S(n + 1)$ is uniquely determined by $S(n)$ for all $n \geq 1$. We often use recursive definitions to define sequences. For example, $x_1 = 1$, $x_{n+1} = \frac{1}{2}(x_n + \frac{2}{x_n})$, $n \geq 1$. The rule used to define x_{n+1} in terms of x_n may involve logical statements and not be given in terms of a simple mathematical formula.

© Springer International Publishing AG 2017
M. Field, *Essential Real Analysis*, Springer Undergraduate Mathematics Series,
https://doi.org/10.1007/978-3-319-67546-6_1

1.2   Sets

Roughly speaking a set is a collection of 'objects'. Each object in the set is regarded as a *member* of the set. If we have a set X and x is an object, then we say x is a member of X if x is one of the objects comprising X. We write this symbolically as "$x \in X$".

Examples 1.2.1

(1) Let $X = \{1, 2, 3\}$ be the set with the members $1, 2, 3$. We have $1 \in X$, $2 \in X$, $3 \in X$, $4 \notin X$, where the last notation means that '4 is not a member of X'.
(2) Let $\mathbb{N} = \{1, 2, \cdots\}$ denote the set of strictly positive integers—the natural numbers. Note the use of the dots to signify that $\mathbb{N}$ consists of all positive integers. We have $10 \in \mathbb{N}$ but $-1, 0, \frac{1}{2} \notin \mathbb{N}$.
(3) Let $\mathbb{Z}$ denote the set of integers: $\mathbb{Z} = \{0, \pm 1, \pm 2, \cdots\}$.
(4) Let $\mathbb{Z}_+$ denote the set of non-negative integers: $\mathbb{Z}_+ = \{0, 1, 2, \cdots\}$.
(5) Let $\mathbb{Q} = \{\frac{r}{s} \mid r, s \in \mathbb{Z}, s \neq 0\}$ denote the set of all rational numbers. We usually assume $s > 0$ and $(r, s) = 1$ (the notation $(r, s) = 1$ signifies that r, s have no common factors).
(6) We let $\mathbb{R}$ denote the set of all real numbers. For the present, we will be imprecise about the exact nature of the members of $\mathbb{R}$. However, if $x \in \mathbb{Z}$ or $x \in \mathbb{Q}$, then $x \in \mathbb{R}$.
(7) Let $[0, 1] = \{x \in \mathbb{R} \mid 0 \le x \le 1\}$. We refer to $[0, 1]$ as the 'closed unit interval.' Observe the logical definition of $[0, 1]$: we impose a condition—$0 \le x \le 1$—on the set of real numbers. The logical condition follows the $\mid$ symbol. In words: $[0, 1]$ is the set of all real numbers x satisfying the condition $0 \le x \le 1$.   ♠

Let $\emptyset$ denote the *empty set*. The empty set is the set with no members.

Example 1.2.2   Consider the sets $\emptyset, \{\emptyset\}, \{\emptyset, \{\emptyset\}\}$. The second set $\{\emptyset\}$ is not empty: it has one member, the empty set $\emptyset$. Similarly, the third set has two members: $\emptyset$ and $\{\emptyset\}$.   ♠

1.2.1   Subsets

Let A, B be sets. We say that A is a *subset* of B, written symbolically as $A \subset B$, if every member of A is a member of B. In terms of the implication symbol $\Longrightarrow$, $A \subset B$ if

$$a \in A \Longrightarrow a \in B.$$

Note that in this definition we allow $A = B$. If we can find $a \in A$ such that $a \notin B$, then A is not a subset of B. We write this as $A \not\subset B$. A consequence of our definitions is that we regard the empty set as a subset of B since we cannot find any member of

$\emptyset$ which is not a member of B (and so $\emptyset \not\subset B$ is false). If $A \subset B$ but $A \neq B$ we write $A \subsetneq B$. If, in addition, $A \neq \emptyset$, we refer to A as a *proper* subset of B. We also allow the notation $A \supset B$ which means that B is a subset of A (or A is a *superset* of B).

Example 1.2.3

$$[0, 1] \subset \mathbb{R}, \quad [0, 1] \not\subset \mathbb{N}, \quad \mathbb{N} \subset \mathbb{Z}_+ \subset \mathbb{Z} \supset \mathbb{Q} \supset \mathbb{R}.$$

As we shall see, $\mathbb{Q}$ is a proper subset of $\mathbb{R}$. ♠

1.2.2 Operations on Sets

Unions and Intersections Let $A, B, \cdots$ be sets. The *union* $A \cup B$ of A and B is defined by

$$A \cup B = \{x \mid x \in A \text{ or } x \in B\}.$$

Observe that $A \subset A \cup B, B \subset A \cup B$ and $A \cup B = B \cup A$.
The *intersection* $A \cap B$ of A and B is defined by

$$A \cap B = \{x \mid x \in A \text{ and } x \in B\}.$$

We have $A \cap B \subset A, B \subset A \cup B$ and $A \cap B = B \cap A$.

Example 1.2.4 $A \cup A = A \cap A = A \cup \emptyset = A$ and $A \cap \emptyset = \emptyset$. ♠

Later we need to look at unions and intersections of families of sets. Suppose then that we are given a set $\mathcal{A} = \{A_i\}$ of sets indexed by a non-empty set I. That is, $\mathcal{A} = \{A_i \mid i \in I\}$ (or, in abbreviated form $\{A_i\}_{i \in I}$). If the index set $I = \mathbb{N}$, then we have a *sequence* of sets $A_1, A_2, \cdots$. We define unions and intersections for the family $\{A_i\}_{i \in I}$ by

$$\cup_{i \in I} A_i = \{x \mid \exists i \in I \text{ with } x \in A_i\},$$

$$\cap_{i \in I} A_i = \{x \mid x \in A_i \ \forall i \in I\}.$$

(Here we have made use of the shorthand symbols $\exists$ ('there exists') and $\forall$ ('for all').)

Examples 1.2.5

(1) $\cup_{n \in \mathbb{N}} \{n\} = \mathbb{N}, \cup_{n \in \mathbb{Z}_+} \{\pm n\} = \mathbb{Z}$.
(2) For $x \in \mathbb{R}$, define $\delta_x = \frac{1}{1+|x|}$, $A_x = [x - \delta_x, x + \delta_x]$. Then $\cup_{x \in \mathbb{R}} A_x = \mathbb{R}$, $\cap_{x \in \mathbb{R}} A_x = \emptyset$. If instead we take $\delta_x = |x|, A_x = [x - \delta_x, x + \delta_x]$, then $\cup_{x \in \mathbb{R}} A_x = \mathbb{R}$, $\cap_{x \in \mathbb{R}} A_x = \{0\}$. ♠

Complements Fix a non-empty set X. If A is a subset of X, we define the *complement* $X \smallsetminus A$ of A (in X) by

$$X \smallsetminus A = \{x \in X \mid x \notin A\}.$$

Remark 1.2.6 There are several other notations commonly in use for the complement $X \smallsetminus A$. For example, $X - A$, A', A^c and $\complement A$. An advantage of our notation is that it enables us to easily write sets built by iterated complementation—for example, $A \smallsetminus (B \smallsetminus C)$.   ✠

Example 1.2.7 We have $X \smallsetminus \emptyset = X$ and $X \smallsetminus X = \emptyset$.   ♠

Lemma 1.2.8 *For all subsets A of X we have*

$$X \smallsetminus (X \smallsetminus A) = A.$$

Proof If $x \in A$, then $x \notin X \smallsetminus A$. Hence x lies in the complement of $X \smallsetminus A$. That is, $x \in X \smallsetminus (X \smallsetminus A)$. We have shown that $A \subset X \smallsetminus (X \smallsetminus A)$. Reversing the argument shows that $X \smallsetminus (X \smallsetminus A) \subset A$. Hence $X \smallsetminus (X \smallsetminus A) = A$.   □

The next result will prove useful when we investigate open and closed sets of metric spaces (Chap. 7).

Proposition 1.2.9 *Let $\{A_i\}_{i \in I}$ be a family of subsets of X. We have*

(1) $X \smallsetminus \bigcap_{i \in I} A_i = \bigcup_{i \in I} (X \smallsetminus A_i)$.
(2) $X \smallsetminus \bigcup_{i \in I} A_i = \bigcap_{i \in I} (X \smallsetminus A_i)$.

Proof The proof is left to the exercises.   □

Example 1.2.10 We define a subset A of $\mathbb{R}$ to be of type **C** if it is either finite, or empty or equal to $\mathbb{R}$. A subset of $\mathbb{R}$ is of type **O** if it is the complement of a subset of type **C**. Since $\mathbb{R}$ and $\emptyset$ are of type **C** it follows (by taking complements) that $\mathbb{R}$ and $\emptyset$ are also are of type **O**. These are the only subsets of $\mathbb{R}$ that are of type **C** *and* type **O**. All other subsets of type **O** are the complement of a finite set and so must be infinite. It follows from Proposition 1.2.9 that the intersection (respectively, union) of any collection of sets of type **C** (respectively, type **O**) is a set of type **C** (respectively, type **O**). On the other hand, only *finite* unions (respectively, intersections) of type **C** (respectively, type **O**) will always be of type **C** (respectively, type **O**).   ♠

The Power Set Let X be a set. We define the *power set of X*, $P(X)$, to be the set of all subsets of X. That is, $P(X) = \{A \mid A \subset X\}$.

Examples 1.2.11

(1) $P(\emptyset) = \{\emptyset\} \neq \emptyset$.
(2) If $X = \{1, 2\}$, then $P(X) = \{\{1\}, \{2\}, \{1, 2\}, \emptyset\}$.   ♠

Note that $P(X)$ always contains $\emptyset$ and X. Hence, provided $X \neq \emptyset$, $P(X)$ must contain at least two members. It is easy to see that if X is finite and contains N members, then $P(X)$ contains exactly 2^N members.

Products of Sets Let X, Y be non-empty sets. We define the (Cartesian) product $X \times Y$ to be the set of ordered pairs of elements of X and Y. That is,

$$X \times Y = \{(x, y) \mid x \in X, \ y \in Y\}.$$

It is straightforward to extend this definition to finite products. For example, given sets $X_1, \cdots, X_N$, we define

$$\Pi_{i=1}^{N} X_i = X_1 \times \cdots \times X_N = \{(x_1, \cdots, x_N) \mid x_i \in X_i, 1 \leq i \leq N\}.$$

Remark 1.2.12 If either X or Y is empty, then $X \times Y = \emptyset$. If X and Y are non-empty, then $X \times Y \neq \emptyset$ since we can pick at least one element x_0 from X, and one element y_0 from Y. Hence, $(x_0, y_0) \in X \times Y$ and $X \times Y \neq \emptyset$. This argument becomes a little dangerous if we try to define the product $\Pi_{i \in I} X_i$ over an *arbitrary* indexing set I. Formally, we can define

$$\Pi_{i \in I} X_i = \{f : I \rightarrow \cup_{i \in I} X_i \mid f(i) \in X_i, \forall i \in I\},$$

where f is a function with domain I and range $\cup_{i \in I} X_i$ which satisfies $f(i) \in X_i$ for all $i \in I$ (see the next section for more on functions). However, with this definition of product, it is not clear that the product is non-empty since, without further assumptions, there seems no obvious way of constructing a function f satisfying $f(i) \in X_i$ for all $i \in I$ (this is not a problem if $I = \mathbb{N}$—use induction). In practice, it is usually *assumed* that $\Pi_{i \in I} X_i \neq \emptyset$ if $X_i \neq \emptyset$, for all $i \in I$, whatever the indexing set I. This assumption is known as the *Axiom of Choice*. The Axiom of Choice, and an equivalent statement called *Zorn's Lemma*, play an important role in many parts of mathematics; in particular, when we require statements that apply in great generality. The need for care was seen early on in the development of set theory because of the appearance of contradictions. Best known is Russell's paradox: if we let $\mathcal{X}$ denote the set of all sets and define $Z \subset \mathcal{X}$ by $Z = \{X \in \mathcal{X} \mid X \notin X\}$, then $Z \in Z$ iff (shorthand for "if and only if") $Z \notin Z$. Russell's and other paradoxes can be avoided by developing axiomatic versions of set theory. Most mathematicians now assume a version of Zermelo–Fraenkel axiomatic set theory (ZF). For more information, we refer the reader to one of the many books on the foundations and history of set theory (personal favourites are [12, 13]).   ✠

EXERCISES 1.2.13

(1) Prove that $A \cap (B \cup C) = (A \cap B) \cup (A \cap C)$ (Distributive law).
(2) True or false for all sets A, B, C? In each case, either prove the statement or provide a counterexample.

    (a) $A = B$ iff $A \subset B$ and $B \subset A$.
    (b) $A \cup B = (A \smallsetminus B) \cup (B \smallsetminus A) \cup (A \cap B)$.
    (c) $A \cup (B \cap C) = (A \cup B) \cap (A \cup C)$.
    (d) $A \subset B$ iff $x \notin B \implies x \notin A$.

(3) Let A, B be subsets of X. Prove that $A \smallsetminus B = (X \smallsetminus B) \cap A$. Deduce that if we use the notation A^c for $X \smallsetminus A$, then every expression involving $A \smallsetminus B$ can be written in terms of A and B^c. If A, B, C are subsets of X, find the simplest expression you can for $((A \smallsetminus B) \smallsetminus C) \cap (A \smallsetminus C)$ in terms of A, B, C, A^c, B^c, C^c.

(4) Complete the proof of Proposition 1.2.9. (To show that $X \smallsetminus \bigcap_{i \in I} X_i = \bigcup_{i \in I} (X \smallsetminus X_i)$), prove that the left-hand side is a subset of the right-hand side and conversely.)

(5) Let X be a non-empty set. Suppose that A, B are proper subsets of X ($A, B \neq \emptyset, X$). Show that we can generate at most four different subsets of X from A and B using the operations of intersection and union. What about if we allow complements? Consider the same question, but now suppose we are given three subsets A, B, C of X. (Hint: Use the result of Q2 and Proposition 1.2.9 for the extension to complements. We remark that the number of subsets we can generate by intersection and union from n subsets of X increases rapidly as a function of n and is closely related to the *Dedekind number* $M(n)$ of n.)

(6) Let A, B, C be subsets of X. Define the *symmetric difference* $A \triangle B$ by

$$A \triangle B = (A \smallsetminus B) \cup (B \smallsetminus A).$$

Complete the sentence '$x \in A \triangle B$ iff $x \in A$ and ... or ... and ... $\notin$ ...'. Prove

(a) $A \triangle B = \emptyset$ iff $A = B$.
(b) $A \smallsetminus B = A \cap (A \triangle B)$.
(c) $A \cap (B \triangle C) = (A \cap B) \triangle (A \cap C)$.
(d) $(A \triangle B) \triangle C = A \triangle (B \triangle C)$ (associativity of symmetric difference).
(e) $A \triangle B = (A \triangle C) \triangle (C \triangle B)$.
(f) Show that $A \triangle B = C \triangle D$ iff $A \triangle C = B \triangle D$.

(7) Let $\{A_i \mid i \in I\}$ and $\{B_j \mid j \in J\}$ be families of subsets of X. Prove that

$$\left(\bigcap_{i \in I} A_i\right) \cup \left(\bigcap_{j \in J} B_j\right) = \bigcap_{i \in I, j \in J} A_i \cup B_j.$$

(The indexing sets I, J are non-empty and may be infinite.)

1.3   Functions

Let X, Y be non-empty sets. A *function*, or *map*, f from X to Y assigns to each $x \in X$ a unique point $f(x)$ in Y. We denote this assignment symbolically by $f : X \to Y$ ("f maps X to Y"). We call $f(x)$ the *value* of f at x. Every function $f : X \to Y$ has a *graph* $\Gamma_f \subset X \times Y$ defined by

$$\Gamma_f = \{(x, f(x)) \mid x \in X\}.$$

Conversely, if $G \subset X \times Y$ has the property that for every $x \in X$, there exists a unique point $y \in Y$ such that $(x, y) \in G$, then G is the graph Γ_f of a function f, where the value $f(x)$ of f at x is y—the unique point in Y such that $(x, y) \in G$.

If $f : X \to Y$, then the *range* or *image* of f is the subset $f(X)$ of Y defined by

$$f(X) = \{f(x) \mid x \in X\}.$$

More generally, if $A \subset X$, then $f(A)$ is the subset of Y defined by $f(A) = \{f(a) \mid a \in A\}$.

If B is a subset of Y, the *inverse image* (by f) of B is the subset $f^{-1}(B)$ of X defined by

$$f^{-1}(B) = \{x \in X \mid f(x) \in B\}.$$

Example 1.3.1 If $B \subset Y \smallsetminus f(X)$, then $f^{-1}(B) = \emptyset$. Conversely, if $B \subset Y$ and $f^{-1}(B) = \emptyset$, then $B \cap f(X) = \emptyset$. ♠

If $f : X \to Y, g : Y \to Z$, then the *composite* gf of f and g is the map $gf : X \to Z$ defined by

$$(gf)(x) = g(f(x)), \quad (x \in X).$$

Remark 1.3.2 The composite gf of f and g is not the multiplicative product of f and g. Of course, if (say) $f, g : X \to \mathbb{R}$, we can form the multiplicative product $f \times g$, defined by $(f \times g)(x) = f(x)g(x)$. As it is natural to abbreviate $f \times g$ as fg (especially if $f, g : \mathbb{R} \to \mathbb{R}$), it is sometimes useful to use a notation like $g \circ f$ for composites so as to make it clear that we are not dealing with $f \times g$. ✳

Example 1.3.3 If $C \subset Z$, then $(gf)^{-1}(C) = f^{-1}(g^{-1}(C))$—note the reverse order. The proof is left to the exercises. ♠

Definition 1.3.4 Let $f : X \to Y$.

(1) f is *onto* (or *surjective*) if $f(X) = Y$.
(2) f is *1:1* (or *injective*) if $f(x) = f(x')$ iff $x = x'$.
(3) f is *1:1 onto* (or *bijective*) if f is 1:1 and onto.

If $f : X \to Y$ is a bijection, then we may define the *inverse map* $f^{-1} : Y \to X$ by defining the value of f^{-1} at $y \in Y$ to be the unique point $f^{-1}(y) \in X$ such that $f(f^{-1}(y)) = y$. Since f is onto, there always exists at least one point $x \in X$ such that $f(x) = y$. Since f is 1:1, the point x is unique. We define $f^{-1}(y) = x$.

Remark 1.3.5 The reader should be aware of the ambiguity caused by using the symbol f^{-1} for inverses of sets and inverse maps. In particular, if $f : X \to Y$, and $b \in Y$, then $f^{-1}(\{b\})$ is a *subset* of X. If f has an inverse, then $f^{-1}(b)$ is a *point* of X. In practice, one writes $f^{-1}(b)$ to cover *both* situations. That is, $f^{-1}(b) = \{x \in X \mid f(x) = b\}$. If f is a bijection, we identify the point $f^{-1}(b) \in X$ with the subset

$\{f^{-1}(b)\}$ of X. If f has inverse function f^{-1}, then the inverse image set $f^{-1}(B)$ is equal to the image of B by the map f^{-1}.   ✸

Example 1.3.6 $f : X \to Y$, $g : Y \to Z$ are bijections, then $gf : X \to Y$ is a bijection and $(gf)^{-1} = f^{-1}g^{-1}$.   ♠

EXERCISES 1.3.7

(1) Let $f : X \to Y$ and $\{U_i \mid i \in I\}$ be a family of subsets of X. Show that $f(\cup_{i \in I} U_i) = \cup_{i \in I} f(U_i)$. What about $f(\cap_{i \in I} U_i)$?

(2) Let $f : X \to Y$ and U, V be subsets of X. Show, by finding an explicit example, that in general, $f(U) \smallsetminus f(V) \neq f(U \smallsetminus V)$.

(3) Let $f : X \to Y$ and $\{U_i \mid i \in I\}$ be a family of subsets of Y. Prove that

  (a) $f^{-1}(\cap_{i \in I} U_i) = \cap_{i \in I} f^{-1}(U_i)$.
  (b) $f^{-1}(\cup_{i \in I} U_i) = \cup_{i \in I} f^{-1}(U_i)$.

(4) Show that if $f : X \to Y$, $g : Y \to Z$ and C is a subset of Z, then $(gf)^{-1}(C) = f^{-1}(g^{-1}(C))$.

(5) Define $f : \mathbb{R} \to \mathbb{R}$ by $f(x) = x^2 + 1$. Find

  (a) $f([0, 2])$.
  (b) $f^{-1}((-1, 1))$.
  (c) $f^{-1}([2, 3])$.

  If $g : \mathbb{R} \to \mathbb{R}$ is the map $g(x) = x^3$, find (a) $(gf)^{-1}([0, 2])$, (b) $gf([0, 2])$, (c) $(fg)^{-1}([0, 2])$, (d) $fg([0, 2])$.

1.4   Countable Sets

1.4.1   *Equivalence of Sets*

Definition 1.4.1 The sets X, Y are *equivalent* if there exists a bijection $f : X \to Y$. We write this symbolically as "$X \sim Y$".

Remark 1.4.2 The relation $\sim$ is *reflexive* $X \sim X$, *symmetric* $X \sim Y \Longrightarrow Y \sim X$, and *transitive* $X \sim Y$, $Y \sim Z \Longrightarrow X \sim Z$. A relation satisfying these properties is called an *equivalence relation*.   ✸

We give a general result on inequivalence that has an intriguing proof discovered by the creator of set theory, Georg Cantor. The proof is reminiscent of the argument used in Russell's paradox (see Remark 1.2.12).

Proposition 1.4.3 *Let X be a set. Then $X \not\sim P(X)$.*

The interest in this result lies in the case when X is not finite (see below). The result is trivially true if $X = \emptyset$ ($P(\emptyset) = \{\emptyset\} \neq \emptyset$) so we assume $X \neq \emptyset$.

Proof We prove that there is no surjection from X to $P(X)$. Our proof goes by contradiction. Suppose that $f : X \to P(X)$ is surjective. Since $f(x)$ is a subset of X for all $x \in X$, we may define the following subset B of X:

$$B = \{x \in X \mid x \notin f(x)\}.$$

Observe that the definition makes sense even if $f(x) = \emptyset$—a possibility since $\emptyset \in P(X)$. Since we assume f is onto, $\exists b \in X$ such that $f(b) = B$. There are exactly two possibilities: $b \in B$, or $b \notin B$. If $b \in B$ then, by definition of B, $b \notin f(b) = B$. Contradiction. Similarly, if $b \notin B = f(b)$, then $b \in B$, by definition of B. Contradiction. Either assumption leads to an absurd conclusion and therefore our original assumption that f is onto must be false. □

Remark 1.4.4 If X is *infinite*, then Proposition 1.4.3 implies that X is not equivalent to the infinite set $P(X)$. In particular, infinite sets need not be equivalent. When Cantor proved this (and more) in 1874, the result was highly controversial and was attacked by many mathematicians, philosophers and theologians. ✱

1.4.2 Finite and Countable Sets

For this section only we use the notation $\mathbf{n}$ for the set $\{1, \cdots, n\}$ of the first n natural numbers. If $m < n \in \mathbb{N}$, then $\mathbf{n} - \mathbf{m}$ will be the set $\{1, \cdots, n - m\}$.

Definition 1.4.5 A set X is *finite* if either $X = \emptyset$ or there exists an $n \in \mathbb{N}$ such that

$$X \sim \mathbf{n}.$$

A set is *infinite* if it is not finite.

We need to check that the n in our definition of finite is uniquely determined by X. For this we need a preliminary result.

Lemma 1.4.6 *Let* X, Y *be equivalent sets containing at least two members. If* $x_0 \in X$ *and* $y_0 \in Y$, *then* $X \smallsetminus \{x_0\} \sim Y \smallsetminus \{y_0\}$.

Proof Let $f : X \to Y$ be a bijection. If $f(x_0) = y_0$, then f restricts to a bijection $f : X \smallsetminus \{x_0\} \to Y \smallsetminus \{y_0\}$ and so $X \smallsetminus \{x_0\} \sim Y \smallsetminus \{y_0\}$. If $f(x_0) \neq y_0$, we may choose $z_0 \in Y$, $z_0 \neq y_0$ (since Y contains at least two members). Now define $g : Y \to Y$ by $g(y) = y$ if $y \notin \{y_0, z_0\}$, $g(z_0) = y_0$, $g(y_0) = z_0$. The composite $gf : X \to Y$ is a bijection and $(gf)(x_0) = y_0$. We proceed as before. □

Lemma 1.4.7 *Let* $n, m \in \mathbb{N}$. *Then* $\mathbf{n} \sim \mathbf{m}$ *iff* $n = m$.

Proof Obviously, if $n = m$, we have $\mathbf{n} \sim \mathbf{m}$. So suppose $\mathbf{n} \sim \mathbf{m}$. Since the result is trivial if $n = 1$ or $m = 1$, we may assume $n, m > 1$. Without loss of generality suppose $m \leq n$. Apply the previous lemma $m - 1$ times to get $\mathbf{1} \sim \mathbf{n} - \mathbf{m} + \mathbf{1}$. It follows that $1 = n - m + 1$. That is, $m = n$. □

As an immediate corollary of Lemma 1.4.7 we have

Corollary 1.4.8 *If X is finite and non-empty, then there exists a unique $n \in \mathbb{N}$ such that $X \sim \mathbf{n}$. The integer n is called the cardinality of X. (If $X = \emptyset$, we regard the cardinality of X as being zero.)*

Example 1.4.9 If the cardinality of X is n, then the cardinality of $P(X)$ is 2^n. Since $2^n > n$, for all $n \in \mathbb{N}$, we see that the cardinality of the power set of a finite set is always greater than the cardinality of the set (this holds true if $X = \emptyset$ as $P(\emptyset) = \{\emptyset\}$).   ♠

Definition 1.4.10 A set X is *countable* if either X is finite or $X \sim \mathbb{N}$. If $X \sim \mathbb{N}$, we sometimes say X is *countably infinite*. If X is not countable, we say X is *uncountable*.

Examples 1.4.11

(1) The set $\mathbb{Z}$ of all integers is countable. For this it is enough to note that the map $f : \mathbb{N} \to \mathbb{Z}$ defined by

$$f(n) = \begin{cases} n/2, & \text{if } n \text{ is even,} \\ -(n+1)/2, & \text{if } n \text{ is odd,} \end{cases}$$

is a bijection.

(2) The set of prime numbers is countably infinite (see the exercises at the end of the section).

(3) Not all sets are countable. For example, $P(\mathbb{N}) \not\sim \mathbb{N}$ (Proposition 1.4.3) and so, since $P(\mathbb{N})$ is infinite ($P(\mathbb{N}) \supset \{\{1\}, \{2\}, \cdots\}$), $P(\mathbb{N})$ cannot be countable.   ♠

Proposition 1.4.12 *A subset of a countable set is countable.*

Proof Let Y be a subset of the countable set X. The result is immediate if X is finite so we shall assume that X is countably infinite. We may write $X = \{x_1, x_2, \cdots\}$. More precisely, there exists a bijection $f : \mathbb{N} \to X$, and so we may define $x_n = f(n)$, $n \in \mathbb{N}$. Let $n_1 \geq 1$ be the smallest integer such that $x_{n_1} \in Y$. Assume we have defined $n_1 < n_2 < \cdots < n_k$ such that $x_{n_j} \in Y$, $1 \leq j \leq k$ and $Y \cap \{x_1, x_2, \cdots, x_{n_k}\} = \{x_{n_1}, \cdots, x_{n_k}\}$. Then either $Y \cap \{x_1, \cdots, x_{n_k}\} = Y$ and so Y is finite or there exists a smallest $n_{k+1} > n_k$ such that $x_{n_{k+1}} \in Y$. It follows that either Y is finite (the process terminates) or we may write $Y = \{x_{n_1}, x_{n_2}, \cdots\}$ and so Y is countable (a bijection $g : \mathbb{N} \to Y$ is defined by $g(k) = x_{n_k}$, $k \geq 1$).   □

Theorem 1.4.13 *Every infinite set X contains a countable subset.*

Proof The proof is similar to that of Proposition 1.4.12 and uses an inductive construction. We construct a 1:1 map $g : \mathbb{N} \to X$. Since $X \neq \emptyset$, we can pick an element $x_1 \in X$. After n steps, suppose we have picked n distinct elements $\{x_1, \cdots, x_n\}$ of X. Since X is not finite, $X \smallsetminus \{x_1, \cdots, x_n\} \neq \emptyset$ and so we can pick $x_{n+1} \in X \smallsetminus \{x_1, \cdots, x_n\}$. The construction defines a 1:1 map $g : \mathbb{N} \to X$ by $g(n) = x_n$, $n \in \mathbb{N}$.   □

Lemma 1.4.14 *Let $f : X \to Y$.*

(1) If f is injective and Y is countable, then X is countable.
(2) If f is surjective and X is countable, then Y is countable.

Proof We prove (1) and leave (2) to the exercises. Since $f(X) \subset Y$ and Y is countable, $f(X)$ is countable by Proposition 1.4.12. The result follows since $X \sim f(X)$ (since f is injective, f defines a bijection of X onto $f(X)$). $\qquad\qquad\square$

Theorem 1.4.15

(1) A finite product of countable sets is countable.
(2) A countable union of countable sets is countable.

Proof (1) We start by proving that $\mathbb{N}^m \sim \mathbb{N}$, for all $m \geq 2$. Our proof is by induction on m. Suppose $m = 2$. Since $\mathbb{N}^2 = \mathbb{N} \times \mathbb{N}$, we may write the points of $\mathbb{N}^2$ as the infinite array

$$
\begin{array}{llll}
(1,1) & (1,2) & (1,3) & \cdots \\
(2,1) & (2,2) & (2,3) & \cdots \\
(3,1) & (3,2) & (3,3) & \cdots \\
\cdots & \cdots & \cdots & \cdots
\end{array}
$$

We give an inductive definition of $F : \mathbb{N} \to \mathbb{N}^2$. We define $F(1) = (1,1)$. Suppose we have defined $F(1), \cdots, F(n-1)$, $n > 1$. We define $F(n)$. Suppose $F(n-1) = (i,j)$. If $j = 1$, we take $F(n) = (1, i+1)$, else we take $F(n) = (i+1, j-1)$. This defines a path through the array which follows the diagonals: $(1,1) \to (1,2) \to (2,1) \to (1,3) \to (2,2) \to \cdots$. The map F defines the required bijection between $\mathbb{N}$ and $\mathbb{N}^2$. Suppose that $m > 2$ and that we have shown $\mathbb{N}^{m-1} \sim \mathbb{N}$. We have

$$\mathbb{N}^m = \mathbb{N}^{m-1} \times \mathbb{N} \sim \mathbb{N} \times \mathbb{N} \sim \mathbb{N},$$

where the first equivalence follows by the inductive hypothesis and the second equivalence is the case $m = 2$. Now suppose $A_1, \cdots, A_m$ are countable sets. For each $j \in \mathbf{m}$, there exists an injection $\alpha_j : A_j \to \mathbb{N}$. Hence we may define the injection $(\alpha_1, \cdots, \alpha_m) : \Pi_{i=1}^m A_i \to \mathbb{N}^m$. Hence $\Pi_{i=1}^m A_i$ is countable by Lemma 1.4.14(1).

It remains to prove (2). Let $\{A_i\}_{i \in I}$ be a countable family of countable sets. We assume that I is infinite and take $I = \mathbb{N}$ (if $I \sim \mathbf{m}$, define $A_j = A_m, j > m$). Since A_j is countable, there exists a surjection $\beta_j : \mathbb{N} \to A_j$ (if A_j is finite with n elements $a_1, \cdots, a_n$, define $\beta_j(m) = a_n, m > n$). Define the surjection $\beta : \mathbb{N}^2 \to \cup_{n \in \mathbb{N}} A_n$ by $\beta(i,j) = \beta_j(i)$. The countability of $\cup_{n \in \mathbb{N}} A_n$ follows by Lemma 1.4.14(2). $\qquad\square$

Examples 1.4.16

(1) For $m \geq 1$, $\mathbb{N}^m$, $\mathbb{Z}^m$ are countable.
(2) The set $\mathbb{Q}$ of rational numbers is countable. Every element of $\mathbb{Q}$ can be represented uniquely in the form r/s, where $s > 0$ and $(r, s) = 1$ (we write $0 = \frac{0}{1}$). Hence we may represent $\mathbb{Q}$ as a subset of $\mathbb{Z}^2 \sim \mathbb{N}^2 \sim \mathbb{N}$. Apply Proposition 1.4.12.

(3) Let $A = \{a \in \mathbb{R} \mid \exists n, p_0 \neq 0, \cdots, p_n \in \mathbb{Z} \text{ with } p_0 a^n + \cdots p_{n-1} a + p_n = 0\}$. Then A—the set of algebraic numbers—is countable. This is a simple consequence of Theorem 1.4.15 and we leave the details to the reader as an exercise. In particular, $\{m^{1/n} \mid m, n \geq 1\} \subset A$ is countable.   ♠

EXERCISES 1.4.17

(1) Prove Lemma 1.4.14(2). (Hint: Show, using an inductive construction, that if $f : X \to Y$ is onto and X is countable, then there exists a subset X' of X such that f maps X' bijectively onto Y. This shows that there exists an injective map $g : Y \to X$—g is the inverse of $f : X' \to Y$.)

(2) Prove that the following sets are countably infinite.

   (a) The set of positive odd integers. (Construct a bijection between the set and $\mathbb{N}$.)

   (b) The set of prime numbers. (Euclid's argument. Suppose the contrary and let $p_1 < \cdots < p_N$ be the set of prime numbers. Derive a contradiction by showing that the prime factorization of $p_1 \cdots p_N + 1$ must have a prime factor bigger than p_N.)

   (c) The set of all real numbers which are roots of an equation of the form $p_0 x^n + \cdots + p_{n-1} x + p_n = 0$, where $n \geq 1$ and $p_0, \cdots, p_n \in \mathbb{Z}$.

   (d) The subset A of $\mathbb{R}$ defined by $a \in A$ iff there exist $n \in \mathbb{N}$ and $p_1, \cdots, p_{n-1} \in \mathbb{Q}$ such that
   $$a^n + p_1 a^{n-1} + \cdots + p_{n-1} a + \sqrt{2} = 0.$$

   (For (c,d) you need to verify that the set can be represented as an *infinite* subset of a (known) countable set.)

(3) Let $p \in \mathbb{N}$, $p > 1$. Define $D_p = \{x \in \mathbb{R} \mid x = \frac{m}{p^n}, \text{ where } m \in \mathbb{Z}, n \in \mathbb{Z}_+\}$. Show that D_p and $\mathbb{Q} \smallsetminus D_p$ are both countably infinite.

(4) Show that if X contains an uncountable subset, then X is uncountable.

(5) Show that if X is an infinite set, then we can find a proper subset Y of X ($Y \neq X$) such that $Y \sim X$. (Hint: use Theorem 1.4.13.)

(6) Let A_0, B_0 be non-empty sets and suppose that $f : A_0 \to B_0$ and $g : B_0 \to A_0$. For $n \geq 0$ define inductively $B_{j+1} = f(A_j)$, $A_{j+1} = g(B_j)$. Show that

   (a) $A_0 \supset A_1 \supset \cdots, B_0 \supset B_1 \supset \cdots$.

   (b) If $E = \cap_{n \geq 0} A_n$, $F = \cap_{n \geq 0} B_n$ and $p \geq 0$, then

   $$A_p = \cup_{n \geq p}(A_n \smallsetminus A_{n+1}) \cup E, \quad B_p = \cup_{n \geq p}(B_n \smallsetminus B_{n+1}) \cup F$$

   as unions of disjoint subsets.

(c) We may write A_0, A_1 as unions of disjoint subsets $A_0 = X \cup Y \cup E$, $A_1 = X \cup Y_1 \cup E$, where

$$X = \cup_{n \geq 1}(A_{2n-1} \smallsetminus A_{2n}),$$

$$Y = \cup_{n \geq 0}(A_{2n} \smallsetminus A_{2n+1}),$$

$$Y_1 = \cup_{n \geq 1}(A_{2n} \smallsetminus A_{2n+1}).$$

Similarly for B_0, B_1.

(d) If f, g are 1:1, then $gf : Y \to Y_1$ is a bijection.

Deduce that if f, g are 1:1, then $A_0 \sim A_1$ and so, since $A_1 \sim B_0$ (by g^{-1}), $A_0 \sim B_0$. (The Cantor–Bernstein theorem: if A is equivalent to a subset of B and B is equivalent to a subset of A, then A is equivalent to B.)

1.5 The Real Numbers

1.5.1 Not All Real Numbers Are Rational

The original formulation of geometry by the Pythagorean school was based on ideas of proportion and tacitly assumed that all numbers were rational.[1] An advantage of this approach was that numbers could, in theory, all be constructed geometrically using ruler and compass. It came as a shock to the Pythagorean school when it was discovered that some numbers that arose geometrically were not rational. The easiest example comes from Pythagoras's theorem: the hypotenuse of an isosceles right angled triangle with side length 1 is $\sqrt{2} \notin \mathbb{Q}$. In most cases, the square root of a positive integer is not rational. Indeed, the only time it is rational is when the integer is the square of another integer.

Proposition 1.5.1 *Let $n \in \mathbb{N}$. Then $\sqrt{n} \in \mathbb{Q}$ iff $\sqrt{n} \in \mathbb{N}$. That is, the set of natural numbers with rational square root is precisely $\{1^2 = 1, 2^2 = 4, 3^2 = 9, \cdots\}$.*

Proof We prove a special case of this result and leave the general case (and extensions) to the exercises. We show that if $p > 1$ is prime then $\sqrt{p} \notin \mathbb{Q}$. Our proof goes by contradiction. Suppose that $\sqrt{p} \in \mathbb{Q}$, then we may write $\sqrt{p} = \frac{r}{s}$, where $r, s \in \mathbb{N}$ and $(r, s) = 1$ (recall that $(r, s) = 1$ means no common factors— the unique factorization of an integer into a product of primes that allows for this representation is used again in the proof). Since we assume $\sqrt{p} = \frac{r}{s}$ we have, on squaring and multiplying by s^2,

$$ps^2 = r^2.$$

---

[1]Strictly speaking, strictly positive numbers; the concepts of negative and zero numbers were developed later in Indian and Arabian mathematics.

It follows that p is a factor of r^2 and so, since $p > 1$ is prime, p must be a factor of r (use the prime factorization of r). Hence we may write $r = pR$, where $R \in \mathbb{N}$. Substituting for r, we get $ps^2 = p^2R^2$ and so, after cancelling p,

$$s^2 = pR^2.$$

Just as before, it follows that p is a factor of s. But we have shown that p is a factor of both r and s. This contradicts our assumption that $(r, s) = 1$. Hence $\sqrt{p}$ cannot be rational.   □

Remark 1.5.2 As remarked above, the discovery that mathematics could not be done within the (countable) framework of rational numbers was of profound significance. It is no coincidence that numbers that are not rational are called *irrational* or that there is the word play between *surd* (root of number) and *absurd*. Irrational numbers cannot be expressed in finite terms—indeed, most irrational numbers correspond (in a sense that can be made very precise) to an infinite sequence of random numbers and so cannot be represented in any finite form. Allowing irrational numbers means the acceptance that randomness can and does play a pivotal role in mathematics—even in a precise and quantitative subject like real analysis.   ✖

EXERCISES 1.5.3

(1) Show that if $p_1, p_2, \ldots, p_n$ are distinct primes then $\sqrt{p_1 p_2 \cdots p_n}$ is irrational.
(2) Complete the proof of Proposition 1.5.1.
(3) Show that $\sqrt[n]{5}$ is irrational for all $n \in \mathbb{N}, n \geq 2$.
(4) Show that if p is prime $\sqrt[n]{p}$ is irrational for all $n \in \mathbb{N}, n \geq 2$.
(5) Show that if $n, m \geq 2$ then $\sqrt[n]{m}$ is rational iff there exists an $\ell \in \mathbb{N}$ such that $m = \ell^n$. (Hints: extend the method of (1) or show directly that if $\sqrt[n]{m} = p/q$, then p/q must be an integer—use prime factorization and raise to the nth power.)

1.5.2   Construction of the Real Numbers

We present an approach to the construction of the real numbers using decimal expansions and approximation. The methods we use originate from the work of the sixteenth century Dutch Mathematician Simon Stevin who developed the foundations of decimal arithmetic and real numbers in a 35-page booklet *De Thiende* ('The art of tenths') published in 1685 (a very readable historical survey on the influence of Stevin's work can be found in the article by Błaszczyk et al. [3, §2]).

Aside from familiarity, the main advantage of our approach is that it leads to an elementary constructive proof of the existence of the least upper bound or supremum (see Chap. 2) and that the methods we use lead naturally to the 'subdivide and conquer' techniques we repeatedly use in Chap. 2. On the other hand there are some technical difficulties to be overcome related to the non-uniqueness of the decimal expansions of some rational numbers (irrational numbers always have a unique decimal expansion). At the end of Chap. 2, we give a more abstract approach in terms of equivalence relations and Cauchy sequences of rational numbers. Even with the general approach there are still many details to be checked.

The reader should view the material in the remainder of the chapter as being for exploration and discussion: starting with the relatively simple idea of a real number as being a decimal expansion, how can one define familiar concepts like order, absolute value, addition and subtraction? How do we represent the rational numbers as a subset of the real numbers and what do we mean by the approximation of a real number by a rational number or truncated decimal? As we shall see, these questions lead naturally to the idea of 'limit'.

1.5.3 Decimal Expansions and Rational Numbers

For us a decimal expansion will be a formal expression

$$x = \pm x_0.x_1x_2\cdots x_n\cdots,$$

where $x_0 \in \mathbb{Z}_+$ (*not* $\mathbb{Z}$) and $x_n \in \{0,\cdots,9\}, n \in \mathbb{N}$.

We regard expansions prefixed by a $+$ as positive, and those prefixed by a $-$ as negative. If we drop the $\pm$ (we often do), we regard the expansion as unsigned (could be either prefixed by $+$ or $-$) or positive (implicit $+$).

A decimal expansion is an infinite string of integers together with a sign. The problem is to give the expansion a useful interpretation. Our first task will be to show that certain types of decimal expansions naturally define rational numbers and, conversely, that rational numbers have a special type of decimal expansion. Our goal is to identify the set of decimal expansions with the set of real numbers. Indeed, we will *define* the real numbers to be the set of all decimal expansions.

Let $N \in \mathbb{N}$. A decimal expansion $x = \pm x_0.x_1x_2\cdots$ is *terminating of length N* if $x_N \neq 0$ and $x_n = 0, n > N$. That is,

$$x = \pm x_0.x_1\cdots x_N\overline{0},$$

where $\overline{0}$ is shorthand for 0 repeated infinitely often. A decimal expansion is *terminating* if it is terminating of length N for some $N \in \mathbb{N}$.

In future we regard the finite expansion $x_0.x_1\cdots x_N$ as being identical to $x_0.x_1\cdots x_N\overline{0}$ and write $x_0.x_1\cdots x_N = x_0.x_1\cdots x_N\overline{0}$. Similarly for negative decimal expansions. We also take $+0.\overline{0}$ as identical to $-0.\overline{0}$ and set $\pm 0.\overline{0} = 0$ (see Example 1.5.8 below).

Definition 1.5.4 (Truncations) If $x = x_0.x_1x_2\cdots x_n\cdots$ is an (unsigned) decimal expansion and $N \in \mathbb{N}$, we set

$$x^N = x_0.x_1x_2\cdots x_N,$$

and call x^N the (decimal) *truncation of x to N-terms*.

Remark 1.5.5 We always use a *capital N superscript* to label the truncation of x to N-terms and generally reserve lower case subscripts to label general terms in the decimal expansion.   ✱

It is easy to identify terminating decimals with rational numbers.

Lemma 1.5.6 *If $x = \pm x_0.x_1x_2 \cdots x_n \cdots$ is a decimal expansion, then the truncation x^N of x to N-terms defines a unique rational number x^N according to the rule*

$$x^N = \pm \left(x_0 + \sum_{n=1}^{N} \frac{x_n}{10^n} \right).$$

Remark 1.5.7 If $x = -x_0.x_1x_2 \cdots x_n \cdots$, then $x^N = -(x_0 + \sum_{n=1}^{N} \frac{x_n}{10^n})$. If instead we had defined decimal expansions to be of the form $x_0.x_1x_2 \cdots$, where $x_0 \in \mathbb{Z}$ (rather than $\mathbb{Z}_+$), then the truncation would fail badly for negative decimal expansions. For example, the truncation x^1 of $x = -1.1$ would give the rational number $x_0 + \frac{x_1}{10} = -1 + \frac{1}{10} = -9/10$.   ✱

Example 1.5.8 For all $N \in \mathbb{N}$, $-0.\bar{0}^N = 0.\bar{0}^N = 0$. This justifies the identification of $0.\bar{0}$ and $-0.\bar{0}$.   ♠

We need an elementary result on geometric series for our study of infinite decimal expansions of rational numbers.

Lemma 1.5.9 *Let $a, r \in \mathbb{Q}$ and suppose $|r| < 1$. Then the geometric series $\sum_{n=0}^{\infty} ar^n$ converges to $a/(1-r) \in \mathbb{Q}$.*

Proof We have $\sum_{n=0}^{m} ar^n = \frac{a(1-r^{m+1})}{1-r}$ and so

$$\left| \sum_{n=0}^{m} ar^n - a/(1-r) \right| = |a||r|^{m+1}/(1-r).$$

Letting $m \to \infty$, the result follows (see Exercise 1.5.17(1) for a proof, not depending on properties of real numbers, that $\lim_{m\to\infty} r^m = 0$).   □

Remark 1.5.10 The formula for the sum of an infinite geometric series is well-known. From our perspective, what is interesting is that provided the constant term a and multiplier r are rational, the infinite sum is *always* rational. This rarely holds for general infinite series. For example, $\sum_{n=1}^{\infty} 1/n^{2k} \notin \mathbb{Q}$, for all $k \in \mathbb{N}$ (see Chap. 5). Without real numbers, the theory of infinite series is effectively restricted to geometric series. In particular, we cannot yet give a meaning to $x_0 + \sum_{n=1}^{\infty} \frac{x_n}{10^n}$ for a general decimal expansion $x_0.x_1 \cdots$ *unless* we can show the series converges to a rational number.   ✱

An unsigned decimal expansion is *eventually periodic* of period $p \geq 1$, if we can find $k \in \mathbb{N}$ and $a_1, \cdots, a_p \in \{0, \cdots, 9\}$ such that

$$x = x_0.x_1 \cdots x_k a_1 a_2 \cdots a_p a_1 a_2 \cdots a_p \cdots$$

We usually write this in abbreviated form as $x = x_0.x_1 \cdots x_k\overline{a_1 a_2 \cdots a_p}$. We also require that p is minimal and that $a_1 \neq 0$ if $p = 1$.

Lemma 1.5.11 *If $x = x_0.x_1 \cdots x_k\overline{a_1 a_2 \cdots a_p}$ is eventually periodic, then*

$$x_0 + \sum_{n=1}^{\infty} \frac{x_n}{10^n} \in \mathbb{Q}.$$

That is, the infinite series $\sum_{n=1}^{\infty} \frac{x_n}{10^n}$ converges to a rational *number.*

Proof For $m \geq 1$ we may write

$$x_0 + \sum_{n=1}^{k+pm} \frac{x_n}{10^n} = x_0 + \sum_{n=1}^{k} \frac{x_n}{10^n} + \frac{A}{10^k} \sum_{j=0}^{m-1} 10^{-pj}, \qquad (1.1)$$

where $A = \sum_{m=1}^{p} \frac{a_m}{10^m}$. Clearly, $x_0 + \sum_{n=1}^{k} \frac{x_n}{10^n} \in \mathbb{Q}$. Since $A, 10^{-p} \in \mathbb{Q}$, Lemma 1.5.9 implies that $\frac{A}{10^k} \sum_{j=0}^{\infty} 10^{-pj}$ converges and is rational. Letting $k \to \infty$ in (1.1), we see that $\sum_{n=1}^{\infty} \frac{x_n}{10^n}$ converges and

$$x_0 + \sum_{n=1}^{\infty} \frac{x_n}{10^n} = x_0 + \sum_{n=1}^{N} \frac{x_n}{10^n} + \frac{A}{10^k} \sum_{j=0}^{\infty} 10^{-pj}$$

is a rational number. $\qquad \square$

Proposition 1.5.12 *Suppose that $x = p/q \in \mathbb{Q}$ where $(p, q) = 1$ and $p, q > 0$. There are two mutually exclusive possibilities.*

(1) *If the prime factorization of q is $2^r 5^s$, $r + s \geq 0$, then the decimal expansion of x is not unique and can be written in precisely two ways: as a terminating decimal $x = x_0.x_1 \cdots x_N$ or as an infinite decimal $x = x_0.x_1 \cdots (x_N - 1)\overline{9}$, where $x_N \in \{1, \cdots, 9\}$.*

(2) *If $q \neq 1$ and the prime factorization of q contains primes other than 2 or 5, then the decimal expansion of x is unique and eventually periodic with period at most q.*

A similar result holds if $p < 0$. If $x = 0$, then x has the unique decimal expansion $0.\overline{0}$ (we regard $\pm 0.\overline{0}$ as being identified).

Proof We leave the proof to the exercises. $\qquad \square$

Remark 1.5.13 Note that in (2) of Proposition 1.5.12, the decimal expansion of x cannot be of period 1 with $a_1 \in \{0, 9\}$. ✱

Examples 1.5.14

(1) If $p \in \mathbb{N}$, $q = 1$, then part (1) of the proposition applies. We may write $p = (p-1).\overline{9}$.

(2) Computing, we find that

$$\frac{3}{34} = 0.0882352941176470588235294 1 \cdots = 0.0\overline{8823529411764705}$$

The decimal expansion is unique and eventually periodic of period $16 \leq 34$. On the other hand, $7/20 = 7/(2^2 5) = 0.35 = 0.34\overline{9}$ and the decimal expansion is not unique.   ♠

1.5.4 Decimal Expansions and Real Numbers

We define the real numbers $\mathbb{R}$ to be the set of all *signed decimal expansions*:

$$\mathbb{R} = \{\pm x_0.x_1 x_2 \cdots x_n \cdots \mid x_0 \in \mathbb{Z}_+, \ x_n \in \{0, 1, \cdots, 9\}, \ n \in \mathbb{N}\}.$$

Let $\mathbb{R}_+ = \{x \in \mathbb{R} \mid x = +x_0.x_1 x_2 \cdots\}$ (the positive real numbers), and $\mathbb{R}_- = \{x \in \mathbb{R} \mid x = -x_0.x_1 x_2 \cdots\}$ (the negative real numbers). So as to simplify notation, we almost always drop the $+$-prefix from decimal expansions in $\mathbb{R}_+$ and write $\mathbb{R}_+ = \{x \in \mathbb{R} \mid x = x_0.x_1 x_2 \cdots\}$. We also identify the zero expansions $\pm 0.\overline{0}$ and denote either expansion by 0. With this convention, $\mathbb{R}_+ \cap \mathbb{R}_- = \{0\}$.

Our first step is to identify the set of rational numbers $\mathbb{Q}$ with a proper subset of $\mathbb{R}$. The only difficulty here is that not all rational numbers have a unique decimal expansion (Proposition 1.5.12). We deal with this the same way we dealt with the two decimal representations $\pm 0.\overline{0}$ of zero. We regard a terminating decimal expansion $\pm x_0.x_1 x_2 \cdots x_N$ (or $\pm x_0.x_1 x_2 \cdots x_N \overline{0}$), with $x_N \in \{1, \cdots, 9\}$, as identified with the decimal expansion $\pm x_0.x_1 x_2 \cdots (x_N - 1)\overline{9}$. If a decimal expansion x does not end with recurring 0's or 9's, then we regard x as uniquely defined by its decimal expansion. It follows from Proposition 1.5.12 that we may regard $\mathbb{Q}$ as identified with a proper subset of $\mathbb{R}$. Specifically, we identify $p/q \in \mathbb{Q}$ with its decimal expansion with the understanding that the non-zero decimal expansion $\pm x_0.x_1 x_2 \cdots x_N$ is identified with $\pm x_0.x_1 x_2 \cdots (x_N - 1)\overline{9}$.

Remark 1.5.15 We can enforce uniqueness of decimal expansions if we insist either that $\mathbb{R}$ contains no terminating decimal expansions (other than $0 = 0.\overline{0}$) or that $\mathbb{R}$ contains no decimal expansions ending in recurring nines. In practice, it is useful to allow both types of expansion.   ✶

It follows from Lemmas 1.5.6, 1.5.11 that if $x \in \mathbb{R}$ is either eventually periodic or terminating, then x is rational. If $x \in \mathbb{Q}$ is eventually periodic and not of the form $\pm x_0.x_1 \cdots x_N \overline{0}$, $x_N \neq 0$, then the decimal expansion of x is unique. If x is neither eventually periodic nor terminating, we say x is *irrational*. Irrational numbers have unique decimal expansions. This is by definition!—we only identify decimal expansions corresponding to rational numbers of the form $p/(2^r 5^s)$, where $r + s \geq 0, p \in \mathbb{Z}$.

Now that we have a minimal description of the real numbers it is easy to prove Cantor's result that $\mathbb{R}$ is uncountable.

Theorem 1.5.16 *The set $\mathbb{R}$ is uncountable.*

Proof We give the proof discovered by Cantor and based on his diagonal method. If $\mathbb{R}$ is countable then certainly the half-open interval $[0, 1) = \{x \in \mathbb{R}_+ \mid x_0 = 0,$ $x \neq 0.\overline{9}\}$ is countable (every subset of a countable set is countable). It is therefore enough to show that $[0, 1)$ is uncountable. Suppose the contrary. Then we may write $[0, 1) = \{x^n \mid n \in \mathbb{N}\}$, where each x^n has decimal expansion not ending in recurring 9's for all $n \in \mathbb{N}$. In terms of decimal expansions, we have

$$x^n = 0.x_1^n x_2^n \cdots x_n^n \cdots ,$$

where the sequence $x_1^n, x_2^n, \cdots$ does not end in recurring 9's. We define $z = 0.z_1 z_2 \cdots \in \mathbb{R}_+$ by

$$z_n = 4, \quad \text{if } x_n^n = 5,$$
$$= 5, \quad \text{if } x_n^n \neq 5$$

Clearly $z \in [0, 1)$ since the decimal expansion of z cannot end in recurring 9's. On the other hand, $z \notin \{x^1, x^2, \cdots\}$ since $z_n \neq x_n^n$, all $n \geq 1$ (decimal expansions are unique granted our condition on recurring 9's). Contradiction. Hence $\mathbb{R}$ cannot be countable. $\Box$

EXERCISES 1.5.17

(1) Find an elementary argument to prove that if r is rational and $|r| < 1$ then $\lim_{n \to \infty} r^n = 0$. (Hints and comments: It is enough to assume $r \in (0, 1)$. We may write $r = 1 - s$, where $1 > s > 0$. Observe that $(1 - s) \leq 1/(1 + s)$. By the binomial theorem $(1 + s)^n \geq 1 + ns > ns$, for all $n \in \mathbb{N}$. Hence $(1 - s)^n \leq an^{-1}$, $s = 1/a$. Now argue using $\lim_{n \to \infty} 1/n = 0$. This argument is elementary and does not rely on using properties of monotone decreasing sequences or the log and exponential functions.)
(2) Show that if $x_N \in \{1, \cdots, 9\}$, then the decimals $x = x_0.x_1 \cdots x_N \overline{0}$ and $x' = x_0.x_1 \cdots (x_N - 1)\overline{9}$ define the same rational number.
(3) Prove Proposition 1.5.12.

1.6   The Structure of the Real Numbers

For the remainder of the chapter we look at the problem of extending order, absolute value, addition and subtraction from the rationals to the real numbers. Order and absolute value are easy and natural to define for real numbers and allow us to make a start on approximating real numbers by rational numbers. Addition and subtraction of *infinite* decimals is trickier as we have to work with rational approximations and be careful about 'bookkeeping'. However, it is easy to add and subtract two decimals if one of the decimals is terminating. As a result, we can approximate a real number by its decimal truncations: for all $x \in \mathbb{R}$, we have $|x - x^N| \leq 10^{-N}$, $N \in \mathbb{N}$. We conclude by defining multiplication and division of real numbers by rational numbers (we defer general multiplication and division of real numbers to Chap. 2).

We suggest a careful reading of the definitions and results on order, absolute value and approximation of real numbers by rational numbers and then skim through the elementary but longer arguments on addition and subtraction of infinite decimals. The alternative more abstract approach we give at the end of Chap. 2 handles the arithmetic properties of real numbers straightforwardly but there are still many details to be checked.

1.6.1   Order on $\mathbb{R}$

Provided we require unique decimal expansions (we deny *either* recurring 0's *or* 9's), it is easy to define an order $<$ on $\mathbb{R}$ that extends the usual order on $\mathbb{Q}$. Suppose we restrict to decimal expansions that do not end with recurring 9's. If $x, y \in \mathbb{R}_+$, we write $x < y$ (equivalently, $y > x$) if there exists an $N \geq 0$ such that $x_n = y_n$, $n < N$, and $x_N < y_N$. Necessarily $x \neq y$ by uniqueness of decimal expansions! If $x \in \mathbb{R}_-$, and $y \in \mathbb{R}_+$ and $x \neq y$ (so x, y are not both equal to zero) we declare that $x < y$, and if $x, y \in \mathbb{R}_-$ then $x < y$ iff $-y < -x$. Since we have unique decimal expansions, this restricts to the usual order on $\mathbb{Q}$ (if we did not have this restriction, then $x = 0.1$, $y = 0.0\overline{9}$ would cause a problem.) We extend the notation in the usual way to $\leq, \geq$. With these conventions we have

$$\mathbb{R}_+ = \{x \in \mathbb{R} \mid x \geq 0\}, \ \mathbb{R}_- = \{x \in \mathbb{R} \mid x \leq 0\}.$$

Remark 1.6.1 If instead we had restricted to decimal expansions of non-zero numbers that do not end with recurring 0's, we would have ended with the same order structure on $\mathbb{R}$. Thus $0.1 < 0.2$ (deny recurring 9's) and $0.0\overline{9} < 0.1\overline{9}$ (deny recurring 0's). Note, however, that if we deny recurring 0's, then we need a special argument for $0 = 0.\overline{0}$. Hence our preference for denying recurring 9's.   ✠

Example 1.6.2 If $x \in \mathbb{R}_+$, there exists an $N \in \mathbb{N}$ such that $x < N$. The proof is immediate: if $x = x_0.x_1 \cdots$ and we deny recurring 9's, define $N = (x_0 + 1).\overline{0}$. This property is known as the *Archimedean property* of real numbers.   ♠

1.6.2   The Absolute Value

Definition 1.6.3 If $x \in \mathbb{R}$, the *absolute value* $|x|$ of x is defined to be x, if $x \geq 0$, and $-x$ if $x < 0$. That is $|x|$ is x 'unsigned'.

Remark 1.6.4 The absolute value restricted to $\mathbb{Q}$ gives the usual absolute value on rationals (same definition).   ✣

Lemma 1.6.5 *If $x \in \mathbb{R}$ and $x_0, \cdots, x_N = 0$, $x_{N+1} \neq 0$, then $10^{-N-1} \leq |x| \leq 10^{-N}$.*

Proof If we replace x_n by 9 for $n > N$, we have

$$|x| = 0.0 \cdots 0 x_{N+1} \cdots \leq 0.0 \cdots 0\overline{9} = 10^{-N-1} \sum_{m=0}^{\infty} \frac{9}{10^m} = 10^{-N}.$$

This shows $|x| \leq 10^{-N}$. On the other hand if $x_{N+1} \neq 0$, we can replace x_{N+1} by 1 and set $x_n = 0, n > N + 1$ to obtain

$$|x| = 0.0 \cdots 0 x_{N+1} \cdots \geq 0.0 \cdots 01\overline{0} = 10^{-N-1}.$$

Hence $10^{-N-1} \leq |x|$.   □

Remark 1.6.6 If x is irrational, we always have strict inequality in Lemma 1.6.5: $10^{-N-1} < |x| < 10^{-N}$.   ✣

Example 1.6.7 We claim that if $x \in \mathbb{R}_+$ is non-zero, then there exists a $z \in \mathbb{R}$ such that $0 < z < x$. Since $x \neq 0$, there exists a least $N \geq 0$ such that $x_N \neq 0$. By Lemma 1.6.5, $x \geq 10^{-N-1}$. Since $10^{-N-1} > 10^{-N-2} > 0$, we may take $z = 10^{-N-2}$. In this case we constructed a rational z. We can find an irrational z by choosing any non-rational decimal expansion, for example define $b = 101^2 01^3 01^4 0 \cdots 01^n 0 \cdots$ (where 1^n is shorthand for n repeated 1's). If we define $z = 0.0^{N+1} b$, then z is positive, irrational and $z < 10^{-N-1} \leq x$ by Lemma 1.6.5. Hence $0 < z < x$. If we assume more structure on the reals (addition and division), it is easy to deduce this result from the Archimedean property of $\mathbb{R}$ (Example 1.6.2). See also Proposition 1.6.20.   ♦

Remark 1.6.8 So far we have made no use of addition and subtraction of real numbers.   ✣

1.6.3 Addition and Subtraction: Terminating Decimals

In this section we review the addition and subtraction of terminating decimals. Necessarily our definition should give the same result as addition of subtraction of rational numbers according to the rule $\frac{p}{q} \pm \frac{r}{s} = \frac{ps \pm rq}{qs}$.

First, some notational conventions. If $n \in \mathbb{N}$, we usually denote the terminating decimals $\pm n.\overline{0}$ by $\pm n$. Note that it follows from our conventions that, as real numbers, we have $n = n.\overline{0} = (n-1).\overline{9}$ and $-n = -n.\overline{0} = -(n-1).\overline{9}$. Let $\mathbb{R}_T \subset \mathbb{Q}$ denote the set of terminating decimals.

Addition of terminating decimals of the same sign follows the standard 'add and carry' rules—we define $x + y = -(-x + -y)$ if $x, y \leq 0$. Suppose $x > 0 > y$. Then

$x + y = x - (-y)$. If $-y < x$, we compute using standard subtract and carry (if $-y > x$, write $x + y = -((-y) - x)$.

Example 1.6.9

$$2.35 + (-1.46) = 2.35 - 1.46 = 0.89,$$

where we carry -1. Note this is correct since $1.46 + 0.89 = 2.35$ (add and carry or compute as a sum of rational numbers). On the other hand, $1.46 - 2.35$ computes to -1.11 if we use subtract and carry and this is incorrect. ♠

These rules give the correct definition of addition for terminating decimals—that is, they give the same result we get using the standard rule for addition of rational numbers.

We define subtraction using addition:

$$x - y \stackrel{\text{def}}{=} x + (-y),\ x, y \in \mathbb{R}_T.$$

Remark 1.6.10 Let $x = 0.x_1 \cdots x_{N-1}x_N$ and suppose $x_N \neq 0$. Define $\bar{x} \in \mathbb{R}_T$ by $\bar{x} = 0.\bar{x}_1 \cdots \bar{x}_{N-1}(\bar{x}_N + 1)$, where $\bar{a} = 9 - a$, $a \in \{0, 1, \cdots, 9\}$. Observe that $\bar{x}_N + 1 \in \{1, \cdots, 9\}$, since $x_N \neq 0$, and $x + \bar{x} = 1$ (add and carry). The difficulties of adding finite decimals of opposite sign occur because if $n \in \mathbb{Z}_+$, and $x = 0.x_1 \cdots x_{N-1}x_N$, then $-n + x = -(n-1).\bar{x}_1 \cdots \bar{x}_{N-1}(\bar{x}_N + 1)$ and this is only equal to $-(n-1).x_1 \cdots x_{N-1}x_N$ if $N = 1$ and $x_1 = 5$. Of course, there is no problem for $n + x$ if $n \geq 0$. ✠

1.6.4 A Special Case of Addition for Infinite Decimals

When we come to the problem of addition and subtraction of infinite decimals, we cannot avoid looking at limits—in this case of sequences of rational approximations. Roughly speaking, if we are given infinite decimals $x = x_0.x_1x_2\cdots$, $y = y_0.y_1y_2\cdots$, we want to define $x + y$ to be the limit as $N \to \infty$ of $x^N + y^N$, where $x^N = x_0.x_1 \cdots x_N$, $y^N = y_0.y_1 \cdots y_N$ are the truncations of x, y to N-terms. The difficulty is that generally we do not know what the limit *is*—as a decimal expansion—and 'add and carry' does *not* work for infinite decimals. However, when we know what the limit is, it is usually easy to prove the convergence of $x^N + y^N$ to the limit as $N \to \infty$. If this sounds tautological it is: the definition of convergence of a sequence assumes we know the limit. Later, in Chap. 2, we introduce the idea of a *Cauchy sequence*, which gives an intrinsic definition of convergence without having to know the limit. In this section we look at the problem of solving the equation $x + y = n$, where x is a given infinite decimal and $n \in \mathbb{Z}$. We show that there is a unique infinite decimal y satisfying the equation and that the decimal expansion of y can be given explicitly in terms of that of x. Once we have this result, it is easy to extend our definition of addition and subtraction to sums and differences when one (not both) of the terms may be an infinite decimal. Although this seems a small step, it allows us to view the truncations x^N, $N \in \mathbb{N}$, as (rational) approximations to an infinite decimal. Specifically, we can easily show that

$$|x - x^N| \leq 10^{-N}, \; N \in \mathbb{N}.$$

It is natural to think of a real number x as the set $\{x^N \mid N \in \mathbb{N}\}$ of all its truncations: this is the way we do computations with irrational numbers in practice. That is, rather than attempting to *define* x by evaluating the infinite sum $\sum_{n=0}^{\infty} \frac{x_n}{10^n}$ (we cannot at this point unless x is rational), we think of x as defined by its set of truncations and then do 'approximate arithmetic' (which will be exact in the limit). Observe that we cannot write down an irrational number in exact form—that requires an infinite string of integers—instead we write down a 'good enough' rational approximation to the number. In order to make this process work and keep control of the errors, we need to introduce ideas based on limits.

Let $x = 0.x_1 x_2 \cdots$ be an infinite (not terminating) decimal and define $\bar{x} = 0.\bar{x}_1 \bar{x}_2 \cdots$ (recall $\bar{a} = 9 - a$ for $a \in \{0, \cdots, 9\}$, see Remark 1.6.10). If we define $z = x + \bar{x}$ by $z_n = x_n + \bar{x}_n$, $n \geq 0$, then $z = 0.\bar{9} = 1$. Unlike what happens for terminating decimals, we will not be able to avoid recurring 9's when we consider addition of infinite decimals.

In terms of the truncations $z^N, x^N, \bar{x}^N$, we have $z^N = x^N + \bar{x}^N = 0.\overline{9}^N$ and so $1 - (x^N + \bar{x}^N) = 10^{-N}$. That is, $\lim_{N \to \infty} x^N + \bar{x}^N = 1$ (this is a statement about rational numbers). In other words, if we define the sum $x + \bar{x}$ by addition of like terms, then the resulting decimal expansion is the limit of the sum of the truncations. This is the key property we need when we come to the sum of general decimal expansions. Observe there is no problem here with the addition as there are no terms to be carried.

Now suppose $x = x_0.x_1 x_2 \cdots$ is an infinite decimal and $n \in \mathbb{Z}$. The general solution of the equation $x + y = n$ is given as follows

$$y = \begin{cases} (n - x_0).\bar{x}_1 \bar{x}_2 \cdots, & \text{if } x_0 < n, \\ -(x_0 - n).x_1 x_2 \cdots, & \text{if } x_0 \geq n. \end{cases} \tag{1.2}$$

We may now easily extend our rules to define $x \pm y$ when one of x, y is an infinite decimal. Addition when x, y are of the same sign follows the pattern given for terminating decimals. For example, if $y = y_0.y_1 \cdots y_M$, and $x, y \geq 0$, then $x + y$ is defined by adding x^M and y^M and appending $x_{M+1} x_{M+2} \cdots$. If $x \geq y$ and x is not terminating, then $x - y$ is defined exactly as we did when both x and y are terminating. If $x < y$, then we may write $x = (x + n) - n$, where $n \in \mathbb{Z}$ is chosen so that $x + n \geq y$. The decimal $z = (x + n) - y$ is well defined $(x + n > y)$ and we have $x - y = z - n = -(n - z)$, which is given by (1.2).

1.6.5 Decimal Approximation of Real Numbers

The results in the previous section allow us to estimate real numbers in terms of rational approximations by finite decimals.

Lemma 1.6.11 *Let $x \in \mathbb{R}$ have decimal expansion $x = x_0.x_1 \cdots$. For all $N \in \mathbb{N}$ we have*

$$0 \le |x - x^N| \le 1/10^N.$$

In particular, $\lim_{N \to \infty} x^N = x$.

Proof Without loss of generality suppose $x \ge 0$. Then

$$0 \le x - x^N = 0.0^N x_{N+1} \cdots \le 0.0^N \overline{9} = 10^{-N}.$$

The result follows since $\lim_{N \to \infty} x^N = x$ iff $\lim_{N \to \infty} |x - x^N| = 0$. □

Remarks 1.6.12

(1) The proof of Lemma 1.6.11 uses only the order structure on $\mathbb{R}$ (elementary) together with arguments involving rational numbers and geometric series (cf. Lemma 1.5.9).

(2) Previously, we have only discussed the convergence of $\sum_{n=0}^{\infty} \frac{x_n}{10^n}$ for eventually periodic decimal expansions. Since $\lim_{N \to \infty} x^N = x$, and $x^N = \sum_{n=0}^{N} \frac{x_n}{10^n}$, we now have the result that

$$\sum_{n=0}^{\infty} \frac{x_n}{10^n} \overset{\text{def}}{=} \lim_{N \to \infty} \sum_{n=0}^{N} \frac{x_n}{10^n} = x.$$

That is, the sequence of partial sums converges to x. ✠

Example 1.6.13 Let x be an infinite decimal and $y \in \mathbb{R}_T$. Set $z = x + y$. Then $\lim_{N \to \infty} (z - (x^N + y^N)) = 0$. If x, y are of the same sign, and y is of length M, then $x^N + y^M = z^N$ for $N \ge M$ and the result follows from Lemma 1.6.11. If $x \ge 0 \ge y$ and $-y > x$, choose $n \in \mathbb{Z}$ so that $n + x > -y$. Set $u = (n + x) + y$. We have $n + x^N + y^N = u^N \to u$ and $u^N - n \to z$. ♠

Remark 1.6.14 If $x \in \mathbb{R}$ is eventually periodic and y is a terminating decimal, then $x \pm y$ is eventually periodic and so rational. We leave it to the exercises for the reader to check that our definition of addition and subtraction gives the same result as when we add/subtract $p/q, r/s$ using $p/q \pm r/s = (ps \pm qr)/qs$. ✠

1.6.6 Addition and Subtraction of Real Numbers

It remains to define addition and subtraction of infinite decimals. Suppose $x, y \in \mathbb{R}$ are infinite decimals. We define $x + y$ to be the limit as $N \to \infty$ of $x^N + y^N$. In order to prove the limit exists, we have to prove that the initial terms of the decimal expansion of $x^N + y^N$ 'stabilize' as $N \to \infty$. To capture this property precisely, we define a new limit operation.

Definition 1.6.15 Let $x_0.x_1x_2\cdots$ be the decimal expansion of $x \in \mathbb{R}$. Suppose that $(x^n)_{n\geq 1}$ is a sequence of decimal expansions. We write $\lim_{n\to\infty}x^n = x$ if, for every $N \in \mathbb{N}$, we can find $M \in \mathbb{N}$ such that

$$x_i^n = x_i, \text{ for all } i \leq N \text{ and } n \geq M.$$

In words, we say that (x^n) converges to x iff for any M, we can find N such that the truncation of x^n to M terms is equal to truncation of x to M terms for all $n \geq N$. Note the purely symbolic character of this definition. There is no use of subtraction or absolute value.

Examples 1.6.16

(1) Suppose that $x = \pm x_0.x_1 \cdots \in \mathbb{R}$ and let $x^N = \pm x_0.x_1\cdots x_N = \pm x_0.x_1\cdots x_N\overline{0}$ denote the truncation of x to N terms. Then $\lim_{N\to\infty}x^N = x$. In this case, given $N \in \mathbb{N}$, we may take $M = N$ in Definition 1.6.15.

(2) Let $x = 1.\overline{0}, y = 0.\overline{9}$. Let (x^N) and (y^N) be the sequences of truncations defined by taking $x^N = 1.\overline{0}^N$, $y^N = 0.\overline{9}^N$. Observe that $\lim_{N\to\infty}x^N = 1.\overline{0} \neq 0.\overline{9} = \lim_{n\to\infty}y^N$, even though the two limits define the same real number. This is only an issue for rational numbers which have a finite decimal expansion.

(3) Suppose $x = 0.1234516\cdots, y = 0.3765484\cdots$. For $N \in \mathbb{N}$, let $z^N = x^N + y^N$. We have

$$z^6 = 0.499999 = 0.49^5,$$

$$z^7 = 0.500000 = 0.50^6.$$

If $N > 7$, it is easy to see that $z_n^N = z_n^7$, if $n \leq 6$, and that $z_7^N \in \{0, 1\}$, whatever the higher-order terms are in the decimal expansions of x and y. For example, consider the 'worst' case $x_n = y_n = 9$, for all $n > 7$. Computing we find that $z^N = 0.50^519^{N-8}8$, for all $N \geq 8$. We see that the initial term of z^N is 0.500000, for $N > 7$, and that $\lim_{N\to\infty}z^N$ exists and is equal to 0.50^519. ◆

Lemma 1.6.17 (Stability Lemma) *Let $x, y \in \mathbb{R}_+$ and set $z^N = x^N + y^N$, $N \in \mathbb{N}$. Given $N_0 > 1$, suppose there exists an $m < N_0$ such that $z_m^{N_0} \leq 8$. Then for $N > N_0$ we have*

$$z_n^N = z_n^{N_0}, \text{ for all } n < m.$$

Proof Let $N > N_0 > m \geq 1$ and suppose that $z_m^{N_0} \leq 8$. Assume first that $x_n = y_n = 9$, $N \geq n > N_0$. Adding, we find that $z_m^{N_0} = 9$ and $z_n^N = z_n^{N_0}$ for all $n < m$. If we vary x_n, y_n, $n > N_0$, we only make z^N smaller. Since $z^N \geq z^{N_0}$, the terms z_n^N are unchanged for all $n < m$.   □

Proposition 1.6.18 *Let* $x, y \geq 0$. *Then* $\lim_{N \to \infty}(x^N + y^N)$ *exists and defines a unique point in* $\mathbb{R}_+$.

Proof The result is trivial if either x or y is zero so suppose $x, y > 0$. If either x or y is a terminating decimal, the result is easy (see Sect. 1.6.4). Hence we may assume that x, y have unique infinite decimal expansions, $x = x_0.x_1 x_2 \cdots$, $y = y_0.y_1 y_2 \cdots$. Set $z^N = x^N + y^N$. Then $z^N = z_0^N . z_1^N \cdots z_N^N$ where the integers $z_0^N, \cdots z_N^N$ may depend on N.

Given $N \geq 2$, we let $m = m(N)$ be the largest value of $m \in \mathbb{N}$ such that $m < N$ and $z_m^N \leq 8$. If $z_m^N = 9$ for $1 \leq m < N$, we set $m(N) = 0$. It follows from Lemma 1.6.17 that

$$z_n^P = z_n^Q, \; P, Q \geq N, \; n < m(N).$$

Set $s_N = m(N) - 1$. Our construction defines an increasing sequence $(s_N) \subset \mathbb{Z}_+$. There are two possibilities: either $s_N \to \infty$ as $N \to \infty$ or there exists a $P \in \mathbb{Z}$ such that $s_N = P$ for all sufficiently large N. If the second condition holds then the $\lim_{N \to \infty}(z^N = x^N + y^N)$ exists and the corresponding decimal expansion ends with recurring 9's. If the first condition holds, then by the definition of **lim**, $\lim_{N \to \infty}(z^N = x^N + y^N)$ exists (given N, take $M = s_N$ in Definition 1.6.15).   □

Granted Proposition 1.6.18 it is now easy to define addition and subtraction of general real numbers.

If $x, y \geq 0$, we define $x + y = \lim_{N \to \infty}(x^N + y^N)$. If both x and y are negative, define $x + y = -(-x + -y)$. If $x \geq 0 \geq y$, choose $n \in \mathbb{N}$ so that $n + y \geq 0$ and then define $x + y = (x + (n + y)) - n$ (using (1.2) as needed). For subtraction, define $x - y = x + (-y)$.

It follows immediately from our constructions and Lemma 1.6.11 that if $x, y \in \mathbb{R}$ and we set $z = x \pm y$ then

$$\lim_{N \to \infty} |z - (x^N \pm y^N)| = 0. \qquad (1.3)$$

That is $\lim_{N \to \infty}(x^N + y^N) = x \pm y$.

Example 1.6.19 Using (1.3), the usual rules for absolute value, such as the triangle inequality, follow immediately from the corresponding rules for rational numbers: if $x, y \in \mathbb{R}$ then

$$|x + y| = \lim_{N \to \infty} |x^N + y^N| \leq \lim_{N \to \infty} (|x^N| + |y^N|) = |x| + |y|.$$

Proposition 1.6.20 *Let $x, y \in \mathbb{R}$ and suppose $x < y$. Then there exists a $z \in \mathbb{R}$ such that $x < z < y$. We may require z to be either rational or irrational.*

Proof Observe that $x < y$ iff $y - x > 0$. By Example 1.6.7, there exists an $a \in \mathbb{R}$ such that $0 < a < y - x$. Hence $x < x + a < y$. If we want z to be irrational, observe that either $x + a$ is irrational (and we done) or $x + a$ is rational. In the latter case we can choose an irrational b satisfying $0 < b < y - x - a$ (for example, $b = 0.0^M 101^2 01^3 \cdots 01^n 0 \cdots$ for large enough M—see Example 1.6.7) and then define $z = b + x + a < y$. If we require z to be rational, take a high enough order truncation of z. Finally, note that we cannot (yet!) take $z = (x + y)/2$ as we have not defined multiplication and division of real numbers. $\square$

Remarks 1.6.21

(1) All the standard rules of addition and subtraction (commutativity, associativity, etc.) are easily seen to hold for real numbers. Indeed they are all inherited through the limit operation from the corresponding properties for rational numbers.

(2) Let $x \in \mathbb{R}$. We have shown that we can view x as the limit of the sequence (x^N) of truncations of x: $x = \lim_{N \to \infty} x^N$. In this sense, we can think of a real number as defined by its set of truncations, all of which are rational. While the rational numbers $\mathbb{Q}$ are naturally defined (in terms of the integers), the restriction to (decimal) truncations depends on the choice of base 10. However, it is not hard to show that changing base does not change the set of real numbers. The problem is avoided by defining a real number in terms of all of its rational approximations—see section "Appendix: Construction of $\mathbb{R}$ Revisited" at the end of Chap. 2.

(3) It is worth emphasizing again the conceptual leap that is required in going from rational to irrational numbers. Rational numbers are given finitely; the specification of an irrational number depends on a limiting process and irrational numbers cannot be described finitely. About the best one can do is define an irrational number by a recursive process. For example, if we take $x_0 = 1$ and define $x_{n+1} = \frac{1}{2}(x_n + \frac{2}{x_n})$, $n \geq 0$, then $(x_n) \subset \mathbb{Q}$ and $\lim_{n \to \infty} x_n = \sqrt{2}$. However, it can be shown that most irrational numbers cannot be specified in this way: there are uncountably many real numbers, but only countably many recursion formulas with rational coefficients! ✠

1.6.7 Multiplication of Real Numbers by Rationals

Let $x, y \in \mathbb{R}$. One way of defining the product xy is to show that $\lim_{N \to \infty} x^N y^N$ exists. However, even more than was the case in the proof of Proposition 1.6.18, bookkeeping is a problem. We give a much simpler and more elegant proof of convergence in Chap. 2 based on results on bounded monotone sequences. On the other hand, if one of x, y is *rational*, it is easy to define the product by making use of our results on addition of real numbers.

Suppose then that $x \in \mathbb{R}$, $p/q \in \mathbb{Q}$. Without loss of generality, assume that $x, p, q > 0$. We outline the main steps in defining the product $\frac{p}{q}x$, leaving the details to the exercises.

(1) Show $px \in \mathbb{R}$. (Since px should be the sum of p copies of x, we define px using addition of real numbers.)
(2) If $q \in \mathbb{N}$ then $\lim_{N \to \infty} \frac{x^N}{q}$ exists. There is no problem with *division* of the decimal expansion x by q—we start with division by q of the initial term x_0 of the decimal expansion and carry to the right. In particular,

$$\left(\frac{x^N}{q} \right)_n = \left(\frac{x^M}{q} \right)_n , \quad \text{all } n \leq \max\{M, N\}.$$

Hence $\lim_{N \to \infty} \frac{x^N}{q}$ exists and we define the limit to be $\frac{p}{q}x$.

Remark 1.6.22 The method gives division of real numbers by rationals: $x/\left(\frac{p}{q}\right) \overset{\text{def}}{=}$ $\frac{q}{p}x$, provided $p \neq 0$.   ✲

EXERCISES 1.6.23

(1) Let x, y be two decimal expansions. Show that $x - y = 0$ (as real numbers) iff either $x = y$ (as decimal expansions) or x and y represent the same rational number which has a terminating decimal expansion. A consequence is that if x, y are real numbers and x is irrational then $x = y$ (as decimal expansions) iff $x - y = 0$.
(2) Show that $\sum_{n=0}^{\infty} 10^{-n^2}$ is irrational. More generally, show that if $p(x) = x^m + a_1 x^{m-1} + \cdots + a_m$ is a polynomial of degree $m \geq 1$ with integer coefficients, then $\sum_{n=0}^{\infty} 10^{-p(n)}$ is rational if and only if $m = 1$.
(3) Verify that we get the same order structure on $\mathbb{R}$ if (a) we deny decimal expansions ending in recurring zeros, or (b) we deny decimal expansions ending in recurring nines.
(4) Let $A > 0$. Show that for $n \geq 0$, it is possible to choose a *unique* finite decimal expansion $X_n = x_0.x_1 \ldots x_n$ such that

(a) $X_n^2 \leq A$.
(b) $(X_n + 10^{-n})^2 > A$.

Show also that the terms $x_0, x_1, \ldots, x_n$ do not depend on n (that is, if $m > n$, then the first $n+1$ terms of X_n, X_m are the same). Deduce that $\lim_{n \to \infty} X_n = X$ exists. Show that $X = \sqrt{A}$. (For the last part, define $Y_n = X_n + 10^{-n}$ and observe that $X_n^2 \leq x < Y_n^2$. Now let $n \to \infty$.)
(5) Extend the method of the previous exercise to show that if $A > 0$ and $p \geq 2$, then the positive pth root of A exists. (We give alternative constructions for rapidly computing roots in terms of rational sequences in Chap. 2.)
(6) Fill in the details for the construction of $\frac{p}{q}x$ in Sect. 1.6.7. (Start by giving an inductive definition of px, $x \in \mathbb{R}_+$, $p \geq 2$—define $px = (p-1)x + x$, $p \geq 2$. Verify that $rx + sx = px$ if $r, s \in \mathbb{N}$, $r + s = p$.)

(7) Show that if $r_1, r_2 \in \mathbb{Q}$, $x \in \mathbb{R}$ then $r_1(r_2 x) = (r_1 r_2)x$ (associative law of multiplication). Deduce that if $r \in \mathbb{Q}$, $r \neq 0$, and $x \in \mathbb{R}$, then $rx \in \mathbb{Q}$ iff $x \in \mathbb{Q}$.

(8) Let $a < b$, $a, b \in \mathbb{R}$. Prove that $(a, b) \cap \mathbb{Q}$ is countably infinite. (Show that all but finitely many members of $\{a + 10^{-n} \mid n \in \mathbb{N}\}$ lie in (a, b).)

(9) Show that

   (a) $(0, 1) \sim \mathbb{R}$. (Look for a bijection of the form $g(x) = A/x + B/(x - 1)$.)

   (b) $(0, 1) \cup C \sim (0, 1)$ for any countable set C disjoint from $(0, 1)$. (Hint: Choose a countable infinite subset K of $(0, 1)$ and observe that $K \cup C \sim K$.) Deduce that $(0, 1) \sim [0, 1]$ (this can never be realized by a continuous map).

   (c) $P(\mathbb{N}) \sim (0, 1)$. (Hint: Let B be the set of all binary expansions $0.b_1 b_2 \cdots$, $b_i \in \{0, 1\}$. Show that every $X \in P(\mathbb{N})$ determines a unique $b \in B$ by $b_n = 1$ iff $n \in X$ and hence show $P(\mathbb{N}) \sim B$. Using (2) show that $B \sim (0, 1)$— you will have to address the non-uniqueness of binary expansions.)

   (d) $P(\mathbb{N}) \sim \mathbb{R}$.

(10) Let $\mathcal{F}$ be the set of all functions $f : [0, 1] \to \mathbb{R}$. Show that $\mathcal{F} \not\sim [0, 1]$. (Hint: use the diagonal method. If we restrict to *continuous* functions, then we do have equivalence—a continuous function on $[0, 1]$ is uniquely determined by its values at the rational points of $[0, 1]$.)

(11) Let X be a non-empty set and $\mathcal{F}$ denote the set of all functions $f : X \to \mathbb{R}$. Show that $\mathcal{F} \not\sim X$.

(12) Let X, Y be non-empty sets and suppose that Y contains at least two points. Let $\mathcal{F}$ denote the set of all functions $f : X \to Y$. Show that $\mathcal{F} \not\sim X$. What happens if Y consists of a single point?

(13) Using decimal expansions, find an *onto* map $F : [0, 1] \to [0, 1]^2$. Is the map F you have constructed 1:1? If not (most likely), show that it is possible to define a *bijection* $G : [0, 1] \to [0, 1]^2$. (Hints and comments for the second part. The new map G is closely related to F. The problem lies with non-uniqueness of decimal expansions. Let D denote the set of *all* decimal expansions $0.x_1 x_2 \cdots$. Show there is a bijection between D and D^2—easy! Then verify $D \sim [0, 1]$, $D^2 \sim [0, 1]^2$; this will require handling countable sets of 'bad' points (use the result of Q9). The maps F, G will not be continuous. Although it is possible to construct continuous maps of $[0, 1]$ *onto* $[0, 1]^2$ (Peano curves), there are no continuous bijections between $[0, 1]$ and $[0, 1]^2$.)

Chapter 2
Basic Properties of Real Numbers, Sequences and Continuous Functions

2.1 Introduction

In this chapter we prove a number of foundational results about real numbers, sequences and continuous functions. Sequences will play a major role throughout. We start by proving key results on the convergence of bounded monotone sequences using methods that develop naturally from our real number constructions in Chap. 1. As an application, we give the general definitions for multiplication and division of real numbers. We then prove the Bolzano–Weierstrass theorem and its important corollary that every bounded sequence has at least one convergent subsequence. We use both results repeatedly in the sequel. Turning next to functions, we verify the equivalence of continuity and sequential continuity, and then use relatively simple sequence-based methods to prove standard results about continuous functions on a closed and bounded interval (boundedness, attainment of bounds, the intermediate value theorem and uniform continuity). Next we define Cauchy sequences and prove the fundamental result that a sequence is convergent if and only if it is Cauchy. As a consequence we obtain an intrinsic definition of convergence that does not explicitly depend on the limit. We devote a section to the definitions and properties of the operations of lim sup and lim inf and show how we may use these concepts to provide alternative proofs of some of our results. After a section reviewing the definition of complex numbers and properties of complex sequences, we conclude the chapter with four appendices. In the first appendix, we review some standard results of the differential calculus. In the second appendix we provide a simple proof that every continuous function on a closed interval has a unique Riemann integral (the proof does not use uniform continuity). In the third appendix, we develop from scratch the theory of the exponential and natural logarithm and prove important and much used growth estimates for $\log x$ and e^x, as $x \to +\infty$, and for $\log x$ as $x \to 0+$. In the final appendix, we outline an approach to the construction of the real number system that is based on Cauchy sequences of rational numbers.

© Springer International Publishing AG 2017
M. Field, *Essential Real Analysis*, Springer Undergraduate Mathematics Series,
https://doi.org/10.1007/978-3-319-67546-6_2

2.2 Sequences

Let Z be a non-empty set. Formally, a *sequence* of points of Z is a function $\mathbf{x}$: $\mathbb{N} \to Z$ (sometimes $\mathbf{x} : \mathbb{Z}_+ \to Z$). We invariably set $\mathbf{x}(n) = x_n, n \geq 1$, denote the sequence by (x_n), or $(x_n)_{n \geq 1}$, and regard (x_n) as being an *ordered* subset of Z—the order being given by $\mathbb{N}$.

Example 2.2.1 If $x \in Z$ and we define $x_n = x$, for all $n \geq 1$, then (x_n) is a constant sequence. In particular, we do not require that the map $\mathbf{x} : \mathbb{N} \to Z$ be 1:1 and (x_n) is not the same as the set $\{x_n \mid n \in \mathbb{N}\}$ which, for this example, is the singleton $\{x\}$. ♠

2.2.1 Sequences of Real Numbers and Convergence

In this chapter we will be mainly interested in sequences (x_n) of real numbers: $x_n \in \mathbb{R}, n \in \mathbb{N}$. We sometimes write $(x_n) \subset \mathbb{R}$ to signify that (x_n) is a sequence of real numbers. That is, $\{x_n \mid n \in \mathbb{N}\} \subset \mathbb{R}$. Similarly, if we write $(x_n) \subset \mathbb{Q}$, then (x_n) will be a sequence of rational numbers.

Example 2.2.2 Since $\mathbb{Q}$ is countable, there is a surjective map $\mathbf{x} : \mathbb{N} \to \mathbb{Q}$. The associated sequence (x_n) has the property that $\cup_{n=1}^{\infty}\{x_n\} = \mathbb{Q}$. We can require that every rational number occurs *infinitely* often in the sequence (x_n). Indeed, this follows since the set of all pairs $(r, s) \in \mathbb{Z}^2$, with $s \neq 0$, is countable and so if $q = r/s$, then $(nr)/(ns) = q$ for all $n \in \mathbb{N}$. Infinity is very elastic. ♠

Definition 2.2.3 The sequence $(x_n) \subset \mathbb{R}$ *is convergent* with limit $x \in \mathbb{R}$ if, for all $\varepsilon > 0$, there exists an $N \in \mathbb{N}$ such that

$$|x - x_n| < \varepsilon, \; n \geq N.$$

We write this as $\lim_{n \to \infty} x_n = x$. We say that x is the *limit* of the sequence (x_n) or that the sequence (x_n) *converges* to x.

Remarks 2.2.4

(1) In limit definitions, we generally use capitals $M, N, \ldots$ for the bounds—this is in contrast to the way we used capitals in the previous chapter for decimal truncations. Nevertheless, we continue to use the notation (x^N) for the sequence of decimal truncations of a real number x.

(2) If the sequence (x_n) is convergent, then the limit is unique. Intuitively this is clear: it is not possible to be arbitrarily close to distinct points. We leave the formal details as an exercise for the reader.

(3) The definition of convergence works perfectly well within the framework of the rational numbers. In this case, we require $(x_n) \subset \mathbb{Q}$ and $x, \varepsilon \in \mathbb{Q}$. We showed in Chap. 1 that the geometric series $\sum ar^n$ always converges in $\mathbb{Q}$ if $a, r \in \mathbb{Q}$ and $|r| < 1$. As we shall see, this is quite exceptional. In general, infinite sequences

or series of rational numbers will not converge in $\mathbb{Q}$ even though they converge in $\mathbb{R}$.

(4) The definition of convergent sequence suffers from the defect that it includes the limit x. Later we shall see that providing we work with the real numbers $\mathbb{R}$ (as opposed to the rationals $\mathbb{Q}$), it is possible to give an *intrinsic* definition of a convergent sequence that does not depend explicitly on the limit x. This is significant as in many cases it is possible to prove convergence without knowing the limit.   ✠

We give some equivalent ways of formulating the limit definition in the next lemma.

Lemma 2.2.5 *Let (x_n) be a sequence of real numbers and $x \in \mathbb{R}$. The following statements are equivalent.*

(1) $\forall \varepsilon > 0, \exists N \in \mathbb{N}$ *such that* $|x - x_n| < \varepsilon$ *for all* $n \geq N$.
(2) $\forall \varepsilon > 0, \exists N \in \mathbb{N}$ *such that* $|x - x_n| \leq \varepsilon$ *for all* $n \geq N$.
(3) $\forall m \in \mathbb{N}, \exists N \in \mathbb{N}$ *such that* $|x - x_n| < 10^{-m}$ *for all* $n \geq N$.
(4) *There exists a sequence (κ_m) of strictly positive numbers converging to zero such that* $\forall m \in \mathbb{N}, \exists N \in \mathbb{N}$ *such that* $|x - x_n| < \kappa_m$ *for all* $n \geq N$.
(5) *For every sequence (κ_m) of strictly positive numbers converging to zero, $\forall m \in \mathbb{N}, \exists N \in \mathbb{N}$ such that $|x - x_n| < \kappa_m$ for all $n \geq N$.*

(In statements (3,4,5), we can replace $<$ by $\leq$ as in (2).)

Proof We need to show that if $p, q \in \{1, \ldots, 5\}, p \neq q$, then $(p) \implies (q)$. That is, if the sequence converges according to (p), then it converges according to (q).

We start by proving the equivalence of (1) and (2). (1) $\implies$ (2) is obvious since $\varepsilon \leq \varepsilon$. For the converse, suppose convergence according to (2). Given $\varepsilon/2 > 0$, we can choose $N \in \mathbb{N}$ such that $|x - x_n| \leq \varepsilon/2$ for all $n \geq N$. Since $\varepsilon/2 < \varepsilon$, we have $|x - x_n| < \varepsilon$ for all $n \geq N$ and so we have convergence according to (1).

Turning to the remaining statements, we have (5) $\implies$ (3,4) and (3) $\implies$ (4) ((5) is the strongest statement, (4) the weakest). Hence, it suffices to show that (1) $\implies$ (4) $\implies$ (5) $\implies$ (1). For (1) $\implies$ (4), take $\kappa_m = \frac{1}{m}$ and apply (1) with $\varepsilon = \kappa_m$. Next suppose (4) holds with the sequence (κ_ℓ). Let (ρ_m) be any sequence of strictly positive numbers converging to zero. Given $m \in \mathbb{N}, \rho_m > 0$ and so, since (κ_ℓ) converges to zero, there exists an $\ell_0 \in \mathbb{N}$ such that $0 < \kappa_{\ell_0} \leq \rho_m$. Hence, by (4), there exists an $N \in \mathbb{N}$ such that $|x - x_n| < \kappa_\ell$ for all $n \geq N$. Since $\kappa_\ell \leq \rho_m$, $|x - x_n| < \rho_m$ for all $n \geq N$, proving that (5) holds. Finally, we show that (5) $\implies$ (1). For this, it is enough to define $\kappa_m = \varepsilon/m$ and apply (5) with $m = 1$.   □

Remarks 2.2.6

(1) Statements (3,4) of the lemma are the easiest to work with as they only require verification of a countable number of conditions. On the other hand, (1,2,5) require verification of an uncountable number of conditions.
(2) There is no loss of generality in requiring in (1,2) that $\varepsilon \in \mathbb{Q}$ and in (4,5) that $(\kappa_m) \subset \mathbb{Q}$.   ✠

Example 2.2.7 Let $x \in \mathbb{R}$ have decimal expansion $x = x_0.x_1x_2\cdots$. For $N \geq 1$, define $x^N = x_0.x_1\cdots x_N$. Then (x^N) is convergent and $\lim_{N\to\infty} x^N = x$. This follows since

$$|x - x_N| \leq 10^{-N}, \ N \geq 1,$$

and so we may use the convergence statement (3) of Lemma 2.2.5 (with $<$ replaced by $\leq$). ♠

In subsequent sections, we often make use of the well-known *squeezing lemma*. We give the statement for reference and leave the straightforward proof to the exercises.

Lemma 2.2.8 *If $(a_n), (b_n), (c_n)$ are sequences of real numbers which satisfy*

(1) $a_n \leq x_n \leq b_n$ *for all $n \geq 1$ (for large enough n suffices);*
(2) $\lim_{n\to\infty} a_n, \lim_{n\to\infty} b_n$ *exist and have the same limit, say $x^\star$,*

then (x_n) is convergent and has limit $x^\star$.

Examples 2.2.9 We give some examples of convergence and applications of the squeezing lemma. Most of the examples require multiplication of real numbers and a very limited knowledge of rational exponents and roots. In every case, the limit will be rational. The examples will not be used in the theoretical developments in this chapter. Indeed, the gaps will be filled in subsequent sections and exercises.

(1) Let $p, q \in \mathbb{N}$, $(p, q) = 1$, and set $\alpha = p/q > 0$. We claim that $(n^{-\alpha})$ converges to zero. Suppose first that $p = q = 1$. Given $\varepsilon > 0$, choose $N \in \mathbb{N}$ such that $N > 1/\varepsilon$ (this uses the Archimedean property of $\mathbb{R}$). We have $|0 - n^{-1}| < \varepsilon$, $n \geq N$, and so $\lim_{n\to\infty} n^{-1} = 0$. Suppose $p > 1$, $q = 1$. Since $n^p \geq n$, $1/n \geq 1/n^p$, for all $n \in \mathbb{N}$. Hence, taking $a_n = 0$, $b_n = n^{-p}$, and $c_n = 1/n$ in the statement of the squeezing lemma, we have $\lim_{N\to\infty} n^{-p} = 0$. Next take $q > 1$, $p = 1$, and recall that $n^{-1/q}$ is the positive qth root of $1/n$. If $n = m^q$, then $n^{-1/q} = 1/m$. Taking $\kappa_m = 1/m$ in Lemma 2.2.5(4), we see that $|0 - n^{-1/q}| < 1/m$, for all $n \geq N = m^q$. Hence $\lim_{n\to\infty} n^{-1/q} = 0$. Finally, $\lim_{n\to\infty} n^{-p/q} = \lim_{n\to\infty} (n^{-1/q})^p = (\lim_{n\to\infty} n^{-1/q})^p = 0$ by standard properties of limits and multiplication of real numbers. (The result extends to strictly positive exponents $\alpha \in \mathbb{R}$. However, for this we need properties of the log and exponential function—see section "Appendix: The Log and Exponential Functions".)

(2) We claim that if $x > 0$, then $(x^{1/n})$ converges to 1. Suppose first that $x > 1$ so that $x^{1/n} > 1$. Set $x_n = x^{1/n} - 1 > 0$. By the binomial theorem $x = (1 + x_n)^n \geq 1 + nx_n$, $n \geq 1$, and so

$$0 < x_n \leq \frac{x-1}{n}.$$

The result follows by (1) and Lemma 2.2.8. If $0 < x < 1$, apply the previous argument to $y = 1/x$.

(3) The sequence $(n^{1/n})$ converges to 1. Set $n^{1/n} = 1 + x_n$. Clearly $x_n \geq 0$. Applying the binomial theorem we have

$$n = (1 + x_n)^n \geq 1 + \frac{n(n-1)}{2} x_n^2,$$

and so

$$0 \leq x_n \leq \sqrt{\frac{2}{n}}.$$

The result follows by (1) and the squeezing lemma.

(4) If $r \in (-1, 1)$, the geometric sequence (r^n) converges and has limit zero. Suppose that $r \in (0, 1)$. Define $x > 0$ by $r = 1/(1 + x)$. Then $r^{-n} = (1 + x)^n \geq nx$ and so $0 \leq r^n \leq x^{-1} n^{-1}$. The result follows by (1) and the squeezing lemma. If $r \in (-1, 0)$, then the same argument shows that $|r|^n \to 0$ and hence $\lim_{n \to \infty} r^n = 0$ (by the definition of the limit). ♠

2.2.2 Subsequences

Definition 2.2.10 Let (x_n) be a sequence of real numbers. A *subsequence* (x_{n_j}) of (x_n) is a sequence of the form $x_{n_1}, x_{n_2}, \cdots$ where $1 \leq n_1 < n_2 < \cdots$. That is, it is a sequence (z_j) where $z_j = x_{n_j}, j \geq 1$.

Remark 2.2.11 If $(x_n) \subset \mathbb{R}$ is a sequence then every countably infinite subset K of $\mathbb{N}$ uniquely determines a subsequence. Indeed, if K is a countably infinite subset of $\mathbb{N}$, we may write K uniquely as $K = \{n_j \mid j \in \mathbb{N}\}$, where $1 \leq n_1 < n_2 < \cdots$. We define the sequence (z_j) by $z_j = x_{n_j}, j \in \mathbb{N}$. ✖
We leave the proof of the next lemma as an exercise.

Lemma 2.2.12 Let (x_n) be a convergent sequence with limit x. Every subsequence (x_{n_j}) of (x_n) is convergent with limit x.

Examples 2.2.13

(1) Define the sequence (x_n) by $x_n = n$, $n \geq 1$. As a subsequence we could take (x_{n_p}) where x_{n_p} denotes the pth prime number (so (x_{n_p}) is the sequence $2, 3, 5, 7, 11, \cdots$). It is clear that (x_n) has no convergent subsequences.

(2) Let (x_n) be a sequence such that $\{x_n \mid n \geq 1\} = \mathbb{Q}$. Obviously, (x_n) is not convergent. However, for every $x \in \mathbb{R}$, we can construct a subsequence (x_{n_j}) of (x_n) which converges to x (this does not contradict Lemma 2.2.12—a non-convergent sequence may have many convergent subsequences). Suppose then that $x \in \mathbb{R}$. We give an inductive construction for a subsequence converging to x which will repeatedly use that $\mathbb{Q} \cap (a, b)$ is countably infinite for all open intervals (a, b), $a < b$ (Exercises 1.6.23(10)). We define x_{n_1} by taking $n_1 \geq 1$ to

be the smallest integer such that $x_{n_1} \in (x - 10^{-1}, x + 10^{-1})$. Suppose we have constructed $x_{n_1}, \cdots, x_{n_{m-1}}$ so that $1 \leq n_1 < \cdots < n_{m-1}$ and

$$x_{n_j} \in (x - 10^{-j}, x + 10^{-j}), \quad j = 1, \cdots, m - 1.$$

Choose n_m to be the smallest integer greater than n_{m-1} such that $x_{n_m} \in (x - 10^{-m}, x + 10^{-m})$. That we can choose n_m follows since $(x - 10^{-m}, x + 10^{-m}) \cap \mathbb{Q}$ contains $(x - 10^{-m}, x + 10^{-m}) \cap (\mathbb{Q} \setminus \{x_1, \cdots, x_{n_{m-1}}\})$, which is countably infinite. This completes the inductive construction of (x_{n_j}). Since $|x - x_{n_m}| < 10^{-m}$, $m \geq 1$, it follows from Lemma 2.2.5 (statement (4) this time) that (x_{n_m}) converges to x.   ♠

EXERCISES 2.2.14

(1) Show that if a sequence is convergent, then the limit is unique.
(2) Prove the squeezing lemma (Lemma 2.2.8). Is the result true if we work over the rational numbers?
(3) Using the squeezing lemma, and a little algebra, show that the sequence (x_n), $x_n = \sqrt{n+1} - \sqrt{n}$ is convergent. What is the limit?
(4) Suppose that the sequences (a_n) and (b_n) are convergent with respective limits $a^\star$ and $b^\star$. Show that if $a_n \leq b_n$ for all $n \in \mathbb{N}$, then $a^\star \leq b^\star$. Can any more be said if $a_n < b_n$ for all $n \in \mathbb{N}$? What about if $a_n \leq b_n$ for all sufficiently large n?
(5) Prove Lemma 2.2.12.
(6) Find a countable infinite subset X of $\mathbb{R}$ such that if $(x_n) \subset X$ is convergent, then (x_n) is eventually constant and the limit of (x_n) lies in X ((x_n) is *eventually constant* if $\exists x, \exists N \in \mathbb{N}$ such that $x_n = x, n \geq N$).
(7) Let X be a non-empty subset of $\mathbb{R}$. A point $x \in \mathbb{R}$ is a *closure point* of X if we can find a sequence $(x_n) \subset X$ which converges to x. Denote the set of closure points of X by $\overline{X}$. Why is it true that $\overline{X} \supset X$?

  (a) Find an example of a countably infinite unbounded set X of $\mathbb{R}$ such that $\overline{X} = X$.
  (b) Find an example of a countably infinite bounded subset X of $\mathbb{R}$ such that $\overline{X} = X$.
  (c) Find an example of a countably infinite bounded subset of $[0, 1]$ such that $\overline{X} \setminus X = \{0, \frac{1}{2}, 1\}$.
  (d) Find an example of a countably infinite subset X of $\mathbb{R}$ such that $\overline{X} = \mathbb{R}$.

(8) Suppose that (x_n) is convergent. Let $\sigma : \mathbb{N} \to \mathbb{N}$ be a bijection. Prove that $(x_{\sigma(n)})$ is convergent and has the same limit as (x_n).
(9) Write the set $\mathbb{Q}$ of all rational numbers as a sequence $(q_n)_{n \geq 1}$ where we assume $q_n \neq q_m$ if $n \neq m$. Given $\varepsilon > 0$, define $I_n = (q_n - \varepsilon 2^{-(n+1)}, q_n + \varepsilon 2^{-(n+1)})$, $n \geq 1$, and set $I = \cup_{n \in \mathbb{N}} I_n$.

  (a) If $|I_n|$ denotes the length of I_n, show that $\sum_{n=1}^{\infty} |I_n| = \varepsilon$.
  (b) Show that the set $X = \mathbb{R} \setminus I$ contains no proper subintervals even though the 'length' of the complement I is at most ε.

(c) The set X consists of irrational numbers. Does X contain all the irrational numbers? Why/Why not?

2.3 Bounded Subsets of $\mathbb{R}$ and the Supremum and Infimum

Definition 2.3.1 Let A be a subset of $\mathbb{R}$.

(1) A is *bounded above* if A is nonempty and $\exists M \in \mathbb{R}$ such that $M \geq x$ for all $x \in A$. We call M an *upper bound* for A.
(2) A is *bounded below* if A is nonempty and $\exists m \in \mathbb{R}$ such that $m \leq x$ for all $x \in A$. We call m a *lower bound* for A.
(3) A is *bounded* if A is bounded above and below.

Examples 2.3.2

(1) $\mathbb{N}$ is bounded below ($m \leq 1$ works) but not bounded above.
(2) $\mathbb{Z}$ is unbounded.
(3) If $a < b$ are real numbers, then (a, b), $[a, b]$ are bounded. In both cases we can take as upper bound any $M \geq b$ and as lower bound any $m \leq a$.
(4) A non-empty subset A of $\mathbb{R}$ is bounded iff $\exists R \geq 0$ such that $A \subset [-R, R]$. ♠

The next two lemmas turn out to be very useful in our discussion of upper and lower bounds and convergence of bounded sequences.

Lemma 2.3.3 *Let A be a nonempty subset of $\mathbb{R}$. If $M + \varepsilon$ is an upper bound for A for all $\varepsilon > 0$, then M is an upper bound for A. An analogous result holds for lower bounds.*

Proof We prove by contradiction. If M is not an upper bound, there exists an $a \in A$ such that $M < a$. Take $\varepsilon = (a - M)/2 > 0$ and observe that $M + \varepsilon = (M + a)/2 < a$, contradicting our assumption that $M + \varepsilon$ is an upper bound for A for all $\varepsilon > 0$. □

Remark 2.3.4 The proof uses Sect. 1.6.7 on multiplication of real numbers by rational numbers—in this case by $\frac{1}{2}$. We can avoid multiplication by choosing $z \in \mathbb{R}$ satisfying $M < z < a$ (Example 1.6.7) and taking $\varepsilon = z - M$. ✠

Lemma 2.3.5 *Let (x_n) be a sequence of real numbers and suppose that $x_n \leq M$ for all $n \in \mathbb{N}$. If $\lim_{n\to\infty} x_n = x^\star$, then $x^\star \leq M$. An analogous result holds for lower bounds.*

Proof We prove by contradiction. Suppose $x^\star > M$. Take $\varepsilon = (x^\star - M)/2 > 0$. Since (x_n) converges to $x^\star$, there exists an $N \in \mathbb{N}$ such that $|x^\star - x_N| < \varepsilon$. Hence $x_N \geq x^\star - \varepsilon > M$. Contradiction. □

Remark 2.3.6 If $x_n < M$ in the statement of Lemma 2.3.5, then we can only infer that $x^\star \leq M$. Typically limits preserve '$\leq$' but *not* strict inequality. ✠

Definition 2.3.7 Let A be a nonempty subset of $\mathbb{R}$.

(1) Suppose A is bounded above. A *least upper bound* for A, or *supremum* for A, is a real number M such that

    (a) M is an upper bound for A.
    (b) If M' is any upper bound for A, then $M \leq M'$.

(2) Suppose A is bounded below. A *greatest lower bound* for A, or *infimum* for A, is a real number m such that

    (a) m is a lower bound for A.
    (b) If m' is any lower bound for A, then $m' \leq m$.

Lemma 2.3.8 *Suppose A is bounded above. If the supremum of A exists, it is unique. Similarly for the infimum of A, if A is bounded below.*

Proof (1) Suppose that M, M' are supremums of A. Then by property (a), both M and M' are upper bounds of A. Since M is a supremum of A, it follows by (b) that $M \leq M'$. Applying the same argument with the roles of M, M' interchanged, we get $M' \leq M$. Hence $M = M'$. We may apply a similar argument to prove the uniqueness of the infimum. $\qquad\square$

Remarks 2.3.9

(1) If they exist, we denote the supremum and infimum of A by $\sup(A)$ and $\inf(A)$ respectively. Alternative and commonly used notations are $\mathrm{lub}(A)$ for $\sup(A)$, and $\mathrm{glb}(A)$ for $\inf(A)$.
(2) In analysis, we only define the maximum of A, $\max(A)$, and minimum of A, $\min(A)$, if A is a finite subset of $\mathbb{R}$. This is a little confusing as we do refer to the maximum and minimum values of a function.   ✳

Lemma 2.3.10 *Suppose $\sup(A)$ exists. Define $-A = \{-a \mid a \in A\}$ and, for $x \in \mathbb{R}$, $A + x = \{a + x \mid a \in A\}$.*

(1) $\inf(-A)$ *exists and equals* $-\sup(A)$.
(2) $\sup(A + x)$ *exists and equals* $\sup(A) + x$.

Similarly, if $\inf(A)$ exists, $\sup(-A) = -\inf(A)$, $\inf(A + x) = \inf(A) + x$, $x \in \mathbb{R}$.

Proof We leave this as an exercise. $\qquad\square$

    We have the following necessary and sufficient condition for the existence of $\sup(A)$.

Lemma 2.3.11 *Let A be a subset of $\mathbb{R}$ which is bounded above and let $M \in \mathbb{R}$. Suppose that*

(1) *M is an upper bound for A.*
(2) *For every $\varepsilon > 0$, there exists an $x \in A$ such that $x > M - \varepsilon$.*

Then $\sup(A) = M$. Conversely, if $M = \sup(A)$, then (1,2) are satisfied. A similar criterion holds for the infimum of A.

Proof If (2) fails then there exists an $\varepsilon > 0$ such that $M - \varepsilon \geq x$ for all $x \in A$. Hence $M - \varepsilon$ is an upper bound for A and M cannot be the supremum of A. Hence conditions (1,2) are necessary for M to be the supremum of A. Conversely, suppose (1,2) hold. Since (2) holds, $M - \varepsilon$ is not an upper bound of A for all $\varepsilon > 0$. Since, by (1), M is an upper bound of A, M must be the least upper bound of A. ☐

Theorem 2.3.12 *Let* $A \subset \mathbb{R}$ *be bounded above. Then* $\sup(A)$ *exists. Similarly, if* A *is bounded below,* $\inf(A)$ *exists.*

Proof Let A be bounded above. It is enough by Lemma 2.3.10 to prove the existence of the supremum of $A + x$ for some $x \in \mathbb{R}$ since $\sup(A) = \sup(A + x) - x$. It follows that there is no loss of generality in requiring that A contains a point $a > 0$ and so we may and shall assume that 0 is not an upper bound of A. This assumption simplifies the proof as we avoid having to make a separate argument to handle negative upper bounds.

We use an inductive technique to construct the decimal expansion $\alpha = \alpha_0.\alpha_1\alpha_2\cdots$ of $\sup(A)$. Specifically, we construct a positive sequence $\alpha^P = \alpha_0.\alpha_1\alpha_2\cdots\alpha_P$, $P \geq 0$, of decimal truncations of α. The truncations will satisfy

(a) α^P is not an upper bound of A, $P \geq 0$.
(b) $\alpha^P + 10^{-P}$ is an upper bound of A, $P \geq 0$.
(c) If $0 \leq P < Q$, then α^P and α^Q agree to the first P decimal places.
(d) The sequence $(\alpha^P)_{P \geq 0}$ is increasing: $\alpha^P \leq \alpha^Q$, if $P < Q$.

Let L be the smallest integer which is an upper bound for A. Since 0 is not an upper bound, $L \in \mathbb{N}$. Define $\alpha^0 = \alpha_0 = L - 1 \geq 0$.

Proceeding inductively, suppose that we have constructed $\alpha^j = \alpha_0.\alpha_1\alpha_2\cdots\alpha_j$ satisfying (a–d) for $j < P$. Consider $Z_p = \alpha^{P-1} + p10^{-P}$, $0 \leq p \leq 10$. We have that Z_{10} is an upper bound of A (using (b) for α^{P-1}), but $Z_0 = \alpha^{P-1}$ is not an upper bound (by (a)). Choose $p \in \{0, \cdots, 9\}$ so that Z_{p+1} is an upper bound but Z_p is not. Define $\alpha_P = p$, $\alpha^P = \alpha_0.\alpha_1\alpha_2\cdots\alpha_P = \alpha^{P-1} + p10^{-P}$. This completes the inductive step.

It is immediate from (c) that $\lim_{n\to\infty} \alpha^P$ converges to the real number $\alpha = \alpha_0.\alpha_1\alpha_2\cdots$. We claim that $\alpha = \sup(A)$.

By property (b), $\alpha^P + 10^{-P}$ is an upper bound for A for all $P \geq 0$. Since (α^P) is an increasing sequence and $\alpha^P \leq \alpha$ for all P, $\alpha + 10^{-P}$ is an upper bound for A for all $P \geq 0$. Hence α is an upper bound of A (Lemma 2.3.3). We need to show α is the least upper bound. Suppose β is an upper bound of A. Then $\beta > \alpha^P$, for all $P \geq 0$ (property (a)). Hence, by Lemma 2.3.5, $\beta \geq \lim_{P\to\infty} \alpha^P = \alpha$ and so $\sup(A) = \alpha$. The result for infimums can be proved along the same lines or, more simply, by using Lemma 2.3.10. ☐

Remarks 2.3.13

(1) The proof of Theorem 2.3.12, which depends on a 'subdivide and conquer' technique, is carefully constructed so as to make transparent the convergence of the sequence (α^P) to the supremum of A. We do this in two ways. First,

the sequence (α^P) is an increasing sequence which obviously converges to α. Secondly, α^P is not an upper bound but $\alpha^P + 10^{-P}$ is an upper bound.

(2) The proof does not work over the rational numbers. There is no reason why the sequence (α^P) should converge to a rational number.

(3) The existence of the supremum is sometimes taken as an Axiom for the real numbers. The point of the proof is that if one thinks of real numbers as being decimal expansions, then it is straightforward to construct the supremum directly. In particular, we construct the supremum as a sequence of rational approximations.   ✹

Examples 2.3.14

(1) Theorem 2.3.12 fails if we work over the rational numbers. The easiest example is found by defining $A = \{x_0.x_1 \cdots x_n \mid n \geq 1\}$, where $\sqrt{2} = x_0.x_1 \cdots$. If $\sup(A)$ existed it would have to be $\sqrt{2}$ but $\sqrt{2} \notin \mathbb{Q}$. Other examples can be found by constructing (according to some simple rule) a non-periodic non-terminating decimal expansion and then taking the set of decimal truncations. For example, define $x^\star = 1.0^1 10^2 10^3 1 \cdots 0^n 1 \cdots$ (here 0^p signifies a string of p zeros).

(2) Let $\{I_j = (a_j, b_j) \mid j \in J\}$ be a set of open intervals of $\mathbb{R}$. Suppose that $\cap_{j \in J} I_i \neq \emptyset$—that is, the intervals I_i share at least one common point, say x_0. We claim that $I = \cup_{j \in J} I_i$ is an open interval (we regard $(-\infty, b)$, (a, ∞) and $\mathbb{R}$ as open intervals). Suppose the sets $\{a_j \mid j \in J\}$, $\{b_j \mid j \in J\}$ are bounded subsets of $\mathbb{R}$ (we leave the case where one or both of these sets is unbounded to the exercises). Set $a^\star = \inf\{a_j \mid j \in J\}$, $b^\star = \sup\{b_j \mid j \in J\}$. It suffices to show that $I = (a^\star, b^\star)$. By definition of the supremum and infimum, given $\varepsilon > 0$, there exist $\ell, m \in J$ such that $a^\star \geq a_\ell - \varepsilon$, $b^\star \leq b_m + \varepsilon$. Since I_ℓ, I_m share the common point x_0, and $a_\ell < x_0 < b_m$, $I_\ell \cup I_m \subset I$ and so $(a^\star + \varepsilon, b^\star - \varepsilon) \subset I$. Since this is true for all $\varepsilon > 0$, we see that $(a^\star, b^\star) \subset I$. On the other hand, $a^\star, b^\star \notin I$. Indeed, if $a^\star \in I$, this would imply that there exists an $m \in J$ such that $a^\star \in I_m$. But then $a_m < a^\star$ and so $a^\star$ could not be a lower bound for $\{a_j \mid j \in J\}$. A similar argument applies to $b^\star$. We have shown that $I = (a^\star, b^\star)$. This result is *false* if we work over the rational numbers (the open interval $(a, b)^{\mathbb{Q}}$ of $\mathbb{Q}$ is defined to be the set $\{x \in \mathbb{Q} \mid a < x < b\}$, where $a, b \in \mathbb{Q}$). We leave the construction of an explicit counterexample to the exercises.   ♠

Remark 2.3.15 It is useful to extend the definition of sup and inf to allow for unbounded sets, Thus, if A is not bounded above, we set $\sup(A) = +\infty$ and if A is not bounded below, we set $\inf(A) = -\infty$.   ✹

2.3.1 Applications to Sequences and Series

Let (a_n) be a sequence of real numbers. Recall that the sequence (a_n) *diverges to* $+\infty$ if for every $M \in \mathbb{R}$, there exists an $N \in \mathbb{N}$ such that $a_n \geq M$ for all $n \geq N$. We often write this as $\lim_{n \to \infty} a_n = +\infty$. We may similarly define divergence to $-\infty$.

We have already made use of increasing sequences in the proof of Theorem 2.3.12. For completeness, we give some formal definitions.

Definition 2.3.16 The sequence (a_n) of real numbers is *increasing* if $a_1 \leq a_2 \leq a_3 \leq \cdots$. That is, $a_n \leq a_m$ whenever $n < m$. The sequence is *strictly increasing* if $a_n < a_m$ whenever $n < m$ and is *eventually increasing* if there exists an $N \in \mathbb{N}$ such that $a_n < a_m$ whenever $N \leq n < m$. We similarly may define *decreasing, strictly decreasing* and *eventually decreasing* sequences.

Definition 2.3.17 The sequence (a_n) of real numbers is *bounded above* if $\{a_n \mid n \in \mathbb{N}\} \subset \mathbb{R}$ is bounded above. The sequence is *bounded below* if $\{a_n \mid n \in \mathbb{N}\}$ is bounded below.

The next result is both simple to state and very special to the real numbers. It provides a gateway into the study of convergence of sequences and series of real numbers. The result is false for sequences of rational numbers.

Theorem 2.3.18 *Let (a_n) be an increasing sequence of real numbers.*

(1) *If (a_n) is not bounded above, then $\lim_{n \to \infty} a_n = +\infty$.*
(2) *If (a_n) is bounded above then (a_n) is convergent and $\lim_{n \to \infty} a_n = \sup\{a_n \mid n \in \mathbb{N}\}$.*

A similar result holds for decreasing sequences. The results also hold for eventually increasing (or decreasing) sequences provided that we take the supremum (or infimum) over the increasing (or decreasing) part of the sequence.

Proof Set $A = \{a_n \mid n \geq 1\}$. Suppose first that A is not bounded. Then for every $M \in \mathbb{R}$, there exists an $N \in \mathbb{N}$ such that $a_N \geq M$. Since (a_n) is increasing, $a_n \geq M$, for all $n \geq N$. Hence $\lim_{n \to \infty} a_n = +\infty$. If $A = \{a_n \mid n \geq 1\}$ is bounded, Theorem 2.3.12 applies and we can define $a^\star = \sup(A)$. We claim (a_n) is convergent with limit $a^\star$. Certainly $a_n \leq a^\star$ for all $n \geq 1$ ($a^\star$ is an upper bound). Further, for every $\varepsilon > 0$, there exists an $N \in \mathbb{N}$ such that $a_N > a^\star - \varepsilon$ (otherwise $a^\star$ would not be the least upper bound). Since (a_n) is increasing, and bounded above by $a^\star$, we have $a^\star \geq a_n > a^\star - \varepsilon$ for all $n \geq N$. That is, $|a^\star - a_n| < \varepsilon$, $n \geq N$. Hence $\lim_{n \to \infty} a_n$ exists and equals $a^\star$. We leave the proofs of the remaining parts of the theorem to the reader. $\qquad \square$

Theorem 2.3.18 has the following important and useful corollary.

Theorem 2.3.19 *Let $\sum_{i=1}^{\infty} a_i$ be a series of (eventually) positive terms. Then either $\sum_{i=1}^{\infty} a_i$ diverges to $+\infty$ or $\sum_{i=1}^{\infty} a_i$ converges. In particular, $\sum_{i=1}^{\infty} a_i$ converges iff the sequence $(S_n = \sum_{i=1}^{n} a_i)$ of partial sums is bounded.*

Proof For $n \geq 1$, define the nth partial sum $S_n = \sum_{i=1}^{n} a_i$. We recall that, by definition, $\sum_{i=1}^{\infty} a_i$ converges iff the sequence (S_n) of partial sums converges. Since it is assumed that the terms in the series are (eventually) positive, it follows that the sequence (S_n) is (eventually) increasing. The result follows by Theorem 2.3.18. $\qquad \square$

Remark 2.3.20 The significance of Theorems 2.3.18 and 2.3.19 is that they give a criterion for convergence that *does not require us to know the limit*. For example,

Theorem 2.3.19 implies that an infinite series of positive terms either converges or diverges to $+\infty$. Aside from the implicit upper bound, nothing is stated or needed about the actual value of the limit.   ✠

2.3.2 Multiplication and Division of Real Numbers

With Theorem 2.3.18, we have all the necessary tools to define multiplication and division of real numbers.

Multiplication Suppose that $x, y \in \mathbb{R}_+$. For $N \geq 1$, let x^N and y^N denote the truncations of x and y to N-terms. Set $z^N = x^N y^N$. Then $(z^N)_{N \geq 1}$ is an increasing sequence of rational numbers bounded above by $(x_0 + 1) \times (y_0 + 1)$. Hence, $\lim_{N \to \infty} z^N$ exists by Theorem 2.3.18(2). We define the product xy to be $\lim_{N \to \infty} z^N$. We extend the definition of product to negative numbers in a way consistent with multiplication of negative rationals: if $x, y < 0$, define $xy = (-x)(-y)$ and if x and y are of opposite sign, define $xy = -(-x)y = -x(-y)$. In all cases, we have $xy = \lim_{N \to \infty} x^N y^N$.

With multiplication defined in terms of rational approximation, it is straightforward to extend standard results on rationals to real numbers.

Examples 2.3.21

(1) For all $x, y, z \in \mathbb{R}$, $x(y + z) = xy + xz$ (distributive law). Indeed,

$$x(y + z) = \lim_{N \to \infty} x^N(y^N + z^N) = \lim_{N \to \infty} x^N y^N + \lim_{N \to \infty} x^N z^N = xy + xz,$$

where we have used the distributive law for rationals, $x^N(y^N + z^N) = x^N y^N + x^N z^N$.

(2) If $x, y \in \mathbb{R}$ and $xy = 0$, then either $x = 0$ or $y = 0$. It is enough to show that if $x, y \neq 0$, then $xy \neq 0$. Without loss of generality take $x, y > 0$. Choose $p, q \in \mathbb{N}$ so that $x_p, y_q \neq 0$. Then $x^N \geq 10^{-p}$, all $N \geq p$, and $y^N \geq 10^{-q}$, all $N \geq q$. Therefore $x^N y^N \geq 10^{-(p+q)}$ for all $N \geq p + q$. Hence $xy = \lim_{N \to \infty} x^N y^N \geq 10^{-(p+q)} > 0$.   ♠

Division Since multiplication of real numbers is now defined, it suffices to define the reciprocal x^{-1} for $x > 0$. The sequence (x^N) is increasing and bounded above by $x_0 + 1$, and so $(\frac{1}{x^N})$ is a decreasing sequence bounded below by $1/(x_0 + 1)$ (the sequence is defined provided $x^N \neq 0$ which is true for large enough N). Hence, by Theorem 2.3.18, we may define $x^{-1} = \lim_{N \to \infty} \frac{1}{x^N}$.

Example 2.3.22 Equipped with division, Examples 2.3.21(2) follows by multiplying the equation $xy = 0$ by x^{-1} if $x \neq 0$.   ♠

Remark 2.3.23 We refer to the exercises for effective ways of computing the reciprocal and the roots $x^{1/p}$ of $x \in \mathbb{R}_+$, $p \geq 2$.   ✠

2.3.3 Examples of Convergent Sequences

For the remainder of this section, we show how we can use Theorem 2.3.18 to give simple proofs of convergence for some basic geometric sequences.

Lemma 2.3.24 *Let $x \in \mathbb{R}$ and consider the sequence (x^n).*

(1) *If $x \in (-1, 1)$, (x^n) converges to 0.*
(2) *If $x = 1$, (x^n) converges to 1.*
(3) *If $x > 1$, (x^n) diverges to $+\infty$.*
(4) *If $x \leq -1$, (x^n) is divergent.*

Proof Statements (2,4) are obvious; we prove (1,3) using Theorem 2.3.18 together with standard facts about limits.

If $x = 0$, (1) is immediate so suppose $x \in (0, 1)$. Then (x^n) is a (strictly) decreasing sequence bounded below by 0. Hence, by Theorem 2.3.18, (x^n) converges with limit $x^\star \geq 0$. We have

$$xx^\star = x \lim_{n \to \infty} x^n = \lim_{n \to \infty} x^{n+1} = x^\star$$

and so $xx^\star = x^\star$. Since $0 < x < 1$, $x^\star = 0$. If $x \in (-1, 0)$, then $|x|^n \to 0$ since $|x| \in (0, 1)$. Hence, by the definition of the limit, $\lim_{n \to \infty} x^n = 0$. We prove (3) by observing that if $x > 1$, then $y = x^{-1} \in (0, 1)$ and so by (1), $\lim_{n \to \infty} y^n = 0$. Hence for any $M > 0$, there exists an $N \in \mathbb{N}$ such that $y^n \leq 1/M$, for all $n \geq N$. That is, $x^n \geq M$, $n \geq N$, and so (x^n) diverges to $+\infty$. □

As an immediate corollary of Lemmas 2.3.24 and 2.2.8, we have

Lemma 2.3.25 *Let (a_n) be a sequence. Suppose that there exist $C \geq 0$ and $r \in (0, 1)$ such that*

$$0 \leq |a_n| \leq Cr^n,$$

for all sufficiently large n. Then (a_n) converges and $\lim_{n \to \infty} a_n = 0$.

Example 2.3.26 If $\alpha \in \mathbb{Q}$ and $r \in (-1, 1)$, then the sequence $(n^\alpha r^n)$ is convergent with limit zero. We may assume that $r \neq 0$ and $\alpha > 0$ (since $0 < n^\alpha \leq 1$ if $\alpha \leq 0$). Set $x_n = n^\alpha r^n$. We have

$$\left| \frac{x_{n+1}}{x_n} \right| = \left(1 + \frac{1}{n} \right)^\alpha |r|, \ n \geq 1.$$

Choose $N \in \mathbb{N}$ so that $(1 + \frac{1}{N})^\alpha |r| \leq (|r| + 1)/2 < 1$. Since $((1 + \frac{1}{n})^\alpha)$ is a decreasing sequence, we have

$$\left(1 + \frac{1}{n} \right)^\alpha |r| \leq (|r| + 1)/2, \ n \geq N.$$

Therefore for $n \geq N$ we have

$$|x_n| \leq |x_N| \left(\frac{|r|+1}{2}\right)^{n-N} = |x_N| 2^N (|r|+1)^{-N} \left(\frac{|r|+1}{2}\right)^n,$$

$$= C \left(\frac{|r|+1}{2}\right)^n.$$

The result follows from Lemma 2.3.24 since $|r| + 1 < 2$.   ♠

The examples we have given so far have a rational limit (zero or one) and follow from relatively simple arguments. We end this section with examples where the limit is not rational and elementary arguments no longer suffice to prove convergence.

Examples 2.3.27

(1) Let $A > 0$. Define the sequence (x_n) inductively by $x_1 = 1$, $x_{n+1} = \frac{1}{2}(x_n + \frac{A}{x_n})$, $n \geq 1$. Clearly (x_n) is a sequence of strictly positive real numbers and if $A \in \mathbb{Q}$, then $x_n \in \mathbb{Q}$ for all $n \in \mathbb{N}$. Recall that if $a, b > 0$, then $(a+b)^2 \geq 4ab$ with equality iff $a = b$. Applying this inequality with $a = x_n$, $b = \frac{2}{x_n}$, we get

$$x_{n+1}^2 \geq A, \ n \geq 1,$$

with equality iff $x_n^2 = A$. Since we assumed $x_1 = 1$, we have $x_n^2 \geq A$ for all $n \geq 2$. A simple computation shows that

$$x_{n+1} - x_n = \frac{x_n - x_{n-1}}{2} \left(1 - \frac{A}{x_n x_{n-1}}\right), \ n \geq 3.$$

Since $x_n^2, x_{n-1}^2 \geq A$, $n \geq 3$, we have $x_n x_{n-1} \geq A$ and so $1 - \frac{A}{x_n x_{n-1}} \geq 0$, for all $n \geq 3$. Hence, the sign of $x_{n+1} - x_n$ is the same as that of $x_n - x_{n-1}$ for all $n \geq 3$. Starting with $x_1 = 1$, we compute that

$$x_3 - x_2 = -\frac{(A-1)^2}{2(1+A)} \leq 0.$$

Hence $(x_n)_{n \geq 3}$ is a decreasing sequence bounded below by 0. It follows from Theorem 2.3.18 that (x_n) converges. If we set $\lim_{n \to \infty} = z$, we have $z^2 \geq A$ (since (x_n^2) is bounded below by A). Since $\lim_{n \to \infty} x_{n+1} = \lim_{n \to \infty} x_n = z$, we have

$$z = \lim_{n \to \infty} \frac{1}{2}\left(x_n + \frac{A}{x_n}\right) = \frac{1}{2}\left(z + \frac{A}{z}\right).$$

That is, $z^2 = A$. Hence $z = \sqrt{A}$—the positive square root of A.

Notice that if $A \in \mathbb{Q}$, then $(x_n) \subset \mathbb{Q}$ but the limit is typically irrational. For example, if $A \in \mathbb{N}$ is not a perfect square.

The convergence to $\sqrt{A}$ is fast—the iteration is based on Newton's method applied to the function $f(x) = x^2 - A$. In the exercises we give similar methods for constructing $x^{1/p}$, $x > 0$, and the reciprocal x^{-1}. Not only do these methods give the existence of general roots of positive reals, they also give rapid numerical computation of the roots and division by real numbers.

(2) Suppose $x_n = (1 + \frac{1}{n})^n$, $n \geq 1$. By the binomial theorem

$$\left(1 + \frac{1}{n}\right)^n = 1 + n\frac{1}{n} + \frac{n(n-1)}{2!}\frac{1}{n^2} + \cdots + \frac{1}{n^n}$$

$$= 1 + 1 + \sum_{j=2}^{n} K_n(j) \geq 2,$$

where

$$K_n(j) = \frac{1}{j!}\left(1 - \frac{1}{n}\right)\left(1 - \frac{2}{n}\right)\cdots\left(1 - \frac{j-1}{n}\right).$$

For fixed j, $K_n(j)$ increases with n as do the number of terms in the expansion of $(1 + \frac{1}{n})^n$. Hence $(1 + \frac{1}{n})^n$ is an increasing sequence and must either converge or diverge to $+\infty$. But since $K_n(j) < \frac{1}{j!}$, we have

$$\left(1 + \frac{1}{n}\right)^n < 1 + 1 + \frac{1}{2!} + \cdots + \frac{1}{n!}$$

$$< 1 + 1 + \frac{1}{2} + \frac{1}{2^2} + \cdots + \frac{1}{2^n} < 3,$$

proving that $(1 + \frac{1}{n})^n$ is bounded and therefore converges with limit in $(2.5, 3]$. In Chap. 3 (Proposition 3.5.7), we show that the limit is e. See also Exercises 2.9.10(5) where we give a less elementary proof that uses properties of the logarithm. ♠

EXERCISES 2.3.28

(1) Prove Lemma 2.3.10.
(2) Construct a countable set of open intervals $(a_j, b_j)^{\mathbb{Q}}$ of rational numbers, $a_j < b_j \in \mathbb{Q}$, with a common point x_0, such that $\cup_{j \geq 1}(a_j, b_j)^{\mathbb{Q}}$ is not an open interval in $\mathbb{Q}$ (see Examples 2.3.14(3) for the definition of $(a, b)^{\mathbb{Q}}$).
(3) Find an example of a *countably infinite* bounded subset A of $\mathbb{R}$ such that

   (a) $\sup(A) \in A$.
   (b) For every $\varepsilon > 0$, $\exists x \in A$ such that $x > \sup(A) - \varepsilon$.
   (c) $\inf(A) \notin A$.
   (d) For every $\varepsilon > 0$, $\exists x \in A$ such that $x < \inf(A) + \varepsilon$.

Which of (a,b,c,d) could hold if A were finite? (Hint: construct (a_n) as the union of two sequences (a_{2n}), (a_{2n-1}). You should give explicit definitions for a_{2n} and a_{2n-1}.)

(4) Find an explicit example of a *countably infinite* subset $A = \{a_n \mid n \in \mathbb{N}\}$ of $\mathbb{R}$ such that the following four properties hold:

(a) $\sup(A) = +\infty$.

(b) $\inf(A) = 0 \notin A$.

(c) For every $\varepsilon > 0$, $\exists x \in A$ such that $x < \inf(A) + \varepsilon$.

(d) If (a_{n_k}) is a subsequence of (a_n), then *either* $\lim_{k\to\infty} a_{n_k} = 0$ *or* $\lim_{k\to\infty} a_{n_k} = +\infty$ *or* (a_{n_k}) is not convergent.

(You should construct A to be the union of two sequences, one diverging to $+\infty$, the other converging to 0—see also the hint for the previous question.)

(5) Construct an explicit example of a *countably infinite* subset $A = \{a_n \mid n \in \mathbb{N}\}$ of $\mathbb{R}$ such that the following four properties hold:

(a) $\inf(A) = -\infty$.

(b) $\sup(A) = 1 \notin A$.

(c) For every $\varepsilon > 0$, $\exists x \in A$ such that $x > \sup(A) - \varepsilon$.

(d) If (a_{n_k}) is a subsequence of (a_n), then *either* $\lim_{k\to\infty} a_{n_k} = -\infty$ *or* $\lim_{k\to\infty} a_{n_k} = 1$ *or* (a_{n_k}) is not convergent. Construct explicit subsequences to show that each of these possibilities can occur.

(You should construct A to be the union of three sequences.)

(6) Construct an explicit example of a *countably infinite* subset $A = \{a_n \mid n \in \mathbb{N}\}$ of $\mathbb{R}$ such that the following three properties hold:

(a) $\sup(A) = +\infty$.

(b) $\inf(A) = -\infty$.

(c) If $p \in \mathbb{Z}$, there exists a sequence of distinct points (a_n) of A such that $\lim_{n\to\infty} a_n = p$.

(d) If (a_n) is a convergent sequence of distinct points of A, then (a_n) converges to an integer.

(You should construct A as a countable union of countable sets. Note the use of the word 'distinct'. Without that we could just take $A = \mathbb{Z}$. Why?)

(7) Let $y > 0$ and $a > 0$ such that $ay < 2$. Define the sequence (x_n) by $x_1 = a$, $x_{n+1} = 2x_n - x_n^2 y$, $n \in \mathbb{N}$. Show that (x_n) converges to y^{-1}. (Hints and comments: The maximum value of $f(x) = 2x - x^2 y$ occurs at $x = y^{-1}$ and $f(0) = f(2y^{-1}) = 0$. Show that $x_n \le y^{-1}$, for all $n \ge 2$, and that $(x_n)_{n \ge 2}$ is an increasing sequence. The construction is based on Newton's method and gives a rapidly convergent method for approximating the reciprocal of a real number and so division by real numbers.)

(8) Let $p \in \mathbb{N}$, $p \ge 2$ and $y \in \mathbb{R}$, $y > 0$. This exercise proves the existence of the positive pth root $y^{1/p}$ of y by giving an explicit method for the computation of $y^{1/p}$. Choose $a > 0$ satisfying $a^p < (p + 1)y$. Define the sequence (x_n)

by $x_1 = a$, $x_{n+1} = x_n \frac{(p+1) - x_n^p y^{-1}}{p}$, $n \in \mathbb{N}$. Show that (x_n) converges and that the limit is $y^{1/p}$ (that is, $(\lim_{n\to\infty} x_n)^p = y$). Show also that this is the unique positive pth root of y.

(9) Show that if the sequence (x_n) converges to a, then the sequence $(\frac{x_1 + \cdots + x_n}{n})$ of *arithmetic means* converges to a. Show also that if (x_n) is a sequence of strictly positive numbers with limit $a > 0$, then we have a similar result for *geometric means*:

$$\lim_{n\to\infty} \sqrt[n]{x_1 x_2 \cdots x_n} = a.$$

What can you say if $a = 0$? (Hint for the first part: start by proving the case $a = 0$. Hint for the second part: use logarithms.)

(10) Using the results of the previous question, show that

(a) $\lim_{n\to\infty} \frac{1 + \frac{1}{2} + \cdots + \frac{1}{n}}{n} = 0$.

(b) $\lim_{n\to\infty} \sqrt[n]{n} = 1$.

(c) $\lim_{n\to\infty} \frac{1 + \sqrt{2} + \cdots + \sqrt[n]{n}}{n} = 1$.

(11) Define the sequence (x_n) by $x_0 = 1$,

$$x_{n+1} = 1 + \frac{1}{x_n}, \quad n \geq 0.$$

Show that

(a) (x_n) is a sequence of rational numbers.

(b) $1 < x_n < 2$ for all $n > 0$.

(c) $x_0 < x_2 < x_4 < \ldots < x_{2n} < \ldots < x_{2m+1} < \ldots < x_5 < x_3 < x_1$, $(n, m \geq 3)$.

Verify $\lim_{n\to\infty} x_n$ exists and is irrational. (Hints for (a,b,c): Induction, induction, induction. The limit is known as the *golden mean* or *golden ratio*. The denominators of x_n give the Fibonacci sequence.)

(12) The examples and exercises we have given so far may suggest that sequences $(x_n)_{n\geq 0}$ defined recursively ($x_{n+1} = f(x_n)$, $n \geq 0$) typically converge. This is far from the case. As an example, suitable for computer experimentation, the reader may investigate the sequence $x_{n+1} = L_\lambda(x_n) \overset{\text{def}}{=} \lambda x_n(1 - x_n)$, where $x_0 \in [0, 1]$ (L_λ is called the *logistic map*). Provided $\lambda \in [0, 4]$, $(x_n) \subset [0, 1]$. The sequence (x_n) converges provided $\lambda \in [0, 3]$. However, for $\lambda \in (3, 4]$, the sequence (x_n) typically does not converge and exhibits ever more complex behaviour, including randomness (or 'chaos'), as λ approaches 4 (for more details and references we refer to the text by Strogatz [28, Chap. 10]).

2.4   The Bolzano–Weierstrass Theorem

Theorem 2.4.1 (Bolzano–Weierstrass Theorem) *If X is an infinite bounded subset of $\mathbb{R}$, then there exists a convergent sequence (x_n) consisting of distinct points of X.*

Proof Since X is bounded, there exists a closed interval $I_0 = [a_0, b_0]$ containing X. We construct a sequence of closed intervals $I_n = [a_n, b_n]$, $n \geq 0$, with the following properties

(1) $I_{n+1} \subset I_n$, $n \geq 0$.
(2) (a_n) is an increasing sequence, (b_n) is a decreasing sequence.
(3) $a_n < b_n$, all $n \geq 0$.
(4) $|I_n| \overset{\text{def}}{=} |b_n - a_n| = 2^{-n}|b_0 - a_0|$, $n \geq 0$.
(5) $X \cap I_n$ is infinite, $n \geq 0$.

Our construction of (I_n) is inductive. When $n = 0$, conditions (3,4,5) are automatically satisfied (conditions (1,2) are empty). So suppose we have constructed intervals $I_0, \cdots, I_n$ satisfying (1–5). Let $J = [a_n, \frac{a_n+b_n}{2}]$, $K = [\frac{a_n+b_n}{2}, b_n]$. Note that $|J| = |K| = 2^{-1}|I_n| = 2^{-(n+1)}|b_0 - a_0|$ by (4). Since $J \cup K = I_n$ and $I_n \cap X$ is infinite, one (at least) of $J \cap X$, $K \cap X$ must be infinite. Choose one of J, K so that the intersection is infinite. Denote the corresponding interval by $I_{n+1} = [a_{n+1}, b_{n+1}]$. Since $[a_{n+1}, b_{n+1}] \subset [a_n, b_n]$, we have $a_n \leq a_{n+1} < b_{n+1} \leq b_n$. This completes the inductive step and the construction of the intervals I_n.

Since (a_n) is bounded above by b_m, $m \geq 0$, and (b_n) is bounded below by a_m, $m \geq 0$, both sequences (a_n), (b_n) converge by Theorem 2.3.18 and $a_m \leq \lim_{n\to\infty} a_n \leq \lim_{n\to\infty} b_n \leq b_m$, for all $m \geq 0$. Applying (4), we see that $\lim_{n\to\infty} a_n = \lim_{n\to\infty} b_n$.

It remains to construct a convergent sequence (x_n) of distinct points of X. Since $I_0 \cap X$ is infinite, we can choose $x_0 \in I_0 \cap X$. Proceeding inductively, suppose we have constructed distinct points $x_j \in I_j \cap X$, $0 \leq j \leq n$. Since $I_{n+1} \cap X$ is infinite, we can choose $x_{n+1} \in (I_{n+1} \cap X) \smallsetminus \{x_0, \cdots, x_n\}$. This completes the inductive construction of (x_n). Since $x_n \in I_n$, $n \geq 0$, we have

$$a_n \leq x_n \leq b_n, \ n \geq 0,$$

and so, by the squeezing lemma, (x_n) is convergent.   □

Remark 2.4.2 Theorem 2.4.1 fails if we work over the rational numbers. The condition that X is bounded is also necessary. A simple counterexample is given by $X = \mathbb{N}$.   ✱

We have a very useful application of Theorem 2.4.1 to sequences.

Proposition 2.4.3 *Let (x_n) be a bounded sequence. Then there exists a convergent subsequence (x_{n_k}) of (x_n).*

Proof An instructive direct proof of the result can be given along the lines of the proof of Theorem 2.4.1—see the exercises at the end of the section. Here we present a proof using Theorem 2.4.1. Set $X = \{x_n \mid n \in \mathbb{N}\}$. First observe that the result is easy if X is finite—we can choose (x_{n_k}) to be a constant sequence. Suppose that X is infinite. By Theorem 2.4.1, we can pick a convergent sequence (z_j) of distinct points of X. For each $j \in \mathbb{N}$ there exists a unique smallest $m_j \in \mathbb{N}$ such that $z_j = x_{m_j}$. If $m_1 < m_2 < \cdots$ we are done. If not, set $\mathfrak{M} = \{m_j \mid j \in \mathbb{N}\}$. Take $n_1 = m_{j(1)} = \min \mathfrak{M}$ and proceed inductively. Assume we have defined $n_1 = m_{j(1)} < \cdots < n_k = m_{j(k)}$, where $j(1) < \cdots < j(k)$. Define $n_{k+1} = m_{j(k+1)} = \min\{m_j \in \mathfrak{M} \mid m_j > n_1, \cdots, n_k, j > j(k)\}$. This defines a subsequence (x_{n_k}) of (x_n) which is also a subsequence of (z_j). Hence, by Lemma 2.2.12, (x_{n_k}) converges (with limit $\lim_{n \to \infty} z_n$). $\qquad\square$

2.4.1 Continuous Functions

We start by recalling the standard definition of a continuous function.

Definition 2.4.4 If $x_0 \in X \subset \mathbb{R}$ and $f : X \to \mathbb{R}$, then f is *continuous* at x_0 if for every $\varepsilon > 0$, there exists a $\delta > 0$ such that

$$|f(x) - f(x_0)| < \varepsilon, \text{ whenever } x \in X, \text{ and } |x - x_0| < \delta.$$

We say f is continuous on X, or just continuous, if f is continuous at every point of X.

Remarks 2.4.5

(1) For most of our initial applications, X will either be an open or closed interval of $\mathbb{R}$. Later we will need to work with more general subsets of $\mathbb{R}$.

(2) The definition of continuity has some unpleasant and subtle features. For example, it appears to require the verification of uncountably many conditions (that is, for each $\varepsilon > 0 \cdots$). However, as in Lemma 2.2.5, we can easily show that it suffices to verify the conditions just for $\varepsilon = 10^{-n}$, $n \geq 1$ (indeed, any sequence converging to zero will do). We give below an alternative, but equivalent, formulation of continuity known as *sequential continuity* that is, in many cases, much easier to work with. In spite of the simplifications obtained either by working with a countable set of conditions or with sequential continuity, the fact remains that the concept of continuity is highly non-intuitive. Contrary to the often made suggestion that the graph of a continuous function is what one gets by 'drawing a line without breaks', the reality is that the graph of a 'typical' continuous function is very jagged on all scales and the function is *nowhere* differentiable. In practice, the functions usually encountered in analysis and its applications have more structure than just continuity. Finally, there is a far more elegant and natural definition of continuity that applies in many contexts (including algebra) and which avoids the arid and uninformative ε, δ notation. This definition does, however, require another significant layer of abstraction. We revisit this issue later when we discuss metric spaces in Chap. 7.

(3) As in Lemma 2.2.5, we can replace $< \varepsilon$ in the definition by $\leq \varepsilon$ (of course we
    cannot replace the condition that δ is *strictly* positive). ✠

Definition 2.4.6 If $x_0 \in X \subset \mathbb{R}$ and $f : X \to \mathbb{R}$, then f is *sequentially continuous* at
x_0 if for every sequence $(x_n) \subset X$ converging to x_0 we have

$$\lim_{n \to \infty} f(x_n) = f(x_0).$$

We say f is sequentially continuous on X if f is sequentially continuous at every
point of X.

Remark 2.4.7 At first sight the definition of sequential continuity requires even
more to be checked than does the definition of continuity. However, the power of
the definition lies in the application to convergent sequences. If f is continuous and
$x_n \to x_0$ then $f(x_n) \to f(x_0)$. As we shall see, this property is very useful, especially
in a context where we can apply the Bolzano–Weierstrass theorem (or its corollary
Proposition 2.4.3). ✠

Example 2.4.8 Let $X = \mathbb{Z} \subset \mathbb{R}$. Every function $f : \mathbb{Z} \to \mathbb{R}$ is sequentially
continuous. This follows since if $(x_n) \subset \mathbb{Z}$ is convergent to $x_0 \in \mathbb{Z}$ then (x_n) is
eventually constant (that is, there exists an $N \in \mathbb{N}$ such that $x_n = x_N$, for all $n \geq N$).
Of course, it is easy to give a direct proof that $f : \mathbb{Z} \to \mathbb{R}$ is continuous—take $\delta < 1$
in Definition 2.4.4. ♠

Theorem 2.4.9 *If $x_0 \in X \subset \mathbb{R}$ and $f : X \to \mathbb{R}$, then f is continuous at x_0 iff f is
sequentially continuous at x_0.*

Proof We start by proving that if f is continuous at x_0 then f is sequentially
continuous at x_0. We have to show that given a sequence (x_n) converging to x_0 and
$\varepsilon > 0$, there exists an $N \in \mathbb{N}$ such that $|f(x_n) - f(x_0)| < \varepsilon$, for all $n \geq N$. Since
f is continuous at x_0, there exists a $\delta > 0$ such that $|f(x) - f(x_0)| < \varepsilon$ whenever
$|x - x_0| < \delta$ (here, and below, we always assume without further comment that
$x \in X$). Since (x_n) converges to x_0, there exists an $N \in \mathbb{N}$ such that $|x_n - x_0| < \delta$, for
all $n \geq N$. But then $|f(x_n) - f(x_0)| < \varepsilon$, for all $n \geq N$. Hence $\lim_{n \to \infty} f(x_n) = f(x_0)$.

It remains to prove the trickier converse that the sequential continuity of f at x_0
implies the continuity of f at x_0. We prove this by contradiction. Suppose that f is
not continuous at x_0. This means that there must be an $\varepsilon_0 > 0$ for which we cannot
find any $\delta > 0$ satisfying the conditions of the continuity definition. Hence, taking
$\delta = 1/n$, we can find an $x_n \in X$ such that $|x_n - x_0| < 1/n$ and $|f(x_0) - f(x_n)| \geq \varepsilon_0$.
By construction $\lim_{n \to \infty} x_n = x_0$ and so, by sequential continuity, $\lim_{n \to \infty} f(x_n) = f(x_0)$. But this implies there exists an $N \in \mathbb{N}$ such that $|f(x_0) - f(x_n)| < \varepsilon_0$, for all
$n \geq N$, and so contradicts our assumption that $|f(x_0) - f(x_n)| \geq \varepsilon_0$, for all $n \in \mathbb{N}$.
Hence f must be continuous at x_0. □

With these preliminaries out of the way, we can now prove a result that gives the
basic properties of a continuous function defined on a closed interval.

Theorem 2.4.10 *Let $f : [a, b] \to \mathbb{R}$ be continuous $(-\infty < a \le b < \infty)$. Then*

(1) *$f([a, b])$ is a bounded subset of $\mathbb{R}$ ("continuous functions are bounded on closed bounded intervals").*

(2) *If $m = \inf(f([a, b]))$, $M = \sup(f([a, b]))$, then there exist $x_m, x_M \in [a, b]$ such that $f(x_m) = m$, $f(x_M) = M$ ("a continuous function on a closed and bounded interval attains its bounds").*

(3) *$f([x_m, x_M]) = [m, M]$. In particular, $f([a, b]) \supset [f(a), f(b)]$ (the intermediate value theorem).*

Proof

(1) Suppose that f is not bounded above on $[a, b]$. Then for each $n \in \mathbb{N}$, there exists an $x_n \in [a, b]$ such that $f(x_n) \ge n$. Applying Proposition 2.4.3, we can choose a convergent subsequence (x_{n_k}) of (x_n). Let $\lim_{k \to \infty} x_{n_k} = x^* \in [a, b]$. By sequential continuity, $\lim_{k \to \infty} f(x_{n_k}) = f(x^*)$. But the sequence $(f(x_{n_k}))$ is unbounded by construction and so cannot converge. Contradiction. Hence f must be bounded above on $[a, b]$. Applying this result to $-f$ shows that f is bounded below on $[a, b]$.

(2) Set $M = \sup(f([a, b]))$. For each $n \in \mathbb{N}$, there exists an $x_n \in [a, b]$ such that $f(x_n) > M - 1/n$ (definition of the supremum). Using Proposition 2.4.3 again, we can pick a convergent subsequence (x_{n_k}) of (x_n). If $\lim_{k \to \infty} x_{n_k} = x^*$, then by sequential continuity we have $f(x^*) = M$. The result for the infimum is obtained by applying the result to $-f$.

(3) We have to prove that $f([x_m, x_M]) = [m, M]$. Without loss of generality assume $x_m < x_M$—if $x_m = x_M$, then $m = M$ and f is constant; if $x_m > x_M$, replace f by $-f$. We will show that for every $z \in (m, M)$ we can find a (least) $x^* \in (m, M)$ such that $f(x^*) = z$.

The basic idea is to look at the set X of points $x \in [x_m, x_M]$ such that $f < z$ on $[x_m, x]$—see Fig. 2.1. Clearly, $X \ne [x_m, x_M]$ (otherwise $m < f(x) < M$ for all $x \in [x_m, x_M]$). So we expect there is a first point $x^* \in [x_m, x_M]$ for which

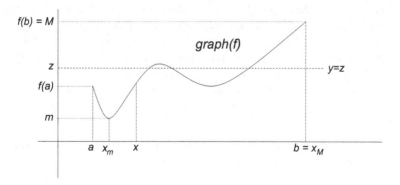

Fig. 2.1 Proving the intermediate value theorem

$f(x^\star) \not< z$. Since $f(x) < z$ for $x < x^\star$, we expect continuity to give us $f(x^\star) = z$. Now for the details.

Given $z \in (m, M)$, define

$$X = \{x \in [x_m, x_M] \mid f(\xi) < z, \text{ all } \xi \in [m, x]\}.$$

Clearly, $X \neq \emptyset$ since $f(x_m) = m < z$ and so $x_m \in X$. Since x_M is an upper bound of X, $x^\star = \sup(X)$ exists and $x^\star \le x_M$. Choose a sequence $(x_n) \subset X$ such that $x_n \to x^\star$. By sequential continuity, $\lim_{k \to \infty} f(x_n) = f(x^\star)$. Since $f(x_n) < z$ for all n, we have $f(x^\star) \le z$. If $f(x^\star) = z$ we are done. If not, then $f(x^\star) < z$, and so $x^\star < x_M$. By the continuity of f, we can find $\delta > 0$ such that $[x^\star - \delta, x^\star + \delta] \subset [x_m, x_M]$ and $f(x) < z$ for all $x \in [x^\star - \delta, x^\star + \delta]$. Since $x^\star - \delta \in X, f(x) < z$ on $[x_m, x^\star - \delta]$ and so $f < z$ on $[x_m, x^\star + \delta]$. Therefore, $x^\star + \delta \in X$, contradicting the definition of $x^\star$ as the supremum of X. Hence $f(x^\star) = z$.

Finally, we need to show that $f([a, b]) \supset [f(a), f(b)]$. This is obvious since $f(a), f(b) \in [m, M]$ and $f([a, b]) = [m, M]$. □

Remarks 2.4.11

(1) The proofs of (1,2) are a little different (and easier) than the proofs given in many texts. We indicate alternative proofs of these results in the exercises.
(2) Note that Theorem 2.4.10 implies that a continuous function $f : \mathbb{R} \to \mathbb{R}$ maps closed bounded intervals to closed bounded intervals. In general, a continuous function $f : \mathbb{R} \to \mathbb{R}$ maps intervals to intervals but does not necessarily map open intervals to open intervals or unbounded closed intervals to closed intervals. See the exercises.   ✱

Examples 2.4.12

(1) All three parts of Theorem 2.4.10 *fail* if we work over the rational numbers or consider real-valued functions defined on intervals of rational numbers. For example, if we set $[0, 1]^\mathbb{Q} = [0, 1] \cap \mathbb{Q}$ and define $f : [0, 1]^\mathbb{Q} \to \mathbb{Q}$ (or $\mathbb{R}$) by $f(x) = 2x^2 - 1$, then $0 \notin f([0, 1]^\mathbb{Q})$ even though $f(0) = -1 < 0 < 1 = f(1)$.
(2) Let $f : [a, b] \to \mathbb{R}$ be continuous and satisfy either $f([a, b]) \subset [a, b]$ or $f([a, b]) \supset [a, b]$. Then f has a *fixed point*. That is, there exists an $x^\star \in [a, b]$ such that $f(x^\star) = x^\star$. To see this, define $g(x) = f(x) - x$. Suppose that $f([a, b]) \subset [a, b]$. Then $f(a) \ge a$ and so $g(a) \ge 0$. Similarly, $g(a) \le 0$. Hence, by the intermediate value theorem $0 \in g([a, b])$ and there exists an $x^\star \in [a, b]$ such that $g(x^\star) = f(x^\star) - x^\star = 0$. We leave the proof of the second statement to the exercises.   ♠

We conclude this review of continuous functions with a definition and result that shows the utility of working with sequential continuity.

Definition 2.4.13 If X is a non-empty subset of $\mathbb{R}$ and $f : X \to \mathbb{R}$, then f is *uniformly continuous* if for every $\varepsilon > 0$, there exists a $\delta > 0$ such that

$$|f(x) - f(y)| < \varepsilon, \text{ whenever } x, y \in X, \text{ and } |x - y| < \delta.$$

Remark 2.4.14 A uniformly continuous function on X is continuous. In terms of the definition of continuity, uniform continuity implies that $\delta > 0$ can be chosen to be independent of $x_0 \in X$. ✠

Theorem 2.4.15 *Every continuous real-valued function defined on a closed and bounded interval* $[a, b]$ *is uniformly continuous.*

Proof Suppose $f : [a, b] \to \mathbb{R}$ is continuous but not uniformly continuous. If f is not uniformly continuous, there exists $\varepsilon_0 > 0$ such that for every $\delta > 0$, there is a pair $x, y \in [a, b]$, with $|x - y| < \delta$ and $|f(x) - f(y)| \geq \varepsilon_0$. Choose $\delta = 1/n$, $n \in \mathbb{N}$. Then for each $n \in \mathbb{N}$, we can find points $x_n, y_n \in [a, b]$ such that

$$|f(x_n) - f(y_n)| \geq \varepsilon_0, \text{ and } |x_n - y_n| < \frac{1}{n}.$$

By Proposition 2.4.3, $(x_n) \subset [a, b]$ has a convergent subsequence, say (x_{n_k}). Let $\lim_{k\to\infty} x_{n_k} = x^\star \in [a, b]$. Since $|x_{n_k} - y_{n_k}| < 1/n_k$, we have

$$\begin{aligned} |x^\star - y_{n_k}| &= |(x^\star - x_{n_k}) + (x_{n_k} - y_{n_k})| \\ &\leq |x^\star - x_{n_k}| + |x_{n_k} - y_{n_k}| \\ &< |x^\star - x_{n_k}| + \frac{1}{n_k}. \end{aligned}$$

Letting $k \to \infty$, we see that (y_{n_k}) is convergent with limit $x^\star$. By the sequential continuity of f, we have $\lim_{k\to\infty} f(x_{n_k}) = \lim_{k\to\infty} f(y_{n_k}) = f(x^\star)$ and hence $\lim_{k\to\infty} |f(x_{n_k}) - f(y_{n_k})| = 0$, contradicting our assumption that $|f(x_n) - f(y_n)| \geq \varepsilon_0$, all $n \in \mathbb{N}$. Hence f must be uniformly continuous. □

EXERCISES 2.4.16

(1) Let (x_n) be a bounded sequence of real numbers. Give a proof based on the subdivision method used in the proof of the Bolzano–Weierstrass theorem to show that (x_n) has a convergent subsequence. (For your proof you should not need to distinguish the cases where $\{x_n \mid n \in \mathbb{N}\}$ is finite or infinite—as a subset of $\mathbb{R}$.)

(2) Find a countable infinite subset X of $\mathbb{R}$ such that if $(x_n) \subset X$ is convergent, then (x_n) is eventually constant and the limit of (x_n) lies in X. ((x_n) is *eventually constant* if $\exists x, \exists N \in \mathbb{N}$ such that $x_n = x, n \geq N$. Eventually constant sequences always converge.)

(3) Let X be a non-empty subset of $\mathbb{R}$. We say that $x \in \mathbb{R}$ is a *closure point* of X if we can find a sequence $(x_n) \subset X$ which converges to x. Denote the set of closure points of X by $\overline{X}$. Why is it true that $\overline{X} \supset X$?

(a) Find an example of a countably infinite unbounded set X of $\mathbb{R}$ such that $\overline{X} = X$.

(b) Find an example of a countably infinite bounded subset X of $\mathbb{R}$ such that $\overline{X} = X$.

(c) Find an example of a countably infinite bounded subset of $[0, 1]$ such that $\overline{X} \smallsetminus X = \{0, \frac{1}{2}, 1\}$.

(d) Find an example of a countably infinite subset X of $\mathbb{R}$ such that $\overline{X} = \mathbb{R}$.

(4) By Theorem 2.4.10, if f is a **continuous** $\mathbb{R}$-valued map on a **closed** and **bounded** interval, then f is *bounded* and *attains its bounds*. Show by means of examples that each of the conditions **continuous, closed,** and **bounded** is necessary for either of the conclusions *bounded, attains its bounds* to hold.

(5) Show that a continuous function $f : \mathbb{R} \to \mathbb{R}$ maps intervals to intervals (an interval may be half-open, open or closed and bounded or bounded). Find an example where a bounded open interval is mapped to a closed interval. Does the closed interval have to be bounded?

(6) Deduce part (2) of Theorem 2.4.10 from part (1) by assuming that the upper bound M of f is not attained and considering the function $1/(M - f(x))$.

(7) Suppose $f : [a, b] \to \mathbb{R}$ is continuous and $f(a) < 0 < f(b)$. Show, using the subdivide and rule method of the proof of Theorem 2.3.12, that there exists a solution of $f(x) = 0$. (Hint and comments. Replacing $f(x)$ by $g(x) = f((b - a)x + a)$, there is no loss of generality in assuming $a = 0, b = 1$. Now construct a sequence $(x^N = 0.x_1 \cdots x_N)$ of decimal truncations such that $f(x^N) < 0 \leq f(x^N + 10^{-N})$. Simon Stevin used a similar method to show that a polynomial which changed sign had a root.)

(8) Suppose that I, J are non-empty closed bounded intervals of $\mathbb{R}, f : I \to \mathbb{R}$ is continuous and $f(I) \supset J$. Show that there exists a closed interval $I^* \subset I$ such that $f(I^*) = J$. Is the result true if I or J are not bounded? Prove it or find counterexample(s).

(9) Is Theorem 2.4.9 true if we work over the rational numbers?

(10) Suppose that $f : \mathbb{R} \to \mathbb{R}$ and that for every sequence (x_n) of real numbers diverging to $+\infty$, we have $\lim_{n \to \infty} f(x_n) = a$. Prove that $\lim_{x \to \infty} f(x) = a$. Is the converse true?

(11) Complete the analysis for the second case in Examples 2.4.12(2).

(12) Show by means of examples that a continuous map $f : [0, 1]^{\mathbb{Q}} \to \mathbb{R}$ need not be uniformly continuous.

(13) Define $f : \mathbb{R} \to \mathbb{R}$ by

$$f(x) = \begin{cases} 10^{-s}, & \text{if } x = \frac{r}{s} \in \mathbb{Q}, \ (r, s) = 1, s > 0, \text{ and } s = 1 \text{ if } r = 0, \\ 0, & \text{if } x \notin \mathbb{Q}. \end{cases}$$

Prove that f is continuous at x iff x is irrational. (Hint: You may assume that every rational r can be approximated by irrationals—$r + 10^{-n}\sqrt{2}$; that will help you prove that f is not continuous at rational points.)

(14) True of false? In each case either *prove* the result or provide a *simple explicit counterexample*.

(a) If $f : [0, 1]^{\mathbb{Q}} \to \mathbb{Q}$ is continuous, then f is bounded.
(b) If $f : [0, 1]^{\mathbb{Q}} \to \mathbb{Q}$ is continuous and bounded, then f attains its bounds.
(c) The intermediate value theorem holds for continuous functions $f : [a, b]^{\mathbb{Q}} \to \mathbb{Q}$.

How would your answers change if instead we looked at continuous maps $f : [0, 1] \to \mathbb{Q}$? (Be advised: the answers change! Hint: since $\mathbb{Q} \subset \mathbb{R}$, every continuous $f : [0, 1] \to \mathbb{Q}$ determines a continuous $\mathbb{R}$-valued map $F : [0, 1] \to \mathbb{R}$ with image $f([0, 1])$ consisting of rational numbers.)

(15) Show that $f(x) = 1/x$ is not uniformly continuous on $(0, 1)$ but is uniformly continuous on $(1, 2)$.

(16) Find examples of functions $f : \mathbb{R} \to \mathbb{R}$ which are (a) uniformly continuous, (b) not uniformly continuous.

(17) A common proof of Theorem 2.4.10(1) proceeds along the following lines. Let $X = \{x \in [a, b] \mid f \text{ bounded on } [a, x]\}$. Show that (a) $X \neq \emptyset$, (b) $\sup(X) \in X$, (c) $\sup(X) = b$. Fill in the details and use similar methods to prove part (2) of Theorem 2.4.10. (Comment: one defect of this approach is that it does not extend well to functions defined on more general sets, for example, subsets of $\mathbb{R}^n$, since it makes use of the order structure on $\mathbb{R}$.)

2.4.2 Cauchy Sequences

Equipped with the Bolzano–Weierstrass theorem we can now give a satisfactory intrinsic definition of a convergent sequence which does not depend on knowing the limit.

Definition 2.4.17 A sequence (x_n) of real numbers is a *Cauchy sequence* if for every $\varepsilon > 0$, there exists an $N \in \mathbb{N}$ such that

$$|x_m - x_n| < \varepsilon, \text{ for all } m, n \geq N.$$

If (x_n) is Cauchy, we write $\lim_{m,n \to \infty} |x_m - x_n| = 0$.

Remarks 2.4.18

(1) Roughly speaking, a Cauchy sequence has the property that terms in the sequence eventually all get arbitrarily close to one another.

(2) As in Lemma 2.2.5, we can replace $< \varepsilon$ by $\leq \varepsilon$ and it is enough to test the truth of the definition for any sequence (κ_m) of strictly positive numbers converging to zero. ✠

Example 2.4.19 Let $x \in \mathbb{R}$. The sequence of decimal truncations $(x^N = x_0.x_1 \cdots x_N)$ to x defines a Cauchy sequence: $|x^M - x^N| \leq 10^{-M}, M \geq N$. ♠

We need the following elementary lemma about Cauchy sequences (this result is also true if we work over $\mathbb{Q}$).

Lemma 2.4.20 *Let (x_n) be a sequence of real numbers.*

(1) *If (x_n) is Cauchy, then $\{x_n \mid n \in \mathbb{N}\}$ is a bounded subset of $\mathbb{R}$.*
(2) *If (x_n) is convergent, then (x_n) is Cauchy.*
(3) *If (x_n) is Cauchy and (x_n) has a convergent subsequence, then (x_n) is convergent.*

Proof

(1) Take $\varepsilon = 1$ in Definition 2.4.17. Then there exists an $N \in \mathbb{N}$ so that $|x_m - x_n| \leq 1$, for all $m, n \geq N$. Taking $m = N$, we see that $|x_n - x_N| \leq 1$ for all $n \geq N$ and so $|x_n| \leq |x_N| + 1, n \geq N$. Hence $|x_n| \leq \max\{|x_1|, \cdots, |x_{N-1}|, |x_N| + 1\}$ for all $n \geq 1$, proving that $\{x_n \mid n \in \mathbb{N}\}$ is a bounded subset of $\mathbb{R}$.
(2) Suppose $\lim_{n\to\infty} x_n = x^\star$. Let $\varepsilon > 0$. Since (x_n) converges to $x^\star$ we can choose $N \in \mathbb{N}$ such that $|x^\star - x_n| < \varepsilon/2, n \geq N$. We have

$$\begin{aligned}
|x_m - x_n| &= |x^\star - x_m + x_n - x^\star| \\
&\leq |x^\star - x_m| + |x^\star - x_n| \\
&< \frac{\varepsilon}{2} + \frac{\varepsilon}{2} = \varepsilon, \text{ if } m, n \geq N.
\end{aligned}$$

(3) Finally, suppose that (x_{n_k}) is a convergent subsequence of (x_n) with limit $x^\star$. Given $\varepsilon > 0$, we can choose $N_1 \in \mathbb{N}$ such that $|x^\star - x_{n_k}| < \varepsilon/2$, provided $n_k \geq N_1$ (it is easier to work with n_k here as opposed to the index k). Since (x_n) is Cauchy, we can choose $N_2 \in \mathbb{N}$ so that $|x_m - x_n| < \varepsilon/2, m, n \geq N_2$. Set $N = \max\{N_1, N_2\}$. For all $n, n_k \in \mathbb{N}$ we have

$$|x^\star - x_n| = |x^\star - x_{n_k} + x_{n_k} - x_n| \leq |x^\star - x_{n_k}| + |x_{n_k} - x_n|.$$

Fix $n_k \geq N$. Then for all $n \geq N$, we have $|x^\star - x_{n_k}| + |x_{n_k} - x_n| < \varepsilon/2 + \varepsilon/2 = \varepsilon$, proving that (x_n) converges to $x^\star$. □

We can now state and prove our main result on Cauchy sequences.

Theorem 2.4.21 *A sequence (x_n) of real numbers is convergent iff (x_n) is Cauchy.*

Proof By Lemma 2.4.20(2), if (x_n) is convergent, then (x_n) is Cauchy. Conversely, if (x_n) is Cauchy then by Lemma 2.4.20(1), (x_n) is bounded and so, by Proposition 2.4.3, (x_n) has a convergent subsequence. Apply Lemma 2.4.20(3). □

Remarks 2.4.22

(1) Theorem 2.4.21 fails over $\mathbb{Q}$. Indeed, the sequence of finite decimal approximations to an irrational number provides an example of a non-convergent Cauchy sequence in $\mathbb{Q}$.

(2) We can use Theorem 2.4.21 as the basis for a more intrinsic (though perhaps less transparent) definition of the real numbers that does not depend on working to a particular base. Specifically, consider the set $\mathcal{C}$ of all Cauchy sequences of rational numbers. We define an equivalence relation $\sim$ on $\mathcal{C}$ by $(x_n) \sim (y_n)$ iff $\lim_{n\to\infty} |x_n - y_n| = 0$. In particular, if one or other sequence converges, then both do with the same limit. We define the set of real numbers as the set of $\sim$ equivalence classes and then prove Theorem 2.4.21 directly without recourse to the Bolzano–Weierstrass theorem. Modulo the abstraction of using equivalence classes, what we are doing with this general construction is defining real numbers by (all of) their rational approximations. For more details, we refer to the appendix at the end of the chapter. ✸

Multiplication and Division Revisited Once we know Cauchy sequences of real numbers converge it is easy to define the operations of multiplication and division on $\mathbb{R}$.

Suppose $x, y \in \mathbb{R}$. Let (x_n), (y_n) be sequences of rational numbers converging to x, y respectively. We want to define $xy = \lim_{n\to\infty} x_n y_n$. For this to work we need to check that (a) $(x_n y_n)$ is convergent and (b) the limit of $(x_n y_n)$ is independent of the choice of sequences (x_n), (y_n) converging to x, y. We verify (a) and leave (b) to the exercises. For (a) we prove that $(x_n y_n)$ is a Cauchy sequence. For this, observe that

$$|x_m y_m - x_n y_n| \le |x_m y_m - x_m y_n| + |x_m y_n - x_n y_n|$$
$$= |x_m||y_m - y_n| + |y_n||x_m - x_n|.$$

By Lemma 2.4.20(1), there exists an $M > 0$ such that $|x_m|, |y_n| \le M$, for all $n, m \in \mathbb{N}$. Given $\varepsilon > 0$, choose $N \in \mathbb{N}$ such that $|y_m - y_n|, |x_m - x_n| < \frac{\varepsilon}{2M}$, $m, n \ge N$. We have

$$|x_m y_m - x_n y_n| \le |x_m||y_m - y_n| + |y_n||x_m - x_n|$$
$$< M\frac{\varepsilon}{2M} + M\frac{\varepsilon}{2M} = \varepsilon, \text{ for } m, n \ge N.$$

Hence $(x_n y_n)$ is Cauchy.

We use exactly the same process to define division of real numbers. Finally, we may deduce all the standard laws of arithmetic for real numbers from the corresponding laws for rational numbers. For example, the distributive law $x(y + z) = xy + xz$ follows from

$$x(y + z) = \lim_{n\to\infty} x_n(y_n + z_n) = \lim_{n\to\infty} x_n y_n + \lim_{n\to\infty} x_n z_n = xy + yz.$$

EXERCISES 2.4.23

(1) Verify that the geometric sequence (r^n) is Cauchy if $|r| < 1$.
(2) Find examples of sequences (x_n) such that
    (a) for every $p \in \mathbb{N}$, $\lim_{n \to \infty} |x_{n+p} - x_n| = 0$, (b) (x_n) is not Cauchy. (Hint: try $x_n = \log(n + 1)$.)
(3) Suppose that (x_n) is a sequence of real numbers and there exists a $k \in (0, 1)$ such that $|x_{n+1} - x_n| < k|x_n - x_{n-1}|$ for all $n \geq 2$. Show that (x_n) is a Cauchy sequence.

     Show by means of an example that if we allow $k \in (0, 1)$ to depend on n, then this result may fail and (x_n) may not converge. (Hint: Look for an increasing sequence (x_n) which diverges to $+\infty$ but for which $x_n/n \to 0$.)
(4) Complete the proof of the definition of multiplication on $\mathbb{R}$ by showing that the limit of $(x_n y_n)$ is the same for all rational sequences (x_n) converging to x and (y_n) converging to y.
(5) Show how we define division by non-zero real numbers.
(6) For $n \geq 1$, define $S_n = \sum_{j=1}^{n} (-1)^{j+1} j^{-2}$. Prove that (S_n) is a Cauchy sequence and hence that the infinite series $\sum_{n=1}^{\infty} (-1)^{n+1} j^{-2}$ converges.
(7) Suppose that $f : \mathbb{Q} \to \mathbb{R}$ is uniformly continuous. Show that there exists a unique continuous function $F : \mathbb{R} \to \mathbb{R}$ such that $F(x) = f(x)$ for all $x \in \mathbb{Q}$ (we say "F is a continuous *extension* of f to $\mathbb{R}$"). Find an example of a continuous (but not uniformly continuous) function $g : \mathbb{Q} \to \mathbb{R}$ which does *not* extend to $\mathbb{R}$. (Hint for first part: Show that every Cauchy sequence $(q_n) \subset \mathbb{Q}$ is mapped by f to a Cauchy sequence $(f(q_n)) \subset \mathbb{R}$. Remember to verify that F is well-defined and does not depend on the particular choice of Cauchy sequence.)

2.5 lim sup and lim inf

2.5.1 *Sequences*

Suppose that $(x_n) \subset \mathbb{R}$ is a bounded sequence. For $n \geq 1$, let $X_n = \{x_m \mid m \geq n\}$. We define

$$\alpha_n = \inf(X_n), \quad \beta_n = \sup(X_n).$$

Since (x_n) is a bounded sequence, we have

$$-\infty < \alpha_n \leq \beta_n < +\infty,$$

for all $n \geq 1$. Moreover, since $X_1 \supset X_2 \supset \cdots$, we have

$$\alpha_1 \leq \alpha_2 \leq \cdots \leq \alpha_n \leq \cdots \leq \beta_n \leq \cdots \leq \beta_2 \leq \beta_1.$$

It follows by Theorem 2.3.18 that $\lim_{n\to\infty} \alpha_n$ and $\lim_{n\to\infty} \beta_n$ exist and that

$$\lim_{n\to\infty} \alpha_n \leq \lim_{n\to\infty} \beta_n. \tag{2.1}$$

We define $\liminf x_n = \lim_{n\to\infty} \alpha_n$ and $\limsup x_n = \lim_{n\to\infty} \beta_n$. Alternative (and commonly used) notations are $\overline{\lim}\, x_n$ for $\limsup x_n$ and $\underline{\lim}\, x_n$ for $\liminf x_n$.

Lemma 2.5.1 *If (x_n) is a bounded sequence of real numbers, then*

(1) $\liminf x_n \leq \limsup x_n$.
(2) $\liminf x_n = \limsup x_n$ *iff (x_n) is convergent. If (x_n) is convergent then the limit of (x_n) must be the common value of $\liminf x_n$ and $\limsup x_n$.*
(3) *There exists a subsequence of (x_n) converging to $\liminf x_n$. Similarly for $\limsup x_n$.*
(4) *If (x_{n_k}) is a convergent subsequence of (x_n), then*

$$\lim_{k\to\infty} x_{n_k} \in [\liminf x_n, \limsup x_n].$$

Proof (1) is immediate from (2.1). The remainder of the proof is left to the exercises. $\square$

Remarks 2.5.2

(1) Lemma 2.5.1(3) gives an alternative proof of Proposition 2.4.3.
(2) We may define lim sup and lim inf for unbounded sequences if we give $\limsup x_n = +\infty$ and $\liminf x_n = -\infty$ the obvious meanings.   ✷

Example 2.5.3 Suppose that (x_n) is a Cauchy sequence. By Lemma 2.4.20, (x_n) is bounded. It follows from the definition of Cauchy sequence that for every $\varepsilon > 0$, we can choose $N \in \mathbb{N}$ so that $|\inf\{x_n \mid n \geq N\} - \sup\{x_n \mid n \geq N\}| < \varepsilon$. Hence $\liminf x_n = \limsup x_n$ and (x_n) is convergent by Lemma 2.5.1(2).   ♠

2.5.2   *Functions, Continuity,* **lim sup** *and* **lim inf**

We start with a review of one-sided limits. Most of this material should be familiar to the reader. For simplicity we usually assume the domain is a closed interval but everything we say extends to general intervals: open or closed, bounded or unbounded. Some of the results we prove can be used to extend the range of applicability of the Riemann integral as well as to develop the theory of the Riemann–Stieltjes integral (see also the exercises at the end of the section and the exercises in section "Appendix: The Riemann Integral").

Let $f : [a, b] \to \mathbb{R}$ be bounded. Given $x_0 \in [a, b]$, let $\lim_{x\to x_0-}$ signify the limit as x approaches x_0 from the left and $\lim_{x\to x_0+}$ denote the limit as x approaches x_0 from the right. If $\lim_{x\to x_0-} f(x)$ exists we set $\lim_{x\to x_0-} f(x) = f(x_0-)$ and similarly define

$f(x_0+)$. If $x_0 = a$ (respectively b), we only consider the limit $\lim_{x \to a+}$ (respectively, $\lim_{x \to b-}$).

Lemma 2.5.4 (Notation and Assumptions as Above) *The function f is continuous at x_0 iff the one-sided limits $\lim_{x \to x_0 \pm} f(x) = f(x_0 \pm)$ both exist and*

$$f(x_0) = f(x_0-) = f(x_0+).$$

(The statement is modified in the obvious way at the end-points of $[a, b]$.)

Proof A standard argument—left to the exercises.   □

Examples 2.5.5

(1) Define $f : [-1, +1] \to \mathbb{R}$ by

$$f(x) = \begin{cases} -1, & x < 0, \\ 0, & x = 0, \\ 1, & x > 0. \end{cases}$$

The map f is continuous except at $x = 0$. We have $f(0-) = -1, f(0+) = 1$ and $f(0) = 0$. We refer to the discontinuity at $x = 0$ as a *jump discontinuity* of f: the limits $\lim_{x \to x_0 \pm} f(x)$ exist but are not equal. Whatever the value of $f(x_0)$, f is not continuous at x_0.

(2) Define $f : [-1, +1] \to \mathbb{R}$ by

$$f(x) = \begin{cases} \sin(1/x), & x \neq 0, \\ 0, & x = 0. \end{cases}$$

The map f is continuous except at $x = 0$. Neither of the limits $\lim_{x \to 0 \pm} f(x)$ exist.

(3) Suppose $f(x) = x^2$, $x \neq 0$ and $f(0) = 1$. In this case $f(0\pm) = 0 \neq f(0)$. We refer to $x = 0$ as a *removable discontinuity* of f: if we redefine f so that $f(0) = 0$, then f will be continuous. Note that neither of the discontinuities in the previous examples are removable.   ♠.

As example (2) above shows, not all discontinuities of a function need be jump discontinuities. In particular, the limits $\lim_{x \to x_0 \pm} f(x)$ need not exist. However, since we are assuming f is bounded, we may use the operations of lim sup and lim inf to define quantities that reflect the variation in f near a discontinuity point x_0. More precisely, let $x_0 \in [a, b)$. Since f is bounded, we may define

$$\bar{f}(x_0+) = \limsup_{x \to x_0+} f(x) \stackrel{\text{def}}{=} \lim_{h \to 0+} \sup f((x_0, x_0 + h]),$$

$$\underline{f}(x_0+) = \liminf_{x \to x_0+} f(x) \stackrel{\text{def}}{=} \lim_{h \to 0+} \inf f((x_0, x_0 + h]).$$

Since f is bounded, we have $-\infty \leq \underline{f}(x_0+) \leq \bar{f}(x_0+) < +\infty$. If $x_0 \in (a, b]$, we may similarly define $\bar{f}(x_0-), \underline{f}(x_0-)$, and if $x_0 \in (a, b)$

$$\underline{f}(x_0) = \liminf_{x \to x_0} f(x), \quad \bar{f}(x_0) = \limsup_{x \to x_0} f(x).$$

It follows easily from the definitions that we have the following relations between these limits.

$$\bar{f}(x_0+) \geq \underline{f}(x_0+),$$

$$\bar{f}(x_0-) \geq \underline{f}(x_0-),$$

$$\bar{f}(x_0) \geq \underline{f}(x_0),$$

$$\bar{f}(x_0) = \max\{f(x_0), \bar{f}(x_0+), \bar{f}(x_0-)\},$$

$$\underline{f}(x_0) = \min\{f(x_0), \underline{f}(x_0+), \underline{f}(x_0-)\}.$$

Example 2.5.6 Define $f : [-1, +1] \to \mathbb{R}$ by

$$f(x) = \begin{cases} 3\max\{0, \sin(1/x)\}, & x > 0, \\ 2 + 5\min\{0, \sin(1/x)\}, & x < 0, \\ 7, & x = 0. \end{cases}$$

In this case we have $\bar{f}(0+) = 3, \underline{f}(0+) = 0, \bar{f}(0-) = -2, \underline{f}(0-) = -3, \bar{f}(0) = 7, \underline{f}(0) = -3$. In general, there are no further relationships we can expect between the various limits. ♠

We define three terms which quantify the 'fluctuation' or 'oscillation' of f at x_0:

$$\omega_f(x_0) = \bar{f}(x_0) - \underline{f}(x_0),$$

$$\omega_f(x_0+) = \bar{f}(x_0+) - \underline{f}(x_0+),$$

$$\omega_f(x_0-) = \bar{f}(x_0-) - \underline{f}(x_0-).$$

Lemma 2.5.7 *(Notation and assumptions as above.)*

(1) *If* $\omega_f(x_0) = 0$, *then* f *is continuous at* x_0.
(2) *If* $\omega_f(x_0+) = 0$, *then* $f(x_0+)$ *exists.*
(3) *If* $\omega_f(x_0-) = 0$, *then* $f(x_0-)$ *exists.*
(4) *If* $\omega_f(x_0\pm) = 0$ *and* $f(x_0\pm) = f(x_0)$, *then* f *is continuous at* x_0.

Proof Straightforward and left to the exercises.   □

Remarks 2.5.8

(1) We briefly mention the important concepts of upper and lower semi-continuity, which play a role in many parts of analysis. In our context, a bounded map $f :$ $[a, b] \to \mathbb{R}$ is *upper semi-continuous* at x_0 if $\bar{f}(x_0) = \limsup_{x \to x_0} f(x) \le f(x_0)$ and *lower semi-continuous* at x_0 if $\underline{f}(x_0) = \liminf_{x \to x_0} f(x) \ge f(x_0)$. It follows from Lemma 2.5.7(1) that if f is upper and lower semi-continuous at x_0, then f is continuous at x_0. If f has a jump discontinuity at x_0, then f will be upper semi-continuous at x_0 if $f(x_0) \ge \max\{f(x_0+), f(x_0-)\}$. It may be shown that if $f : [a, b] \to \mathbb{R}$ is bounded, then ω_f is upper semi-continuous.
(2) Given $f : [a, b] \to \mathbb{R}$, it was shown by Young [31] that the subset of points of $[a, b]$ where either $\bar{f}(x_0-) \ne \bar{f}(x_0+)$ or $\underline{f}(x_0+) \ne \underline{f}(x_0+)$ is countable. (See Exercises 7.11.10(16,17) for an outline proof of Young's theorem.) ✱

Theorem 2.5.9 *Let $I \subset \mathbb{R}$ be an interval and $f : I \to \mathbb{R}$ be monotone and bounded. Then*

(1) *For all $x_0 \in I$, $f(x_0+)$ and $f(x_0-)$ exist. In particular, all discontinuities of f are jump discontinuities.*
(2) *The set of points where f is discontinuous is a countable subset of I.*

Proof Without loss of generality assume f is monotone increasing. Since f is increasing, we have $\liminf_{x \to x_0-} f(x) = \lim_{x \to x_0-} f(x) = f(x_0-)$ and $\limsup_{x \to x_0+} f(x) = \lim_{x \to x_0+} f(x) = f(x_0+)$, proving (1). Let $\mathcal{D}_f \subset I$ denote the set of points of discontinuity of f. Since f is increasing, it follows from (1) and Lemma 2.5.7 that $\mathcal{D}_f = \{x \in I \mid \omega_f(x) = f(x+) - f(x-) > 0\}$. Given $n \in \mathbb{N}$, define

$$D_n = \{x \in I \mid \omega_f(x) \ge 1/n\}.$$

Since f is bounded and increasing, D_n is finite for all $n \in \mathbb{N}$ ($f(b) - f(a) \ge \sum_{x \in D_n} \omega_f(x)$). The result follows since $\mathcal{D}_f = \cup_{n \ge 1} D_n$. □

EXERCISES 2.5.10

(1) Complete the proof of Lemma 2.5.1.
(2) Provide the proof of Lemma 2.5.4.
(3) Provide the details of the proof of Lemma 2.5.7 and verify that the conditions are all necessary. (For example, find a function f which satisfies $\omega_f(x_0\pm) = 0$ and $f(x_0+) = f(x_0)$, but is not continuous at x_0.)
(4) Show that $f : [a, b] \to \mathbb{R}$ is upper semi-continuous at x_0 iff $-f : [a, b] \to \mathbb{R}$ is lower semi-continuous at x_0.
(5) Show that $f : [a, b] \to \mathbb{R}$ is upper semi-continuous at x_0 iff for all sequences $(x_n) \subset [a, b]$ converging to x_0 we have $\limsup_{n \to \infty} f(x_n) \le f(x_0)$. Formulate and prove the analogous statements for lower semi-continuity.
(6) Show that

   (a) the *floor function* $f(x) = \lfloor x \rfloor$, which returns the greatest integer $\le x$, is upper semi-continuous,
   (b) the *ceiling function* $f(x) = \lceil x \rceil$, which returns the smallest integer $\ge x$, is lower semi-continuous.

(7) Show that if $f : [a, b] \to \mathbb{R}$ is upper semi-continuous, then f is bounded above on $[a, b]$ and attains it upper bound. Formulate and prove a corresponding result for lower semi-continuous functions. (Hint: Use the sequence method used in the proof of Theorem 2.4.10.)

(8) A function $f : [a, b] \to \mathbb{R}$ is of *bounded variation* if there exists an $M \geq 0$ such that $\sum_{j=0}^{n-1} |f(x_j) - f(x_{j+1})| \leq M$ for all finite partitions $\mathcal{P} = \{x_j \mid a = x_0 \leq x_1 \leq \cdots \leq x_n = b$ of $[a, b]$. Show that

(a) If f is monotone or continuously differentiable, then f is of bounded variation.

(b) If $f(x) = \sin(1/x)$, $x \in (0, 1]$, $f(0) = 0$, then f is not of bounded variation. Show also that $xf(x)$ is not of bounded variation (note that $xf(x)$ is differentiable but not continuously differentiable).

(c) If f is of bounded variation, $f(x_0\pm)$ exist for all $x_0 \in [a, b]$ and the set of discontinuities is countable.

(9) Show that if $f, g : [a, b] \to \mathbb{R}$ are of bounded variable so are $f \pm g, f \times g$.

(10) Let $f : [a, b] \to \mathbb{R}$ be of bounded variation and $x \in [a, b]$. If $\mathcal{P}$ is a finite partition of $[a, x]$, let $V(\mathcal{P}) = \sum_{j=0}^{n-1} |f(x_j) - f(x_{j+1})|$ and define $V_a^x(f) = \sup_{\mathcal{P}} V(\mathcal{P})$ (taken over all finite partitions of $[a, x]$). Show that

(a) $V(x) = V_a^x(f)$ is monotone increasing on $[a, b]$.

(b) $W = V - f$ is monotone increasing on $[a, b]$.

(c) f is of bounded variation if and only if f can be written as the difference of two monotone functions (either both strictly increasing or both strictly decreasing).

Can you prove (c) more geometrically if f is C^1?

2.6 Complex Numbers

We recall the definitions and elementary properties of complex numbers and extend results on sequences of real numbers to complex numbers.

2.6.1 Review of Complex Numbers

A *complex number* $z = x + \imath y$ may be identified with the point $(x, y) \in \mathbb{R}^2$. Addition and subtraction of the complex numbers $z_1 = x_1 + \imath y_1$ and $z_2 = x_2 + \imath y_2$ corresponds to vector addition in $\mathbb{R}^2$:

$$z_1 \pm z_2 = (x_1 \pm x_2) + \imath (y_1 \pm y_2).$$

Multiplication of complex numbers is defined by

$$z_1 z_2 = (x_1 + \imath y_1)(x_2 + \imath y_2) = (x_1 x_2 - y_1 y_2) + \imath (x_1 y_2 + x_2 y_1).$$

On $\mathbb{R}^2$, multiplication is given by $(x_1, y_1)(x_2, y_2) = (x_1 x_2 - y_1 y_2, x_1 y_2 + x_2 y_1)$. It is easy to check that multiplication is commutative ($z_1 z_2 = z_2 z_1$), associative ($z_1(z_2 z_3) = (z_1 z_2)z_3$) and that the distributive law holds

$$z_1(z_2 + z_3) = z_1 z_2 + z_1 z_3.$$

If we let $\mathbb{C}$ denote the set of complex numbers and identify $\mathbb{C}$ with $\mathbb{R}^2$ by $z = x + \imath y \leftrightarrow (x, y)$, then the real numbers $\mathbb{R}$ are naturally defined as the subset $\{(x, 0) \mid x \in \mathbb{R}\}$ of $\mathbb{C}$. We say the complex number z is real if $z = x + \imath 0$. We define $1 = (1, 0) = 1 + \imath 0$, and $0 = (0, 0)$. We then have $1z = z1 = z$, and $z + 0 = 0 + z = z$ for all $z \in \mathbb{C}$.

Since $\imath^2 = (0, 1) \times (0, 1) = -(1, 0) = -1$, we have $\imath^2 = -1$. A complex number z is *imaginary* if $z = \imath y$ for some $y \in \mathbb{R}$. The square of every imaginary number is negative.

We define the *modulus* $|z|$ of $z = x + \imath y \in \mathbb{C}$ by

$$|z| = \sqrt{x^2 + y^2}.$$

Of course, $|z|$ is the Euclidean length of the vector $(x, y) \in \mathbb{R}^2$. It is straightforward to verify that $|z_1 z_2| = |z_1||z_2|$, for all $z_1, z_2 \in \mathbb{C}$. If $z \in \mathbb{C}$ is real, then $|z|$ is the absolute value of z. We have the important triangle-inequality

$$|z_1 + z_2| \le |z_1| + |z_2|, \text{ for all } z_1, z_2 \in \mathbb{C}.$$

Let $c : \mathbb{C} \to \mathbb{C}$ be the real linear map defined by

$$c(x + \imath y) = x - \imath y.$$

We refer to c as *complex conjugation* and write $c(z) = \bar{z}$. Observe that $\bar{z} = z$ iff z is real and $\bar{z} = -z$ iff z is imaginary. Since $z\bar{z} = (x + \imath y)(x - \imath y) = x^2 - \imath^2 y^2 = x^2 + y^2$, we have

$$|z|^2 = z\bar{z}.$$

If $|z| = 1$, then z is a point on the unit circle $x^2 + y^2 = 1$.

Referring to Fig. 2.2, z defines a unique $\theta \in [0, 2\pi)$ such that $z = \cos\theta + \imath \sin\theta$ (that is, the Cartesian coordinates of z are $(\cos\theta, \sin\theta)$). If we *define* $e^{\imath\theta} = \cos\theta + \imath \sin\theta$, then $e^{-\imath\theta} = \cos\theta - \imath \sin\theta$ and so

$$\cos\theta = \frac{e^{\imath\theta} + e^{-\imath\theta}}{2}, \quad \sin\theta = \frac{e^{\imath\theta} - e^{-\imath\theta}}{2\imath}.$$

Fig. 2.2 Complex number of unit modulus

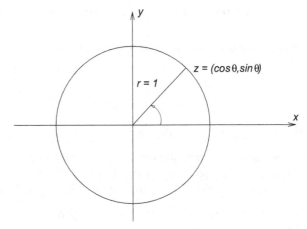

Remark 2.6.1 If we substitute $x = \imath\theta$ in the exponential series $e^x = \sum_{n=0}^{\infty} \frac{x^n}{n!}$, then we obtain the well-known infinite series for $\cos\theta$ and $\sin\theta$ (see Chap. 5). ✱
We may use standard trigonometric identities to verify

$$e^{\imath 0} = 1, \quad \overline{e^{\imath\theta}} = e^{-\imath\theta}, \quad e^{\imath\theta}e^{\imath\phi} = e^{\imath(\theta+\phi)}.$$

In particular, for $n \in \mathbb{Z}$ we have De Moivre's formula

$$(e^{\imath\theta})^n = e^{\imath n\theta}.$$

We leave to the exercises the proof that if $a \in \mathbb{C}$ and $ae^{\imath\theta} \neq 1$, then

$$\sum_{p=0}^{n} a^p e^{\imath p\theta} = \frac{(1 - a^{n+1}e^{\imath(n+1)\theta})}{1 - ae^{\imath\theta}}.$$

If $z \neq 0$, there exists a unique $\theta \in [0, 2\pi)$ such that

$$z = |z|e^{\imath\theta}.$$

For this we observe that $u = z/|z|$ lies on the unit circle and so defines a unique $\theta \in [0, 2\pi)$ as described above. We call $z = |z|e^{\imath\theta}$ the *modulus and argument* form of z. If $z = x + \imath y$, then $r = |z|$, θ are the polar coordinates of (x, y).

Multiplication takes a particularly simple form if we use the modulus and argument representation of complex numbers. If $z_1 = |z_1|e^{\imath\theta_1}$ and $z_2 = |z_1|e^{\imath\theta_1}$, then

$$z_1 z_2 = |z_1||z_2|e^{\imath\theta_1}e^{\imath\theta_1} = |z_1 z_2|e^{\imath(\theta_1+\theta_2)}.$$

2.6.2 Sequences of Complex Numbers

Definition 2.6.2 A sequence (z_n) of complex numbers is *convergent* if there exists a $z \in \mathbb{C}$ such that

$$\lim_{n \to \infty} |z - z_n| = 0.$$

We call z the *limit* of the sequence (z_n) and write $\lim_{n \to \infty} z_n = z$.

Remark 2.6.3 The definition is formally identical to that of convergence of a real sequence with the proviso that we replace the absolute value by the modulus.   ✶
The next lemma allows us to switch easily between real and complex sequences.

Lemma 2.6.4 *Let (z_n) be a sequence of complex numbers. If we write $z_n = x_n + \iota y_n$, then (z_n) is convergent iff both the real sequences (x_n) and (y_n) are convergent.*

Proof Observe that if $z = x + \iota y$, then

$$|x|, |y| \le |z| \le |x| + |y|. \tag{2.2}$$

Suppose that (z_n) is convergent with limit $z = x + \iota y$. By definition, $\lim_{n \to \infty} |z - z_n| = 0$. By the left-hand inequality of (2.2), we have $|x - x_n|, |y - y_n| \le |z - z_n|$, for all $n \in \mathbb{N}$. Hence, by the squeezing lemma, $\lim_{n \to \infty} |x - x_n|, |y - y_n| = 0$ and the sequences (x_n), (y_n) converge with respective limits x and y. The converse is equally simple using the right-hand inequality of (2.2).   ☐

Definition 2.6.5 A sequence (z_n) of complex numbers is a *Cauchy sequence* if $\lim_{m,n \to \infty} |z_m - z_n| = 0$.

Theorem 2.6.6 *A sequence (z_n) is Cauchy iff it is convergent.*

Proof We leave the proof, which is an easy consequence of Theorem 2.4.21 and Lemma 2.6.4, to the exercises.   ☐

Remark 2.6.7 Lemma 2.6.4 and Theorem 2.4.21 allow us to extend many results on real sequences and series to complex sequences and series. Subsequently, we usually indicate these extensions in remarks rather than developing the complex theory separately.   ✶

EXERCISES 2.6.8

(1) Verify that $|z_1 z_2| = |z_1||z_2|$ and $\overline{z_1 z_2} = \bar{z}_1 \bar{z}_2$ for all $z_1, z_2 \in \mathbb{C}$.
(2) Complete the proof of Theorem 2.6.6.
(3) Verify the formula for the sum of a geometric series. (Hint: multiply both sides by $1 - a e^{\iota \theta}$.)
(4) A subset A of $\mathbb{C}$ is *bounded* if there exists a $C \ge 0$ such that $|z| \le C$ for all $z \in A$. Show that if A is an infinite bounded subset of $\mathbb{C}$ then there exists a

convergent subsequence (z_n) consisting of distinct points of A. (Hint: Use the Bolzano–Weierstrass theorem twice and Lemma 2.6.4.)

(5) Show that every bounded sequence of complex numbers has a convergent subsequence.

(6) Show that a continuous function $f : [a, b] \to \mathbb{C}$ is bounded and attains its bounds.

2.7 Appendix: Results from the Differential Calculus

We review some definitions and results from the differential calculus of functions of one variable. For the results on Taylor's theorem, we only use an elementary result from the theory of Riemann integration. Namely that if f has an anti-derivative F ($F' = f$), then $\int_a^b f(t)\, dt = F(b) - F(a)$ (see the second appendix for additional comments).

Definition 2.7.1 Let I be an interval (open or closed, bounded or unbounded). If $f : I \to \mathbb{R}$ is continuous and $x_0 \in I$, then f is *differentiable* at x_0 if $\lim_{h \to 0} \frac{f(x_0+h)-f(x_0)}{h}$ exists. We denote the value of the limit by $f'(x_0)$ and call $f'(x_0)$ the *derivative* of f at x_0. If f is differentiable at every point of I, we say f is differentiable on I.

Remarks 2.7.2

(1) If $x_0 \in I$ is an end-point of I, then we take the appropriate one-sided limit. For example, if $I = [a, b]$, then $f'(a) = \lim_{h \to 0+} \frac{f(a+h)-f(a)}{h}$.

(2) Continuity of f at x_0 is implied by the existence of the limit $\lim_{h \to 0} \frac{f(x_0+h)-f(x_0)}{h}$. The verification is routine.   ✸

Easily the most important foundational theorem in the differential calculus is the *mean value theorem*. The mean value theorem follows simply from Rolle's theorem which we state and prove first.

Theorem 2.7.3 (Bhaskara (1114–1185), Rolle 1691) *Let $f : [a, b] \to \mathbb{R}$ be continuous on $[a, b]$ and differentiable on (a, b). If $f(a) = f(b) = 0$, there exists a $z \in (a, b)$ such that $f'(z) = 0$.*

Proof Either f is constant, in which case $f' \equiv 0$ and we may take $z = (a+b)/2$, or not. If not, then by the continuity of f, $f(x)$ attains minimum and maximum values $m < M$ on $[a, b]$. Since f is not constant, at least one of m, M is non-zero. Without loss of generality, suppose $M > 0$ and that $M = f(z)$ where necessarily $z \in (a, b)$. It is a simple consequence of the definition of derivative that $f'(z) = 0$ (if not, f would have to take values greater than M close to z).   □

Theorem 2.7.4 (Mean Value Theorem) *Let $f : [a, b] \to \mathbb{R}$ be continuous on $[a, b]$ and differentiable on (a, b). Then there exists a $z \in (a, b)$ such that*

$$f(b) - f(a) = f'(z)(b - a).$$

Proof Define $G(x) = (f(b) - f(a))(x - a) - (f(x) - f(a))(b - a)$, $x \in [a, b]$. Then $G(a) = G(b) = 0$ and G satisfies the conditions of Rolle's theorem. Therefore there exists a $z \in (a, b)$ such that $G'(z) = 0$. Observe that $G'(z) = (f(b) - f(a)) - f'(z)(b - a)$. □

Remark 2.7.5 The precise value of z given by the mean value theorem is rarely of interest. If we set $M = \sup_{x \in (a,b)} |f'(x)|$, then provided $M < \infty$, we obtain the very useful estimate

$$|f(b) - f(a)| \leq M|b - a|. \tag{2.3}$$

This is the form in which we make most use of the mean value theorem and it is also the form in which it generalizes to functions of several variables (see Chap. 9). ✸

Corollary 2.7.6 *Suppose that* $f : [a, b] \to \mathbb{R}$ *is continuous and differentiable on* (a, b)*. If* $f' = 0$ *on* (a, b)*, then* f *is constant.*

Proof If $x \in [a, b]$, then $|f(x) - f(a)| \leq 0$, by (2.3). □

2.7.1 Higher Derivatives and Taylor's Theorem

Let I be an interval. If $f : I \to \mathbb{R}$ is differentiable, let $f' : I \to \mathbb{R}$ denote the derivative map. We say f is *continuously differentiable*, or C^1, if f is differentiable and $f' : I \to \mathbb{R}$ is continuous. Proceeding inductively, $f : I \to \mathbb{R}$ is *r-times continuously differentiable*, or just C^r, if the derivative maps $f', \ldots, f^{(r-1)}$ exist and are continuous on I and the derivative map $f^{(r-1)} : I \to \mathbb{R}$ is differentiable with derivative map $(f^{(r-1)})' : I \to \mathbb{R}$ continuous. Set $f^{(r)} = (f^{(r-1)})'$. If f is r-times continuously differentiable for all $r \geq 1$, f is said to be *smooth* or C^∞. We make a special study of smooth functions later in Chap. 5.

Theorem 2.7.7 (Taylors Theorem: Integral Remainder) *Let* I *be a non-empty open or closed interval,* $r \in \mathbb{N}$ *and* $f : I \to \mathbb{R}$ *be* $(r + 1)$*-times continuously differentiable. Given* $a, x \in I$*, we have*

$$f(x) = f(a) + \frac{f'(a)}{1!}(x - a) + \cdots + \frac{f^{(r)}(a)}{r!}(x - a)^r + R_r(a, x),$$

where the remainder term $R_r(a, x)$ *is given explicitly by*

$$R_r(a, x) = \frac{1}{r!} \int_a^x f^{(r+1)}(t)(x - t)^r \, dt$$

$$= \frac{(x - a)^{r+1}}{r!} \int_0^1 (1 - s)^r f^{(r+1)}(a + s(x - a)) \, ds.$$

Proof The proof is by induction. The result is trivially true when $r = 0$ (the integral is defined in terms of the anti-derivative f of f'). So suppose we have shown

$$f(x) = \sum_{i=0}^{k} \frac{f^{(i)}(a)}{i!}(x-a)^i + \frac{1}{k!} \int_a^x f^{(k+1)}(t)(x-t)^k \, dt,$$

where $0 < k < r$. Integrating by parts we have

$$\int_a^x f^{(k+1)}(t)(x-t)^k \, dt = -\frac{1}{k+1} f^{(k+1)}(t)(x-t)^{k+1} \big|_{t=a}^x$$

$$+ \frac{1}{k+1} \int_a^x f^{(k+2)}(t)(x-t)^{k+1} \, dt$$

$$= \frac{1}{k+1} f^{(k+1)}(a)(x-a)^{k+1}$$

$$+ \frac{1}{k+1} \int_a^x f^{(k+2)}(t)(x-t)^{k+1} \, dt.$$

Dividing by $k!$, we see that $R_{k+1}(a,x) = \frac{1}{(k+1)!} \int_a^x f^{(k+2)}(t)(x-t)^k \, dt$, completing the inductive step. It remains to prove that the two versions of the remainder term are equal. This is easily done by means of the substitution $t = a + s(x-a)$ and we leave the details to the reader. □

Taylor's theorem with integral remainder will suffice for our later applications. We remark that if $[a - \delta, a + \delta] \subset I$, and we set $M_{r+1} = \sup_{t \in [a-\delta,a+\delta]} |f^{(r+1)}(t)|$, then we have the estimate

$$\left| f(x) - \sum_{i=0}^{r} \frac{f^{(i)}(a)}{i!}(x-a)^i \right| \leq M_{r+1} \frac{|x-a|^{r+1}}{r!}, \quad |x-a| \leq \delta. \qquad (2.4)$$

The estimate follows easily from the second form for the remainder.

Definition 2.7.8 If $f : I \to \mathbb{R}$ is C^∞, then the *Taylor series* $T_a f$ of f at a is defined by

$$T_a f(x) = \sum_{n=0}^{\infty} \frac{f^{(n)}(a)}{n!}(x-a)^n.$$

Remark 2.7.9 This is a formal definition and we caution the reader that the Taylor series, *even if it converges*, may bear little, if any, relation to the values of $f(x)$, when $x \neq a$ (see Chap. 5). However, the Taylor series does encode information about f— all the derivatives of f at $x = a$. ✱

2.7.2   Other Forms of the Remainder in Taylor's Theorem

It is possible to express the remainder in a form that assumes weaker conditions on f. In this section we state some characteristic results.

Given that $f : I \to \mathbb{R}$ is r-times differentiable at a, recall that if $x \in I$, then the remainder $R_r(a, x)$ is defined by

$$R_r(a, x) = f(x) - \sum_{i=0}^{r} \frac{f^{(i)}(a)}{i!}(x - a)^i.$$

We start by giving a variant of the remainder estimate (2.4).

Theorem 2.7.10 *If* $f : I \to \mathbb{R}$ *is* r *times differentiable at* a, *then*

$$\lim_{x \to a} \frac{|R_r(a, x)|}{|x - a|^r} = 0.$$

(Limit through points of I.)

Proof The proof is by induction on r and uses Rolle's theorem (see Exercise 2.7.12(5)). Note that the case $r = 1$ is the definition of differentiability at $x = a$.   □

Theorem 2.7.11 (Classical Taylor's Theorem) *Suppose that* $f : I \to \mathbb{R}$.

(a) *If* f *is* $(r+1)$-*times differentiable on* I, *then for each* $a < x \in I$, $\exists \xi \in (a, x)$ *such that*

$$R_r(a, x) = \frac{f^{(r+1)}(\xi)}{(r+1)!}(x - a)^{r+1}.$$

   (Lagrange form of the remainder.)

(b) *If* f *is* C^{r+1}, *then for each* $x \in I$, $\exists \xi \in (a, x)$ *such that*

$$R_r(a, x) = \frac{f^{(r+1)}(\xi)}{r!}(x - \xi)^r(x - a).$$

   (Cauchy form of the remainder.)

Proof We give the proof of the Cauchy remainder leaving the Lagrange form to the exercises. Define $g(y) = \frac{1}{r!} \int_a^y f^{(r+1)}(t)(x-t)^r \, dt$, $y \in [a, x]$. Since g is differentiable (fundamental theorem of calculus), the mean value theorem implies there exists an $\xi \in (a, x)$ such that $R_n(a, x) = g(x) - g(a) = (x - a)g'(\xi) = (x - a)\frac{1}{r!}f^{(r+1)}(\xi)(x - \xi)^r$.   □

EXERCISES 2.7.12

(1) Suppose that $h : [a, x] \to \mathbb{R}$ is $(r + 1)$ times differentiable and $h^{(j)}(a) = h(x) = 0$, $0 \leq j \leq r$. Using induction and Rolle's theorem, show that there exists $\xi \in (a, x)$ such that $h^{(r+1)}(\xi) = 0$.

(2) Assume the conditions of Theorem 2.7.11(a). Set $T(t) = \sum_{j=0}^{r} \frac{f^{(j)}(a)}{j!}(t - a)^j$ and define $h : [a, x] \to \mathbb{R}$ by

$$h(t) = f(t) - T(t) - \left[\frac{f(x) - T(x)}{(x - a)^{r+1}}\right](t - a)^{r+1}, \; t \in [a, x].$$

Show that h satisfies the conditions of (1) and deduce the Lagrange form of the remainder in Taylor's theorem.

(3) Using the Lagrange form of the remainder, show that if $n \in \mathbb{N}$, then $e = \sum_{j=0}^{n} \frac{1}{j!} + \frac{1}{(n+1)!}e^\theta$, where $0 < \theta < 1$. Deduce that e is irrational. (Hint: suppose $e = p/q$. Choose $n > q$.)

(4) Show that Theorem 2.7.10 follows from Theorem 2.7.11 if f is C^{r+1}.

(5) Prove Theorem 2.7.10 (assume only that f is C^r).
    (Hint. The result is true for $r = 1$. Let $a < b \in I$. Given $x \in [a, b]$, define

$$G(x) = R_r(a, b)(x - a)^r - R_r(a, x)(b - a)^r.$$

Verify $G(a), G(b) = 0$ and use Rolle's Theorem and induction on r.)

2.8 Appendix: The Riemann Integral

Suppose that f is a *continuous* function defined on a closed and bounded interval. In this appendix we show how to define, construct and compute the Riemann integral of f. We make use of two results: Theorem 2.4.10(1,2) (f is bounded on a closed bounded interval and attains its bounds) and the mean value theorem.

Rather than defining the integral of f by approximating upper and lower sums, we instead state two simple properties that the integral should possess. We show these properties are reasonable by verifying that they give the correct areas under the graph of a constant function (area of a rectangle) and under the graph of $y = x$ (area of a triangle). We then prove that these properties *uniquely determine* the integral if it exists. It is then almost a *triviality* to observe that if f has an anti-derivative F ($F' = f$) then the integral from a to x of f exists and is equal to $F(x) - F(a)$ (the fundamental theorem of calculus). We conclude with an elementary proof of the main theoretical result that every continuous function defined on a closed interval has an anti-derivative. This result amounts to an existence theorem for solutions of the ordinary differential equation $\frac{dy}{dx} = f(x)$. We briefly indicate how to extend our definition of the integral to include bounded functions with at most countably many discontinuities.

If f is everywhere positive then we think of the integral of f as the area under the graph of f. If f is everywhere negative, then the corresponding integral will be negative and equal to minus the area under the graph of $-f$.

2.8.1 Two Basic Properties Required of the Integral

Let f be a real-valued function with domain $\mathcal{D} \subset \mathbb{R}$, where $\mathcal{D}$ is an *interval* which may be open, closed, half-open or unbounded. We assume f is bounded on all closed intervals $[a, b] \subset \mathcal{D}$. Every continuous function f satisfies this condition by Theorem 2.4.10.

Definition 2.8.1 A function $\mathcal{I}(x, y)$, with domain $\mathcal{D} \times \mathcal{D}$, is a (definite) integral for f if

(1) Given $a \leq b \leq c, a, b, c \in \mathcal{D}$, we have

$$\mathcal{I}(a, c) = \mathcal{I}(a, b) + \mathcal{I}(b, c). \tag{2.5}$$

(2) Given $a < b, a, b \in \mathcal{D}$,

$$m(b - a) \leq \mathcal{I}(a, b) \leq M(b - a), \tag{2.6}$$

where m is any lower bound for f on $[a, b]$ and M is any upper bound for f on $[a, b]$.

Remark 2.8.2 We should emphasize that $\mathcal{I}(x, y)$ depends on f—we could have written $\mathcal{I}_f$ rather than $\mathcal{I}$ but we prefer the simpler notation with the understanding that the function f remains fixed.

In Fig. 2.3 we show the meaning of (2.5). The condition implies that if we choose any finite sequence $a = x_0 < x_1 \ldots < x_N = b$, then

$$\mathcal{I}(a, b) = \sum_{n=0}^{N-1} \mathcal{I}(x_n, x_{n+1}). \tag{2.7}$$

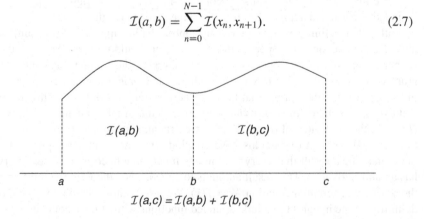

$$\mathcal{I}(a,c) = \mathcal{I}(a,b) + \mathcal{I}(b,c)$$

Fig. 2.3 Condition (2.5)

Turning to the second condition, assume for the moment that f is positive (as in Fig. 2.4). The first inequality $m(b-a) \le \mathcal{I}(a,b)$ of (2.6) says that whatever $\mathcal{I}(a,b)$ is, it cannot be smaller than the area of the *largest* rectangle with base $[a,b]$ that we can fit *under* the graph of f. Similarly, the second inequality $\mathcal{I}(a,b) \le M(b-a)$ implies that $\mathcal{I}(a,b)$ can be no larger than the area of the smallest rectangle with base $[a,b]$ that contains the graph of f.

Examples 2.8.3

(1) Let $f(x) = C$ be a constant function. Then $\mathcal{I}(a,b) = C(b-a)$: take $m = M = C$ and the result is immediate from (2.6). Notice that this gives the (signed) area of the rectangle with base $[a,b]$ and height C. Of course, if $C > 0$ we get the usual unsigned area. If f is negative, then $\mathcal{I}(a,b)$ is 'signed' and negative.

(2) If f takes positive and negative values there can be cancellation between the positive and negative parts of $\mathcal{I}(a,b)$ and so the inequality (2.6) is weaker—see Fig. 2.5 where $m = -1, M = +1$ and $\mathcal{I}(a,b) = 0$ (using (2.5) and the previous example). ♠

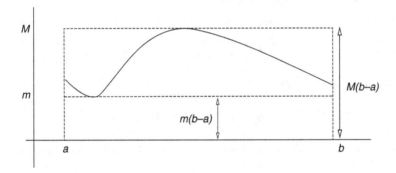

Fig. 2.4 Condition (2.6)

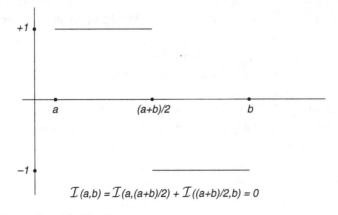

$\mathcal{I}(a,b) = \mathcal{I}(a,(a+b)/2) + \mathcal{I}((a+b)/2,b) = 0$

Fig. 2.5 A case where $\mathcal{I}(a,b) = 0$

As we shall soon see, (2.5), (2.6) uniquely characterize the function $\mathcal{I}(x, y)$ when f is continuous.

In practice, it is useful to extend (2.5) to allow for arbitrary triples $a, b, c \in \mathcal{D}$. First note that if $a = b = c$ we get

$$\mathcal{I}(a, a) = \mathcal{I}(a, a) + \mathcal{I}(a, a)$$

and so $\mathcal{I}(a, a) = 0$. It follows that if we want to define $\mathcal{I}(a, b)$ when $b < a$ we must take $\mathcal{I}(a, b) = -\mathcal{I}(a, b)$ since for (2.5) to hold (for a, b, a) we need

$$0 = \mathcal{I}(a, a) = \mathcal{I}(a, b) + \mathcal{I}(b, a)$$

With these conventions, it is easy to check that if (2.5) holds then we have

$$\mathcal{I}(a, c) = \mathcal{I}(a, b) + \mathcal{I}(b, c),$$

for *all* $a, b, c \in \mathcal{D}$.

Summarizing, we henceforth suppose that $\mathcal{I}(x, y)$ satisfies

(I) For all $a, b, c \in \mathcal{D}$ we have

$$\mathcal{I}(a, c) = \mathcal{I}(a, b) + \mathcal{I}(b, c).$$

(II) Given $a < b, a, b \in \mathcal{D}$,

$$m(b - a) \leq \mathcal{I}(a, b) \leq M(b - a),$$

where m is any lower bound for f on $[a, b]$ and M is any upper bound for f on $[a, b]$.

Example 2.8.4 Conditions **(I,II)** allow us to do some simple computations. For example, suppose $f(x) = x$ and we take $a = 0$, $b = 1$. We compute $\mathcal{I}(0, 1)$. Take the subdivision $0, 1/N, 2/N, \ldots, (N - 1)/N, 1$ of $[0, 1]$. Applying (2.7), we have

$$\mathcal{I}(0, 1) = \sum_{n=0}^{N-1} \mathcal{I}\left(\frac{n}{N}, \frac{n+1}{N}\right).$$

On the interval $[\frac{n}{N}, \frac{n+1}{N}]$ we have the bounds $\frac{n}{N} \leq x \leq \frac{n+1}{N}$. Hence $n/N^2 \leq \mathcal{I}(\frac{n}{N}, \frac{n+1}{N}) \leq (n + 1)/N^2$. Summing from $n = 0$ to $N = 1$, we obtain the estimate

$$\frac{1}{N^2} \sum_{n=0}^{N-1} n \leq \mathcal{I}(0, 1) \leq \frac{1}{N^2} \sum_{n=0}^{N-1} n + 1.$$

The arithmetic progressions $0, 1, \ldots, N - 1$ and $1, 2, \ldots, N$ have respective sums $N(N-1)/2$ and $N(N+1)/2$ and so

$$\frac{N-1}{2N} \leq \mathcal{I}(0, 1) \leq \frac{N+1}{2N}.$$

This estimate holds for all $N \geq 1$. Letting $N \to \infty$, the squeezing lemma implies that $\mathcal{I}(0, 1) = \frac{1}{2}$ (the area of the triangle of base 1 and height 1). $\spadesuit$

2.8.2 Existence of $\mathcal{I}(x, y)$, Part 1

We now assume that f is continuous. Recall that $f : \mathcal{D} \to \mathbb{R}$ has an *anti-derivative* F if there exists a differentiable function $F : \mathcal{D} \to \mathbb{R}$ such that $F' = f$ on $\mathcal{D}$.

Lemma 2.8.5 *If f is continuous and has an anti-derivative F, then $\mathcal{I}(a, b) = F(b) - F(a)$ satisfies properties I, II.*

Proof Define

$$\mathcal{I}(a, b) = F(b) - F(a), \quad a, b \in \mathcal{D}.$$

Since $(F(b) - F(a)) + (F(c) - F(b)) = (F(c) - F(a))$ (for all $a, b, c \in \mathcal{D}$), it is obvious that $\mathcal{I}(a, b)$ satisfies **I**. It remains to show that $\mathcal{I}$ satisfies **II**. Suppose $a, b \in \mathcal{D}$, $a < b$. Let M, m be upper and lower bounds for f on $[a, b]$. By the mean value theorem, we can find $z \in (a, b)$ so that

$$\mathcal{I}(a, b) = F(b) - F(a) = F'(z)(b - a) = f(z)(b - a).$$

Since $m \leq f(x) \leq M$ on $[a, b]$, it follows immediately that

$$m(b - a) \leq f(z)(b - a) = \mathcal{I}(a, b) \leq M(b - a),$$

proving **II**. $\square$

2.8.3 Uniqueness of $\mathcal{I}(x, y)$, f Continuous

Theorem 2.8.6 *Let $f : \mathcal{D} \to \mathbb{R}$ be continuous. If we can find $\mathcal{I}$ satisfying I, II, then*

(1) *For all $a \in \mathcal{D}$, $F_a(x) = \mathcal{I}(a, x)$ is an anti-derivative of f.*
(2) *$\mathcal{I}$ is unique.*

Proof Fix $a \in \mathcal{D}$ and set $F(x) = \mathcal{I}(a, x)$, $x \in \mathcal{D}$. We claim that F is an anti-derivative of f: $F'(x) = f(x)$, $x \in \mathcal{D}$.

Fix $x \in \mathcal{D}$ and choose $h \in \mathbb{R}$ so that $x + h \in \mathcal{D}$. (If x is not an end-point of $\mathcal{D}$, we have $x + h \in \mathcal{D}$ for sufficiently small h. Otherwise we restrict to positive or negative values of h as appropriate.) By **I**, we have

$$\mathcal{I}(a, x) + \mathcal{I}(x, x + h) = \mathcal{I}(a, x + h).$$

Hence

$$\mathcal{I}(a, x + h) - \mathcal{I}(a, x) = \mathcal{I}(x, x + h).$$

Therefore, if $h \neq 0$,

$$\frac{\mathcal{I}(a, x + h) - \mathcal{I}(a, x)}{h} = \frac{\mathcal{I}(x, x + h)}{h}.$$

Let m_h and M_h respectively denote the infimum and supremum of f on $[x, x + h]$. Since f is continuous, $-\infty < m_h \leq M_h < \infty$ and $\lim_{h \to 0} M_h, m_h = f(x)$. Suppose first that $h > 0$. From **II**, we have

$$m_h h \leq \mathcal{I}(x, x + h) \leq M_h h,$$

and so

$$m_h \leq \mathcal{I}(x, x + h)/h \leq M_h.$$

Letting $h \to 0+$, we see

$$\lim_{h \to 0+} \mathcal{I}(x, x + h)/h = f(x).$$

If $h < 0$, then $\frac{\mathcal{I}(x, x+h)}{h} = \frac{\mathcal{I}(x+h, x)}{-h}$ (since $\mathcal{I}(x, x + h) = -\mathcal{I}(x + h, x)$). The argument now proceeds as before, using **II** applied to the interval $[x + h, x]$ (note $x + h < x$) to give

$$\lim_{h \to 0-} \mathcal{I}(x, x + h)/h = f(x).$$

Hence we have shown that

$$\lim_{h \to 0} \frac{\mathcal{I}(a, x + h) - \mathcal{I}(a, x)}{h} = f(x)$$

and $F(x) = \mathcal{I}(a, x)$ is an anti-derivative of f, proving (1).

Since any two anti-derivatives of f differ by a constant,[1] (2) follows from (1). $\square$

---

[1] By the mean value theorem, $F' - G' = 0$ implies $G = F + c$ if the common domain is an interval.

Remark 2.8.7 Note that Lemma 2.8.5 and Theorem 2.8.6 suffice for all the standard applications and examples in a first calculus course: all the functions considered invariably have an anti-derivative and so the definite (or indefinite) integral is given by the anti-derivative. No arguments needing approximating sums are needed. ✱
In future we adopt the usual notation and set $\mathcal{I}(a,b) = \int_a^b f(t)\, dt$. We refer to $\int_a^b f(t)\, dt$ as the *Riemann or definite integral of f from a to b*. If $x \in [a,b]$, then Theorem 2.8.6 implies the fundamental theorem of calculus

$$\frac{d}{dx}\left(\int_a^x f(t)\, dt\right) = f(x). \tag{2.8}$$

2.8.4 Existence of the Integral, Part 2

In this section we prove

Theorem 2.8.8 *Every continuous function $f : \mathcal{D} \to \mathbb{R}$ has an anti-derivative.*
Our proof of Theorem 2.8.8 proceeds by constructing a function $L(x,y)$ that satisfies conditions **I, II**. This construction is quite straightforward and uses only parts (1,2) of Theorem 2.4.10 (in particular, no use is made of results on uniform continuity).

Proof of Theorem 2.8.8 Fix an interval $[a,b]$ and suppose that f is continuous on $[a,b]$. A *partition* $\mathcal{P}$ of $[a,b]$ consists of a finite number of points $t_0, \ldots, t_N$ satisfying

$$a = t_0 \le t_1 \le \ldots \le t_N = b.$$

Given a partition $\mathcal{P}$, set $m_j = \inf\{f(s) \mid s \in [t_j, t_{j+1}]\}, 0 \le j < N$. Define the (lower) sum $L(\mathcal{P},f)$ by

$$L(\mathcal{P},f) = \sum_{j=0}^{N-1} m_j(t_{j+1} - t_j).$$

If m, M denote lower and upper bounds for f on $[a,b]$ then $m \le m_j \le M$ and so

$$m(b-a) \le L(\mathcal{P},f) \le M(b-a). \tag{2.9}$$

If we add new points to $\mathcal{P}$, say to form $\mathcal{P}'$, then the reader may easily check that $L(\mathcal{P}',f) \ge L(\mathcal{P},f)$. It follows from (2.9), that $m(b-a)$ and $M(b-a)$ are lower and upper bounds respectively for $L(\mathcal{P},f)$ for all partitions $\mathcal{P}$ of $[a,b]$. Hence if we define

$$L(a,b) = \sup\{L(\mathcal{P},f) \mid \text{all partitions } \mathcal{P} \text{ of } [a,b]\},$$

we have

$$m(b - a) \le L(a, b) \le M(b - a).$$

Hence $L(a, b)$ satisfies **II**. Since we can always add a point b to a partition of $[a, c]$ $(a \le b \le c)$, it is easy to see that

$$L(a, b) + L(b, c) = L(a, c),$$

and so $L(a, b)$ satisfies **I**. It follows by Theorem 2.8.6 that $L(a, x)$ is an anti-derivative of f.   □

Remarks 2.8.9

(1) We could have done the construction using 'upper' sums $U(\mathcal{P}, f)$. Since we have proved already that the integral is unique, we get for free that $\sup_{\mathcal{P}} L(\mathcal{P}, f) = \inf_{\mathcal{P}} U(\mathcal{P}, f) = \int_a^b f(x)\, dx$. We can similarly use approximating sums that lie between $L(\mathcal{P}, f)$ and $U(\mathcal{P}, f)$. For example, $\sum_{j=0}^{N-1} f(t_j)(t_{j+1} - t_j)$ (see the exercises below).

(2) If we use uniform partitions $\mathcal{P}$ $(t_{j+1} - t_j = (b - a)/N$ is independent of j), then we need the uniform continuity of f in order to prove that $\lim_{N\to\infty} U(\mathcal{P}, f), L(\mathcal{P}, f) = \int_a^b f(x)\, dx$. But this assumption is *not* needed to prove the existence of the integral. Indeed, a virtue of the Riemann integral is that once you know the integrand f has an anti-derivative F, then you can write down the integral in terms of F. Nothing is needed about approximating sums. The only technical difficulty is proving the existence of the Riemann integral for general continuous functions. This may be regarded as an existence theorem for ordinary differential equations: given a continuous function f, the ordinary differential equation

$$\frac{dy}{dx} = f(x)$$

has a C^1 solution $y = F(x)$. (Later we address the existence theorem when f is a function of y, rather than x.)   ✠

2.8.5   Methods of Integration

Once we know the integral can be given in terms of an anti-derivative, all the standard results from calculus follow more or less immediately. For example, suppose that f is continuous (and so has an anti-derivative) and g is any differentiable function with continuous derivative. We have

$$\int_{g(a)}^{g(b)} f(s)\, ds = \int_a^b f(g(t))g'(t)\, dt.$$

Indeed, since f is continuous, there exists an F such that $F' = f$. Hence $(F \circ g)'(t) = F'(g(t))g'(t) = f(g(t))g'(t)$ and the result follows. If g is invertible, we obtain the *integration by substitution formula*:

$$\int_a^b f(s)\, ds = \int_{g^{-1}(a)}^{g^{-1}(b)} f(g(t))g'(t)\, dt.$$

(Substitute $s = g(t)$ and $ds = g'(t)\, dt$.)

2.8.6 Extensions

We can prove the existence of an integral satisfying **I, II** for any *bounded* function on a domain $\mathcal{D}$ which has at most countably many discontinuities. This is easy to do when there are finitely many discontinuities of f and more challenging when there are countably infinitely many discontinuities (see the exercises). We can also weaken the boundedness condition to allow for functions that grow slowly enough near singular points (for example $1/\sqrt{|x|}$ near zero) as well as to allow for the definition of the integral on unbounded domains. We address these issues in more detail as and when they arise in the text (see, in particular, the first section of Chap. 6).

EXERCISES 2.8.10

(1) Suppose that the function $\mathcal{I}(x, y)$, $x, y \in \mathcal{D}$, satisfies condition **I** whenever $a \leq b \leq c$. Show that if we define $\mathcal{I}(a, b) = -\mathcal{I}(b, a)$ for $b < a$, then **I** holds for all $a, b, c \in \mathcal{D}$.

(2) Show that if $f : \mathcal{D} \to \mathbb{R}$ is bounded and $\mathcal{I}(a, b)$ exists and satisfies **I, II**, then $\mathcal{I}(a, x)$ is a continuous function of $x \in \mathcal{D}$, a a fixed point of $\mathcal{D}$.

(3) Let $\mathcal{P}, \mathcal{P}'$ be two (finite) partitions of $[a, b]$. Show that if we define $\mathcal{Q} = \mathcal{P} \cup \mathcal{P}'$, then $L(\mathcal{Q}, f) \geq \max\{L(\mathcal{P}, f), L(\mathcal{P}', f)\}$.

(4) Let $n \in \mathbb{N}$ and let $\mathcal{P}_n$ be the partition of $[a, b]$ defined by $t_j = a + \frac{j(b-a)}{n}$, $0 \leq j \leq n$. Using the uniform continuity of f (Theorem 2.4.15), verify that $\lim_{n \to \infty} L(\mathcal{P}_n, f) = \int_a^b f(t)\, dt$. (Hint: Show $\lim_{n \to \infty} U(\mathcal{P}_n, f) - L(\mathcal{P}_n, f) = 0$.)

(5) Suppose $f : [a, b] \to \mathbb{R}$ has finitely many discontinuities. Show how we may define the definite integral $\int_a^b f(t)\, dt$ and verify that the derivative of $\int_a^x f(t)\, dt$ exists and equals $f(x)$ at all points x where $f(x)$ is continuous.

(6) Let $f, g : [a, b] \to \mathbb{R}$ be continuous and suppose that g is of constant sign (that is, either positive or negative). Show that there exists an $x \in (a, b)$ such that

$$\int_a^b f(t)g(t)\, dt = f(x) \int_a^b g(t)\, dt.$$

Hint: Show that $(\int_a^b f(t)g(t)\,dt/\int_a^b g(t)\,dt) \in [m, M]$, where m, M respectively denote the infimum and supremum of f on $[a, b]$. (The result is known as the *first mean value theorem for integrals*. Note that if $g \equiv 1$, then we obtain $\int_a^b f(t)\,dt = f(x)(b - a)$—often called the mean value theorem for integrals.)

(7) Let $f, g : [a, b] \to \mathbb{R}$ be continuous and suppose that g is positive and monotone decreasing. There exists an $x \in (a, b]$ such that

$$\int_a^b f(t)g(t)\,dt = g(a) \int_a^x f(t)\,dt.$$

(This result is known as the *second mean value theorem for integrals*.) Find an example to show that we may have $x = b$ and so deduce that we cannot use the condition $x \in (a, b)$.

(8) Let $\mathcal{P} = \{t_j \mid a = t_0 < t_1 < \cdots < t_N = b\}$ be a partition of $[a, b]$ and define $\delta(\mathcal{P}) = \max\{t_{j+1} - t_j \mid j = 0, \cdots, N - 1\}$. If $f : [a, b] \to \mathbb{R}$ is continuous, show that $L(\mathcal{P}, f), U(\mathcal{P}, f) \to \int_a^b f(t)\,dt$ as $\delta(\mathcal{P}) \to 0$. Deduce that if we define $V(\mathcal{P}, f) = \sum_{j=0}^{N-1} \xi_j(t_{j+1} - t_j)$, where $\xi_j \in [m_j, M_j]$, and m_j, M_j are respectively the infimum and supremum of $f|[t_j, t_{j+1}]$, then $V(\mathcal{P}, f) \to \int_a^b f(t)\,dt$ as $\delta(\mathcal{P}) \to 0$. (Hint: use the uniform continuity of f. Note that you are expected to show that if $\varepsilon > 0$, then there exists a $\delta_0 > 0$ such that if $\delta(\mathcal{P}) < \delta_0$, then $|V(\mathcal{P}, f) - \int_a^b f(t)\,dt| < \varepsilon$.)

(9) In this extended exercise we consider the problem of defining $\int_a^b f(t)\,dt$ when f is bounded and the discontinuity set $\mathcal{D}$ of f is countable. When there are countably infinitely many discontinuities, we use the definition of integrability given by equality of integrals defined by upper and lower sums: $\overline{\int_a^b} f(t)\,dt = \underline{\int_a^b} f(t)\,dt$. We use the results and notation from Sect. 2.5.2.

(a) Given $\ell > 0$, let $G_\ell = \{x \mid \omega_f(x) \geq \ell\}$. Show that given $\delta > 0$, we can choose a finite set of open intervals $\{J_i \mid i \in \mathbf{k}\}$ of total length at most ε such that $\cup_{i \in \mathbf{k}} J_i \supset G_\ell$. (Hints: Since the discontinuity set $\mathcal{D}$ is countable and $\mathcal{D} \subset G_\ell$, we can choose a countable set of open intervals I_i such that $\cup_i I_i \supset G_\ell$. Now show, using the result of Exercises 2.5.10(7), that we can pick a finite number of the intervals I_i with union containing G_ℓ.)

(b) Let $\varepsilon > 0$. Show that by choosing $\delta, \ell > 0$ sufficiently small we have $\overline{\int_a^b} f(t)\,dt - \underline{\int_a^b} f(t)\,dt < \varepsilon$. (Hint: Use the result of (a) together with the definition of $\omega_f(x)$.)

(c) Deduce that $\overline{\int_a^b} f(t)\,dt = \underline{\int_a^b} f(t)\,dt$.

(d) Prove that if f is continuous at $x \in [a, b]$ then $\int_a^x f(t)\,dt$ is differentiable at x with derivative $f(x)$.

We remark that this result requires more serious analysis than what is required if f is continuous or has an anti-derivative.

(10) This is an extended exercise that may be used for discussion and projects. The aim is to define integrals of continuous functions on rectangular domains in $\mathbb{R}^2$. Recall that a (bounded) *rectangle R* is a subset of $\mathbb{R}^2$ that can be written as a product of closed and bounded intervals: $R = [a, b] \times [c, d] = \{(x, y) \in \mathbb{R}^2 \mid x \in [a, b], y \in [c, d]\}$, where $-\infty < a \leq b < \infty$, $-\infty < c \leq d < \infty$. If $f : \mathbb{R}^2 \to \mathbb{R}$, then f is continuous at (x_0, y_0) if, given $\varepsilon > 0$, there exists a $\delta > 0$ such that $|f(x, y) - f(x_0, y_0)| < \varepsilon$, whenever $|x - x_0|, |y - y_0| < \delta$. We assume here that continuous functions on a rectangle are bounded (see (a) below and also Chap. 7). In what follows f will be fixed, continuous and have domain $\mathbb{R}^2$ and we consider the problem of defining the integral $\int_R f$ of f over rectangles $R = [a, b] \times [c, d]$. Just as in the 1-variable case, we approach the problem by stating the properties we require of the integral and then prove existence and uniqueness.

(a) (Preliminaries on continuity.) If $f : [a, b] \times [c, d] \to \mathbb{R}$ is continuous, show that f is (a) sequentially continuous and (b) uniformly continuous. (Hint and comments: use the same method as in the 1-variable case. Given sequential continuity, it is easy to show that f is bounded and attains its bounds.)

(b) Generalize conditions **I, II** so that they apply to bounded functions on a rectangle $R = [a, b] \times [c, d]$. (Let $\mathcal{I}((a, b), (c, d))$ denote the candidate for $\int_R f$. For **II**, we allow the rectangle R to be written as a finite union of rectangles which only meet along their boundaries.) In (c–f) below the aim is to prove that **I, II** uniquely characterize the integral of f. We then give two solutions to finding $\mathcal{I}((a, b), (c, d))$—which must be equal, by uniqueness.

(c) Assume $f : R \to \mathbb{R}$ is independent of y. Show that if **I, II** hold, then $\mathcal{I}((a, b), (c, d)) = (d - c)^{-1} \int_a^b f(x)\, dx$ for all $a < b, c < d$. Verify the similar statement if f is independent of x. (Hint: fix c, d and show that $\mathcal{I}(a, b) = (d - c)^{-1}\mathcal{I}((a, b), (c, d))$ satisfies conditions **I, II** for functions of one variable.)

(d) Let $(x, y) \in R = [a, b] \times [c, d]$ and define $V(x, y) = \mathcal{I}((a, x), (c, y))$. Verify that (i) V is continuous, (ii) $\frac{\partial V}{\partial x}(x, y) = \int_c^y f(x, y)\, dy$, (iii) $\frac{\partial V}{\partial y}(x, y) = \int_a^x f(x, y)\, dx$. (Hint: for the proof of (ii,iii) use uniform continuity—see (a) above.)

(e) Using (d), show that if $\mathcal{I}((a, x), (c, y))$ exists, $x \in [a, b], y \in [c, d]$, then it is unique.

(f) Show that $\mathcal{I}((a, c), (c, d)) = \int_c^d \int_a^b f(x, y)\, dxdy = \int_a^b \int_c^d f(x, y)\, dydx$. This not only gives the existence of the integral but also gives Fubini's theorem.

(g) What is $\frac{\partial^2 V}{\partial x \partial y}(x, y)$?

The arguments above allow us to give an elementary definition of double integrals on rectangles. The results suffice for all but one of our applications of double integrals in Chap. 6 as well as our construction of uniform approximations in Chap. 9. The arguments easily generalize to unbounded rectangles (for example, quadrants of $\mathbb{R}^2$) as well as integrals over rectangular regions in $\mathbb{R}^n$, $n > 2$. However, much extra work has to be done to rigorously establish integrals on general non-rectangular domains and, in particular, to prove the change of variables formula for multiple integrals. This is best done in the framework of Lebesgue integration though the one application of a linear change of variables we make in Chap. 6 can be done fairly easily using direct arguments.

2.9   Appendix: The Log and Exponential Functions

In this section we give a summary, with proofs, of the main properties of the logarithm and exponential functions. Particularly important for us will be the result that as $x \to +\infty$, $\log x$ grows more slowly than x^a, for any $a > 0$, and e^x grows faster than any power of x.

2.9.1   The Logarithm

For $x > 0$, we define the *natural* or *Napierian* logarithm by

$$\log x = \int_1^x \frac{dt}{t}.$$

Remark 2.9.1 We avoid the alternative notation $\ln x$ for the logarithm of x on the grounds that the base 10 logarithm is rarely used these days and so there is no longer a good reason to use an unpronounceable notation for the natural logarithm. It is immediate from the fundamental theorem of calculus (2.8) that log is continuously differentiable on $(0, \infty)$ with derivative given by

$$\frac{d}{dx} \log(x) = \frac{1}{x}.$$

From this it follows that $\log : (0, \infty) \to \mathbb{R}$ is C^∞ (infinitely differentiable). Since $\log'(x) = 1/x > 0$ for all $x > 0$, log is a strictly increasing function of x.

Proposition 2.9.2 *We have*

(1) $\log xy = \log x + \log y$, *for all* $x, y > 0$.
(2) $\log 1 = 0$.
(3) $\log x^{-1} = -\log x$, *for all* $x > 0$.
(4) $\log x^{\frac{p}{q}} = \frac{p}{q} \log x$, *for all* $\frac{p}{q} \in \mathbb{Q}$.

Proof We have

$$\log xy = \int_1^{xy} \frac{dt}{t} = \int_1^x \frac{dt}{t} + \int_x^{xy} \frac{dt}{t} = \log x + \int_x^{xy} \frac{dt}{t}.$$

Hence to prove (1), it suffices to show that $\int_x^{xy} \frac{dt}{t} = \log y$. For this, we make the substitution $t = ux$, to obtain $\int_x^{xy} \frac{dt}{t} = \int_1^y \frac{du}{u} = \log y$.

Statement (2) follows from (1) by taking $x = y = 1$. Alternatively, take $x = 1$ in the definition of $\log x$. Statement (3) follows from (2) by taking $y = x^{-1}$ in (1).

Finally, (1) and (3) imply that $\log x^n = n \log x$, $n \in \mathbb{Z}$. Therefore, for $q \in \mathbb{N}$, we have $\log(x^{\frac{1}{q}})^q = q \log x^{\frac{1}{q}}$, and so $\log x^{\frac{1}{q}} = \frac{1}{q} \log x$. Hence $\log x^{\frac{\ell}{q}} = \log(x^{\frac{1}{q}})^p = \frac{\ell}{q} \log x$. $\qquad\square$

Proposition 2.9.3 *The logarithm maps* $(0, \infty)$ *bijectively onto* $\mathbb{R}$. *In particular,* $\lim_{x \to 0+} \log x = -\infty$, $\lim_{x \to +\infty} \log x = +\infty$.

Proof Since log is strictly increasing, log is a bijection onto its image. Since $\log 2 > 0$ ($\log 1 = 0$ and log is strictly increasing), $\lim_{n \to \infty} \log 2^n = n \log 2 = +\infty$. Hence $\lim_{x \to +\infty} \log x = +\infty$. On the other hand $\lim_{x \to 0+} \log x = \lim_{y \to +\infty} \log y^{-1} = -\lim_{y \to +\infty} \log y = -\infty$. It remains to show that log maps $(0, \infty)$ onto $\mathbb{R}$. Let $y \in \mathbb{R}$. Choose $n \in \mathbb{N}$ so that $-n \log 2 \leq y \leq n \log 2$. Since $-n \log 2 = \log 2^{-n}$, $n \log 2 = \log 2^n$, the intermediate value theorem implies there exists an $x \in [2^{-n}, 2^n]$ such that $\log x = y$. $\qquad\square$

Remarks 2.9.4

(1) By Proposition 2.9.3, we may define $e > 1$ to be the unique real number such that $\log e = 1$.

(2) We may use Proposition 2.9.3 to *define* x^a for all $x > 0$, $a \in \mathbb{R}$. Thus we define x^a to be the unique positive real number with logarithm $a \log x$. Granted Proposition 2.9.2(4), this definition of x^a coincides with the usual one when a is rational. We also have the obvious extension of Proposition 2.9.2(4): $\log x^a = a \log x$ for all $x > 0$, $a \in \mathbb{R}$. For further properties of a^x, see the exercises at the end of the section. ✱

2.9.2 The Exponential Function

We define the *exponential* function $\exp : \mathbb{R} \to (0, \infty)$ to be the inverse of $\log : (0, \infty) \to \mathbb{R}$. As is customary, we often use the notation e^x for $\exp(x)$. This is justified by (2,3,4) of the next proposition.

Proposition 2.9.5 *We have*

(1) $e^{\log x} = x$, *for all* $x > 0$, $\log(e^x) = x$, *for all* $x \in \mathbb{R}$.

(2) $e^{x+y} = e^x e^y$, *for all* $x, y \in \mathbb{R}$.

(3) $e^0 = 1$.

(4) $e^{-x} = 1/e^x$, for all $x > 0$.

Proof (1) follows since exp is the inverse of log. The remaining properties follow easily from (1) and Proposition 2.9.2. For example, $x + y = \log(e^x) + \log(e^y) = \log(e^x e^y)$. Exponentiate to get (2).   □

Proposition 2.9.6

(1) $\exp : \mathbb{R} \to (0, \infty)$ *is strictly increasing.*

(2) exp *is continuous.*

(3) exp *is* C^∞ *and* $\exp'(x) = \exp(x)$, *for all* $x \in \mathbb{R}$.

Proof Since $\log : (0, \infty) \to \mathbb{R}$ is strictly increasing, $\exp : \mathbb{R} \to (0, \infty)$ is strictly increasing.

We prove the continuity of exp. Let $\varepsilon > 0$, and $x_0 \in \mathbb{R}$. We must find $\delta > 0$ such that $|\exp(x) - \exp(x_0)| < \varepsilon$, if $|x - x_0| < \delta$. Set $y_0 = \exp(x_0)$ and suppose $0 < \varepsilon < y_0$. Set $a = \log(y_0 - \varepsilon)$, $b = \log(y_0 + \varepsilon)$ and note that $a < x_0 < b$. We have $\exp(a, b) = (y_0 - \varepsilon, y_0 + \varepsilon)$ and so if we take $\delta = \min\{x_0 - a, b - x_0\}$, we have $|\exp(x) - \exp(x_0)| < \varepsilon$ if $|x - x_0| < \delta$.

It remains to prove that exp is C^∞. We start by proving that exp is differentiable at $x = 0$ with derivative 1. That is, we claim $\lim_{h \to 0}(e^h - 1)/h = 1$. Setting $h = \log x$, we have

$$\lim_{h \to 0} \frac{e^h - 1}{h} = \lim_{x \to 1} \frac{e^{\log x} - e^{\log 1}}{\log x}$$

$$= \lim_{x \to 1} \frac{x - 1}{\log x - \log 1}$$

$$= 1,$$

since $\lim_{x \to 1} \frac{\log x - \log 1}{x - 1} = \log'(1) = 1$. The derivative of exp at x is defined by

$$\exp'(x) = \lim_{h \to 0} \frac{e^{x+h} - e^x}{h}$$

$$= \exp(x) \lim_{h \to 0} \frac{e^h - 1}{h}$$

$$= \exp(x),$$

where we have used $\exp'(0) = 1$. Hence for all $x \in \mathbb{R}$, $\exp'(x) = \exp(x)$. Since exp is continuous, $\exp' = \exp$ is continuous and so exp is C^1. Proceeding inductively, we have for all $n \in \mathbb{N}$, $\exp^{(n)} = \exp$ and so exp is C^∞.   □

Remark 2.9.7 As a corollary of Proposition 2.9.6, we see that $x^a = \exp(a \log x)$ is differentiable with derivative ax^{a-1}.   ✖

2.9.3 Estimates

Proposition 2.9.8 *Let* $a, b > 0$.

(1) $\lim_{x\to\infty} x^{-a} (\log x)^b = 0$.
(2) $\lim_{x\to 0+} x^a (\log x)^b = 0$.

Proof We prove (1) ((2) follows from (1) by replacing x by x^{-1}). Since $x^{-a}(\log x)^b = (x^{-a/b}(\log x))^b$, there is no loss of generality in taking $b = 1$ and verifying that $\lim_{x\to\infty} x^{-a} \log x = 0$ for all $a > 0$. Computing the derivative of $f(x) = x^{-a} \log x$, we find that $f'(x) < 0$ if $a \log x > 1$. Hence, $x^{-a} \log x$ is monotone decreasing for sufficiently large x and so $\lim_{x\to\infty} x^{-a} \log x$ exists and is greater than or equal to zero. Now $(2^n)^{-a} \log 2^n = (2^{-a})^n n \log 2 \to 0$ as $n \to \infty$, since $2^{-a} < 1$ (Example 2.3.26). Hence $\lim_{x\to\infty} x^{-a} \log x = 0$. $\qquad\square$

Proposition 2.9.9 *Let* $a \in \mathbb{R}$, $c > 0$. *Then* $\lim_{x\to +\infty} x^a e^{-cx} = 0$.

Proof We leave this as an exercise, using Proposition 2.9.8. $\qquad\square$

EXERCISES 2.9.10

(1) For $a > 0$, show that $\lim_{x\to\infty} (\log x)^{-a} \log\log x = 0$. State and prove an analogous result that applies as $x \to 0+$.
(2) Provide the proof of Proposition 2.9.9.
(3) Show that $\log 3 > 1 > \log 2$ and deduce that $e \in (2, 3)$.
(4) Using calculus, show that

   (a) $x - \frac{x^2}{2} \le \log(1 + x) \le x$, for all $x \ge 0$.
   (b) $-x \ge \log(1 - x) \ge -x - x^2$, for all $x \in [0, 1/2]$.

(5) Using the results of the previous exercise show that for all $x \in \mathbb{R}$

$$\lim_{n\to\infty} \left(1 + \frac{x}{n}\right)^n = e^x.$$

(6) Show that

   (a) $\lim_{n\to\infty} \left(\frac{2n^2-1}{2n^2+1}\right)^{n^2} = e^{-1}$.
   (b) $\lim_{n\to\infty} \left(\frac{3n}{3n-1}\right)^n = \sqrt[3]{e}$.

(7) Let $p, q \in \mathbb{N}$. Find $\lim_{n\to\infty} \left(\frac{qn+p}{qn}\right)^n$.
(8) For $a > 0$, $x \in \mathbb{R}$, define $a^x = \exp(x \log a)$. Verify that

   (a) $a^x a^y = a^{x+y}$, all $x, y \in \mathbb{R}$.
   (b) $a^0 = 1$, $a^1 = a$.
   (c) $a^{-x} = 1/a^x$.
   (d) a^x is infinitely differentiable and the derivative of a^x is $(\log a)a^x$.

   Show also that if $a > 1$ (respectively, $a < 1$) then a^x defines a monotone strictly increasing (respectively, decreasing) bijection of $\mathbb{R}$ onto $(0, \infty)$.

(9) Let $a \in (0, 1)$. Verify that $\lim_{x \to \infty} a^x x^b = 0$ for all $b \in \mathbb{R}$.

(10) Let $\alpha > 0$. Find $\lim_{n \to \infty} n^{n^{-\alpha}}$.

2.10   Appendix: Construction of $\mathbb{R}$ Revisited

We look at the construction of the real numbers using Cauchy sequences of rational numbers rather than decimal expansions. Most of this appendix should be regarded as being for group discussion—at most we give brief proofs, preferring instead to make precise the results that need to be proved.

Let C denote the set of all Cauchy sequences of *rational* numbers. Our aim is to show that there is a natural way to partition C as $\{C_\alpha \mid \alpha \in \mathbb{R}\}$. Rather than thinking of a real number as a single 'point', we view the real number α as the set of all possible rational approximations to α. That is, we think of each partition set C_α as defining a real number. Practically speaking, this is the way we handle irrational numbers—we compute using rational approximations. The devil is in the details—though nothing is hard, there are *many* points to be checked. One advantage of the approach is that we avoid the problems of addition, subtraction and multiplication of decimal expansions as well as issues about whether or not a rational number has more than one decimal expansion. This time we just use the standard and simple arithmetic properties of rational numbers: $\frac{p}{q} \pm \frac{r}{s} = \frac{ps \pm rq}{qs}$, $\frac{p}{q} \times \frac{r}{s} = \frac{pr}{qs}$. Disadvantages are that we work at a more abstract level and that the arguments verifying the existence of an order on the real numbers are a little harder than what we sketched in Chap. 1. Also the methods used in Chap. 1 lead to natural and constructive proofs of, for example, the existence of the supremum of a bounded set.

If $s = (x_n) \in C$, we define

$$C_s = \{t = (y_n) \in C \mid \lim_{n \to \infty} |x_n - y_n| = 0\}.$$

Since $s \in C_s$, $C_s \neq \emptyset$. The next lemma gives a natural partition of C.

Lemma 2.10.1 *If $s, t \in C$, then either $C_s = C_t$ or $C_s \cap C_t = \emptyset$. In particular, $C_s = C_t$ iff $t \in C_s$ and $\{C_s \mid s \in C\}$ defines a partition of C.*

Let $\mathbb{R} = \{C_s \mid s \in C\}$ denote the partition of C given by the lemma. There is a natural way to embed the rational numbers $\mathbb{Q}$ in $\mathbb{R}$. Given $q \in \mathbb{Q}$, let $C_{\mathbf{q}} \in \mathbb{R}$ be defined by the constant Cauchy sequence $\mathbf{q} = (q)$. By Lemma 2.10.1, if $q, r \in \mathbb{Q}$, then $C_{\mathbf{q}} = C_{\mathbf{r}}$ iff $q = r$.

2.10.1   Arithmetic

Let $s = (x_n), t = (y_n) \in C$. We define $s \pm t = (x_n \pm t_n)$, $st = s \times t = (x_n y_n)$. We also let $\mathbf{0} = (0)$ denote the Cauchy sequence all of whose terms are zero and $\mathbf{1} = (1)$ denote the Cauchy sequence all of whose terms are one.

Lemma 2.10.2 *Let $s, t \in C$.*

(1) $s \pm t$, $st \in C$.
(2) $s \pm 0 = s$, $1s = s$, $0s = 0$.

We need to be careful when it comes to division; more precisely, the definition of the reciprocal. Let $C^\star$ denote the subset of C consisting of Cauchy sequences (x_n) for which $x_n \neq 0$, all $n \in \mathbb{N}$. Given $s \in C$, set $C_s^\star = C^\star \cap C_s$. We remark that $C_s^\star \neq \emptyset$ (if $s = (x_n) \in C$ replace every term x_n which is zero by $1/n^2$ to get a sequence $s' \in C_s^\star$).

Lemma 2.10.3 *Suppose that $s = (x_n) \in C^\star$ and that $s \notin C_0$. Then $s^{-1} = (x_n^{-1}) \in C$.*

Now the idea is to extend Lemmas 2.10.2, 2.10.3 to $\mathbb{R}$. Suppose $C_s, C_t \in \mathbb{R}$. We define

$$C_s \pm C_t = C_{s \pm t},$$

$$C_s C_t = C_{st},$$

$$C_s^{-1} = C_{\bar{s}^{-1}}, \text{ where } \bar{s} \in C_s^\star, \text{ and } s \notin C_0.$$

Lemma 2.10.4 *Our definitions of $\pm$, $\times$ and the reciprocal on $\mathbb{R}$ are well defined and are compatible with the usual definitions of $\pm$, $\times$, and the reciprocal on $\mathbb{Q} \subset \mathbb{R}$.*

Proof To verify that the definition of $\pm$ on $\mathbb{R}$ is well defined, we have to show that $C_{s \pm t}$ depends only on the partition sets C_s, C_t and not on the particular choices of s and t. That is, if $s' \in C_s$ and $t' \in C_t$, we have to show $C_{s \pm t} = C_{s' \pm t'}$. This follows from Lemma 2.10.1. Similar arguments hold for multiplication and the reciprocal.
□

Proposition 2.10.5 *With our definition of $\pm$, $\times$, and reciprocal, $\mathbb{R}$ inherits all of the standard laws of arithmetic from $\mathbb{Q}$. In particular, zero is represented by C_0, 1 by C_1 and we have*

(1) $C_s + C_t = C_t + C_s$, $C_s C_t = C_t C_s$ *(commutativity)*.
(2) $C_s + C_0 = C_s$, $C_s C_0 = C_0$, $C_s C_1 = C_s$.
(3) $(C_s + C_t) + C_u = C_s + (C_t + C_u)$, $(C_s C_t) C_u = C_s (C_t C_u)$ *(associativity)*.
(4) $C_s (C_t + C_u) = C_s C_t + C_s C_u$ *(distributivity)*.

The additive inverse of $-C_s$ is defined to be C_{-s} and the multiplicative inverse C_s^{-1} of C_s is defined for $s \notin C_0$ by $C_{\bar{s}^{-1}}$, where $\bar{s} \in C_s^\star$.

Remark 2.10.6 Setting up the basic arithmetic is easier when we work with Cauchy sequences of rational numbers as opposed to the decimal expansions used in Chap. 1. ✱

2.10.2 Order Structure on $\mathbb{R}$

Definition 2.10.7 Given $C_s \in \mathbb{R}$, we write $C_s > C_0$ if there exists $t = (t_n) \in C_s$, $\delta \in \mathbb{Q}$, with $\delta > 0$, and $N \in \mathbb{N}$ such that

$$t_n \geq \delta, \text{ for all } n \geq N.$$

We write $C_s < C_0$ if $-C_s > C_0$.

Lemma 2.10.8 *Let $C_s > C_0$. Then for every $u = (u_n) \in C_s$, there exists a $\delta \in \mathbb{Q}$, $\delta > 0$, and $N \in \mathbb{N}$ such that*

$$u_n \geq \delta, \text{ for all } n \geq N.$$

(Both δ and N will depend on the choice of $u \in C_s$.)

Remark 2.10.9 It follows from Lemma 2.10.8 that if $q \in \mathbb{Q}$ then $C_q > C_0$ iff $q > 0$ (standard order on $\mathbb{Q}$).   ✣

Lemma 2.10.10 *Let $C_s \in \mathbb{R}$. Then exactly one of the following statements holds:*

$$C_s = C_0, \ C_s > C_0, \ C_s < C_0.$$

In particular, $<$ is well defined.

Remark 2.10.11 The issue here is to that show if $C_s \neq C_0$, then either $C_s > C_0$ or $C_s < C_0$. Note that the theory here is harder than what we did in Chap. 1 using decimal expansion. The reason is that when we used decimal expansions, the sequences of approximating rationals were monotone increasing (and naturally defined).   ✣

Using Lemma 2.10.10 and our results on the arithmetic on $\mathbb{R}$, we can define an order on $\mathbb{R}$ by $C_s > C_t$ if $C_s - C_t = C_{s-t} > C_0$.

Proposition 2.10.12 *With our definition of $<$, $\mathbb{R}$ inherits all of the standard properties of $<$ holding on $\mathbb{Q}$. In particular,*

(1) *If $C_s < C_t$ and $C_t < C_u$, then $C_s < C_u$.*
(2) *If $C_s < C_t$, then If $C_s + C_u < C_t + C_u$ for all $C_u \in \mathbb{R}$.*
(3) *If $C_s < C_t$ and $C_u > C_0$, then $C_s C_u < C_t C_u$.*
(4) *$C_s < C_t$ iff $-C_s > -C_t$.*

Remark 2.10.13 A consequence of Lemma 2.10.10 and the definition of order is that if $C_s > C_0$ then there exists a $q \in \mathbb{Q}$ such that $C_s > C_q > 0$. This property is equivalent to the *Archimedean* property of $\mathbb{R}$: if $C_s > C_0$, there exists an $n \in \mathbb{N}$ such that $C_n > C_s$.   ✣

2.10.3   Absolute Value

Just as for decimal expansions, it is easy to define the absolute value once we have the order structure on $\mathbb{R}$.

Definition 2.10.14 Given $C_s \in \mathbb{R}$ we define $|C_s| = C_s$ if $C_s \geq C_0$ and $|C_s| = -C_s$ if $C_s < C_0$.

It is clear that $| \cdot |$ is compatible with the absolute value defined on $\mathbb{Q}$. That is, if $q \in \mathbb{Q}$, we have $|C_q| = C_{|q|}$.

Lemma 2.10.15 (The Triangle Inequality) *For all $C_s, C_t \in \mathbb{R}$, we have*

$$|C_s + C_t| \leq |C_s| + |C_t|.$$

Proof Given $s = (s_n) \in C$, define $|s| = (|s_n|)$ and note that $|C_s| = C_{|s|}$. In order to prove the triangle inequality we must show

$$C_{|s+t|} \leq C_{|s|} + C_{|t|}.$$

If $s = (s_n)$, $t = (t_n)$, then s, t are Cauchy sequences of rational numbers and so by the triangle inequality on $\mathbb{Q}$, we have

$$|s_n + t_n| \leq |s_n| + |t_n|, \ n \in \mathbb{N}. \tag{2.10}$$

Now argue by contradiction: suppose $C_{|s+t|} > C_{|s|} + C_{|t|}$. Then $|s_n + t_n| > |s_n| + |t_n|$ for sufficiently large n (Lemma 2.10.8), contradicting (2.10). □

2.10.4   Limits, Density and Completeness

Definition 2.10.16 A sequence (C_{s_n}) in $\mathbb{R}$ is convergent if there exists a $C_s \in \mathbb{R}$ such that $|C_{s_n} - C_s| \to C_0$ as $n \to \infty$. That is, given $C_\varepsilon > C_0$, there exists an $N \in \mathbb{N}$ such that $C_0 \leq |C_{s_n} - C_s| < C_\varepsilon$ for all $n \geq N$.

Theorem 2.10.17 (Density of Rational Numbers) *Every $C_s \in C$ is the limit of a sequence of rational numbers. That is, given $C_s \in C$, there exists a sequence $(q_n) \subset \mathbb{Q}$ such that*

$$\lim_{n \to \infty} C_{q_n} = C_s.$$

(For $n \geq 1$, C_{q_n} is defined using the constant Cauchy sequence, all terms of which equal q_n.)

Proof Suppose $s = (s_n)$. We define $q_n = s_n, n \geq 1$. □

Theorem 2.10.18 (Cauchy Sequences Converge in $\mathbb{R}$) *Every Cauchy sequence* (C_{s^n}) *in* $\mathbb{R}$ *is convergent.*

Proof Define the sequence $(S_n) \subset \mathbb{Q}$ by

$$S_n = s_n^n, \ n \in \mathbb{N}.$$

Then (S_n) is Cauchy and if we set $S = (S_n)$, $\lim_{n \to \infty} C_{s^n} = C_S$.   □

Chapter 3
Infinite Series

3.1 Introduction

In this chapter we undertake a detailed study of the convergence of infinite series. This work forms an essential foundation for the construction and analysis of functions that we give in Chaps. 4–6.

We start by looking at general infinite series, then specialize to series of positive terms and give a number of criteria for convergence. Next, using our results on Cauchy sequences, we consider absolutely and conditionally convergent series and find conditions for convergence. As an illustration of the care that needs to be taken, we prove Riemann's rearrangement theorem: if an infinite series is convergent but not absolutely convergent, then we can add the terms in a different order so as to make the series converge to any preassigned number or not converge at all. We conclude with definitions and results on doubly infinite series and infinite products and prove the infinite product formula for $\sin x$.

3.2 Generalities

First, we recall some definitions and results from Chap. 2. Let (a_n) be a sequence of real numbers. For $n \in \mathbb{N}$, we define the *partial sum* $S_n = \sum_{i=1}^{n} a_i$.

Definition 3.2.1 The infinite series $\sum_{n=1}^{\infty} a_n$ is *convergent* if the sequence (S_n) of partial sums is convergent. If (S_n) is convergent, we define $\sum_{n=1}^{\infty} a_n$ to be equal to $\lim_{n \to \infty} S_n$.

Remarks 3.2.2

(1) The infinite series $\sum_{n=1}^{\infty} a_n$ should be thought of symbolically—as shorthand for the sequence of partial sums. When (and only when) the sequence is known

© Springer International Publishing AG 2017
M. Field, *Essential Real Analysis*, Springer Undergraduate Mathematics Series,
https://doi.org/10.1007/978-3-319-67546-6_3

to be convergent, we identify $\sum_{n=1}^{\infty} a_n$ with the limit of the corresponding sequence of partial sums. Of course, this is what we did previously in our description of real numbers. If x has decimal expansion $\pm x_0.x_1 \cdots$, then we identify x with the infinite sum $\pm \sum_{n=0}^{\infty} x_n 10^{-n}$.

(2) We sometimes write "$\sum_{n=1}^{\infty} a_n < \infty$" to signify that the infinite series $\sum_{n=1}^{\infty} a_n$ is convergent. A statement like "$\sum_{n=1}^{\infty} a_n = 5$" should be interpreted as saying that the infinite series $\sum_{n=1}^{\infty} a_n$ is convergent and that the limit (of the sequence of partial sums) is equal to 5. We say "the sum of the infinite series is 5".

(3) Although we shall not spell out the details, all the usual limit laws for sequences carry over to infinite series. For example, if^1 the infinite series $\sum_{n=1}^{\infty} a_n$, $\sum_{n=1}^{\infty} b_n$ are both convergent, then so is the infinite series $\sum_{n=1}^{\infty} (a_n + b_n)$ and $\sum_{n=1}^{\infty} (a_n + b_n) = \sum_{n=1}^{\infty} a_n + \sum_{n=1}^{\infty} b_n$.   ✱

There is precisely one general necessary condition for convergence of an infinite series.

Lemma 3.2.3 *If the infinite series $\sum_{n=1}^{\infty} a_n$ is convergent, then $\lim_{n \to \infty} a_n = 0$. (No restrictions on the signs of the a_n.)*

Proof If $\sum_{n=1}^{\infty} a_n$ is convergent, then $\lim_{n \to \infty} S_n = \lim_{n \to \infty} S_{n-1}$. Hence $\lim_{n \to \infty} a_n = \lim_{n \to \infty} (S_n - S_{n-1}) = 0$.   □

Remark 3.2.4 Everything above extends immediately to infinite series of complex terms. In the next section we study series of positive terms. The results we obtain have no analogue for complex series (there is no natural order relation on the complex numbers).   ✱

3.3   Series of Eventually Positive Terms

If the terms of (a_n) are all positive (respectively, eventually positive), then (S_n) is increasing (respectively, eventually increasing). As a consequence of Theorem 2.3.18, we see that the infinite series $\sum_{n=1}^{\infty} a_n$ is convergent iff the sequence (S_n) of partial sums is bounded.

Examples 3.3.1

(1) The infinite series $\sum_{n=2}^{\infty} \frac{1}{n(n-1)}$ is convergent and $\sum_{n=2}^{\infty} \frac{1}{n(n-1)} = 1$. For $n \geq 2$ we have $\frac{1}{n(n-1)} = \frac{1}{n-1} - \frac{1}{n} > 0$. Hence

$$\sum_{n=2}^{N} \frac{1}{n(n-1)} = \sum_{n=2}^{N} \left(\frac{1}{n-1} - \frac{1}{n} \right) = 1 - \frac{1}{N}.$$

---

[1] An essential 'if'.

Letting $N \to \infty$, we see that $\sum_{n=2}^{\infty} \frac{1}{n(n-1)}$ converges to 1.

(2) The infinite series $\sum_{n=1}^{\infty} \frac{1}{n}$ diverges to $+\infty$. For $n \geq 1$, set $N = 2^n = 1 + \sum_{i=0}^{n-1} 2^i$. We have

$$\sum_{i=1}^{N} \frac{1}{i} = 1 + \frac{1}{2} + \left(\frac{1}{3} + \frac{1}{4}\right) + \left(\frac{1}{5} + \cdots + \frac{1}{8}\right) + \cdots$$

$$+ \left(\frac{1}{2^{n-1} + 1} + \cdots + \frac{1}{2^n}\right)$$

$$\geq 1 + \frac{1}{2} + 2\frac{1}{2^2} + \cdots + 2^j \frac{1}{2^{j+1}} + \cdots + 2^{n-1} \frac{1}{2^n}$$

$$= 1 + n\frac{1}{2} = \frac{n+2}{2}.$$

This estimate shows that the increasing sequence (S_N) is not bounded above and so $\sum_{n=1}^{\infty} \frac{1}{n}$ diverges to $+\infty$. ♠

Our aim in the remainder of this section is to develop some convergence tests for infinite series of (eventually) positive terms. These tests range from the highly practical (comparison, ratio and Cauchy integral tests) to the more theoretical D'Alembert and Cauchy tests. For most practical examples, readers are advised not to use the theoretical tests—at least until simpler tests have been tried. They rarely work better and it is easy to make errors when applying them.

3.3.1   The Comparison Test

Proposition 3.3.2 (The Comparison Test) *Let $(u_n), (v_n)$ be sequences of real numbers satisfying $0 \leq u_n \leq v_n$, for all $n \in \mathbb{N}$.*

(1) *If $\sum_{n=1}^{\infty} v_n$ is convergent, then (a) $\sum_{n=1}^{\infty} u_n$ is convergent, and (b) $0 \leq \sum_{n=1}^{\infty} u_n \leq \sum_{n=1}^{\infty} v_n$.*
(2) *If $\sum_{n=1}^{\infty} u_n$ is divergent, then $\sum_{n=1}^{\infty} v_n$ is divergent (in either case to $+\infty$).*

(The result applies with minor changes in the statements if (u_n) and (v_n) are eventually positive and $u_n \leq v_n$ for all sufficiently large n.)

Proof For $n \in \mathbb{N}$, define $S_n = \sum_{i=1}^{n} u_i$, $T_n = \sum_{i=1}^{n} v_i$. Since $0 \leq u_i \leq v_i$, we have

$$0 \leq S_n \leq T_n, \text{ for all } n \in \mathbb{N}.$$

Suppose that $\sum_{n=1}^{\infty} v_n$ is convergent, with limit T. Then $0 \leq S_n \leq T_n \leq T$, for all $n \in \mathbb{N}$. Hence the increasing sequence (S_n) is bounded above by T and so $\sum_{n=1}^{\infty} u_n$ is convergent and $\sum_{n=1}^{\infty} u_n \leq \sum_{n=1}^{\infty} v_n$, proving (1).

If $\sum_{n=1}^{\infty} u_n$ is divergent, then the series must diverge to $+\infty$ (Theorem 2.3.19). Hence for all $K \geq 0$, there exists an $N \in \mathbb{N}$ such that $\sum_{n=1}^{m} u_n \geq K, m \geq N$. Hence $\sum_{n=1}^{m} v_n \geq \sum_{n=1}^{m} u_n \geq K$, for all $m \geq N$, and so $\sum_{n=1}^{\infty} v_n$ diverges to $+\infty$. $\quad\square$

Examples 3.3.3

(1) Using the comparison test, we show that $\sum_{n=1}^{\infty} n^{-p}$ is convergent for $p \geq 2$. If $p > 2$, then $n^{-p} \leq n^{-2}$ for all $n \in \mathbb{N}$ and so, by (1) of the comparison test, it suffices to show that $\sum_{n=1}^{\infty} n^{-2}$ is convergent. Observe that for $n \geq 2$ we have

$$\frac{1}{n^2} < \frac{1}{n(n-1)}.$$

The series $\sum_{n=2}^{\infty} \frac{1}{n(n-1)}$ is convergent, with sum 1, by Examples 3.3.1(1). Hence, by the comparison test, $\sum_{n=1}^{\infty} n^{-2} = 1 + \sum_{n=2}^{\infty} n^{-2}$ is convergent with sum at most 2.

(2) Using the comparison test, we show that $\sum_{n=1}^{\infty} n^{-p}$ is divergent for $p \leq 1$. If $p \leq 1$, we have $n^{-p} \geq n^{-1}$, for all $n \in \mathbb{N}$. Since $\sum_{n=1}^{\infty} n^{-1}$ is divergent by Examples 3.3.1(2), the divergence of $n^{-p} \geq n^{-1}$ is immediate from (2) of the comparison test. $\quad\spadesuit$

3.3.2 The Ratio Test

Proposition 3.3.4 (The Ratio Test) *Let (a_n) be a sequence of positive real numbers and suppose that $\lim_{n\to\infty} \frac{a_{n+1}}{a_n}$ exists.*

(1) *If $\lim_{n\to\infty} \frac{a_{n+1}}{a_n} < 1$, the series $\sum_{n=1}^{\infty} a_n$ is convergent.*
(2) *If $\lim_{n\to\infty} \frac{a_{n+1}}{a_n} > 1$, the series $\sum_{n=1}^{\infty} a_n$ is divergent.*

Proof We prove (1) and leave (2) to the exercises. If $\lim_{n\to\infty} \frac{a_{n+1}}{a_n} = s < 1$, then there exists an $N \in \mathbb{N}$ such that $\frac{a_{n+1}}{a_n} \leq r = (s+1)/2 < 1$ for all $n \geq N$. Consequently, $a_{N+p} \leq r a_{N+p-1} \leq \cdots \leq r^p a_N$ for all $p \in \mathbb{N}$. The series $\sum_{p=0}^{\infty} a_{N+p}$ therefore converges by comparison with the geometric series $\sum_{p=0}^{\infty} a_N r^p$. If $\sum_{p=0}^{\infty} a_{N+p}$ converges, then obviously $\sum_{n=1}^{\infty} a_n$ converges. $\quad\square$

Remark 3.3.5 We emphasize that for the ratio test to apply, it is necessary that $\lim_{n\to\infty} \frac{a_{n+1}}{a_n}$ exists. $\quad\maltese$

Examples 3.3.6

(1) Convergence *does not* follow if $a_{n+1}/a_n < 1$ for all $n \in \mathbb{N}$. It is essential to compute the limit (if it exists). As a simple example, take $a_n = 1/n$. Then $a_{n+1}/a_n = n/(n+1) < 1$ for all $n \in \mathbb{N}$, yet $\sum_{n=1}^{\infty} 1/n$ diverges (Examples 3.3.1(2)).

(2) The classic area of application of the ratio test is to *power series*. As an example, consider the exponential series $\sum_{n=0}^{\infty} \frac{x^n}{n!}$, where $x \in \mathbb{R}^+$. If $x = 0$, $\sum_{n=0}^{\infty} \frac{x^n}{n!} = 1$ and there is nothing to prove. If we fix $x > 0$, and define $a_n = x^n/n!$, we have

$$\frac{a_{n+1}}{a_n} = \frac{x^{n+1} n!}{x^n (n+1)!} = \frac{x}{n+1}.$$

Since $\lim_{n\to\infty} \frac{x}{n+1} = 0 < 1$, the ratio test applies and so $\sum_{n=0}^{\infty} \frac{x^n}{n!}$ is convergent.
♠

3.3.3   D'Alembert's Test

Proposition 3.3.7 (D'Alembert's Test) *Let (a_n) be a sequence of positive real numbers.*

(1) *If* $\limsup_{n\to\infty} \frac{a_{n+1}}{a_n} < 1$, *the series* $\sum_{n=1}^{\infty} a_n$ *is convergent.*
(2) *If* $\liminf_{n\to\infty} \frac{a_{n+1}}{a_n} > 1$, *the series* $\sum_{n=1}^{\infty} a_n$ *is divergent.*

Proof The proof is almost identical to that of the ratio test. For example, the condition $\limsup_{n\to\infty} \frac{a_{n+1}}{a_n} < 1$ implies that there exists $0 < r < 1$, $N \in \mathbb{N}$ such that $\frac{a_{n+1}}{a_n} \le r < 1$ for all $n \ge N$. The proof then proceeds exactly as in the ratio test.   □

Example 3.3.8 It is quite difficult to find interesting examples of series where the ratio test fails to apply but D'Alembert's test is applicable. As a somewhat contrived example, define (a_n) by

$$a_{2n} = \left(\frac{1}{2}\right)^n \left(\frac{1}{3}\right)^{n-1}, \; n \ge 1,$$

$$a_{2n-1} = \left(\frac{1}{2}\right)^{n-1} \left(\frac{1}{3}\right)^{n-1}, \; n \ge 1.$$

We have $a_{2n}/a_{2n-1} = 1/2 \ne 1/3 = a_{2n+1}/a_{2n}$ and so the limit as $n \to \infty$ of a_{n+1}/a_n does not exist. However, D'Alembert's test applies, since $\limsup_{n\to\infty} \frac{a_{n+1}}{a_n} = 1/2$, and so the infinite series $\sum_{n=1}^{\infty} a_n$ converges. Of course, the convergence is easily seen by comparison with the geometric series $\sum_{n=1}^{\infty} 2^{-n}$.
♠

3.3.4 Cauchy's Test

Proposition 3.3.9 (Cauchy's Test) *Let (a_n) be a sequence of positive real numbers.*

(1) *If* $\limsup a_n^{\frac{1}{n}} < 1$, *the series* $\sum_{n=1}^{\infty} a_n$ *is convergent.*
(2) *If* $\limsup a_n^{\frac{1}{n}} > 1$, *the series* $\sum_{n=1}^{\infty} a_n$ *is divergent.*

Proof We prove (1) and leave (2) to the exercises. If $\limsup a_n^{\frac{1}{n}} = s < 1$, then there exists an $N \in \mathbb{N}$ such that $\sup\{a_n^{\frac{1}{n}} \mid n \geq N\} \leq r = (s+1)/2 < 1$. We therefore have $a_n \leq r^n$, all $n \geq N$. Hence $\sum_{n=1}^{\infty} a_n$ converges by comparison with the geometric series $\sum_{n=1}^{\infty} r^n$.   □

Remarks 3.3.10

(1) The Cauchy test is of great theoretical importance, as we see when we look at power series. However, in most practical applications it is usually best to start by trying the ratio test.
(2) If $\lim a_n^{\frac{1}{n}} = \lambda$ exists, then we have the simpler form of Cauchy's test: the series converges if $\lambda < 1$ and diverges if $\lambda > 1$. This is the form that is used in most of the exercises at the end of the section.
(3) The divergence condition for Cauchy's test uses the $\limsup$, not (the weaker) $\liminf$.   ✠

Examples 3.3.11

(1) We examine the convergence of the series $\sum_{n=1}^{\infty} nx^n$, $x \in \mathbb{R}^+$. We have $\lim_{n \to \infty} (nx^n)^{1/n} = \lim_{n \to \infty} n^{1/n} x = x$, by Examples 2.2.9(3). Hence, by Cauchy's test, $\sum_{n=1}^{\infty} nx^n$ converges if $x \in [0, 1)$. We see by inspection that the series diverges if $x \geq 1$ (Lemma 3.2.3). These results follow more easily using the ratio test.
(2) Consider the series $\sum_{n=1}^{\infty} 2^{-n} \left(1 + \frac{1}{n}\right)^{n^2}$. We have

$$\lim_{n \to \infty} \left(2^{-n}(1 + \frac{1}{n})^{n^2}\right)^{1/n} = 2^{-1} \lim_{n \to \infty} (1 + \frac{1}{n})^n > 1.25,$$

where the last inequality follows from Examples 2.3.27(2). Hence the series diverges. The result is not so easy if we try the ratio test.   ♠

3.3.5 Cauchy's Integral Test

If $f : [a, \infty) \to \mathbb{R}$ is continuous, then we define the infinite integral $\int_a^{\infty} f(t)\, dt$ to be $\lim_{x \to +\infty} \int_a^x f(t)\, dt$ if the limit exists (and is finite). A necessary (but not sufficient) condition for the existence of $\int_a^\infty f(t)\, dt$ is that $\lim_{x \to +\infty} f(x) = 0$. We refer to $\int_a^\infty f(t)\, dt$ as an *improper integral* (see the exercises for more definitions and examples). When $\int_a^\infty f(t)\, dt$ exists, we say the integral converges and write $\int_a^\infty f(t)\, dt < \infty$.

Proposition 3.3.12 (Cauchy's Integral Test) *Let $f : [1, \infty) \to \mathbb{R}$ be a positive, continuous and monotone decreasing function. A necessary and sufficient condition for the convergence of $\sum_{n=1}^\infty f(n)$ is the convergence of the improper integral $\int_1^\infty f(t)\, dt$. For all $n \in \mathbb{N}$ we have the estimate*

$$f(n) + \int_1^n f(t)\, dt \leq \sum_{j=1}^n f(j) \leq f(1) + \int_1^n f(t)\, dt, \ n > 1. \tag{3.1}$$

If either the series or the integral converges, then we have the estimate

$$\int_1^\infty f(t)\, dt \leq \sum_{j=1}^\infty f(j) \leq f(1) + \int_1^\infty f(t)\, dt. \tag{3.2}$$

Proof For $n > 1$, we have (property **I** of the integral)

$$\int_1^n f(t)\, dt = \sum_{j=1}^{n-1} \int_j^{j+1} f(t)\, dt.$$

Since f is monotone decreasing, we have (property **II** of the integral)

$$f(j) \geq \int_j^{j+1} f(t)\, dt \geq f(j+1).$$

Using these estimates we easily verify (3.1). Since $f \geq 0$, the sequence $(\sum_{j=1}^n f(j))$ of partial sums is increasing and so, by Theorem 2.3.18, $\sum_{j=1}^\infty f(j)$ converges iff $\int_1^\infty f(t)\, dt$ converges. Finally, (3.2) follows by letting $n \to \infty$ in (3.1) and using $\lim_{n \to \infty} f(n) = 0$ (Lemma 3.2.3). $\qquad \square$

Remark 3.3.13 Cauchy's integral test was originally found by Maclaurin in 1742 and rediscovered later by Cauchy. Early versions of the test were used in the fourteenth century by the Kerala school of mathematics in India. ✳

Example 3.3.14 We consider the convergence of $\sum_{n=1}^\infty \frac{1}{n^p}$, where $p \in \mathbb{R}^+$. Define the continuous function $f(x) = 1/x^p$, $x \in [1, \infty)$. Since $p \geq 0$, f is monotone decreasing and so Cauchy's integral test applies. If $p \neq 1$, we have

$$\int_1^n \frac{1}{t^p}\, dt = \frac{1}{p-1}\left(1 - \frac{1}{n^{p-1}}\right).$$

If $p > 1$, $\lim_{n\to\infty} \int_1^n \frac{1}{t^p}\, d = (p-1)^{-1}$ and so the improper integral converges. Hence $\sum_{n=1}^{\infty} \frac{1}{n^p}$ converges by Cauchy's integral test and

$$(p-1)^{-1} \le \sum_{n=1}^{\infty} \frac{1}{n^p} \le 1 + (p-1)^{-1}.$$

On the other hand, if $p < 1$, the improper integral diverges and so $\sum_{n=1}^{\infty} \frac{1}{n^p}$ diverges by Cauchy's integral test. There remains the case $p = 1$. We have

$$\int_1^n \frac{1}{t}\, dt = \log n - 1.$$

Since $\lim_{n\to+\infty} \log n = +\infty$ as $n \to \infty$, the improper integral diverges and so $\sum_{n=1}^{\infty} \frac{1}{n}$ diverges by Cauchy's integral test. We note for future reference the useful estimate

$$\frac{1}{n} + \log n \le \sum_{j=1}^{n} \frac{1}{j} \le 1 + \log n. \qquad (3.3)$$

We provide a number of other examples of applications of Cauchy's integral test in the exercises.   ♠

EXERCISES 3.3.15

(1) Complete the proofs of D'Alembert's and Cauchy's test—take particular care with the divergence statement (2) in Cauchy's test.

(2) Let $T_n = \sum_{j=1}^{n} 1/(2j-1)$ and $S_n = \sum_{j=1}^{n} 1/j$, $n \ge 1$. Show that $T_n > S_n/2$ and deduce that the series $\sum_{j=1}^{\infty} 1/(2j-1)$ is divergent. Show also how this result can be derived from Cauchy's integral test.

(3) Cauchy's test is stronger than D'Alembert's test, which is stronger than the ratio test. For each of the following series, determine the weakest test that proves convergence (implicit in the question is showing why the weaker tests fail; you do not have to prove that tests stronger than the weakest test that works also work.)

   (a) $1 + \frac{(a+1)}{b+1} + \frac{(a+1)(2a+1)}{(b+1)(2b+1)} + \cdots + \frac{(a+1)\cdots(na+1)}{(b+1)\cdots(nb+1)} + \cdots$, where $b > a > 0$.
   (b) $1 + \alpha + \beta^2 + \alpha^3 + \beta^4 + \cdots$, where $0 < \alpha < \beta < 1$.

(4) Show that $\sum_{n=1}^{\infty} q^{n^2} x^n$ converges for all positive values of x if $0 < q < 1$. What happens if $q > 1$?

(5) Show that $\sum_{n=2}^{\infty} \frac{1}{(\log n)^n}$ is convergent.

(6) Determine whether or not the followings series converge

   (a) $\sum_{n=1}^{\infty} n^{-1/2}(\sqrt{n+1} - \sqrt{n})$.
   (b) $\sum_{n=1}^{\infty} n^{-2/3}(\sqrt{n+1} - \sqrt{n})$.

(c) $\sum_{n=1}^{\infty} \left(\frac{n+1}{n}\right)^{-n^2}$.

(d) $\sum_{n=1}^{\infty} \left(\frac{n+1}{n}\right)^{n^2} 5^{-n}$.

(7) Show that

(a) $\sum_{n=2}^{\infty} \frac{1}{n \log n}$ is divergent. (Start at $n = 2$ as $\log 1 = 0$.)

(b) $\sum_{n=2}^{\infty} \frac{1}{n(\log n)^p}$ is convergent if $p > 1$. (Start at $n = 2$ as $\log 1 = 0$.)

(c) $\sum_{n=3}^{\infty} \frac{1}{n \log n \log \log n}$ is divergent.

($\log n$ is the logarithm to base e, also denoted by $\ln x$.) Show also that

$$\sum_{n=2}^{\infty} \frac{1}{n(\log n)^2} \in \left[\frac{1}{\log 2}, \frac{1}{\log 2} + \frac{1}{2(\log 2)^2}\right].$$

(8) Show that $\sum_{n=2}^{\infty} \frac{1}{(\log n)^{\log n}}$ is convergent but $\sum_{n=2}^{\infty} \frac{1}{(\log n)^{\log \log n}}$ is divergent.

(9) For what values of $x \geq 0$ are the following series convergent?

(a) $\sum_{n=1}^{\infty} n^{\log n} x^n$.

(b) $\sum_{n=1}^{\infty} n^{\sqrt{n}} x^n$.

(c) $\sum_{n=1}^{\infty} n^n x^{n^2}$.

(10) Show that if $\sum_{n=1}^{\infty} a_n$ is convergent, then $\sum_{n=1}^{\infty} \sqrt{a_n a_{n+1}}$ converges (it is always assumed that (a_n) is a sequence of positive numbers).

(a) Show that if (a_n) is a decreasing sequence, then the convergence of $\sum_{n=1}^{\infty} \sqrt{a_n a_{n+1}}$ implies the convergence of $\sum_{n=1}^{\infty} a_n$.

(b) Find an example where $\sum_{n=1}^{\infty} \sqrt{a_n a_{n+1}}$ converges but $\sum_{n=1}^{\infty} a_n$ diverges (by (a), (a_n) cannot be decreasing).

(11) Suppose that $\sum_{n=1}^{\infty} a_n$ and $\sum_{n=1}^{\infty} b_n$ are series of positive terms.

(a) Show that if $\sum_{n=1}^{\infty} a_n$ and $\sum_{n=1}^{\infty} b_n$ are convergent, then $\sum_{n=1}^{\infty} \sqrt{a_n b_n}$ converges.

(b) Show that if $\sum_{n=1}^{\infty} a_n^2$ converges then so does $\sum_{n=1}^{\infty} a_n/n$.

(c) Show, by means of an example, that the converse of (b) is false, even if (a_n) is decreasing.

(12) Show that

(a) $S_n = 1 + \frac{1}{2} + \cdots + \frac{1}{n} - \log n \in [0, 1]$, $n \geq 1$.

(b) (S_n) is an increasing sequence.

Deduce that $\lim_{n\to\infty}(\sum_{j=1}^{n} 1/j - \log n)$ exists and lies in $[0, 1]$. (The limit is usually denoted by γ and referred to as *Euler's constant*. The value of γ is approximately $0.5772 \cdots$ (see Chap. 6, Sect. 6.3.5). It is not yet (2017) known whether γ is rational or irrational.

(13) Show that $\sum_{n=1}^{\infty} \left(\frac{1}{n} - \log\left(\frac{n+1}{n}\right)\right)$ is convergent.

(14) Suppose $a_n > 0$ and $\sum a_n$ diverges (to $+\infty$). What can be said about the convergence or divergence of $\sum \frac{a_n}{1+na_n}$, $\sum \frac{a_n}{1+n^2a_n}$?

3.4   General Principle of Convergence

In the next four sections, we study series where the terms are not necessarily all of the same sign. The sequence of partial sums will no longer be monotone and so we will not be able to apply Theorems 2.3.18, 2.3.19. Instead, we will need to use the result that if the sequence of partial sums is Cauchy, then it converges.

Theorem 3.4.1 (General Principle of Convergence) *Let* (a_n) *be a sequence of real numbers. Then* $\sum_{n=1}^{\infty} a_n$ *is convergent iff for every* $\varepsilon > 0$, *there exists an* $N \in \mathbb{N}$ *such that*

$$|x_m + x_{m+1} + \cdots + x_n| < \varepsilon, \text{ for all } n \geq m \geq N.$$

Proof The sequence (S_n) of partial sums is convergent iff it is a Cauchy sequence. That is, given $\varepsilon > 0$, there exists an $N \in \mathbb{N}$ such that $|S_n - S_{m-1}| = |x_m + \cdots + x_n| < \varepsilon$, for all $n \geq m \geq N$.   □

Remark 3.4.2 Theorem 3.4.1 extends to infinite series of complex numbers. The proof is formally the same as that of Theorem 3.4.1 but using Theorem 2.6.6.   ✱

3.5   Absolute Convergence

Definition 3.5.1 Let $(a_n) \subset \mathbb{R}$. The infinite series $\sum_{n=1}^{\infty} a_n$ is *absolutely convergent* if $\sum_{n=1}^{\infty} |a_n|$ is convergent.

Remark 3.5.2 The results for sums and differences of convergent series extend to absolutely convergent series using the corresponding results for series of positive terms. For example, if $\sum a_n$ and $\sum b_n$ are absolutely convergent then so are the series $\sum (a_n \pm b_n)$.   ✱

Theorem 3.5.3 *Every absolutely convergent series is convergent.*

Proof Our proof makes essential use of Theorem 3.4.1 (general principle of convergence). Suppose that $\sum_{n=1}^{\infty} |a_n| < \infty$. Then, by Theorem 3.4.1, given $\varepsilon > 0$, there exists an $N \in \mathbb{N}$ such that

$$|a_m| + |a_{m+1}| + \cdots + |a_n| < \varepsilon, \text{ for all } n \geq m \geq N.$$

But $|a_m + a_{m+1} + \cdots + a_n| \leq |a_m| + |a_{m+1}| + \cdots + |a_n|$ and so

$$|a_m + a_{m+1} + \cdots + a_n| < \varepsilon, \text{ for all } n \geq m \geq N.$$

Hence, $\sum_{n=1}^{\infty} a_n$ is convergent by Theorem 3.4.1.   □

Remarks 3.5.4

(1) The reader is cautioned that the converse to Theorem 3.5.3 is *false*: a convergent series need not be absolutely convergent. We give examples shortly.

(2) If $\sum_{n=1}^{\infty} a_n$ is an infinite series of complex numbers, then the series is *absolutely convergent* if $\sum_{n=1}^{\infty} |a_n| < \infty$ (where $|\cdot|$ now denotes the modulus of a complex number). Without exception, all of our results on absolutely convergent real series extend to absolutely convergent complex series.   ✳

Theorem 3.5.3 allows us to translate results on convergent series of positive terms to absolutely convergent series.

Example 3.5.5 The series $\sum_{n=1}^{\infty}(-1)^{n+1}n^{-p}$ is absolutely convergent iff $p > 1$, Example 3.3.14. Hence $\sum_{n=1}^{\infty}(-1)^{n+1}n^{-p}$ is convergent, $p > 1$. Later in this chapter we prove $\sum_{n=1}^{\infty}(-1)^{n+1}n^{-p}$ is convergent for $p > 0$ even though absolute convergence fails if $p \leq 1$.   ♠

We conclude this section with an important and practical result called *Tannery's theorem* that allows us to interchange limit operations in a countable set of absolutely convergent series. Specifically, given sequences $(a_n(p))$, for $p = 1, 2, \cdots$, Tannery's theorem gives easily verifiable conditions under which

$$\lim_{p \to \infty} \sum_{n=1}^{\infty} a_n(p) = \sum_{n=1}^{\infty} \lim_{p \to \infty} a_j(p).$$

Theorem 3.5.6 (Tannery's Theorem) *Suppose we are given sequences* $(a_n(p)) \subset \mathbb{R}$ *depending on* $p \in \mathbb{N}$. *Assume*

(1) $\lim_{p \to \infty} a_n(p) = a_n$, $n \in \mathbb{N}$.
(2) $|a_n(p)| \leq M_n$, *for all* $n, p \in \mathbb{N}$, *where* $\sum_{n=1}^{\infty} M_n < \infty$.

Then

$$\lim_{p \to \infty} \sum_{n=1}^{p} a_n(p) = \lim_{p \to \infty} \sum_{n=1}^{\infty} a_n(p) = \sum_{n=1}^{\infty} a_n.$$

The result continues to hold if $(a_n(p)) \subset \mathbb{C}$.

Proof We have to show that given $\varepsilon > 0$, there exists an $N \in \mathbb{N}$ such that if $p \geq N$ then

$$\left| \sum_{n=1}^{p} a_n(p) - \sum_{n=1}^{\infty} a_n \right| < \varepsilon.$$

It follows from (1,2) that $|a_n(p)|$, $|a_n| \le M_n$, for all $p, n \in \mathbb{N}$ and so $\sum_{n=1}^{\infty} a_n(p)$ and $\sum_{n=1}^{\infty} a_n$ are absolutely convergent. Moreover, we may choose $N_1 \in \mathbb{N}$ such that $\sum_{n=N_1}^{m} |a_n|$, $\sum_{n=N_1}^{m} |a_n(p)| < \varepsilon/3$, for $\infty \ge m \ge N_1$ and $p \in \mathbb{N}$. Hence for $m \ge N_1$ we have

$$\left| \sum_{n=1}^{m} a_n(p) - \sum_{n=1}^{\infty} a_n \right| \le \sum_{n=1}^{N_1} |a_n(p) - a_n| + \sum_{n=N_1+1}^{\infty} |a_n| + \sum_{n=N_1+1}^{m} |a_n(p)|$$

$$< \sum_{n=1}^{N_1} |a_n(p) - a_n| + 2\varepsilon/3.$$

By (1), we can choose $N \ge N_1$ such that $|a_n(p) - a_n| < \varepsilon/(3N_1)$, for all $p \ge N$ and $1 \le n \le N_1$. Hence, taking $m = p$,

$$\left| \sum_{n=1}^{p} a_n(p) - \sum_{n=1}^{\infty} a_n \right| < N_1 \frac{\varepsilon}{3N_1} + \frac{2\varepsilon}{3} = \varepsilon, \text{ for all } p \ge N.$$

The proof extends immediately to the case of complex series—absolute value is replaced by the modulus of a complex number. $\square$

3.5.1   The Exponential Series

The exponential series $\sum_{n=0}^{\infty} \frac{x^n}{n!}$ is absolutely convergent for all $x \in \mathbb{R}$ by Examples 3.3.6(2). Hence, $\sum_{n=0}^{\infty} \frac{x^n}{n!}$ converges for all $x \in \mathbb{R}$. We may give an infinite series definition of the *exponential function* e^x or $\exp(x)$ by

$$e^x = \sum_{n=0}^{\infty} \frac{x^n}{n!}, \ x \in \mathbb{R}.$$

The next result shows that the series definition of exp gives the same function as that defined by the inverse of the logarithm in section "Appendix: The Log and Exponential Functions".

Proposition 3.5.7 *For all $x \in \mathbb{R}$,*

$$\lim_{n \to \infty} \left(1 + \frac{x}{n} \right)^n = \sum_{n=0}^{\infty} \frac{x^n}{n!}$$

$$= e^x.$$

Proof We refer to Exercises 2.9.10(5) for the proof that e^x equals $\lim_{n\to\infty}\left(1+\frac{x}{n}\right)^n$, where e^x is defined as the inverse of $\log x$. To simplify notation, we define $e^x = \sum_{n=0}^{\infty}\frac{x^n}{n!}$ and show that $\lim_{n\to\infty}\left(1+\frac{x}{n}\right)^n = e^x$.

Fix $x \in \mathbb{R}$. By the binomial theorem, we have

$$(1+\frac{x}{p})^p = 1 + p\frac{x}{p} + \frac{p(p-1)}{2}\frac{x^2}{p^2} + \cdots + \frac{x^p}{p^p}$$

$$= 1 + \sum_{n=1}^{p}\frac{x^n}{n!}K_p(n),$$

where $K_p(1) = 1$ and $K_p(n) = (1-\frac{1}{p})(1-\frac{2}{p})\cdots(1-\frac{n-1}{p}) < 1, p \geq n > 1$. We use Tannery's theorem to complete the proof. Following the notation of Theorem 3.5.6, define $a_n = x^n/n!$ and

$$a_n(p) = \begin{cases} 1, & n = 0, \\ \frac{x^n}{n!}K_p(n), & p \geq n, \\ 0, & p < n. \end{cases}$$

We have $\lim_{p\to\infty} a_n(p) = \frac{x^n}{n!}$, verifying (1) of Theorem 3.5.6. Condition (2) holds since $|\frac{x^n}{n!}K_p(n)| \leq \frac{|x|^n}{n!}$ and $\sum_{n=0}^{\infty}\frac{x^n}{n!}$ is absolutely convergent. Applying Tannery's theorem we have

$$\lim_{n\to\infty}\left(1+\frac{x}{n}\right)^n = \lim_{n\to\infty}\sum_{j=0}^{n}a_j(n) = \sum_{n=0}^{\infty}\frac{x^n}{n!}. \qquad \square$$

Remarks 3.5.8

(1) This result continues to hold if we allow for complex variables: $\lim_{n\to\infty}(1 + \frac{z}{n})^n = e^z$, for all $z \in \mathbb{C}$. The proof is exactly the same with 'absolute value' replaced everywhere by 'modulus'. (It is not straightforward to define $\exp(z)$ in terms of the inverse of $\log z$, if z complex.)
(2) A proof of Proposition 3.5.7 can be based on the method of Examples 2.3.27(2), avoiding Tannery's theorem. But the argument still has to address the issues that arise in the proof of Tannery's theorem.   ✳

3.5.2   Tests for Absolutely Convergent Series

We present versions of the comparison, ratio and Cauchy test appropriate for proving absolute convergence. Proofs all use Theorem 3.5.3 together with the corresponding result for series of positive terms. The results also apply to complex series with 'absolute value' replaced by the 'modulus'.

Proposition 3.5.9 (The Comparison Test) *Let $(u_n), (v_n)$ be sequences of real numbers satisfying $|u_n| \leq |v_n|$, for all $n \in \mathbb{N}$.*

(1) *If $\sum_{n=1}^{\infty} v_n$ is absolutely convergent, then $\sum_{n=1}^{\infty} u_n$ is absolutely convergent (and so convergent).*
(2) *If $\sum_{n=1}^{\infty} u_n$ is not absolutely convergent then $\sum_{n=1}^{\infty} v_n$ is not absolutely convergent.*

Remark 3.5.10 For the second statement, $\sum_{n=1}^{\infty} v_n$ may converge even if $\sum_{n=1}^{\infty} u_n$ is divergent. ✠

Proposition 3.5.11 (The Ratio Test) *Let (a_n) be a sequence of real numbers and suppose that $\lim_{n\to\infty} |\frac{a_{n+1}}{a_n}| = \ell$.*

(1) *If $\ell < 1$, the series $\sum_{n=1}^{\infty} a_n$ is absolutely convergent (and so convergent).*
(2) *If $\ell > 1$, the series $\sum_{n=1}^{\infty} a_n$ is divergent.*

Proposition 3.5.12 (Cauchy's Test) *Let (a_n) be a sequence of real numbers.*

(1) *If $\limsup |a_n|^{\frac{1}{n}} < 1$, the series $\sum_{n=1}^{\infty} a_n$ is absolutely convergent (and so convergent).*
(2) *If $\limsup |a_n|^{\frac{1}{n}} > 1$, the series $\sum_{n=1}^{\infty} a_n$ is divergent.*

Definition 3.5.13 Let (a_n) be a sequence of real numbers. The series $\sum_{n=1}^{\infty} b_n$ is a *rearrangement* of $\sum_{n=1}^{\infty} a_n$ if there exists a bijection $\sigma : \mathbb{N} \to \mathbb{N}$ such that $b_n = a_{\sigma(n)}$, for all $n \in \mathbb{N}$. That is, a rearrangement of $\sum_{n=1}^{\infty} a_n$ is a series $\sum_{n=1}^{\infty} a_{\sigma(n)}$ where σ is a "permutation" of $\mathbb{N}$.

Example 3.5.14 The series $x + x^3 - x^2 + x^5 + x^7 - x^4 + \cdots$ is a rearrangement of $\sum_{n=1}^{\infty}(-1)^{n+1}x^n$. ♠

We end this section with an important result that fails dramatically when the series is convergent but not absolutely convergent.

Theorem 3.5.15 *Every rearrangement $\sum_{n=1}^{\infty} a_{\sigma(n)}$ of an absolutely convergent series $\sum_{n=1}^{\infty} a_n$ is convergent and*

$$\sum_{n=1}^{\infty} a_{\sigma(n)} = \sum_{n=1}^{\infty} a_n.$$

Proof If $\sum_{n=1}^{\infty} a_n$ is absolutely convergent, it follows from the general principle of convergence that given $\varepsilon > 0$, there exists an $N \in \mathbb{N}$ such that $|a_m| + \cdots + |a_n| < \varepsilon$, for all $n > m \geq N$. Let $M = \max\{\sigma^{-1}(1), \cdots \sigma^{-1}(N-1)\} \in \mathbb{N}$. Observe that if $n \geq M$, then $\sigma(n) \geq N$. Let $n > m \geq M$ and set $n' = \max\{\sigma(m), \cdots, \sigma(n)\}$ and $m' = \min\{\sigma(m), \cdots, \sigma(n)\}$. We have $n' > m' \geq N$ and so

$$|a_{\sigma(m)}| + \cdots + |a_{\sigma(n)}| \leq |a_{m'}| + |a_{m'+1}| + \cdots + |a_{n'}| < \varepsilon.$$

Hence by the general principle of convergence $\sum_{n=1}^{\infty} a_{\sigma(n)}$ is absolutely convergent. It remains to prove that the series have the same sum. Let $\sum_{n=1}^{\infty} a_n = \ell$. With the same notation used above, suppose $p > M$ and set $q = \max\{\sigma^{-1}(N), \cdots \sigma^{-1}(p)\}$. We have $\left|\sum_{n=1}^{N-1} a_n - \ell\right| \le \varepsilon$ and so

$$\left|\sum_{n=1}^{p} a_{\sigma(n)} - \ell\right| \le \left|\sum_{n=1}^{N-1} a_n - \ell\right| + |a_N + \cdots + a_q| < 2\varepsilon.$$

This estimate holds for all $p > M$ and so $\sum_{n=1}^{\infty} a_{\sigma(n)} = \ell$. □

EXERCISES 3.5.16

(1) Suppose that (a_n) is a sequence of real numbers such that (a) $\sum_{n=1}^{\infty} a_n$ is convergent and (b) the a_n are eventually of the same sign. Show that $\sum_{n=1}^{\infty} a_n$ is absolutely convergent.

(2) For each of the following series find $R > 0$ such that the series converges if $|x| < R$ and diverges if $|x| > R$.

(a) $\sum_{n=1}^{\infty} n^2 x^n$.

(b) $\sum_{n=1}^{\infty} n^{-2} x^{n^2}$.

(c) $\sum_{n=1}^{\infty} \frac{n^n}{n!} x^n$.

(d) $\sum_{n=0}^{\infty} \frac{(2n)!}{(n!)^2} x^n$.

(e) $\sum_{n=0}^{\infty} \frac{(2n)!^2}{(4n)!} x^{2n}$.

(f) $\sum_{n=1}^{\infty} n^{\sqrt{n}} x^n$.

(3) For what values of x are

(a) $\sum_{n=1}^{\infty} \left(1 + \frac{x^2}{2n}\right)^{-n^2} e^{2n}$,

(b) $\sum_{n=1}^{\infty} \left(1 + \frac{x^2}{n}\right)^{-n^2} x^{2n}$,

convergent?

(4) Suppose that $p : \mathbb{N} \to \mathbb{N}$ satisfies $a \le \frac{p(n)}{n} \le A$, where $A \ge a > 0$. Show that if we assume conditions (1,2) of Tannery's theorem (Theorem 3.5.6) then $\lim_{n\to\infty} \sum_{j=1}^{n} a_j(p(n)) = \sum_{m=1}^{\infty} a_m$.

(5) Using Tannery's theorem, prove that

$$\frac{e}{e-1} = \lim_{n\to\infty} \left[\sum_{j=0}^{n-1} \left(\frac{n-j}{n}\right)^n\right].$$

(Hint: Let $a_n(p) = (\frac{p-n}{p})^p$ if $n < p$ and be zero otherwise.)

(6) Show that the rearrangement theorem also holds for absolutely convergent series of complex terms.

3.6 Conditionally Convergent Series

In this section we look at convergent real series that are not absolutely convergent. The results we obtain do not have (simple) extensions to complex series.

Definition 3.6.1 Let (a_n) be a sequence of real numbers. The series $\sum_{n=1}^{\infty} a_n$ is *conditionally convergent* if it is convergent but not absolutely convergent.

Remark 3.6.2 If $\sum_{n=1}^{\infty} a_n$ is conditionally convergent, then the series must have infinitely many positive and negative terms. Else the terms would either be eventually positive or eventually negative and the series would be absolutely convergent.   ✠

Example 3.6.3 The series $\sum_{n=1}^{\infty} (-1)^{n+1}/n$ is conditionally convergent. Since

$$S_{2n} = \left(1 - \frac{1}{2}\right) + \cdots + \left(\frac{1}{2n-1} - \frac{1}{2n}\right) \geq 0, \ n \geq 1,$$

the sequence (S_{2n}) of partial sums is an increasing sequence of positive numbers. On the other hand,

$$S_{2n+1} = 1 - \left(\frac{1}{2} - \frac{1}{3}\right) - \left(\frac{1}{2n} - \frac{1}{2n+1}\right) = S_{2n} - \frac{1}{2n+1},$$

is a decreasing sequence bounded above by 1. Now $S_{2n} = S_{2n+1} + \frac{1}{2n+1} < 1$, $n \geq 1$, and so (S_{2n}) is an increasing sequence of positive numbers bounded above by $1 + 1/(2m + 1)$, for all $m \geq 1$. Hence (S_{2n}) converges to $\ell \in (0, 1]$. Since $S_{2n+1} = S_{2n} - \frac{1}{2n+1}$, (S_{2n+1}) also converges to ℓ. Therefore, $\lim_{n\to\infty} S_n = \ell \in (0, 1]$. (In the next chapter we show that $\ell = \log 2$.)   ♠

Let $\sum_{n=1}^{\infty} a_n$ be a conditionally convergent series. Define sequences (u_n), (v_n) of positive real numbers by

$$u_n = \max\{0, a_n\}, \quad v_n = -\min\{0, a_n\}.$$

Observe that for all $n \in \mathbb{N}$ we have

$$a_n = u_n - v_n, \quad |a_n| = u_n + v_n.$$

The next result will be useful when we prove Riemann's theorem on rearrangements of a conditionally convergent series.

Proposition 3.6.4 *If $\sum_{n=1}^{\infty} a_n$ is a conditionally convergent series and we define the sequences (u_n), (v_n) as above, then $\sum_{n=1}^{\infty} u_n$, $\sum_{n=1}^{\infty} v_n$ both diverge to $+\infty$.*

Proof If $\sum_{n=1}^{\infty} v_n$ converges then $\sum_{n=1}^{\infty} (a_n + v_n) = \sum_{n=1}^{\infty} u_n$ converges. Therefore, $\sum_{n=1}^{\infty} (u_n + v_n) = \sum_{n=1}^{\infty} |a_n|$ converges, contradicting the conditional convergence of $\sum_{n=1}^{\infty} a_n$. Hence $\sum_{n=1}^{\infty} v_n = +\infty$. A similar argument shows that $\sum_{n=1}^{\infty} u_n = +\infty$. □

Example 3.6.5 The series $1 - \frac{1}{2^2} + \frac{1}{3} - \frac{1}{4^2} + \frac{1}{5} + \cdots$ is divergent. For this series, $\sum_{n=1}^{\infty} u_n$ is divergent (Exercises 3.3.15(2)) but $\sum_{n=1}^{\infty} v_n$ is convergent (by comparison with $\sum 1/n^2$). Hence the series cannot be absolutely or conditionally convergent (Proposition 3.6.4) and therefore must diverge (in this case to $+\infty$). ♠

3.6.1 Alternating Series

Definition 3.6.6 The series $\sum_{n=1}^{\infty} (-1)^{n+1} a_n$ is called an *alternating series* if (a_n) is a sequence of positive real numbers.

Proposition 3.6.7 (Leibniz Alternating Series Test) *Let (a_n) be a sequence of positive numbers. The alternating series $\sum_{n=1}^{\infty} (-1)^{n+1} a_n$ converges if*

(a) (a_n) *is a decreasing sequence.*
(b) $\lim_{n \to \infty} a_n = 0$.

If (a,b) hold then $\sum_{n=1}^{\infty} (-1)^{n+1} a_n \in [0, a_1]$.

Proof The proof is formally identical to the argument used in Example 3.6.3 and we leave the details to the reader. □

Example 3.6.8 The alternating series $\sum_{n=2}^{\infty} \frac{(-1)^n}{\log n}$, $\sum_{n=3}^{\infty} \frac{(-1)^n}{\log \log n}$ are convergent. Note that the convergence of these series is very slow. For example, we need to take n greater than 10^{64} to ensure that the nth term of the second series is less than $1/5$. ♠

3.6.2 Riemann's Theorem

In this section we look at rearrangements of a conditionally convergent series. We start with a simple example.

Example 3.6.9 Consider the conditionally convergent series $1 - 1 + \frac{1}{2} - \frac{1}{2} + \cdots + \frac{1}{n} - \frac{1}{n} + \cdots$. The series trivially converges to zero. Take the rearrangement

$$1 + \frac{1}{2} - 1 + \frac{1}{3} + \frac{1}{4} - \frac{1}{2} + \cdots$$

It is easy to see that $S_{3n} = \sum_{j=1}^{n} \frac{1}{2j(2j-1)}$. Using the identity $\frac{1}{2j(2j-1)} = \frac{1}{2j-1} - \frac{1}{2j}$, we find that $S_{3n} = 1 - \frac{1}{2} + \frac{1}{3} - \cdots - \frac{1}{2n}$. We deduce easily that the rearranged series converges with sum equal to $\log 2 > 0$. This example shows the failure of the rearrangement theorem when the series is not absolutely convergent. ♠

Theorem 3.6.10 (Riemann's Rearrangement Theorem) *Let $\sum_{n=1}^{\infty} a_n$ be a conditionally convergent series.*

(a) *For every $x \in \mathbb{R} \cup \{-\infty, +\infty\}$ there exists a rearrangement σ such that*

$$\sum_{n=1}^{\infty} a_{\sigma(n)} = x.$$

(b) *There exist rearrangements σ such that $\sum_{n=1}^{\infty} a_{\sigma(n)}$ does not converge (even to $\pm\infty$).*

Proof We prove (a) and leave (b) to the exercises. Our first step is to define sequences (p_k) and (q_k) by requiring that (p_k) is the subsequence of (a_n) defined by the positive terms and (q_k) is the subsequence of (a_n) defined by the strictly negative terms. Note that the sequence (p_k) may contain zeros and that for each $k \in \mathbb{N}$, there exist unique $n_k, m_k \in \mathbb{N}$ such that $p_k = a_{n_k}, q_k = a_{m_k}$.

Let $x \in \mathbb{R}$. For simplicity, suppose $x \geq 0$. Since $\sum_{k=1}^{\infty} p_k$ diverges to $+\infty$, there exists a unique $k_1 \geq 1$ such that $p_1 + \cdots + p_{k_1-1} \leq x < p_1 + \cdots + p_{k_1} = P_1$. Since $\sum_{k=1}^{\infty} q_k$ diverges to $-\infty$, there exists a unique $\ell_1 \geq 1$ such that $Q_1 = P_1 + q_1 + \cdots + q_{\ell_1} < x \leq P_1 + q_1 + \cdots + q_{\ell_1-1} = Q_1 - q_{\ell_1}$. In the obvious way, we may inductively define increasing sequences (k_n), (ℓ_n) and sequences (P_n), (Q_n) so that $P_n = \sum_{j=k_{n-1}+1}^{k_n} p_j + Q_{n-1}$, $Q_n = P_n + \sum_{j=\ell_{n-1}+1}^{\ell_n} q_j$ and $P_n - p_{k_n-1} \leq x < P_n$, $Q_n < x \leq Q_n - q_{\ell_n-1}$, for all $n \geq 1$. Since $k_j, \ell_j \geq 1$, there are at least $2n - 1$ terms from the series $\sum a_n$ in P_n, and at least $2n$ terms in Q_n. This construction defines the rearrangement

$$\sum_{n=1}^{\infty} a_{\sigma(n)} = a_{n_1} + \cdots + a_{n_{k_1}} + a_{m_1} + \cdots + a_{m_{\ell_1}} + a_{n_{k_1+1}} + \cdots$$

We claim the rearranged series converges to x. Since $\sum a_n$ is convergent, $\lim_{n \to \infty} a_n = 0$ and so $\lim_{n \to \infty} p_n = \lim_{n \to \infty} q_n = 0$. Hence there exists an $N \geq 1$ so that $|p_n|, |q_n| < \varepsilon$ for all $n \geq N$. Choose $M \in \mathbb{N}$ so that if $n \geq M$, then $a_{\sigma(n)}$ is either p_k, with $k \geq N$ or q_k with $k \geq N$. It follows from the construction of the rearrangement that $|\sum_{n=1}^{m} a_{\sigma(n)} - x| < \varepsilon$ for all $m \geq M$. Hence $\sum_{n=1}^{\infty} a_{\sigma(n)} = x$.

If $x = +\infty$, we modify the construction by requiring that $P_n - p_{k_n-1} \leq n < P_n$ and $Q_n < n - 1 \leq Q_n - q_{\ell_n-1}$. The argument when $x = -\infty$ is similar. □

Remark 3.6.11 The statement of Riemann's rearrangement theorem can be strengthened along the following lines: let $-\infty \leq x_1 < x_2 < \cdots < x_N \leq +\infty$. Then there exists a rearrangement σ of $\sum_{n=1}^{\infty} a_n$ such that for each x_j there exists a subsequence (S_{n_k}) of (S_n) which converges to x_j. (See also the exercises.) ✱

EXERCISES 3.6.12

(1) Determine the convergence of the series

(a) $\sum_{n=1}^{\infty} \frac{(-1)^n n}{n+1}$.

(b) $\sum_{n=1}^{\infty} \frac{\cos n\pi}{\sqrt{n}}$.

(c) $\sum_{n=1}^{\infty} 3^n e^{-n}$.

(d) $\sum_{n=2}^{\infty} \frac{(-1)^n}{\log n}$.

(2) Suppose $\sum_{n=1}^{\infty} a_n$ and $\sum_{n=1}^{\infty} b_n$ both converge. Must $\sum_{n=1}^{\infty} a_n b_n$ converge? Either prove it or find a counterexample.

(3) Show that the rearrangement $1 - \frac{1}{2} - \frac{1}{4} + \frac{1}{3} - \frac{1}{6} - \frac{1}{8} + \frac{1}{5} - \frac{1}{10} + \cdots$ of $\sum_{n=1}^{\infty} (-1)^{n+1}/n$ converges to $\frac{1}{2} \log 2$. (Hint: Look at the partial sum to $3n$ terms of the series.)

(4) Show that the rearrangement $1 + \frac{1}{3} - \frac{1}{2} + \frac{1}{5} + \frac{1}{7} - \frac{1}{4} + \frac{1}{9} + \frac{1}{11} - \cdots$ of $\sum_{n=1}^{\infty} (-1)^{n+1}/n$ converges to $\frac{3}{2} \log 2$. (Hint: Work with the partial sum to $3n$ terms. At some point you will need to show that $\frac{1}{n+1} + \cdots + \frac{1}{2n}$ is equal to the partial sum to $2n$ terms of the series $\sum_{n=1}^{\infty} (-1)^{n+1}/n$.)

(5) Show that $1 - \frac{1}{2} - \frac{1}{4} + \frac{1}{5} + \frac{1}{7} - \frac{1}{8} - \frac{1}{10} + + \cdots = \frac{2}{3} \log 2$.

(6) Show that $1 + \frac{1}{3} + \frac{1}{5} - \frac{1}{2} - \frac{1}{4} - \frac{1}{6} + + + \cdots = \log 2$.

(7) Prove part (b) of Riemann's rearrangement theorem.

(8) Prove the result indicated in Remark 3.6.11.

3.7 Abel's and Dirichlet's Tests

In this section we state and prove two powerful tests that can be used to determine the convergence of non-absolutely convergent series. Both results depend on a simple but subtle inequality due to Abel.

Lemma 3.7.1 (Abel's Lemma) *If the sequence (S_n) of partial sums of the infinite series $\sum_{n=1}^{\infty} a_n$ satisfies the bounds*

$$m \leq S_n \leq M, \ n \in \mathbb{N},$$

then for any decreasing sequence (u_n) of positive real numbers we have the bounds

$$mu_1 \leq \sum_{j=1}^{n} a_j u_j \leq Mu_1, \ n \in \mathbb{N}. \tag{3.4}$$

Proof Since $a_n = S_n - S_{n-1}, n \geq 2$, we have

$$\sum_{j=1}^{n} a_j u_j = S_1 u_1 + (S_2 - S_1)u_2 + \cdots + (S_n - S_{n-1})u_n$$

$$= S_1(u_1 - u_2) + \cdots + S_{n-1}(u_{n-1} - u_n) + S_n u_n.$$

Since (u_n) is a decreasing sequence of positive numbers, we have $u_1 - u_2, \cdots, u_{n-1} - u_n, u_n \geq 0$. Hence, upper and lower bounds for $\sum_{j=1}^{n} a_j u_j$ are given respectively by

$$M((u_1 - u_2) + (u_2 - u_3) + \cdots + (u_{n-1} - u_n) + u_n) = M u_1,$$

$$m((u_1 - u_2) + (u_2 - u_3) + \cdots + (u_{n-1} - u_n) + u_n) = m u_1,$$

and so $m u_1 \leq \sum_{j=1}^{n} a_j u_j \leq M u_1$. □

Proposition 3.7.2 (Abel's Test) *If the infinite series $\sum_{n=1}^{\infty} a_n$ is convergent and (v_n) is a bounded monotone sequence of real numbers, then $\sum_{n=1}^{\infty} a_n v_n$ is convergent.*

Proof Since (v_n) is monotone and bounded, (v_n) is convergent, say to v. If (v_n) is increasing, set $u_n = v - v_n$ and if (v_n) is decreasing set $u_n = v_n - v$. In both cases (u_n) is a monotone decreasing sequence of positive numbers. Since $a_n v_n = a_n v - a_n u_n$ or $a_n v_n = a_n u_n - a_n v$ and $\sum a_n v = v \sum a_n$ is convergent, it is enough to prove that $\sum_{n=1}^{\infty} a_n u_n$ converges.

Fix $m \geq 2$ and define $K_m = \sup\{|S_n - S_{m-1}| \mid n \geq m\}$. By the general principle of convergence, $\lim_{m \to \infty} K_m = 0$. Applying Abel's lemma to $\sum_{j=m}^{n} a_j u_j$ gives

$$\left| \sum_{j=m}^{n} a_j u_j \right| \leq K_m u_m \leq K_m u_1, \ n \geq m.$$

Let $\varepsilon > 0$. Since $\lim_{m \to \infty} K_m = 0$, there exists an $N \in \mathbb{N}$ such that $|K_m| < \varepsilon / u_1$. Therefore,

$$\left| \sum_{j=m}^{n} a_j u_j \right| \leq K_m v_1 < \varepsilon, \ \text{if } n \geq m \geq N.$$

The result follows by the general principle of convergence. □

Examples 3.7.3

(1) Consider the series $1 - 1 + \frac{1}{2} - \frac{1}{2} + \frac{1}{3} - \cdots$, which trivially converges to zero (see Example 3.6.9). Take the bounded increasing sequence $1, \frac{1}{2}, \frac{1}{2}, \frac{2}{3}, \frac{2}{3}, \frac{3}{4}, \cdots$. Multiplying the terms of the series by the corresponding term of the decreasing sequence yields the infinite series

$$1 - \frac{1}{2} + \frac{1}{2^2} - \frac{1}{3} + \frac{2}{3^2} - \frac{1}{4} + \frac{3}{4^2} - \cdots$$

Abel's test implies that this series converges. Note that the alternating series test does not apply to this series as the terms in the series do not define a decreasing sequence. (It is not too difficult to show that the series converges to $2 - \sum_{n=1}^{\infty} n^{-2}$.)

(2) If $\sum_{n=1}^{\infty} a_n$ converges, then $\sum_{n=1}^{\infty} \frac{a_n}{n^x}$ converges if $x \geq 0$.   ♠

Proposition 3.7.4 (Dirichlet's Test) *Suppose that the sequence (S_n) of partial sums of the infinite series $\sum_{n=1}^{\infty} a_n$ is bounded and (u_n) is a decreasing sequence of positive numbers. If $\lim_{n \to \infty} u_n = 0$, then $\sum_{n=1}^{\infty} a_n u_n$ is convergent.*

Proof Suppose that $\alpha \leq S_n \leq \beta$ for all $n \in \mathbb{N}$. If we set $K = \max\{|\alpha|, |\beta|\}$, then $|S_n| \leq K$ for all $n \geq 1$. We have $|S_n - S_m| \leq |S_n| + |S_m|$ and so $|S_n - S_m| \leq 2K$ for all $n \geq m \geq 1$. Applying Abel's lemma gives the estimate

$$\left| \sum_{j=m}^{n} a_j u_j \right| \leq 2K u_m, \text{ for all } n \geq m \geq 1.$$

Given $\varepsilon > 0$, choose $N \in \mathbb{N}$ such that $u_m < \varepsilon/2K$ for all $m \geq N$. We have

$$\left| \sum_{j=m}^{n} a_j u_j \right| \leq 2K u_m < \varepsilon, \ n \geq m \geq N.$$

It follows from the general principle of convergence that $\sum_{n=1}^{\infty} a_n u_n$ is convergent.   □

Examples 3.7.5

(1) The series $\sum_{n=1}^{\infty} (-1)^{n+1}$ is not convergent but the partial sums are bounded. If (u_n) is a decreasing sequence of positive numbers converging to zero then Dirichlet's test implies that $\sum_{n=1}^{\infty} (-1)^{n+1} u_n$ converges (Leibniz test).

(2) If the partial sums of $\sum_{n=1}^{\infty} a_n$ are bounded then $\sum_{n=1}^{\infty} \frac{a_n}{n^x}$ converges if $x > 0$.

(3) Dirichlet's test is often useful for the study of trigonometric series. For example, consider the series $\sum_{n=1}^{\infty} \frac{\cos n\theta}{n}$. Using Dirichlet's test we show that the series converges provided that θ is not an integer multiple of 2π. We start by noting that if θ is an integer multiple of 2π then the series is the harmonic series $\sum_{n=1}^{\infty} \frac{1}{n}$ which is divergent. If θ is an odd multiple of π, then the series is $\sum_{n=1}^{\infty} \frac{(-1)^n}{n}$ which converges by the alternating series test. For other values of θ, in particular irrational multiples of π, the issue of convergence is quite subtle.

The main ingredient in the proof of convergence of $\sum_{n=1}^{\infty} \frac{\cos n\theta}{n}$ is the trigonometric identity

$$\sum_{j=1}^{n} \cos(n\theta) = \frac{\cos\left(\frac{n+1}{2}\theta\right) \sin\left(\frac{n}{2}\theta\right)}{\sin\left(\frac{\theta}{2}\right)}, \ \theta \neq 2n\pi. \tag{3.5}$$

(We give the proof of this identity in an appendix at the end of the chapter.) Provided θ is not an integral multiple of 2π, (3.5) gives the estimate

$$\left| \sum_{j=1}^{n} \cos(n\theta) \right| \le \frac{\left| \cos\left(\frac{n+1}{2}\theta\right) \sin\left(\frac{n}{2}\theta\right) \right|}{\left| \sin\left(\frac{\theta}{2}\right) \right|} \le \frac{1}{\left| \sin\left(\frac{\theta}{2}\right) \right|}, \quad n \ge 1.$$

Take $a_n = \cos(n\theta)$, $u_n = 1/n$ in Dirichlet's test. ♠

EXERCISES 3.7.6

(1) Suppose that $\sum_{n=1}^{\infty} na_n$ converges. Show that $\sum_{n=1}^{\infty} a_n$ converges. What about $\sum_{n=1}^{\infty} \sqrt{n}a_n$? $\sum(-1)^{n+1}a_n$? (You may not assume all the terms are of the same sign. Either prove it or find a counterexample.)

(2) Suppose that $\sum_{n=1}^{\infty} a_n$ converges. Show that $\sum_{n=1}^{\infty} \sqrt[n]{n}a_n$ converges.

(3) For what values of $\theta \in \mathbb{R}$ is $\sum_{n=1}^{\infty} \frac{\sin(n\theta)}{n}$ convergent?

(4) Prove that $\sum_{n=0}^{\infty} \frac{\sin((2n+1)\theta)}{(2n+1)}$ converges for all $\theta \in \mathbb{R}$.

(5) For what values of $\theta \in \mathbb{R}$ is the series with nth term

$$\left(1 + \frac{1}{2} + \cdots + \frac{1}{n}\right) \frac{\sin(n\theta)}{n}$$

convergent? (You may assume $(1 + \frac{1}{2} + \cdots + \frac{1}{n})/n \to 0$.)

(6) Prove the following extensions of Abel's and Dirichlet's tests (due to Dedekind).

  (a) Let $\sum_{n=1}^{\infty} a_n$ be convergent and suppose (v_n) is a bounded sequence such that $\sum_{n=1}^{\infty} |v_{n+1} - v_n|$ converges. Then $\sum_{n=1}^{\infty} v_n a_n$ converges.

  (b) Let (a_n) and (v_n) be such that the sequences of partial sums $(\sum_{j=1}^{n} a_j)$ and $(\sum_{j=1}^{n} |v_{j+1} - v_j|)$ are bounded and $\lim_{n\to\infty} v_n = 0$. Then $\sum_{n=1}^{\infty} v_n a_n$ converges.

(7) Prove that $\sum_{n=1}^{\infty} (-1)^n \frac{\sin(\log n)}{n^a}$ is convergent if $a > 0$. (Hint: Use the result of the previous question with $v_n = \frac{\sin(\log n)}{n^a}$. Estimate $|v_{n+1} - v_n|$ using the mean value theorem.)

3.8 Double Series

A *double series* is an infinite series of the form

$$\sum_{m,n=1}^{\infty} a_{m,n},$$

where $a_{m,n} \in \mathbb{R}$ (or $\mathbb{C}$ for a complex double series). Given $m, n \in \mathbb{N}$, we define the partial sum $S_{m,n}$ by

$$S_{m,n} = \sum_{i \leq m, j \leq n} a_{i,j} = \sum_{i=1}^{m} \sum_{j=1}^{n} a_{i,j}.$$

Definition 3.8.1 The double series $\sum_{m,n=1}^{\infty} a_{m,n}$ is convergent, with sum S, if there exists an $S \in \mathbb{R}$ such that for every $\varepsilon > 0$, there exists an $N \in \mathbb{N}$ such that

$$|S_{m,n} - S| < \varepsilon, \ m, n \geq N.$$

That is, if $\lim_{m,n \to \infty} S_{m,n} = S$.

Examples 3.8.2

(1) The double series $\sum_{m,n=1}^{\infty} 2^{-m} 3^{-n}$ is convergent with sum $1/2$. Since $S_{m,n} = (\sum_{i=1}^{m} 2^{-i})(\sum_{j=1}^{n} 3^{-j})$, we easily compute that

$$S_{m,n} = (1 - 2^{-m})(1 - 3^{-n})/2.$$

The result follows since $\lim_{m \to \infty} 1 - 2^{-m} = 1$ and $\lim_{n \to \infty} 1 - 3^{-n} = 1$.

(2) It is not enough in Definition 3.8.1 to require that $\lim_{n \to \infty} S_{n,n} = 0$. For example, define $a_{i,i} = i^{-2}$, $i \in \mathbb{N}$, and $a_{i,j} = -a_{j,i} = 1$, if $i < j$. Then $S_{n,n} = \sum_{i=1}^{n} i^{-2}$ but $a_{m,n} \nrightarrow 0$ as $n, m \to \infty$! (cf. Lemma 3.2.3). The definition we have given is simple and leads quickly to results on repeated series, but alternative definitions are possible that define the partial sums on non-rectangular regions. ♠

Definition 3.8.3 The double series $\sum_{m,n=1}^{\infty} a_{m,n}$ is *absolutely convergent* if $\sum_{m,n=1}^{\infty} |a_{m,n}|$ is convergent.

Proposition 3.8.4 *An absolutely convergent double series is convergent.*

Proof Suppose that $\sum_{m,n=1}^{\infty} a_{m,n}$ is absolutely convergent and that $\sum_{m,n=1}^{\infty} |a_{m,n}| = \hat{S}$. Let $\hat{S}_{m,n}$ and $S_{m,n}$ denote the partial sums for $\sum_{m,n=1}^{\infty} |a_{m,n}|$ and $\sum_{m,n=1}^{\infty} a_{m,n}$, respectively. Since $\sum_{m,n=1}^{\infty} a_{m,n}$ is absolutely convergent, $(\hat{S}_{n,n})$ is a Cauchy sequence. Just as in the proof of Theorem 3.5.3, it follows easily that $(S_{n,n})$ is Cauchy and so $(S_{n,n})$ converges, say to S. Suppose $m \geq n$. We have

$$|S - S_{m,n}| \leq |S - S_{n,n}| + |S_{n,n} - S_{m,n}| \leq |S - S_{n,n}| + |\hat{S}_{m,n} - \hat{S}_{n,n}|.$$

Letting $m \geq n \to \infty$, we see that $\lim_{m,n \to \infty} S_{m,n} = S$. The same argument applies if $m \leq n$. □

Proposition 3.8.5 *Let $\sigma : \mathbb{N}^2 \to \mathbb{N}^2$ be a bijection. If $\sum_{m,n=1}^{\infty} a_{m,n}$ is absolutely convergent then so is $\sum_{m,n=1}^{\infty} a_{\sigma(m,n)}$ and the two sums are equal.*

Proof The proof is similar to that of Theorem 3.5.15; we leave the details to the exercises.   □

3.8.1 Repeated Series

A *repeated series* is an infinite series which is of either of the forms

$$\sum_{m=1}^{\infty}\left(\sum_{n=1}^{\infty}a_{m,n}\right) \text{ or } \sum_{n=1}^{\infty}\left(\sum_{m=1}^{\infty}a_{m,n}\right).$$

If we think of the terms $a_{m,n}$ as defining an infinite matrix $[a_{m,n}]$, then the first repeated sum is naturally called the *sum by rows* of the double series $\sum_{m,n=1}^{\infty}a_{m,n}$ and the second repeated sum is called the *sum by columns*.

Proposition 3.8.6 *Suppose that $\sum_{m,n=1}^{\infty}a_{m,n}$ converges and that for all $m,n \in \mathbb{N}$, the series $\sum_{n=1}^{\infty}a_{m,n}$ and $\sum_{m=1}^{\infty}a_{m,n}$ converge, then the repeated series both converge and we have*

$$\sum_{m,n=1}^{\infty}a_{m,n} = \sum_{m=1}^{\infty}\left(\sum_{n=1}^{\infty}a_{m,n}\right) = \sum_{n=1}^{\infty}\left(\sum_{m=1}^{\infty}a_{m,n}\right).$$

Proof Set $\sum_{m,n=1}^{\infty}a_{m,n} = S$. Given $\varepsilon > 0$, there exists an $N \in \mathbb{N}$ such that

$$|S_{m,n} - S| < \varepsilon, \; m,n \geq N.$$

Hence $|\lim_{n\to\infty}S_{m,n} - S| \leq \varepsilon$ for all $m \geq N$ since $\sum_{n=1}^{\infty}a_{m,n}$ converges for all $m \in \mathbb{N}$. It follows that $\lim_{m\to\infty}(\lim_{n\to\infty}S_{m,n}) = S$. The same argument proves that $\lim_{m\to\infty}(\lim_{n\to\infty}S_{m,n}) = S$.   □

Example 3.8.7 If both repeated series converge but the double series is divergent then the repeated sums may be different. We give an example due to Arndt [5, Chap. V]. If we define

$$a_{m,n} = \frac{1}{m+1}\left(\frac{m}{m+1}\right)^{n} - \frac{1}{m+2}\left(\frac{m+1}{m+2}\right)^{n},$$

then after some computation we find that

$$S_{m,n} = \left(\frac{1}{2} - \frac{1}{2^{n+1}}\right) - \left[\frac{m+1}{m+2} - \left(\frac{m+1}{m+2}\right)^{n+1}\right].$$

It follows that

$$\sum_{m=1}^{\infty}\left(\sum_{n=1}^{\infty}a_{m,n}\right) = \lim_{m\to\infty}\left(\lim_{n\to\infty}S_{m,n}\right) = -\frac{1}{2},$$

$$\sum_{n=1}^{\infty}\left(\sum_{m=1}^{\infty}a_{m,n}\right) = \lim_{n\to\infty}\left(\lim_{m\to\infty}S_{m,n}\right) = \frac{1}{2}.$$

It is clear that $\lim_{m,n\to\infty}S_{m,n}$ does not exist. ♠

Proposition 3.8.8 *Suppose that one of*

$$\sum_{m,n=1}^{\infty}|a_{m,n}|,\quad \sum_{m=1}^{\infty}\left(\sum_{n=1}^{\infty}|a_{m,n}|\right),\quad \sum_{n=1}^{\infty}\left(\sum_{m=1}^{\infty}|a_{m,n}|\right)$$

is convergent, then all three series are convergent with the same sum and we have

$$\sum_{m,n=1}^{\infty}a_{m,n} = \sum_{m=1}^{\infty}\left(\sum_{n=1}^{\infty}a_{m,n}\right) = \sum_{n=1}^{\infty}\left(\sum_{m=1}^{\infty}a_{m,n}\right).$$

Proof The result follows straightforwardly from Propositions 3.8.6 and 3.8.4 and we leave the details to the exercises. □

EXERCISES 3.8.9

(1) Let $a_{m,n} = 1/(\alpha^n + \beta^m)$. Show that if $\alpha, \beta > 1$, then the double series $\sum_{m,n=1}^{\infty}a_{m,n}$ is convergent.

(2) State and prove a version of the comparison test that is applicable to double series of positive terms.

(3) Prove Proposition 3.8.5.

(4) Prove Proposition 3.8.8.

(5) Given the series $\sum_{n=0}^{\infty}a_n$, $\sum_{m=0}^{\infty}b_m$, define $c_{m,n} = a_m b_n$, $m, n \geq 0$. Show that if $\sum_{n=0}^{\infty}a_n$, $\sum_{m=0}^{\infty}b_m$ are convergent, then the double series $\sum_{m,n=1}^{\infty}a_m b_n$ is convergent with sum equal to $(\sum_{n=1}^{\infty}a_n)(\sum_{m=1}^{\infty}b_m)$.

(6) Suppose that $f : [1,\infty)^2 \to \mathbb{R}$ is continuous, monotone decreasing (in the sense that $f(x',y') \leq f(x,y)$, $x' \leq x$, $y' \leq y$) and $\lim_{(x,y)\to+\infty}f(x,y) = 0$. Show that the double series $\sum_{m,n=1}^{\infty}f(m,n)$ converges if and only if $\int_1^{\infty}\int_1^{\infty}f(x,y)\,dxdy$ converges.

(7) Let $f : [0,\infty)^2 \to \mathbb{R}$ be continuous and positive. For $s \geq 0$, define $g(s) = \inf_{x\in[0,s]}f(x,s-x)$, $G(s) = \sup_{x\in[0,s]}f(x,s-x)$. Suppose that $sg(s)$ and $sG(s)$ are both monotone decreasing and converge to zero as $s \to \infty$. Show that the double series $\sum_{m,n=1}^{\infty}f(m,n)$ converges if $\int_1^{\infty}sG(s)\,ds$ converges and diverges if $\int_1^{\infty}sg(s)\,ds$ diverges. (Hint: Given n, estimate the sum of terms on the diagonal $x + y = n$.)

(8) Show that

(a) $\sum_{m,n=1}^{\infty} (m+n)^{-a}$ converges iff $a > 2$.

(b) $\sum_{m,n=1}^{\infty} (Am^2 + 2Bmn + Cn^2)^{-a}$ converges if $a > 2, A, C > 0$ and $AC > B^2$ (if $B < 0$).

(c) If $f : [1, \infty) \to \mathbb{R}$ is continuous and monotone decreasing to zero, then $\sum_{m,n=1}^{\infty} f(Am^2 + 2Bmn + Cn^2)$ converges iff $\int_1^{\infty} f(x)\, dx$ converges (assume the coefficients A, B, C satisfy the conditions of the previous question).

(9) Show that if $a_{nm} = (-1)^{m+n}/mn$, then the double series and associated repeated series are convergent with common sum $(\log 2)^2$ but the series is not absolutely convergent

(10) Show that $\sum_{m,n=1}^{\infty} x^{mn} = \sum_{n=1}^{\infty} \frac{x^n}{1-x^n}$, $|x| < 1$ (Lambert's series).

(11) Let $a_{m,n} = (-1)^{m+n} mn/(m+n)^2$. Show that

$$\sum_{m=1}^{\infty} \left(\sum_{n=1}^{\infty} a_{m,n} \right) = \sum_{n=1}^{\infty} \left(\sum_{m=1}^{\infty} a_{m,n} \right) = \frac{1}{6} \left(\log 2 - \frac{1}{4} \right) = L$$

but that the double series $\sum_{m,n=1}^{\infty} a_{m,n}$ is not convergent (it oscillates between $L - \frac{1}{16}$ and $L + -\frac{1}{16}$).

3.9   Infinite Products

Suppose that (a_n) is a sequence of real numbers. The infinite product $\prod_{n=1}^{\infty} a_n$ is defined to be the sequence (P_n) of partial products where $P_n = a_1 \times a_2 \times \cdots \times a_n$, $n \in \mathbb{N}$. Roughly speaking, the infinite product $\prod_{n=1}^{\infty} a_n$ *converges* if the sequence (P_n) converges. In practice, it is useful to avoid situations where $\lim_{n\to\infty} P_n = 0$. If $\lim_{n\to\infty} P_n$ exists and is either 0 or $\pm\infty$, the infinite product is said to diverge (we refine this definition later). As we shall soon see, there is a close connection between the theories of infinite series and infinite products. This relationship is best seen by working with infinite products of the form $\prod_{n=1}^{\infty}(1 + a_n)$ rather than $\prod_{n=1}^{\infty} a_n$.

Definition 3.9.1 Let (a_n) be a sequence of real numbers. The infinite product $\prod_{n=1}^{\infty}(1 + a_n)$ is *convergent* if the sequence (P_n) of partial products is convergent and does not converge to either 0 or $+\infty$. If $\prod_{n=1}^{\infty}(1 + a_n)$ is not convergent, it is *divergent*.

Remarks 3.9.2

(1) Provided that the terms $1 + a_n$ are all positive, the infinite product $\prod_{n=1}^{\infty}(1 + a_n)$ converges iff the infinite sum $\sum_{n=1}^{\infty} \log(1 + a_n)$ converges.

(2) We can define infinite products $\prod_{n=1}^{\infty}(1 + a_n)$ with $a_n \in \mathbb{C}$. The definition of convergence is the same though we no can longer relate the convergence of the infinite product with the convergence of $\sum_{n=1}^{\infty} \log(1 + a_n)$. With the exception

of Proposition 3.9.10, many of the results and tests we give below do not apply
to the complex case.

(3) In practice, it is prudent to slightly modify the definition of convergence. We
say that the infinite product $\prod_{n=1}^{\infty}(1 + a_n)$ is convergent if there exists an $N \in \mathbb{N}$ such that $a_n \neq -1$, $n \geq N$, and $\prod_{n=N}^{\infty}(1 + a_n)$ converges in the sense of
Definition 3.9.1. Otherwise we say the product diverges. The reason for this
variation will be clearer when we look at the infinite product formula for the
sine function.   ✠

Examples 3.9.3

(1) The infinite product $\prod_{n=1}^{\infty}(1 + \frac{1}{n})$ diverges. We have $(1 + 1)(1 + 1/2) \cdots (1 + 1/n) \geq 1 + 1/2 + \cdots + 1/n$. Since the series $\sum_{n=1}^{\infty} 1/n$ is divergent to $+\infty$,
$\prod_{n=1}^{\infty}(1 + \frac{1}{n})$ diverges to $+\infty$.

(2) The infinite product $\prod_{n=2}^{\infty}(1 - \frac{1}{n^2})$ is convergent. Observe that $1 - \frac{1}{n^2} = \frac{(n-1)(n+1)}{n^2}$. Consequently, $P_n = \prod_{j=2}^{n} \frac{(j-1)(j+1)}{j^2} = \frac{1}{2}\frac{n+1}{n}$. Hence $\prod_{n=2}^{\infty}(1 - \frac{1}{n^2})$
is convergent and $\prod_{n=2}^{\infty}(1 - \frac{1}{n^2}) = \frac{1}{2}$.   ♠

If we assume all the terms a_n are positive then it is easy to give necessary and
sufficient conditions for convergence of $\prod_{n=1}^{\infty}(1 + a_n)$.

Lemma 3.9.4 *Assume* $a_n \geq 0$ *for all* $n \in \mathbb{N}$. *Then* $\prod_{n=1}^{\infty}(1 + a_n)$ *converges iff*
$\sum_{n=1}^{\infty} a_n < \infty$ *and then*

$$\sum_{n=1}^{\infty} a_n \leq \prod_{n=1}^{\infty}(1 + a_n) \leq \exp\left(\sum_{n=1}^{\infty} a_n\right).$$

Proof For $n \in \mathbb{N}$,

$$\sum_{i=1}^{n} a_i \leq \prod_{i=1}^{n}(1 + a_i) \leq e^{\sum_{i=1}^{n} a_i},$$

where the last inequality follows from $1 + a_i \leq e^{a_i}$. The result follows from
Theorem 2.3.18.   □

Example 3.9.5 As an immediate consequence of Lemma 3.9.4 and our results on
series, we have $\prod_{n=1}^{\infty}(1 + \frac{1}{n^p})$ converges iff $p > 1$.   ♠

We have a useful variation of Lemma 3.9.4 that allows for all the terms a_i to be
negative.

Lemma 3.9.6 *Assume* $a_n \in [0, 1)$ *for all* $n \in \mathbb{N}$. *Then* $\prod_{n=1}^{\infty}(1 - a_n)$ *converges iff*
$\sum_{n=1}^{\infty} a_n < \infty$.

Proof Since $(1 - a) \leq (1 + a)^{-1}$ if $a \in [0, 1)$, we have $\prod_{n=1}^{m}(1 - a_n) \leq (\prod_{n=1}^{m}(1 + a_n))^{-1}$ for all $m \in \mathbb{N}$. By Lemma 3.9.4 it follows that if $\sum_{n=1}^{\infty} a_n$ diverges to $+\infty$,
then $\prod_{n=1}^{\infty}(1 - a_n)$ diverges to zero. Conversely, suppose that $\sum_{n=1}^{\infty} a_n$ is convergent.

An easy induction on n verifies that for $n \geq m \geq 1$ we have

$$\prod_{j=m}^{n}(1 - a_j) \geq 1 - \sum_{j=m}^{n} a_j.$$

Since $\sum_{n=1}^{\infty} a_n$ is convergent, there exists an $N \in \mathbb{N}$ such that $\sum_{j=m}^{n} a_j < 1/2$ for all $n \geq m \geq N$. Hence

$$\prod_{j=m}^{n}(1 - a_j) > \frac{1}{2}, \quad n \geq m \geq N.$$

Consequently, $P_n = \prod_{j=1}^{n}(1 - a_j) \geq \frac{1}{2}\prod_{j=1}^{N}(1 - a_j) = C > 0$, for all $n \in \mathbb{N}$. Therefore the decreasing sequence (P_n) is bounded below by $C > 0$ and therefore $\prod_{n=1}^{\infty}(1 - a_n)$ converges.   □

Remark 3.9.7 We refer the reader to Exercises 3.9.19(4) for the *Weierstrass inequalities* which we have made use of in the proofs of Lemmas 3.9.4, 3.9.6.   ✲

Lemmas 3.9.4, 3.9.6 will suffice for most of our intended applications to infinite products (in particular, our Fourier series proof in Chap. 5 of the infinite product formula for the sine function). In the remainder of the section, we develop some more advanced topics from the theory of infinite products that parallels our previous work on conditional and absolute convergence for infinite series.

3.9.1   Tests for Convergence of an Infinite Product

We start with the definition of absolute convergence for infinite products.

Definition 3.9.8 The infinite product $\prod_{n=1}^{\infty}(1 + a_n)$ is *absolutely convergent* if $\prod_{n=1}^{\infty}(1 + |a_n|)$ is convergent.
Before giving our main result, we prove a lemma that is useful for estimating products.

Lemma 3.9.9 *Let (a_n) be a sequence of real or complex numbers. Then*

(1) $\left|\prod_{j=1}^{n}(1 + a_j)\right| \leq e^{\sum_{j=1}^{n}|a_j|}.$

(2) $\left|\prod_{j=1}^{n}(1 + a_j) - 1\right| \leq e^{\sum_{j=1}^{n}|a_j|} - 1.$

Proof Estimate (1) follows easily from $|1 + x| \leq 1 + |x| \leq e^{|x|}$ ($x \in \mathbb{R}$ or $x \in \mathbb{C}$). For estimate (2), observe that $\left|\prod_{j=1}^{n}(1 + a_j) - 1\right| \leq \left|\prod_{j=1}^{n}(1 + |a_j|) - 1\right|$ and then use (1).   ⊔

Proposition 3.9.10 *Let (a_n) be a sequence of real (or complex) numbers none of which equals -1.*

(a) $\prod_{n=1}^{\infty}(1 + a_n)$ *is absolutely convergent iff* $\sum_{n=1}^{\infty} a_n$ *is absolutely convergent.*
(b) *If* $\prod_{n=1}^{\infty}(1 + a_n)$ *is absolutely convergent, then* $\prod_{n=1}^{\infty}(1 + a_n)$ *is convergent.*

Proof (a) Lemma 3.9.4 implies that $\prod_{n=1}^{\infty}(1 + |a_n|)$ is convergent iff $\sum_{n=1}^{\infty} |a_n|$ is convergent. It remains to prove (b). Set $P_n = \prod_{j=1}^{n}(1 + a_j)$. We prove that the sequence (P_n) of partial products is a Cauchy sequence. If $n > m$, then

$$P_n - P_m = P_m \left(\prod_{j=m+1}^{n} (1 + a_j) - 1 \right).$$

In order to estimate $|P_n - P_m|$, we make use of the inequalities (a) $|1 + x| \leq 1 + |x|$, $x \in \mathbb{R}$ (or $\mathbb{C}$), and (b) $1 + x \leq e^x$ for all $x \geq 0$. We have

$$|P_n - P_m| = |P_m| \left| \prod_{j=m+1}^{n} (1 + a_j) - 1 \right|$$

$$\leq e^{\sum_{j=1}^{m} |a_j|} \left(e^{\sum_{j=m+1}^{n} |a_j|} - 1 \right), \text{ by Lemma 3.9.9,}$$

$$= e^{\sum_{j=1}^{n} |a_j|} - e^{\sum_{j=1}^{m} |a_j|}.$$

Since $\sum_{n=1}^{\infty} |a_n| < \infty$, $(\sum_{j=1}^{n} |a_j|)$ is a Cauchy sequence and so therefore is $(e^{\sum_{j=1}^{n} |a_j|})$. Hence $|P_n - P_m| \to 0$ as $m, n \to \infty$ and (P_n) is a Cauchy sequence. This proves that $\prod_{n=1}^{\infty}(1 + a_n)$ exists. It remains to show that $\prod_{n=1}^{\infty}(1 + a_n) = L \neq 0$ if $a_n \neq -1$ for all $n \in \mathbb{N}$. Now if $a_n \neq -1$, for all $n \in \mathbb{N}$, then $P_m \neq 0$ for all $m \geq 1$. From our previous estimates for $|P_n - P_m|$ we see that

$$|P_n - P_m| \leq |P_m| \left(e^{\sum_{j=m+1}^{n} |a_j|} - 1 \right).$$

Since $\sum_{n=1}^{\infty} |a_n| < \infty$, there exists an $N \in \mathbb{N}$ such that $e^{\sum_{j=m+1}^{n} |a_j|} - 1 \leq \frac{1}{2}$ for all $n > m \geq N$ and so

$$|P_n - P_m| \leq |P_m|/2, \ n > m \geq N.$$

Letting $n \to \infty$, we get $|L - P_m| \leq |P_m|/2$, $m \geq N$. In particular, if $L = 0$, this implies $|P_N| \leq |P_N|/2$ and so $P_N = 0$. Contradiction. $\square$

Example 3.9.11 The infinite products $\prod_{n=2}^{\infty}(1 \pm n^{-p})$ converge if $p > 1$. ♠

Theorem 3.9.12 (General Principle of Convergence for Products) *Let* (a_n) *be a sequence of real (or complex) numbers none of which equals* -1. *The infinite product* $\prod_{n=1}^{\infty}(1 + a_n)$ *converges iff for every* $\varepsilon > 0$ *there exists an* $N \in \mathbb{N}$ *such that*

$$\left| \prod_{j=m}^{n}(1 + a_n) - 1 \right| < \varepsilon, \text{ for all } m, n \geq N.$$

Proof We leave the proof to the exercises.   □

We conclude with some additional tests for convergence which only apply to infinite products of real numbers and give necessary and sufficient conditions for convergence in terms of infinite sums.

Lemma 3.9.13 *Let (a_n) be a sequence of real numbers none of which equals -1. If $\sum_{n=1}^{\infty} a_n^2 < \infty$, then*

(a) $\prod_{n=1}^{\infty}(1 + a_n)$ *converges if $\sum_{n=1}^{\infty} a_n$ converges.*
(b) $\prod_{n=1}^{\infty}(1 + a_n)$ *diverges to $+\infty$ if $\sum_{n=1}^{\infty} a_n$ diverges to $+\infty$.*
(c) $\prod_{n=1}^{\infty}(1 + a_n)$ *diverges to 0 if $\sum_{n=1}^{\infty} a_n$ diverges to $-\infty$.*

If $\sum_{n=1}^{\infty} a_n^2$ diverges and $\sum_{n=1}^{\infty} a_n$ converges, then $\prod_{n=1}^{\infty}(1 + a_n)$ diverges to zero.

Proof A straightforward application of the calculus shows that

$$
u^2/4 \le u - \log(1 + u) \le \begin{cases} u^2, & \text{if } 0 \le u \le 1, \\ \frac{1}{2}u^2/(1 + u), & \text{if } 0 > u > -1. \end{cases}
$$

Since one of $\sum_{n=1}^{\infty} a_n$, $\sum_{n=1}^{\infty} a_n^2 < \infty$ converges, $a_n \to 0$. Hence we may choose $N \in \mathbb{N}$ such that $|a_n| \le 1/2, n \ge N$. We have $|(1 + a_n)| \ge 1 - |a_n| \ge 1/2$, for all $n \ge N$. Using the inequalities above, we have for $n > m \ge N$

$$
\frac{1}{4} \sum_{i=m+1}^{n} a_i^2 \le \sum_{i=m+1}^{n} a_i - \log\left[\prod_{i=m+1}^{n}(1 + a_i)\right]
$$

$$
\le \sum_{i=m+1}^{n} a_i^2.
$$

Hence if $\sum a_n^2$ is convergent then $(a_{m+1} + \cdots + a_n) - \log\left[\prod_{i=m+1}^{n}(1 + a_i)\right]$ converges to zero as $n \ge m \to \infty$. Statement (a,b,c) now follow by the general principle of convergence. If $\sum_{n=1}^{\infty} a_n^2$ diverges and $\sum_{n=1}^{\infty} a_n$ converges, then $\log[\prod_{i=m+1}^{n}(1 + a_i)]$ must diverge to $-\infty$ proving the final statement.   □

Example 3.9.14 Lemma 3.9.13 implies that $\prod_{n=1}^{\infty}(1 + \frac{1}{n})$ is divergent while $\prod_{n=1}^{\infty}(1 + \frac{(-1)^n}{n})$ is convergent.   ♠

3.9.2   Tannery's Theorem and an Infinite Product for $\sin x$

We have a version of Tannery's theorem for infinite products.

Theorem 3.9.15 *Suppose we are given a sequence $(a_n(p)) \subset \mathbb{R}$ depending on $p \in \mathbb{N}$. Assume*

(1) $\lim_{p \to \infty} a_n(p) = a_n, n \in \mathbb{N}$.
(2) $|a_n(p)| \le M_n$, *for all $n, p \in \mathbb{N}$, where $\sum_{n=1}^{\infty} M_n < \infty$.*

Then

$$\lim_{p\to\infty}\prod_{n=1}^{p}(1 + a_n(p)) = \lim_{p\to\infty}\prod_{n=1}^{\infty}(1 + a_n(p)) = \prod_{n=1}^{\infty}(1 + a_n).$$

The result continues to hold if $(a_n(p)) \subset \mathbb{C}$.

Proof Suppose first that $(a_n(p)) \subset \mathbb{R}$. For sufficiently large N, we can assume $|a_n(p)| \leq 1/2, n, p \geq N$. We have

$$\prod_{j=N}^{n}(1 + a_j(p)) = \exp\left(\sum_{j=N}^{n}\log(1 + a_j(p))\right).$$

Now $A_j(p) = \log(1 + a_j(p))$ satisfies the conditions of Tannery's theorem for series since if $|a_j(p)| \leq 1/2$, we have $|A_j(p)| = |\log(1 + a_j(p))| \leq 2|a_j(p)|$ (use Exercise 2.9.9(4)). Now apply Tannery's theorem for series (Theorem 3.5.6). If $(a_n(p)) \subset \mathbb{C}$, we reduce to the series case using the method of Proposition 3.9.10 (see Remark 3.9.16 below). □

Remark 3.9.16 In the complex case we take as definitions $\exp(z) = e^z = \sum_{n=0}^{\infty}\frac{z^n}{n!}$, $z \in \mathbb{C}$, and $\log(1 + z) = \sum_{n=1}^{\infty}(-1)^{n+1}\frac{z^n}{n}$, for $|z| < 1$. It is not hard to show (using absolute convergence) that $\exp(\log(1 + z)) = 1 + z$, if $|z| < 1$, which is needed for the proof of Tannery's theorem for complex infinite products.   ✲

Proposition 3.9.17 *For all $z \in \mathbb{C}$*

$$\sin z = z\prod_{n=1}^{\infty}\left(1 - \frac{z^2}{n^2\pi^2}\right).$$

Proof We give a proof of Proposition 3.9.17 that uses Tannery's theorem for infinite products and a minimal amount of complex variable theory. (We give an alternative and simpler real variable proof based on Fourier series in Chap. 5.)

For $z \in \mathbb{C}$ we have (by definition)

$$\sin z = \frac{e^{iz} - e^{-iz}}{2i}. \tag{3.6}$$

Applying the complex version of Proposition 3.5.7 to $e^{\pm iz}$ gives

$$\sin z = \lim_{n\to\infty}\left[\frac{(1 + \frac{iz}{n})^n - (1 - \frac{iz}{n})^n}{2i}\right]$$

$$= \lim_{n\to\infty}P_n(z),$$

where $P_n(z)$ is a polynomial of degree (at most) n. Our approach will be to factorize $P_n(z)$ and for this we need to find the solutions of $P_n(z) = 0$. Observe that

$$P_n(z) = 0 \iff \left(1 + \frac{\iota z}{n}\right)^n = \left(1 - \frac{\iota z}{n}\right)^n$$

$$\iff 1 + \frac{\iota z}{n} = u\left(1 - \frac{\iota z}{n}\right), \text{ where } u^n = 1$$

$$\iff z = \frac{n}{\iota}\left(\frac{u-1}{u+1}\right).$$

From now we assume that n is odd and so $u \neq -1$. The solutions of $u^n = 1$ are given by

$$u = e^{\frac{2k\pi\iota}{n}}, \ k = -\frac{n-2}{2}, \cdots, -1, 0, 1, \cdots, \frac{n-1}{2}.$$

For $k \in \{-\frac{n-2}{2}, \cdots, -1, 0, 1, \cdots, \frac{n-1}{2}\}$ we have $P(z_k) = 0$, where

$$z_k = \frac{n}{\iota}\left(\frac{e^{\frac{2k\pi\iota}{n}} - 1}{e^{\frac{2k\pi\iota}{n}} + 1}\right)$$

$$= n\frac{(e^{\frac{k\pi\iota}{n}} - e^{\frac{-k\pi\iota}{n}})/2\iota}{(e^{\frac{k\pi\iota}{n}} + e^{\frac{k\pi\iota}{n}})/2}$$

$$= n\tan\left(\frac{k\pi}{n}\right).$$

Since $\tan x$ is an odd function, the roots of $P_n(z) = 0$ are

$$0, \pm n\tan z\left(\frac{\pi}{n}\right), \pm n\tan\left(\frac{2\pi}{n}\right), \cdots, \pm n\tan\left(\frac{\frac{n-1}{2}\pi}{n}\right),$$

and so (for n odd) we have $P_n(z) = Cz\prod_{j=1}^{\frac{n-1}{2}}\left(1 - \frac{z^2}{n^2\tan^2(\frac{j\pi}{n})}\right)$. The coefficient of z in $P_n(z)$ is easily verified to be 1 and so $C = 1$. Hence

$$P_n(z) = z\prod_{j=1}^{\frac{n-1}{2}}\left(1 - \frac{z^2}{n^2\tan^2(\frac{j\pi}{n})}\right).$$

For fixed j, we have $\lim_{n\to\infty} n^2\tan^2(\frac{j\pi}{n}) = j^2\pi^2$. Now we apply Tannery's theorem for infinite products with

$$a_n(p) = \begin{cases} 1 - \frac{z^2}{p^2\tan^2(\frac{n\pi}{p})}, & n \geq \frac{p-1}{2}, \\ 0, & n < \frac{p-1}{2}. \end{cases}$$

Noting that $\tan x \geq x$ for $x \in [0, \pi/2)$, we see that $\left| \frac{z^2}{p^2 \tan^2(\frac{n\pi}{p})} \right| \leq \frac{|z|^2}{n^2\pi^2}$, $1 \leq p \leq$
$(n-1)/2$. It follows by Proposition 3.9.10 that condition (2) of Tannery's theorem
for infinite products is satisfied. Condition (1) is immediate since $\lim_{p\to\infty} a_n(p) =$
$(1 - \frac{z^2}{n^2})$.   □

Remark 3.9.18 The most famous infinite product formula is that found by Euler for
the *Riemann zeta-function* (see Exercises 3.9.19(6c)). For many other examples of
infinite products we refer the reader to

www-elsa.physik.uni-bonn.de/~dieckman/InfProd/InfProd.html

for the encyclopedic list compiled by Andreas Dieckmann.   ✱

EXERCISES 3.9.19

(1) Show that

  (a) $\prod_{n=2}^{\infty} \left(1 - \frac{2}{n(n+1)} \right) = \frac{1}{3}$.
  (b) $\prod_{n=2}^{\infty} \left(1 + \frac{(-1)^n}{n} \right) = 1$.
  (c) $\prod_{n=2}^{\infty} \left(1 + \frac{1}{n} \right) = 0$ (in particular, the product diverges).
  (d) $\prod_{n=3}^{\infty} \left(1 + \frac{4}{n^2-4} \right) = 6$.
  (e) $\prod_{n=2}^{\infty} \left(1 + \frac{2n+1}{n^2-1} \right) = \frac{1}{3}$.

(2) Prove Theorem 3.9.12.
(3) Evaluate $\prod_{n=1}^{N}(1 + x^{2^{n-1}})$ and hence show that $\prod_{n=1}^{\infty}(1 + x^{2^{n-1}})$ converges to
    $(1-x)^{-1}$, for $x \in (-1, 1)$.
(4) Suppose that $(a_n) \subset (0, 1)$. Show that for $n > N \geq 1$ we have the Weierstrass
    inequalities

  (a)

$$\prod_{j=N}^{n}(1 + a_j) \geq 1 + \sum_{j=N}^{n} a_j, \quad \prod_{j=N}^{n}(1 - a_j) \geq 1 - \sum_{j=N}^{n} a_j.$$

  (b)

$$\prod_{j=N}^{n}(1 + a_j) \leq \left(1 - \sum_{j=N}^{n} a_j \right)^{-1}, \quad \prod_{j=N}^{n}(1 - a_j) \leq \left(1 + \sum_{j=N}^{n} a_j \right)^{-1},$$

  provided $\sum_{j=N}^{n} a_j < 1$.

Deduce that provided $\sum_{j=N}^{n} a_j < 1$ we have for all $n > N$ the estimates

$$\left(1 - \sum_{j=N}^{n} a_j\right)^{-1} \geq \prod_{j=N}^{n}(1 + a_j) \geq 1 + \sum_{j=N}^{n} a_j$$

$$\left(1 + \sum_{j=N}^{n} a_j\right)^{-1} \geq \prod_{j=N}^{n}(1 - a_j) \geq 1 - \sum_{j=N}^{n} a_j.$$

As a corollary, show that $\sum_{n=1}^{\infty} a_n$ converges iff either the infinite product $\prod_{n=1}^{\infty}(1 + a_n)$ converges or the infinite product $\prod_{n=1}^{\infty}(1 - a_n)$ converges.

(5) Show that

(a) $\prod_{n=1}^{\infty}\left(1 + \frac{(-1)^{n-1}}{\log n \sqrt{n}}\right)$ converges.

(b) $\prod_{n=1}^{\infty}\left(1 + \frac{(-1)^{n-1}}{\sqrt{n}}\right)$ diverges to zero.

(6) Let (p_n) denote the sequence of prime numbers > 1 written in ascending order: $2 = p_2 < p_3 < \cdots$. Show that

(a) $\sum_{n=2}^{\infty} \frac{1}{p_n^x}$ converges for $x > 1$.

(b) $\prod_{n=2}^{\infty}(1 - \frac{1}{p_n^x})$ converges for $x > 1$ and, in particular, is non-zero.

(c) $\prod_{n=2}^{\infty}(1 - \frac{1}{p_n^x})^{-1} = \sum_{n=1}^{\infty} \frac{1}{n^x}$, $x > 1$ (Euler product for the zeta-function).

(d) $\sum_{n=2}^{\infty} \frac{1}{p_n}$ is divergent.

(Hints for parts (c,d): Every $n \in \mathbb{N}$ can be written uniquely as a product of primes. Given $N \geq 2$, let $P(N) \subset \mathbb{N}$ be the subset of all positive integers whose prime factors are $p_2, \cdots, p_N$. We regard $1 \in P(N)$. Verify that for $x > 0$, $\prod_{n=2}^{N}(1 - \frac{1}{p_n^x})^{-1} = \sum_{k \in P(N)} \frac{1}{k^x}$. Use Lemma 3.9.6 for (d).)

(7) Show that if there are only finitely many primes $p_1, \cdots, p_N$, then we would have $\sum_{n=1}^{\infty} 1/n = \prod_{n=1}^{N}(1 - p_n^{-1})^{-1} < \infty$, and so deduce that there are infinitely many primes.

(8) Prove that if $\prod(1 + a_n)$ is absolutely convergent, then the value of the product is independent of the order of the factors.

(9) State and prove an analogue of Riemann's rearrangement theorem for infinite products that are not absolutely convergent.

(10) Taking $z = \iota \pi$ in the product formula for $\sin z$ verify that

$$\frac{e^\pi - e^{-\pi}}{2\pi} = \prod_{n=1}^{\infty}\left(1 + \frac{1}{n^2}\right).$$

(Assume $\sin z$ is defined as in (3.6).) Hence, using Examples 3.9.3(2), find $\prod_{n=1}^{\infty}(1 - \frac{1}{n^4})$.

(11) Show that $\frac{e^{\pi/2} - e^{-\pi/2}}{\pi} = \prod_{n=1}^{\infty}(1 + \frac{1}{4n^2})$.

(12) Prove the infinite product formula for $\cos z$

$$\cos z = \prod_{n=1}^{\infty} \left(1 - \frac{4z^2}{(2n-1)^2 \pi^2} \right).$$

(Hint: Use the product formula for $\sin z$ together with the trigonometric identity $\sin 2z = 2 \sin z \cos z$.)

3.10   Appendix: Trigonometric Identities

In this appendix we prove some very useful trigonometric identities.

Theorem 3.10.1 *Let $\alpha, \beta \in \mathbb{R}$ and suppose that β is not an integer multiple of 2π. For $n \geq 0$ we have*

$$\sum_{k=0}^{n} \cos(\alpha + k\beta) = \frac{\sin\left(\frac{(n+1)\beta}{2}\right) \cos\left(\alpha + \frac{n\beta}{2}\right)}{\sin\left(\frac{\beta}{2}\right)},$$

$$\sum_{k=0}^{n} \sin(\alpha + k\beta) = \frac{\sin\left(\frac{(n+1)\beta}{2}\right) \sin\left(\alpha + \frac{n\beta}{2}\right)}{\sin\left(\frac{\beta}{2}\right)},$$

$$\sum_{k=1}^{n} \cos(k\beta) = \frac{\cos\left(\frac{(n+1)\beta}{2}\right) \sin\left(\frac{n\beta}{2}\right)}{\sin\left(\frac{\beta}{2}\right)},$$

$$\sum_{k=1}^{n} \sin(k\beta) = \frac{\sin\left(\frac{(n+1)\beta}{2}\right) \sin\left(\frac{n\beta}{2}\right)}{\sin\left(\frac{\beta}{2}\right)}.$$

Proof By DeMoivre's theorem we have

$$\cos(\alpha + k\beta) + \imath \sin(\alpha + k\beta) = e^{\imath\alpha + \imath k\beta} = e^{\imath\alpha} e^{\imath k\beta}.$$

Therefore

$$\sum_{k=0}^{n} \cos(\alpha + k\beta) + \imath \sin(\alpha + k\beta) = e^{\imath\alpha} \sum_{k=0}^{n} e^{\imath k\beta}.$$

Provided that β is not an integer multiple of 2π, we have

$$\sum_{k=0}^{n} e^{\imath k\beta} = \frac{1 - e^{\imath(n+1)\beta}}{1 - e^{\imath\beta}}.$$

(This is most easily verified by multiplying both sides by $1 - e^{i\beta}$. Alternatively, divide.) Taking real and imaginary parts gives us

$$\sum_{k=0}^{n} \cos(\alpha + k\beta) = \text{Real}\left(e^{i\alpha}\frac{1 - e^{i(n+1)\beta}}{1 - e^{i\beta}}\right),$$

$$\sum_{k=0}^{n} \sin(\alpha + k\beta) = \text{Im}\left(e^{i\alpha}\frac{1 - e^{i(n+1)\beta}}{1 - e^{i\beta}}\right).$$

We have

$$e^{i\alpha}\frac{1 - e^{i(n+1)\beta}}{1 - e^{i\beta}} = e^{i\alpha}\frac{(1 - e^{i(n+1)\beta})(1 - e^{-i\beta})}{2 - e^{i\beta} - e^{-i\beta}}$$

$$= e^{i\alpha}\frac{(1 - e^{i(n+1)\beta})(1 - e^{-i\beta})}{2 - 2\cos\beta}$$

$$= \frac{A + iB}{4\sin^2(\frac{\beta}{2})},$$

where

$$A = \cos\alpha + \cos(n\beta + \alpha) - \cos((n+1)\beta + \alpha) - \cos(\alpha - \beta),$$

$$B = \sin\alpha + \sin(n\beta + \alpha) - \sin((n+1)\beta + \alpha) - \sin(\alpha - \beta).$$

Using the trigonometric identities $\cos a + \cos b = 2\cos(\frac{a+b}{2})\cos(\frac{a-b}{2})$ and $\cos a - \cos b = 2\sin(\frac{a+b}{2})\sin(\frac{b-a}{2})$, it is straightforward to show that

$$A = 4\cos(\alpha + \frac{n\beta}{2})\sin(\frac{(n+1)\beta}{2})\sin(\beta/2).$$

Hence $\sum_{k=0}^{n}\cos(\alpha + k\beta) = A/4\sin^2(\frac{\beta}{2}) = \frac{\sin\left(\frac{(n+1)\beta}{2}\right)\cos\left(\alpha + \frac{n\beta}{2}\right)}{\sin(\frac{\beta}{2})}$. A similar analysis using the identities $\sin a \pm \sin b = 2\sin(\frac{a\pm b}{2})\cos(\frac{a\mp b}{2})$ gives the result for the sum of sines. Alternatively, replace α by $\alpha - \pi/2$ in the cosine sum formula.

Finally, we need to show $\sum_{k=1}^{n}\cos(k\beta) = \frac{\cos\left(\frac{(n+1)\beta}{2}\right)\sin\left(\frac{n\beta}{2}\right)}{\sin(\frac{\beta}{2})}$. This follows from the expression for the sum from $k = 0$ to n (with $\alpha = 0$) if we subtract the initial term 1 (cos 0) and then use the formula $\sin(\frac{(n+1)\beta}{2})\cos(\frac{n\beta}{2}) - \cos(\frac{(n+1)\beta}{2})\sin(\frac{n\beta}{2}) = \sin(\frac{\beta}{2})$. □

EXERCISES 3.10.2

(1) Show that provided x is not an odd multiple of π we have

$$\sum_{k=1}^{n}(-1)^{k+1}\cos(kx) = \frac{\cos(\frac{x}{2}) + (-1)^{k+1}\cos((k+\frac{1}{2})x)}{2\cos(\frac{x}{2})}.$$

(2) Find formulas for $\sum_{k=0}^{n}(-1)^{k+1}\cos(kx)$ and $\sum_{k=1}^{n}(-1)^{k+1}\sin(kx)$. (Hint: To get the alternating sum, replace x by $x + \pi$ in the original formulas.)

(3) Find formulas for $\sum_{k=0}^{n}(-1)^{k+1}\cos(\alpha + kx)$ and $\sum_{k=0}^{n}(-1)^{k+1}\sin(\alpha + kx)$, $\alpha \in \mathbb{R}$.

Chapter 4
Uniform Convergence

4.1 Introduction

In this chapter we begin our study of continuous and differentiable functions. We focus on construction and properties. Our main strategy will be to build functions as infinite series (or products) of elementary functions such as x^n or $\sin nx$ and $\cos nx$. For example, we develop techniques that enable us to give conditions for a *power series* $\sum_{n=0}^{\infty} a_n x^n$ to converge to an infinitely differentiable function. We also investigate continuity properties of *trigonometric* or *Fourier series* such as the sine series $\sum_{n=1}^{\infty} b_n \sin(nx)$. We conclude the chapter with an example of a trigonometric series that converges to a continuous function on $\mathbb{R}$ that is *nowhere* differentiable. Overall, the aim in this chapter is to develop the tools—which are largely based on the concept of *uniform convergence*. In the next chapter, we use these tools to study several important classes of functions. Although in this and the following chapter we work almost exclusively with real-valued functions defined on subsets, usually subintervals of the real line, the ideas and methods we develop have general applicability and most of the results apply to complex or vector valued functions.

We start by looking at convergence of *sequences* of functions. We then apply our results to the partial sums of infinite series of functions. All of this is along the lines developed in the previous chapter and indeed much of our work will be making the translation from sequences/series of real numbers to sequences/series of functions. A new and important issue will be the validity of term-by-term integration and differentiation of infinite series. For example, when can we find the integral of a function defined as an infinite series by integrating term-by-term? Many foundational theorems in analysis are about precisely this problem of interchanging the order of limiting operations.

© Springer International Publishing AG 2017

M. Field, *Essential Real Analysis*, Springer Undergraduate Mathematics Series,
https://doi.org/10.1007/978-3-319-67546-6_4

4.2 Pointwise Convergence

We always assume that I is a non-empty subset of $\mathbb{R}$. Typically, I might be an interval, possibly unbounded, which may be open, closed, or half-open. However, all of what we say works perfectly well if I is any non-empty subset of $\mathbb{R}$. Suppose that we are given a sequence (u_n) of real-valued functions on I. That is, for each $n \in \mathbb{N}$, $u_n : I \to \mathbb{R}$. At this point we do not assume any additional properties of the functions u_n (such as continuity). Observe that for each $x \in I$, $(u_n(x))$ is a sequence of real numbers. The next definition gives a natural definition of convergence of the sequence of functions (u_n) in terms of the sequences $(u_n(x))$, $x \in I$.

Definition 4.2.1 (Notation and Assumptions as Above) The sequence (u_n) of functions on I is *pointwise convergent* (on I) if there exists a function $u : I \to \mathbb{R}$ such that for every $x \in I$ we have

$$\lim_{n \to \infty} u_n(x) = u(x).$$

We refer to u as the *pointwise limit* of the sequence (u_n).

Examples 4.2.2

(1) Take $I = [0, 1]$, let $f : I \to \mathbb{R}$ be any function and define $u_n = f/n$, $n \in \mathbb{N}$. That is, for each $x \in I$, $n \in \mathbb{N}$, $u_n(x) = f(x)/n$. Although f may not be bounded on I (we are not assuming f is continuous), it is true that for every (fixed) $x \in I$, $f(x) \in \mathbb{R}$, and so $\lim_{n \to \infty} u_n(x) = \lim_{n \to \infty} f(x)/n = 0$. Hence (u_n) is pointwise convergent on I with pointwise limit the zero function. In this case, the pointwise limit is continuous even though the terms in the sequence might be discontinuous at every point of I.

(2) Take $I = [0, 1]$ and let $u_n(x) = x^n$, $x \in I$, $n \in \mathbb{N}$. If $0 \le x < 1$, we have $\lim_{n \to \infty} u_n(x) = \lim_{n \to \infty} x^n = 0$. On the other hand, $\lim_{n \to \infty} u_n(1) = 1$. The pointwise limit u is continuous on $[0, 1)$ but has a discontinuity at $x = 1$: without further conditions, the pointwise limit of continuous functions need not be continuous. A feature of this example is that as x gets close to 1, convergence to $u(x)$ is slow. More specifically, given $x \in [0, 1)$, $1 > \varepsilon > 0$, let $N(x) \in \mathbb{N}$ be the smallest integer such that $u_n(x) = x^n < \varepsilon$. Clearly $N(0) = 1$ and if $0 < x < 1$, $N(x)$ is the smallest integer bigger than $\frac{\log \varepsilon}{\log x}$. Consequently, $\lim_{x \to 1-} N(x) = +\infty$ and convergence is slow when x is close to 1.

(3) Even if the pointwise limit of a sequence of continuous functions is continuous, the convergence can have unpleasant features. For example, take $I = [0, 1]$, $p \in \mathbb{R}$ and define

$$u_n(x) = n^p x^n(1 - x), \; x \in [0, 1], \; n \in \mathbb{N}.$$

Since $\lim_{n \to \infty} n^p x^n(1 - x) = 0$, if $x \in [0, 1]$, we see that (u_n) is pointwise convergent on I with pointwise limit the zero function (note that $u_n(1) = 0$, all n). A straightforward application of the differential calculus shows that the maximum value of u_n on I is $n^{p-1}(n/(n+1))^{n+1}$ and is attained when $x = n/(n+1)$. We see that if $p < 1$, then $\lim_{n\to\infty} \sup_{x\in I} u_n(x) = \lim_{n\to\infty} n^{p-1}(n/(n+1))^n = 0$. If $p = 1$, then $\lim_{n\to\infty} \sup_{x\in I} u_n(x) = e^{-1}$ (where we have used $\lim_{n\to\infty}(n/(n+1))^n = 1/(1+1/n)^n = e^{-1}$). If $p > 1$, then $\lim_{n\to\infty} \sup_{x\in I} u_n(x) = +\infty$. If $p \geq 1$, then even though (u_n) converges pointwise to the zero function, the graph of u_n does not approach that of the zero function. It is also natural to consider the area under the graph of u_n. We have

$$\int_0^1 u_n(x)\, dx = \int_0^1 n^p x^n (1-x)\, dx = \frac{n^p}{(n+1)(n+2)}, \quad n \geq 1.$$

Clearly $\lim_{n\to\infty} \int_0^1 u_n(x)\, dx = 0 = \int_0^1 \lim_{n\to\infty} u_n(x)\, dx$ iff $p < 2$. If $p = 2$, $\lim_{n\to\infty} \int_0^1 u_n(x) = 1$ and if $p > 2$, then $\lim_{n\to\infty} \int_0^1 u_n(x) = +\infty$. This shows that without further conditions on the convergence of functions we cannot interchange the order of limit and integration.   ♠

4.3   Uniform Convergence of Sequences

The examples in the previous section show that pointwise convergence of functions does not handle continuity well and can lead to some nasty pathology (as shown in Examples 4.2.2(3)). We seek a definition of convergence of functions that behaves well with respect to continuity and basic operations of analysis such as integration and differentiation.

Suppose that $f, g : I \subset \mathbb{R} \to \mathbb{R}$. What does it mean for f and g to be 'close'? One natural approach is to require that $|f(x) - g(x)|$ is small for *all* $x \in I$. That is, we are asking that the graphs of f and g are close as subsets of $\mathbb{R}^2$. More formally, given $\varepsilon > 0$, let $T(f, \varepsilon) \subset \mathbb{R}^2$ denote the tube of width 2ε centred on the graph of f. That is,

$$T(f, \varepsilon) = \{(x, y) \mid x \in I, |f(x) - y| < \varepsilon\}.$$

See Fig. 4.1. In order that f and g are ε-close, we require that graph$(g) \subset T(f, \varepsilon)$. Obviously, graph$(g) \subset T(f, \varepsilon)$ iff $|f(x) - g(x)| < \varepsilon$ for all $x \in I$. Hence graph$(g) \subset T(f, \varepsilon)$ iff graph$(f) \subset T(g, \varepsilon)$ and so the condition is symmetric in f and g.

In the remainder of this section we formalize this idea of closeness or *uniform approximation*. We do this by first restricting to the class of bounded functions (defined on any subset of $\mathbb{R}$) and then giving a precise definition of what we mean by the distance between two functions f, g such that $f - g$ is bounded. This will enable us to give a good definition of convergence for sequences of continuous functions. We develop these ideas further in the next chapter where we show how we can

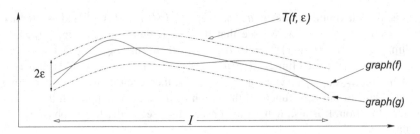

Fig. 4.1 Graphs of f and g that are ε-close to each other

approximate a continuous function, which may be nowhere differentiable, by more regular functions, such as polynomials.

4.3.1   Spaces of Bounded Functions

We continue to assume that I is a non-empty subset of $\mathbb{R}$. A function $f : I \to \mathbb{R}$ is *bounded* if there exists an $M \geq 0$ such that

$$|f(x)| \leq M, \text{ for all } x \in I.$$

We do not assume yet that f is continuous. If $f : I \to \mathbb{R}$ is bounded, we define

$$\|f\| = \sup\{|f(x)| \mid x \in I\} < \infty.$$

Remark 4.3.1 The number $\|f\|$ is often called the *uniform*-norm of f (also the C^0-norm, ∞-norm or supremum norm). It is commonly denoted by $\|f\|_\infty$.   ✱

Definition 4.3.2 Let $B(I)$ denote the set of all bounded functions $f : I \to \mathbb{R}$.

Example 4.3.3 Constant functions are bounded and so $B(I)$ contains all the constant functions, including the zero function.   ♠

Lemma 4.3.4 (Notation as Above) *Let $f, g \in B(I)$.*

(1) *For all $c \in \mathbb{R}$, we have $f + cg \in B(I)$.*
(2) *$\|f + g\| \leq \|f\| + \|g\|$, and $\|cf\| = |c|\|f\|$, all $c \in \mathbb{R}$.*
(3) *$\|f\| = 0$ iff $f \equiv 0$.*

In particular, $B(I)$ is a vector space: for all $f, g \in B(I)$, $c, d \in \mathbb{R}$, we have $cf + dg \in B(I)$.

Proof We start by showing that if $f \in B(I)$, $c \in \mathbb{R}$, then $cf \in B(I)$ and $\|cf\| = |c|\|f\|$. Since $|f(x)| \le \|f\|$, we have $|cf(x)| \le |c|\|f\|$, all $x \in I$, and so $|c|\|f\|$ is an upper bound for cf. Hence $cf \in B(I)$. We claim that $\|cf\| = |c|\|f\|$. If not, there exists an $M < |c|\|f\|$ such that M is an upper bound for cf. But then $M/|c| < \|f\|$ would be an upper bound for f, contradicting the definition of $\|f\|$.

Next we prove that if $f, g \in B(I)$ then $f + g \in B(I)$ and $\|f + g\| \le \|f\| + \|g\|$ (this will complete the proof of (1,2)). For all $x \in I$, we have

$$|f(x) + g(x)| \le |f(x)| + |g(x)| \le \|f\| + \|g\|.$$

Therefore $\|f\| + \|g\|$ is an upper bound for $\{|f(x)+g(x)| \mid x \in I\}$ and so $f+g \in B(I)$ and $\|f + g\| \le \|f\| + \|g\|$.

Finally, suppose $\|f\| = 0$. Then $\sup\{|f(x)| \mid x \in I\} = 0$. Hence $f(x) = 0$, for all $x \in I$, and so $f \equiv 0$. The converse is trivial.   □

Definition 4.3.5 Suppose that $f, g : I \to \mathbb{R}$ and $f - g \in B(I)$. We define the distance between f and g, $\rho(f, g)$, by

$$\rho(f, g) = \|f - g\|.$$

Lemma 4.3.6 (Notation as Above) *Suppose that $f, g, h \in B(I)$. We have*

(1) $\rho(f, g) \ge 0$ *and* $\rho(f, g) = 0$ *iff* $f = g$.
(2) $\rho(f, g) = \rho(g, f)$.
(3) $\rho(f, h) \le \rho(f, g) + \rho(g, h)$ *(triangle inequality)*.

Proof The result is immediate from Lemma 4.3.4.   □

Remark 4.3.7 For the previous lemma to hold it suffices that $f - g, g - h$, $f - h \in B(I)$.   ✶

4.3.2   Spaces of Continuous Functions

Let $C^0(I)$ denote the space of continuous real-valued functions on $I \subset \mathbb{R}$. In general, $C^0(I) \not\subset B(I)$ (take $I = (0, 1)$ and $f(x) = x^{-1}$). However, there is a large class of subsets I of $\mathbb{R}$ for which $C^0(I) \subset B(I)$. We concentrate on the best known case.

Theorem 4.3.8 *If I is a closed and bounded interval, then $C^0(I) \subset B(I)$.*

Proof This is a restatement of Theorem 2.4.10(1): continuous functions on a closed and bounded interval are bounded.   □

4.3.3 Convergence of Functions

Definition 4.3.9 Let $I \subset \mathbb{R}$. If (u_n) is a sequence of functions on I, then (u_n) *converges uniformly* to $u : I \to \mathbb{R}$ if

$$\lim_{n\to\infty} \rho(u, u_n) = 0.$$

Remarks 4.3.10

(1) If (u_n) converges uniformly to $u : I \to \mathbb{R}$, then we must have $u - u_n \in B(I)$, at least for large enough n. In particular, if $(u_n) \subset B(I)$ then $u \in B(I)$ since $\rho(u, u_n) < \infty$ implies that $u - u_n \in B(I)$ and so, $u = (u - u_n) + u_n \in B(I)$ (Lemma 4.3.4).
(2) The use of the term 'uniform' in the definition should be clear. The sequence (u_n) converges uniformly to u if for every $\varepsilon > 0$, we can find an $N \in \mathbb{N}$ such that if $n \geq N$ then $|u_n(x) - u(x)| < \varepsilon$ for all $x \in I$. This is a much stronger condition than pointwise convergence, where N may depend strongly on x—see Example 4.2.2. ✠

Proposition 4.3.11 (Notation as Above) *If (u_n) converges uniformly to u, then (u_n) converges pointwise to u.*

Proof We must prove that for each $x \in I$, $\lim_{n\to\infty} u_n(x) = u(x)$. Let $\varepsilon > 0$. Since $\lim_{n\to\infty} \rho(u_n, u) = 0$, there exists an $N \in \mathbb{N}$ such that $\rho(u_n, u) < \varepsilon$, for all $n \geq N$. That is,

$$\rho(u_n, u) = \sup\{|u_n(y) - u(y)| \mid y \in I\} < \varepsilon,\ n \geq N.$$

Since $|u_n(x) - u(x)| \leq \rho(u_n, u)$, we have $|u_n(x) - u(x)| < \varepsilon$ for all $n \geq N$ and so $\lim_{n\to\infty} u_n(x) = u(x)$. □

The next result shows that uniform convergence behaves well with respect to both continuity and boundedness and so avoids the problems we have seen with pointwise convergence.

Theorem 4.3.12 *Let $I \subset \mathbb{R}$ and (u_n) be a sequence of continuous (respectively, bounded) functions on I which converges uniformly to u. Then u is continuous (respectively, bounded).*

Proof Suppose that $(u_n) \subset C^0(I)$ converges uniformly to u. We are required to prove that if $x_0 \in I$ and $\varepsilon > 0$, then there exists a $\delta > 0$ such that $|u(x_0) - u(x)| < \varepsilon$, for all $x \in I$ such that $|x_0 - x| < \delta$. The idea of the proof is to approximate u sufficiently closely by a continuous function u_N (how large we need to take N depends on ε) and then use the continuity of u_N to deduce the estimate we require on u. In more detail, choose $N \in \mathbb{N}$ such that $\rho(u_N, u) < \varepsilon/3$. By definition of

$\rho(u_N, u)$, we have

$$|u_N(y) - u(y)| < \varepsilon/3, \text{ for all } y \in I.$$

Since u_N is continuous on I, there exists a $\delta > 0$ such that

$$|u_N(x_0) - u_N(x)| < \varepsilon/3, \text{ for all } x \in I \text{ such that } |x_0 - x| < \delta.$$

Now we use the triangle inequality. Suppose $x \in I$, then

$$
\begin{aligned}
|u(x_0) - u(x)| &= |u(x_0) - u_N(x_0) + u_N(x_0) - u_N(x) + u_N(x) - u(x)| \\
&\leq |u(x_0) - u_N(x_0)| + |u_N(x_0) - u_N(x)| + |u_N(x) - u(x)| \\
&< \varepsilon/3 + \varepsilon/3 + \varepsilon/3 = \varepsilon,
\end{aligned}
$$

where the last inequality holds provided $|x_0 - x| < \delta$.
The final statement follows from Remarks 4.3.10(1). $\qquad\square$

Corollary 4.3.13 *Let (u_n) be a sequence of continuous functions on the closed and bounded interval $I = [a, b]$. Suppose that (u_n) converges uniformly to u, then u is continuous and bounded.*

Proof An immediate corollary of Theorem 4.3.12 since every continuous function on $[a, b]$ is bounded. $\qquad\square$

Examples 4.3.14

(1) Take $I = [0, 1]$ and let $u_n(x) = x^n$, $x \in I$ (as in Examples 4.2.2(2)). Recall that the pointwise limit u of (u_n) is the function which is equal to zero on $[0, 1)$ and 1 at $x = 1$. We claim that $\rho(u, u_n) = 1$ for all $n \in \mathbb{N}$ and so the convergence is not uniform. It suffices to show that for every $\varepsilon > 0$, there exists an $x \in [0, 1)$ such that $|u_n(x) - u(x)| = |u_n(x)| = x^n > 1 - \varepsilon$. This is immediate from the continuity of u_n at $x = 1$. (Of course, since u is not continuous, we can deduce that (u_n) does not converge uniformly to u using Corollary 4.3.13.)

(2) Take $I = [0, 1]$, $p \in \mathbb{R}$ and let $u_n(x) = n^p x^n (1 - x)$, $x \in I$ (as in Examples 4.2.2(3)). Recall that the pointwise limit u of (u_n) is identically zero. We have (see Examples 4.2.2(3)), $\rho(u, u_n) = n^{p-1}(n/(n + 1))^n$, $n \in \mathbb{N}$. Hence $\rho(u, u_n) \to 0$ iff $p < 1$. If $p = 1$, $\rho(u, u_n) \to e^{-1}$, and if $p > 1$, $\rho(u, u_n) \to +\infty$. Hence we only have uniform convergence when $p < 1$. $\qquad\spadesuit$

4.3.4 General Principle of Convergence

Just as we did for sequences of real numbers we may define Cauchy sequences of functions. The Cauchy sequence definition has the merit of not requiring knowledge of the actual limit.

Definition 4.3.15 If (u_n) is a sequence of functions on the non-empty subset I of $\mathbb{R}$, then (u_n) is a *Cauchy sequence* if $\rho(u_m, u_n) \to 0$ as $m, n \to \infty$. That is, if for every $\varepsilon > 0$, there exists an $N \in \mathbb{N}$ such that

$$\rho(u_m, u_n) < \varepsilon, \text{ for all } m, n \geq N.$$

Theorem 4.3.16 (General Principle of Uniform Convergence) *Let $I \subset \mathbb{R}$ and (u_n) be a sequence of functions on I. Then (u_n) is uniformly convergent on I iff (u_n) is a Cauchy sequence. If either condition holds and the limit function is u, then u will be bounded (respectively, continuous) if $(u_n) \subset B(I)$ (respectively $(u_n) \subset C^0(I)$).*

Proof Suppose that (u_n) is a Cauchy sequence. We start by verifying that (u_n) is pointwise convergent. Let $\varepsilon > 0$ and choose $N \in \mathbb{N}$ so that $\rho(u_m, u_n) < \varepsilon$ for all $m, n \geq N$. If $x \in I$, we have $|u_m(x) - u_n(x)| \leq \rho(u_m, u_n) < \varepsilon$, for all $m, n \geq N$. Hence $(u_n(x))$ is a Cauchy sequence and by the general principle of convergence for sequences of real numbers, there exists a $u(x) \in \mathbb{R}$ such that $\lim_{n \to \infty} u_n(x) = u(x)$. This construction defines a function $u : I \to \mathbb{R}$. Observe that u is the pointwise limit of the sequence (u_n). The estimate

$$|u_m(x) - u_n(x)| < \varepsilon, \ m, n \geq N, \tag{4.1}$$

holds for all $x \in I$. That is, the integer N does not depend on the choice of $x \in I$. Letting $m \to \infty$ in (4.1) gives

$$|u(x) - u_n(x)| \leq \varepsilon, \ n \geq N, \text{ for all } x \in I,$$

and so $\rho(u, u_n) \leq \varepsilon$, for all $n \geq N$. Hence (u_n) converges uniformly to u. We leave the proof that a uniformly convergent sequence is Cauchy to the exercises. The final statements follow from Remarks 4.3.10(1) and Theorem 4.3.12. $\qquad\square$

Our main applications of the general principle of uniform convergence will be to infinite series and are described in the next section.

EXERCISES 4.3.17

(1) Complete the proof of Theorem 4.3.16 by showing that a uniformly convergent sequence of functions is a Cauchy sequence.
(2) Show that if (f_n) converges uniformly to f on $I \subset \mathbb{R}$, and (f_n) converges uniformly to g on $J \subset \mathbb{R}$, then (a) $f|I \cap J = g|I \cap J$ ("$f|I \cap J$" means f restricted to $I \cap J$), (b) (f_n) converges uniformly to a function $F : I \cup J \to \mathbb{R}$ where $F|I = f$, $F|J = g$.
(3) Show that if (f_n), (g_n) respectively converge uniformly to f, g on I, then $(f_n \pm g_n)$ converges uniformly to $f \pm g$ on I. Show that if $(f_n), (g_n) \in B(I)$, then uniform convergence of (f_n), (g_n) implies $(f_n g_n)$ converges uniformly to fg. Show by means of examples that this result may fail if either (f_n) or (g_n) consists of unbounded functions.

(4) Find the pointwise limit of the following sequences of functions on the specified domain. In each case describe the continuity properties of the limit function.

(a) $f_n(x) = \tan^{-1}(nx), x \geq 0$.
(b) $f_n(x) = \frac{nx}{1+n^2x^2}, x \in \mathbb{R}$.

Is the convergence for either of these sequences uniform? Why/Why not?

(5) Suppose $u_n(x) = x^n(1-x^n), x \in [0,1]$. Is (u_n) pointwise convergent on $[0,1]$? uniformly convergent on $[0,1]$?

(6) Let $p, q \in \mathbb{Z}_+$. Let $v_n(x) = x^{pn}(1-x^{qn}), x \in [0.1]$. Show that (v_n) is pointwise convergent on $[0,1]$? Can we choose p, q so that (v_n) is uniformly convergent on $[0,1]$?

(7) Let $u_n(x) = x^n(1-x^{n^2}), x \in [0,1]$. Investigate the pointwise and uniform convergence of the sequence (u_n) on $[0,1]$.

(8) Let $u_n(x) = x^{n^2}(1-x^n)$. Show that the sequence (u_n) converges uniformly on $[0,1]$. What is the limit?

(9) Suppose $v_n(x) = nxe^{-nx}, x \in [0,1]$. Is (v_n) pointwise convergent on $[0,1]$? uniformly convergent on $[0,1]$? Would the answer change if we took $v_n(x) = nxe^{-nx^2}$? $v_n(x) = nxe^{-n\sqrt{x}}$?

(10) Let $f_n(x) = n^p x^n(1-x)^2$. Show that (f_n) is uniformly convergent on $[0,1]$ iff $p < 2$. (Hint: Proposition 3.5.7.) What about if $f_n(x) = n^p x^n(1-x)^q, q > 2$?

(11) Suppose that (f_n) is a sequence of continuous functions which is pointwise convergent to f on the open interval (a, b). Suppose that the convergence of (f_n) to f is *uniform* on every closed subinterval of (a, b). Prove that f is continuous on (a, b).

(12) The sequences (f_n) on $[0,1]$, (g_n) on $[0,100]$, and (h_n) on $\mathbb{R}$ are defined by

(a) $f_n(x) = x^n(1-x)$.
(b) $g_n(x) = \frac{nx^3}{1+nx}$.
(c) $h_n(x) = \frac{nx^4}{1+nx^2}$.

Find the pointwise limits of these sequences and prove that the convergence is uniform.

(13) Determine whether or not the following sequences converge uniformly on the specified domains. It is a good idea to start by finding pointwise limits.

(a) $f_n(x) = \frac{1}{n+x}, x \geq 0, n \geq 1$.
(b) $f_n(x) = \frac{x^n}{1+x^n}, x \in [0,1], n \geq 1$.

(14) Let (q_n) be a sequence consisting of all the rational numbers with $q_n \neq q_m$, $n \neq m$. Let $C > 0$ and (a_n) be any sequence of real numbers such that $|a_n| \geq C > 0$ for all $n \in \mathbb{N}$. For $n \in \mathbb{N}$ define

$$f_n(x) = \begin{cases} a_n, & \text{if } x = q_n, \\ 0, & \text{otherwise.} \end{cases}$$

Show that (a) (f_n) is pointwise convergent, (b) (f_n) is not uniformly convergent on any closed interval $[a, b]$, $a \neq b$.

(15) Suppose that (u_n) converges uniformly to u on I. Show that if $I^* \subset I$ is such that u_n is continuous on I^* for all n, then u is continuous on I^* (this is a slight extension of Theorem 4.3.12).

(16) Following Exercises 2.4.16(6), define $f : \mathbb{R} \to \mathbb{R}$ by $f(x) = 0$ if $x \notin \mathbb{Q}$ and $f(x) = 10^{-s}$ if $x = r/s$, where $(r, s) = 1$ and $s > 0$ and we take $s = 1$ if $x = 0$. Let (q_n) be a sequence consisting of all the rational numbers and suppose $q_n \neq q_m$, $n \neq m$. For $n \in \mathbb{N}$ define

$$f_n(x) = \begin{cases} 10^{-s}, & \text{if } x = r/s \in \{q_1, \cdots, q_n\}, \\ 0, & \text{otherwise.} \end{cases}$$

Show that (f_n) converges uniformly to f on $\mathbb{R}$. Deduce, using the previous exercise, that f is continuous on $\mathbb{R} \smallsetminus \mathbb{Q}$.

(17) Let (f_n) be uniformly bounded sequence of continuous functions on $[a, b]$ such that for each $x \in [a, b]$, $(f_n(x))$ is monotone.

(a) Show that (f_n) converges pointwise to a function $f : [a, b] \to \mathbb{R}$.
(b) Show that f need not be continuous (construct an example).
(c) Show that if f is continuous, then the convergence of (f_n) to f is *uniform*.

(Remark and hints. Result (c) is *Dini's theorem*. In order to prove (c), fix $\varepsilon > 0$ and define $\Delta_n = \{x \mid |f(x) - f_n(x)| \geq \varepsilon\}$. Since $(f_n(x))$ is monotone, we have $\Delta_1 \supset \cdots \supset \Delta_n \supset \cdots$. It suffices to prove that there exists an $N \in \mathbb{N}$ such that $\Delta_N = \emptyset$. Prove by contradiction. Useful observations are (1) $\cap_{n \geq 1} \Delta_n = \emptyset$; (2) if $x \notin \Delta_n$, then $\exists \delta > 0$ such that $(x - \delta, x + \delta) \cap \Delta_m = \emptyset$, all $m \geq n$.)

4.4   Uniform Convergence of Infinite Series

In this section I will always denote a subinterval of $\mathbb{R}$ (open, closed, half-open, bounded or unbounded). However, the results we give easily extend to functions defined on an arbitrary non-empty subset of $\mathbb{R}$.

Let (u_n) be a sequence of functions defined on I. For $n \geq 1$, we define the sequence (S_n) of partial sums by

$$S_n(x) = \sum_{j=1}^{n} u_j(x), \quad x \in I.$$

Note that (S_n) is a sequence of functions defined on I.

Definition 4.4.1 (Notation as Above)

(a) The infinite series $\sum_{n=1}^{\infty} u_n$ is *pointwise convergent* (on I) to the function $S :$ $I \to \mathbb{R}$ if the sequence of partial sums (S_n) is pointwise convergent to S. (That is, $\lim_{n \to \infty} S_n(x) = S(x)$, for all $x \in I$).

(b) The infinite series $\sum_{n=1}^{\infty} u_n$ is *uniformly convergent* (on I) to the function $S :$ $I \to \mathbb{R}$ if the sequence of partial sums (S_n) is uniformly convergent to S. (That is, $\lim_{n \to \infty} \rho(S, S_n) = 0$.)

Example 4.4.2 We claim that the series $\sum_{n=1}^{\infty} \frac{x^n}{n^2}$ is uniformly convergent on $[-1, 1]$. For $x \in [-1, 1]$, $m < n \in \mathbb{N}$, we have

$$|S_n(x) - S_m(x)| = \left| \sum_{j=m+1}^{n} \frac{x^j}{j^2} \right| \le \sum_{j=m+1}^{n} \frac{|x|^j}{j^2} \le \sum_{j=m+1}^{n} \frac{1}{j^2},$$

with equality if $x = 1$. Hence $\rho(S_n, S_m) = \sum_{j=m+1}^{n} \frac{1}{j^2} \to 0$, as $m, n \to \infty$. It follows by the general principal of uniform convergence that $\sum_{n=1}^{\infty} \frac{x^n}{n^2}$ is uniformly convergent on $[-1, 1]$. ♠

As an immediate consequence of our results on the uniform convergence of sequences, we have the first of our main results on uniform convergence of infinite series.

Theorem 4.4.3 *Let (u_n) be a sequence of continuous (respectively, bounded) functions on I. If the infinite series $\sum_{n=1}^{\infty} u_n$ is uniformly convergent to the function $S : I \to \mathbb{R}$, then S is continuous (respectively, bounded).*

For applications, it is useful to have a slightly stronger version of the continuity statement in Theorem 4.4.3.

Theorem 4.4.4 *Let (u_n) be a sequence of continuous functions on I. If the infinite series $\sum_{n=1}^{\infty} u_n$ is uniformly convergent on every closed and bounded subinterval of I, then $\sum_{n=1}^{\infty} u_n$ converges to a continuous function on I.*

Proof The result is immediate from the previous theorem if I is a closed and bounded interval. So assume that I is not a closed and bounded interval. The hypotheses of the theorem imply that $\sum_{n=1}^{\infty} u_n$ converges pointwise on I to a function $S : I \to \mathbb{R}$. Indeed, given $x \in I$, apply the uniform convergence hypothesis of the theorem to the closed interval $[x, x]$. In order to show that S is continuous it is enough to prove that S is continuous on every closed and bounded subinterval of I. But this follows from the hypotheses of the theorem and Theorem 4.4.3. (If $I = [a, +\infty)$ then we prove S continuous on any bounded interval $[a, b]$, $b > a$, and that suffices for continuity at a. For all other points $x \in I$, choose $a < x < b$ so that $[a, b] \subset I$. Then S is continuous on $[a, b]$ and certainly continuous at x.) □

Remarks 4.4.5

(1) We do not claim in Theorem 4.4.4 that S is bounded.
(2) If I is an arbitrary non-empty subset of $\mathbb{R}$, then Theorem 4.4.4 continues to apply provided that $\sum_{n=1}^{\infty} u_n$ converges uniformly on $[a, b] \cap I$ for all $-\infty < a \le b < +\infty$.   ✠

One last, but key, result before we give some examples.

Theorem 4.4.6 (General Principle of Uniform Convergence for Series) *Let (u_n) be a sequence of functions on I. The infinite series $\sum_{n=1}^{\infty} u_n$ is uniformly convergent on I iff the sequence of partial sums (S_n) is Cauchy. More formally, if for every $\varepsilon > 0$, there exists an $N \in \mathbb{N}$ such that*

$$\|u_m + \cdots + u_n\| < \varepsilon, \text{ for all } n \ge m \ge N,$$

then there exists a function $S : I \to \mathbb{R}$ such that $\sum_{n=1}^{\infty} u_n$ converges uniformly to S. If the sequence (u_n) consist of continuous functions, then S is continuous.

Proof Apply Theorem 4.3.16 to the sequence of partial sums.   □

Examples 4.4.7

(1) Let $u_n(x) = \sin nx/n^2$, $n \ge 1$. We claim that $\sum_{n=1}^{\infty} \frac{\sin nx}{n^2}$ is uniformly convergent on $\mathbb{R}$ and $S(x) = \sum_{n=1}^{\infty} \frac{\sin nx}{n^2}$ is continuous on $\mathbb{R}$. To see this, observe that for all $x \in \mathbb{R}$, we have $|u_n(x)| \le 1/n^2$. We know that $\sum_{n=1}^{\infty} 1/n^2 < \infty$ and so, by the general principle of convergence for series (of real numbers) given $\varepsilon > 0$, there exists an $N \in \mathbb{N}$ such that $|\frac{1}{m^2} + \cdots + \frac{1}{n^2}| < \varepsilon$, for all $n \ge m \ge N$. Hence

$$\left| \sum_{j=m}^{n} \frac{\sin jx}{j^2} \right| \le \sum_{j=m}^{n} \frac{1}{j^2} < \varepsilon, \text{ for all } m \ge n \ge N.$$

Therefore $\rho(S_n, S_{m-1}) = \| \sum_{j=m}^{n} u_j \| < \varepsilon$, for all $n \ge m \ge N$, and the sequence of partial sums is Cauchy. The result follows from Theorem 4.4.6. The reader should note how this proof is a mix of the theory of series of real numbers (using in this case the convergence of $\sum 1/n^2$) and results on uniform convergence. The method we used to deduce uniform convergence by comparing with a 'known' series of real numbers is very powerful and due to Weierstrass (it is a special case of the Weierstrass M-test—see below).

(2) Define $u_n(x) = x^n$, $n \ge 0$. Let $a \in (0, 1)$ and regard u_n as defined on $[-a, a]$. For all $x \in [-a, +a]$, $n \ge m$, we have

$$|x^m + \cdots + x^n| = |x|^m |1 + \cdots + x^{n-m}| \le a^m \sum_{j=0}^{\infty} a^j = a^m/(1 - a).$$

Hence $\|u_m + \cdots + u_n\| \leq a^m/(1 - a) \to 0$ as $n \geq m \to \infty$. By the general principle of uniform convergence for series (Theorem 4.4.3), $\sum_{n=0}^{\infty} x^n$ is uniformly convergent to a continuous function on $[-a, a]$. This argument is valid for all $a \in (0, 1)$ and so $\sum_{n=0}^{\infty} x^n$ converges to a continuous function S on $(-1, 1)$ (Theorem 4.4.4). In this case we know that $S(x) = 1/(1-x)$. The reader should note that $\sum_{n=0}^{\infty} x^n$ is *not* uniformly convergent to $1/(1 - x)$ on $(-1, 1)$. This is easily seen since given any $n \geq m \geq 1$, we can make $a^m + \cdots + a^n > 1/2$ by taking a sufficiently close to 1. In particular, the sequence of partial sums cannot be Cauchy on $(-1, 1)$.   ♠

Theorem 4.4.8 (Weierstrass M-Test) *Suppose that (u_n) is a sequence of functions defined on I and that there exists a sequence (M_n) of positive real numbers such that*

(a) $|u_n(x)| \leq M_n$ *for all* $x \in I$, $n \in \mathbb{N}$.
(b) $\sum_{n=1}^{\infty} M_n < \infty$.

Then $\sum_{n=1}^{\infty} u_n$ is uniformly convergent on I. If the (u_n) are all continuous, so is $S = \sum_{n=1}^{\infty} u_n$.

Proof The proof is the same as that used in Examples 4.4.7(1). We prove that the sequence (S_n) of partial sums is Cauchy. Let $\varepsilon > 0$. Since $\sum_{n=1}^{\infty} M_n < \infty$, there exists an $N \in \mathbb{N}$ such that $M_m + \cdots + M_n < \varepsilon$ for all $n \geq m \geq N$. It follows from assumption (a) that for $n \geq m \geq N$ and $x \in I$ we have

$$\left| \sum_{j=m}^{n} u_j(x) \right| \leq \sum_{j=n}^{m} M_j < \varepsilon.$$

Hence $\rho(S_n, S_{m-1}) = \|\sum_{j=m}^{n} u_j\| \leq \sum_{j=m}^{n} M_j < \varepsilon$, for all $n \geq m \geq N$, and so (S_n) is a Cauchy sequence. Now apply Theorem 4.4.6.   □

We give some characteristic applications of the M-test in the next set of examples (more examples appear in the following section).

Examples 4.4.9

(1) Consider the series $\sum_{n=1}^{\infty} \frac{n}{1+x^2+n^3}$. We have $0 < \frac{n}{1+x^2+n^3} \leq \frac{n}{1+n^3} < \frac{1}{n^2}$ for all $x \in \mathbb{R}$, $n \in \mathbb{N}$. Taking $M_n = \frac{1}{n^2}$ in the M-test, we see that $\sum_{n=1}^{\infty} \frac{n}{1+x^2+n^3}$ converges uniformly to a continuous function on $\mathbb{R}$.

(2) Let (a_n) be any sequence of real numbers and $p > 1$. The infinite series $\sum_{n=1}^{\infty} \frac{\sin(a_n x)}{n^p}$ converges uniformly to a continuous function on $\mathbb{R}$. For this, we note that $|\frac{\sin(a_n x)}{n^p}| \leq n^{-p}$ and take $M_n = n^{-p}$ in the M-test (since $p > 1$, $\sum n^{-p} < \infty$).

(3) Consider the exponential series $\sum_{n=0}^{\infty} \frac{x^n}{n!}$. This series does not converge uniformly on $\mathbb{R}$. We show that the series converges uniformly on every closed and bounded interval $[-R, R]$, $R \geq 0$. Certainly $|\frac{x^n}{n!}| \leq \frac{R^n}{n!}$ for all $x \in [-R, R]$. We take $M_n = \frac{R^n}{n!}$ in the M-test. Since $\sum_{n=0}^{\infty} \frac{R^n}{n!} < \infty$, it follows by the M-test

that $\sum_{n=0}^{\infty} \frac{x^n}{n!}$ converges uniformly on $[-R, R]$ for all $R \geq 0$. As a consequence $\exp(x) = \sum_{n=0}^{\infty} \frac{x^n}{n!}$ defines a continuous function on $\mathbb{R}$.   ♠

EXERCISES 4.4.10

(1) Consider the infinite series $\sum_{n=1}^{\infty} \frac{1}{1+n^2 x}$.

    (a) For what values of x is the series convergent?

    (b) On what closed intervals does it converge uniformly.

    (c) Is $f(x) = \sum_{n=1}^{\infty} \frac{1}{1+n^2 x}$ continuous on the set of points where the series converges?

(2) Show that the series $\sum_{n=1}^{\infty} \frac{x}{n(1+nx^2)}$ is uniformly convergent on $\mathbb{R}$.

(3) Show that the series $\sum_{n=1}^{\infty} \frac{1}{n^3+n^4 x^2}$ is uniformly convergent on $\mathbb{R}$.

(4) Show that the series $\sum_{n=1}^{\infty} x^n(1 - x^n)$ converges pointwise but not uniformly on $[0, 1]$.

(5) Show that the series $\sum_{n=1}^{\infty} x^{n^2}(1 - x^n)/n$ converges uniformly on $[0, 1]$. What about $\sum_{n=1}^{\infty} x^n(1 - x^{n^2})/n$?

(6) Show that the infinite series $\sum_{n=0}^{\infty} \frac{x^2}{(1+x^2)^n}$ converges pointwise on $\mathbb{R}$ and find the sum. Show that the series converges uniformly on any closed and bounded interval I not containing $x = 0$. What happens if $0 \in I$?

(7) Show that $\sum_{n=1}^{\infty} (-1)^{n+1}/(n + x)$ is uniformly convergent on $[0, \infty)$ but that the M-test does not apply to give uniform convergence on *any* closed interval $J \subset [0, \infty)$.

(8) Let u be defined by the geometric series $u(x) = \sum_{n=0}^{\infty} \frac{x}{(1+x)^n}$, $x \geq 0$.

    (a) Find $u(x)$, $x \geq 0$. Is u continuous?

    (b) Is the convergence uniform on $[0, \infty)$? Why/Why not?

    (c) Show that the convergence of $\sum_{n=0}^{\infty} \frac{x}{(1+x)^n}$ is uniform on $[X, \infty)$, provided $X > 0$.

(9) We say that $[a, b]$ is a proper closed interval if $a \neq b$. Prove the following extension of Theorem 4.4.4: if $\{I_i \mid i \in I\}$ is a family of proper closed subintervals of I such that (a) $\sum_{n=1}^{\infty} u_n$ is uniformly convergent on I_i for all $i \in I$, and (b) $\cup_{i \in I} I_i = I$, then $\sum_{n=1}^{\infty} u_n$ converges to a continuous function on I.

(10) Show that $\sum_{n=1}^{\infty} a_n \sin nx$ converges uniformly on $\mathbb{R}$ if $\sum a_n$ is absolutely convergent.

4.5   Power Series

In this section we consider the convergence properties of series of the form $\sum_{n=0}^{\infty} a_n x^n$ (more generally, $\sum_{n=0}^{\infty} a_n(x - x_0)^n$). This type of series is called a *power series*.

Lemma 4.5.1 *Suppose that the power series $\sum_{n=0}^{\infty} a_n x^n$ converges for either $x = r$ or $x = -r$. Then there exists a $C = C(r) > 0$ such that*

$$|a_n| \leq Cr^{-n}, \ n \geq 0.$$

Proof Suppose that $|x| = r$ and $\sum_{n=0}^{\infty} a_n x^n$ converges. Now $\lim_{n \to \infty} a_n x^n = 0$ (Lemma 3.2.3) and so the sequence $(|a_n x^n|) = (|a_n| r^n)$ is bounded. Hence there exists a $C > 0$ such that $|a_n| r^n \leq C$ for all $n \geq 0$. □

Lemma 4.5.2 *Suppose that the power series $\sum_{n=0}^{\infty} a_n x^n$ converges if $x = z \neq 0$. Then $\sum_{n=0}^{\infty} a_n x^n$ is absolutely convergent for all $x \in (-|z|, |z|)$. If $\sum_{n=0}^{\infty} a_n x^n$ diverges for $x = z$, then $\sum_{n=0}^{\infty} a_n x^n$ diverges if $|x| > |z|$.*

Proof Suppose that $\sum_{n=0}^{\infty} a_n z^n$ is convergent. By Lemma 4.5.1, there exists a $C \geq 0$ such that

$$|a_n| \leq C|z|^{-n}, \ n \geq 0.$$

Therefore

$$|a_n x^n| \leq C \left(\frac{|x|}{|z|} \right)^n, \ n \geq 0,$$

and $\sum_{n=0}^{\infty} a_n x^n$ is absolutely convergent if $|x| < |z|$ by comparison with the geometric series $C \sum_{n=0}^{\infty} \left(\frac{|x|}{|z|} \right)^n$.

Suppose that $\sum_{n=0}^{\infty} a_n z^n$ is divergent and that there exists an x, $|x| > |z|$, such that $\sum_{n=0}^{\infty} a_n x^n$ is convergent. Then by the first part of the lemma, $\sum_{n=0}^{\infty} a_n y^n$ will be convergent if $|y| < |x|$, contradicting the divergence of the series at z. Therefore, the series is divergent for all x satisfying $|x| > |z|$. □

Definition 4.5.3 The *radius of convergence R of $\sum_{n=0}^{\infty} a_n x^n$* is defined by

$$R = \sup\{|x| \mid \sum_{n=0}^{\infty} a_n x^n \text{ converges}\}.$$

Examples 4.5.4

(1) The exponential series $\sum_{n=0}^{\infty} \frac{x^n}{n!}$ has radius of convergence $R = +\infty$.

(2) Using the ratio test, the series $\sum_{n=1}^{\infty} (-1)^{n+1} \frac{x^n}{n}$ converges if $|x| < 1$ and diverges if $|x| > 1$. Hence the radius of convergence $R = 1$. The series converges if $x = 1$ and diverges if $x = -1$. ♠

Proposition 4.5.5 *Suppose that $\sum_{n=0}^{\infty} a_n x^n$ has radius of convergence $R > 0$. Then $S(x) = \sum_{n=0}^{\infty} a_n x^n$ defines a continuous function on $(-R, R)$.*

Proof Let $a \in (0, R)$ and choose b, $a < b < R$. By Lemma 4.5.1, there exists a $C \geq 0$ such that $|a_n| \leq Cb^{-n}$. Therefore $|a_n a^n| \leq C(\frac{a}{b})^n$, for all $n \geq 0$. Since $0 \leq b/a < 1$, $\sum_{n=0}^{\infty} C(\frac{a}{b})^n < \infty$. Take $M_n = C(\frac{a}{b})^n$. Then $|a_n x^n| \leq M_n$ for all $x \in [-a, a]$ and so, by the M-test, $\sum_{n=0}^{\infty} a_n x^n$ is uniformly convergent on $[-a, a]$. Since the functions $a_n x^n$ are continuous, it follows that $\sum_{n=0}^{\infty} a_n x^n$ converges to a continuous function on $[-a, a]$. This holds for all $a \in (-R, R)$, and so $\sum_{n=0}^{\infty} a_n x^n$ is continuous on $(-R, R)$ (see Theorem 4.4.4).   □

Example 4.5.6 The series $\sum_{n=1}^{\infty} (-1)^{n+1} \frac{x^n}{n}$ defines a continuous function on $(-1, 1)$ since the radius of convergence is 1 by Examples 4.5.4(2). This series converges at $x = 1$ and we shall show in the next section that the series converges uniformly on $[0, 1]$ (this uses Abel's test for uniformly convergent series).   ♠

Using Cauchy's test we can give an explicit formula for the radius of convergence of a power series.

Proposition 4.5.7 *The radius of convergence of $\sum_{n=0}^{\infty} a_n x^n$ is equal to* $1/(\limsup |a_n|^{1/n})$.

Proof Set $\ell = \limsup |a_n|^{1/n}$. Then $\limsup |a_n x^n|^{1/n} = \ell|x|$, and so $\sum_{n=0}^{\infty} a_n x^n$ converges if $|x|\ell < 1$ and diverges if $|x|\ell > 1$ by Cauchy's test. Hence $R = 1/(\limsup |a_n|^{1/n})$.   □

4.5.1   Sums, Products and Quotients of Power Series

We define the sum $\sum_{n=0}^{\infty} a_n x^n + \sum_{n=0}^{\infty} b_n x^n$ of two power series to be the power series

$$\sum_{n=0}^{\infty} (a_n + b_n)x^n.$$

We may similarly define the difference of two power series.

Proposition 4.5.8 *If $\sum_{n=0}^{\infty} a_n x^n$ has radius of convergence R and $\sum_{n=0}^{\infty} b_n x^n$ has radius convergence S, then the radius of convergence of $\sum_{n=0}^{\infty} (a_n \pm b_n)x^n$ is at least* $\min\{R, S\}$.

Proof We may assume $\min\{R, S\} > 0$, else the result is trivial. Let $0 < |x| < \min\{R, S\}$, then $\sum_{n=0}^{\infty} a_n x^n$ and $\sum_{n=0}^{\infty} b_n x^n$ both converge and so $\sum_{n=0}^{\infty} (a_n \pm b_n)x^n$ converges by standard properties of convergent series. Since this is so for all $|x| < \min\{R, S\}$, the radius of convergence of $\sum_{n=0}^{\infty} (a_n \pm b_n)x^n$ is at least $\min\{R, S\}$.   □

Definition 4.5.9 The *product* $(\sum_{n=0}^{\infty} a_n x^n)(\sum_{n=0}^{\infty} b_n x^n)$ of two power series is defined to be the power series $\sum_{n=0}^{\infty} c_n x^n$, where

$$c_n = \sum_{j=0}^{n} a_j b_{n-j}, \ n \geq 0.$$

Proposition 4.5.10 *If the power series $\sum_{n=0}^{\infty} a_n x^n$ and $\sum_{n=0}^{\infty} b_n x^n$ have radii of convergence R and S respectively, then the product of the series has radius of convergence at least $\min\{R, S\}$. In particular, if we set $f(x) = \sum_{n=0}^{\infty} a_n x^n$, $x \in (-R, R)$, and $g(x) = \sum_{n=0}^{\infty} b_n x^n$, $x \in (-S, S)$, then*

$$f(x)g(x) = \sum_{n=0}^{\infty} c_n x^n, \quad x \in (-\min\{R, S\}, +\min\{R, S\}).$$

Proof We assume $\min\{R, S\} > 0$, otherwise the result is trivial. Fix $0 < r < \min\{R, S\}$ and choose $s \in (r, \min\{R, S\})$. By Lemma 4.5.1, there exists a $C > 0$ such that $|a_n|, |b_n| \leq Cs^{-n}$ for all $n \geq 0$. It follows that $|c_n| \leq \sum_{j=0}^{n} |a_j||b_{n-j}| \leq C^2 s^{-n}(n + 1)$, $n \geq 0$. Hence $|c_n x^n| \leq C^2 (n + 1)(\frac{r}{s})^n = M_n$, if $|x| \leq r$. Since $\sum M_n < \infty$, it follows by the M-test that $\sum_{n=0}^{\infty} c_n x^n$ converges uniformly on $[-r, r]$. This holds for all $0 < r < \min\{R, S\}$ and so the radius of convergence of the product is at least $\min\{R, S\}$. □

Let $\sum_{n=0}^{\infty} a_n x^n$ have radius of convergence $R > 0$ and let $f(x) = \sum_{n=0}^{\infty} a_n x^n$, $x \in (-R, R)$. We recall that f is continuous on $(-R, R)$ by Proposition 4.5.5. Suppose that $f(0) \neq 0$—that is, $a_0 \neq 0$. Since f is continuous, f will be non-vanishing near $x = 0$. We define the sequence $(d_n) \subset \mathbb{R}$ recursively by $d_0 = a_0^{-1}$ and

$$d_n = -\frac{1}{a_0} \left(\sum_{j=0}^{n-1} a_{n-j} d_j \right), \quad n \geq 1.$$

We refer to $\sum_{n=0}^{\infty} d_n x^n$ as the *reciprocal* power series $\sum_{n=0}^{\infty} a_n x^n$.

Proposition 4.5.11 (Notation and Assumptions as Above) *If we denote the radius of convergence of $\sum_{n=0}^{\infty} d_n x^n$ by S, then $S > 0$ and*

$$\sum_{n=0}^{\infty} d_n x^n = 1/f(x), \quad x \in (-\min\{R, S\}, +\min\{R, S\}).$$

Proof If $\sum_{n=0}^{\infty} d_n x^n$ has non-zero radius of convergence S, then the radius of convergence of $(\sum_{n=0}^{\infty} d_n x^n)(\sum_{n=0}^{\infty} a_n x^n)$ is at least $r = \min\{R, S\}$ by Proposition 4.5.10. Since $\sum_{j=0}^{n} a_j d_{n-j} = 0$, $n > 0$, and $a_0 d_0 = 1$, the product is identically one on $(-r, r)$.

Let $r \in (-R, R)$. By Lemma 4.5.1, there exists a $C = C(r) > 0$ such that $|a_n| \leq Cr^{-n}$, $n \geq 0$. Define

$$s = \frac{r|a_0|}{C + |a_0|}.$$

Observe that $s \in (0, r)$ and for future reference note that

$$\frac{C}{|a_0|} \frac{s}{r - s} = 1. \tag{4.2}$$

Set $D = |a_0|^{-1}$. We claim that $|d_n| \le Ds^{-n}$, all $n \ge 0$. This shows that the radius of convergence of $\sum_{n=0}^{\infty} d_n x^n$ is at least $s > 0$. Our proof is by induction. The result is trivial if $n = 0$. Suppose we have proved the estimate for $j = 0, \cdots, n - 1$. Since $d_n = -\frac{1}{a_0} \left(\sum_{j=1}^{n} a_j d_{n-j} \right)$, we have

$$|d_n| \le \frac{DC}{|a_0|} \left(\sum_{j=1}^{n} r^{-j} s^{-n+j} \right)$$

$$= \frac{DC}{|a_0|} s^{-n} \sum_{j=1}^{n} \left(\frac{s}{r} \right)^j$$

$$= \frac{DC}{|a_0|} s^{-n} \left(\frac{s}{r} \right) \sum_{j=0}^{n-1} \left(\frac{s}{r} \right)^j$$

$$\le \frac{DC}{|a_0|} s^{-n} \left(\frac{s}{r} \right) \frac{1}{1 - s/r}$$

$$= \frac{DC}{|a_0|} s^{-n} \frac{s}{r - s}$$

$$= Ds^{-n},$$

where the last statement follows by (4.2). Since $D = |a_0|^{-1}$ works for $|d_0| = D$, the induction shows that this value of D works for all $n \ge 0$, granted our choice of s. $\square$

Suppose that $\sum_{n=0}^{\infty} a_n x^n$, $\sum_{n=0}^{\infty} b_n x^n$ are power series with non-zero radius of convergence R and S, respectively. Let $f(x) = \sum_{n=0}^{\infty} a_n x^n$, $x \in (-R, R)$, and $g(x) = \sum_{n=0}^{\infty} b_n x^n$, $x \in (-S, S)$ be the continuous functions defined by the power series. If $g(0) \ne 0$, then it follows from propositions 4.5.10, 4.5.11 that there exists an $s > 0$ such that the quotient $f(x)/g(x)$ has a power series representation on $(-s, s)$.

Let $\mathbb{R}\{x\}$ denote the set of all power series with strictly positive radius of convergence. Our results show that $\mathbb{R}\{x\}$ is closed under addition and multiplication ($\mathbb{R}\{x\}$ is a *ring*). Let $\mathbb{R}^*\{x\} \subset \mathbb{R}\{x\}$ be the set of all power series with non-vanishing constant coefficient. If $u \in \mathbb{R}^*\{x\}$, then the reciprocal $u^{-1} \in \mathbb{R}^*\{x\}$ and $\mathbb{R}^*\{x\}$ is closed under multiplication and division ($\mathbb{R}^*\{x\}$ is a group—the group of units of $\mathbb{R}\{x\}$). In Chap. 5, we shall use these properties of power series as part of a study of *real analytic functions*—functions which have a power series representation at every point in their domain.

EXERCISES 4.5.12

(1) Find the radius of convergence of $\sum_{n=0}^{\infty} \frac{n!(2n)!}{(3n)!} x^{2n}$.

(2) Find the radius of convergence of $\sum_{n=0}^{\infty} \frac{(2n)!(3n)!}{(5n)!} x^n$. More generally, suppose $p+q = N$, where $p, q \in \mathbb{N}$. Find the radius of convergence of $\sum_{n=0}^{\infty} \frac{(pn)!(qn)!}{(Nn)!} x^n$. How would the result change if you replaced x^n by x^{rn} for some fixed $r \in \mathbb{N}$?

(3) Find the radius of convergence of $\sum_{n=1}^{\infty} n^n x^{n^2}$. (Hint: $R \neq 0$.)

(4) For $\alpha \in \mathbb{R}$, let $\binom{\alpha}{n}$ denote the generalized binomial coefficient $\frac{1}{n!} \prod_{j=0}^{n-1} (\alpha - j)$. Find the radius of convergence of the binomial series $\sum_{n=0}^{\infty} \binom{\alpha}{n} x^n$. (A special argument is needed if $\alpha \in \mathbb{Z}_+$.) Using Taylor's theorem with integral remainder (Theorem 2.7.7), show that $\sum_{n=0}^{\infty} \binom{\alpha}{n} x^n = (1+x)^{\alpha}$ if $|x| < 1$.

(5) In Proposition 4.5.10, it is stated that the radius of convergence is at least $\min\{R, S\}$. Show by means of an example that the radius of convergence may be strictly greater than $\min\{R, S\}$.

(6) Use Proposition 4.5.11 to find a power series for $(1 + x^2)^{-1}$. What is the radius of convergence?

(7) Find the first four terms in the power series expansion of $(2 + x + x^2)^{-1}$. Using the result of (1), find the first four terms in the power series expansion of $(1 + x^2)^{-1}(2 + x + x^2)^{-1}$.

(8) Using Proposition 4.5.10, prove that $e^x e^y = e^{x+y}$, where e^x is assumed to be defined by the power series $\sum_{n=0}^{\infty} \frac{x^n}{n!}$.

4.6 Abel and Dirichlet's Test for Uniform Convergence

Proposition 4.6.1 (Abel's Test for Uniform Convergence) *Given sequences* (a_n), (u_n) *of functions defined on* $I \subset \mathbb{R}$, *the series* $\sum_{n=1}^{\infty} a_n u_n$ *is uniformly convergent on* I *if*

(1) *The series* $\sum_{n=1}^{\infty} a_n$ *is uniformly convergent on* I.
(2) $\exists K \geq 0$ *such that* $\|u_n\| \leq K$, *for all* $n \in \mathbb{N}$.
(3) $(u_n(x))$ *is either decreasing for all* $x \in I$ *or increasing for all* $x \in I$.

In particular, if a_n, u_n *are continuous,* $n \geq 1$, *then* $U = \sum_{n=1}^{\infty} a_n u_n$ *is continuous on* I.

Proof The proof is obtained by using the argument of the proof of Abel's test for infinite series. We leave the proof to the exercises (see also the proof of the Dirichlet test for uniform convergence which we give below). □

Proposition 4.6.2 (Dirichlet's Test for Uniform Convergence) *Given sequences* (a_n), (u_n) *of functions defined on* $I \subset \mathbb{R}$, *the series* $\sum_{n=1}^{\infty} a_n u_n$ *is uniformly convergent on* I *if*

(1) $\exists K \geq 0$ *such that* $\|a_1 + \cdots + a_n\| \leq K$ *for all* $n \geq 1$.
(2) $(u_n(x))$ *is decreasing for all* $x \in I$.
(3) (u_n) *is uniformly convergent to the zero function on* I.

In particular, if a_n, u_n are continuous, $n \geq 1$, then $S = \sum_{n=1}^{\infty} a_n u_n$ is continuous on I.

Proof We apply the argument of the proof of Dirichlet's test for infinite series pointwise to the infinite series $\sum_{n=1}^{\infty} a_n u_n$. Thus, using Abel's lemma, we have the estimate

$$\left| \sum_{j=m}^{n} a_j(x) u_j(x) \right| \leq 2K u_m(x), \text{ for all } n \geq m \geq 1, \, x \in I.$$

Hence $|\sum_{j=m}^{n} a_j(x) u_j(x)| \leq 2K \|u_m\|$, for all $n \geq m \geq 1$, $x \in I$, and so

$$\left\| \sum_{j=m}^{n} a_j u_j \right\| \leq 2K \|u_m\|, \text{ for all } n \geq m \geq 1.$$

Given $\varepsilon > 0$, choose $N \in \mathbb{N}$ such that $\|u_m\| < \varepsilon/2K$, all $m \geq N$. We have

$$\left\| \sum_{j=m}^{n} a_j u_j \right\| \leq \varepsilon, \text{ for all } n \geq m \geq N.$$

It follows from the general principle of uniform convergence (Theorem 4.4.6) that $\sum_{n=1}^{\infty} a_n u_n$ is uniformly convergent on I.   □

Examples 4.6.3

(1) $\sum_{n=1}^{\infty} \frac{(-1)^{n+1} x^n}{n}$ is uniformly convergent on $[0, 1]$. In particular, $\lim_{x \to 1-} \sum_{n=1}^{\infty} \frac{(-1)^{n+1} x^n}{n} = \sum_{n=1}^{\infty} \frac{(-1)^{n+1}}{n}$. This is an application of Abel's test with $a_n = (-1)^{n+1}/n$ and $u_n(x) = x^n$.

(2) We claim that for all $\varepsilon \in (0, \pi)$ and $m \in \mathbb{Z}$, the series $\sum_{n=1}^{\infty} \frac{\cos(nx)}{n}$ is uniformly convergent on $[2m\pi + \varepsilon, 2(m+1)\pi - \varepsilon]$. In particular, $\sum_{n=1}^{\infty} \frac{\cos(nx)}{n}$ defines a continuous function on $(2m\pi, 2(m+1)\pi)$ for all $m \in \mathbb{Z}$. Similar results hold for $\sum_{n=1}^{\infty} \frac{\sin(nx)}{n}$. To prove the claim, suppose that x is not an integer multiple of 2π. We have

$$\sum_{k=1}^{n} \cos(kx) = \frac{\cos\left(\frac{(n+1)x}{2}\right) \sin\left(\frac{nx}{2}\right)}{\sin(\frac{x}{2})}.$$

If we suppose $x \in [2m\pi + \varepsilon, 2(m+1)\pi - \varepsilon]$, where $\varepsilon \in (0, \pi)$ and $m \in \mathbb{Z}$, then the minimum value of $|\sin(\frac{x}{2})|$ is taken at $x = 2m\pi + \varepsilon$ (or $2(m+1)\pi - \varepsilon$).

Hence we have the estimate

$$\left|\sum_{k=1}^{n} \cos(kx)\right| \le 1/|\sin(\tfrac{\varepsilon}{2})|, \text{ for all } x \in [2m\pi + \varepsilon, 2(m+1)\pi - \varepsilon].$$

We now apply Dirichlet's test with $K = 1/|\sin(\tfrac{\varepsilon}{2})|$ and $u_n = 1/n$. ♠

Examples 4.6.3(1) is a special case of Abel's theorem, which we now state and prove.

Theorem 4.6.4 (Abel's Theorem) *If the infinite series $\sum_{n=0}^{\infty} a_n$ is convergent, then $\sum_{n=0}^{\infty} a_n x^n$ is uniformly convergent on $[0, 1]$ and*

$$\lim_{x \to 1-} \sum_{n=0}^{\infty} a_n x^n = \sum_{n=0}^{\infty} a_n.$$

Proof The method is the same as that used for Examples 4.6.3(1) and depends on Abel's test. □

EXERCISES 4.6.5

(1) Show that if $\sum_{n=1}^{\infty} a_n$ converges then

  (a) $\sum_{n=1}^{\infty} a_n \frac{x^n}{1+x^n}$ converges uniformly on $[0, 1]$,
  (b) $\sum_{n=1}^{\infty} a_n \frac{nx^n(1-x)}{1-x^n}$ converges uniformly on $[0, 1]$
    (for (b), the x-dependent terms are defined to be equal to 1 at $x = 1$).
    Deduce that these infinite series define continuous functions on $[0, 1]$.

(2) Show that if the partial sums of $\sum_{n=1}^{\infty} a_n$ are bounded then $\sum_{n=1}^{\infty} \frac{a_n}{n^x}$ defines a continuous function on $(0, \infty)$.

(3) Show that if $\sum_{n=1}^{\infty} a_n$ converges, then $\sum_{n=1}^{\infty} e^{-nx^2} a_n$ is uniformly convergent on $\mathbb{R}$. Why is the result easier if $a_n \ge 0$ for all $n \in \mathbb{N}$?

(4) Show that if $\sum_{n=1}^{\infty} a_n$ is convergent then $\sum_{n=1}^{\infty} \frac{a_n}{n^x}$ defines a continuous function on $[0, \infty)$.

(5) Show that $\sum_{n=1}^{\infty} \frac{\sin(nx)}{\sqrt{n}}$ defines a continuous function on $(2m\pi, 2(m+1)\pi)$ for all $m \in \mathbb{Z}$.

(6) Show that $\sum_{n=0}^{\infty} \frac{\sin((2n+1)x)}{2n+1}$ defines a continuous function on $(m\pi, (m+1)\pi)$ for all $m \in \mathbb{Z}$.

(7) Show that $\sum_{n=1}^{\infty} (-1)^{n+1} \frac{\cos(nx)}{n}$ defines a continuous function on $((2m - 1)\pi, (2m+1)\pi)$ for all $m \in \mathbb{Z}$. (Hint: use Exercises 3.10.2(1).)

(8) Suppose that (a_n) is a monotone decreasing sequence of positive numbers. Show that $\sum_{n=1}^{\infty} a_n \sin nx$ converges uniformly on $\mathbb{R}$ only if $\lim_{n \to \infty} na_n = 0$. (Hint: take $x = \pi(2p + 1)$ and show that $|\sum_{n=m}^{p} a_n \sin nx| > 0.4 p a_p$ if $p > 2m - 1$. Note that $\sqrt{2}/\pi > 0.4$. It can be shown that the condition is also sufficient.)

4.7   Integrating and Differentiating Term-by-Term

In this section we address the question of when we can interchange the order of integration or differentiation with summation. We start with a special case of our main result on interchanging the order of summation and integration.

Proposition 4.7.1 *Let* (u_n) *be a sequence of continuous functions defined on the closed and bounded interval I and suppose that* $\sum_{n=1}^{\infty} u_n$ *converges uniformly on I to* $U : I \to \mathbb{R}$. *Given* $a \in I$, *we have for all* $x \in I$,

$$\int_a^x U(t)\, dt = \int_a^x \left(\sum_{n=1}^{\infty} u_n(t) \right) dt = \sum_{n=1}^{\infty} \int_a^x u_n(t)\, dt, \tag{4.3}$$

and the series $\sum_{n=1}^{\infty} \int_a^x u_n(t)\, dt$ *is uniformly convergent on I.*

Remark 4.7.2 Since $\sum_{n=1}^{\infty} u_n$ is uniformly convergent on I, and the terms u_n are all continuous, $U = \sum_{n=1}^{\infty} u_n$ is continuous on I (Theorem 4.4.3) and therefore integrable.

Proof of Proposition 4.7.1 Let $a, x \in I$. Since I is an interval, $[a, x] \subset I$ (abusing notation, we allow $x < a$). Set $S_n = \sum_{j=1}^{n} u_j$, $n \geq 1$. In order to prove (4.3), it suffices to show that given $\varepsilon > 0$, we can find $N \in \mathbb{N}$ such that

$$\left| \int_a^x U(t)\, dt - \int_a^x S_n(t)\, dt \right| = \left| \int_a^x U(t)\, dt - \sum_{j=1}^{n} \int_a^x u_j(t)\, dt \right| < \varepsilon,$$

for all $n \geq N$. Since the result is trivial if $a = x$, we may assume $x \neq a$. Denote the length of the interval I by $|I|$. Since $\sum_{n=1}^{\infty} u_n$ is uniformly convergent on I, we can choose $N \geq 1$ such that $\|S_n - U\| < \varepsilon/|I|$ for all $n \geq N$. Integrating from a to x we have, for $n \geq N$,

$$\left| \int_a^x U(t)\, dt - \int_a^x S_n(t)\, dt \right| \leq \left| \int_a^x |U(t) - S_n(t)|\, dt \right|$$

$$\leq \left| \int_a^x \|S_n - U\|\, dt \right|$$

$$= |a - x|\, \|S_n - U\|$$

$$\leq |I|\, \|S_n - U\|$$

$$< \varepsilon.$$

Hence $\int_a^x \left(\sum_{n=1}^{\infty} u_n(t) \right) dt = \sum_{n=1}^{\infty} \int_a^x u_n(t)\, dt$. The uniform convergence of $\sum_{n=1}^{\infty} \int_a^x u_n(t)\, dt$ is immediate from the estimate $| \int_a^x U(t)\, dt - \int_a^x S_n(t)\, dt | \leq |I|\, \|S_n - U\|$, $x \in I$.   □

Remark 4.7.3 Proposition 4.7.1 holds for uniformly convergent sequences of continuous functions—indeed, that is exactly how the proposition was proved. ✸

For applications, we need a stronger version of Proposition 4.7.4 that applies when I is not necessarily closed or bounded.

Theorem 4.7.4 *Given a sequence (u_n) of continuous functions defined on the interval $I \subset \mathbb{R}$, suppose that $\sum_{n=1}^{\infty} u_n$ converges uniformly on closed and bounded subintervals of I to $U : I \to \mathbb{R}$. Given $a \in I$, we have for all $x \in I$,*

$$\int_a^x U(t)\, dt = \int_a^x \left(\sum_{n=1}^{\infty} u_n(t) \right) dt = \sum_{n=1}^{\infty} \int_a^x u_n(t)\, dt,$$

and the series $\sum_{n=1}^{\infty} \int_a^x u_n(t)\, dt$ is uniformly convergent on every closed and bounded subinterval of I.

Proof Since $\sum_{n=1}^{\infty} u_n$ is uniformly convergent on all closed and bounded subintervals of I and the terms u_n are all continuous, $U = \sum_{n=1}^{\infty} u_n$ is continuous on I (Theorem 4.4.4) and therefore integrable. The first part of the theorem follows from Proposition 4.7.1 with $I = [a, x]$. For the uniform convergence statement observe that every closed and bounded subinterval J of I is contained in an interval $[x, a] \cup [a, y]$ where $x \le a \le y$. By Proposition 4.7.1, we have uniform convergence on $[x, a]$ and $[a, y]$ and hence on $[x, a] \cup [a, y]$ and therefore on J. □

Examples 4.7.5

(1) We have $(1 + x)^{-1} = \sum_{n=0}^{\infty} (-1)^n x^n$, $x \in (-1, 1)$. Convergence is uniform on every closed subinterval $[-x, x] \subset (-1, 1)$. Take $a = 0$ in the statement of Theorem 4.7.4. We have $U(x) = 1/(1 + x)$, $x \in (-1, 1)$, and

$$\log(1 + x) = \int_0^x \frac{1}{1+t}\, dt = \int_0^x \left(\sum_{n=0}^{\infty} (-1)^n t^n \right) dt$$

$$= \sum_{n=0}^{\infty} \int_0^x (-1)^n t^n\, dt$$

$$= \sum_{n=0}^{\infty} (-1)^n \frac{x^{n+1}}{n}.$$

It follows from Abel's theorem (see also Examples 4.6.3(1)) that convergence of $\sum_{n=0}^{\infty} (-1)^n \frac{x^{n+1}}{n}$ is uniform on $[0, 1]$ and so $\log(1 + x) = \sum_{n=0}^{\infty} (-1)^n \frac{x^{n+1}}{n}$ for $x \in (-1, 1]$. Taking $x = 1$ (strictly, taking the limit of both sides as $x \to 1-$) we get the series formula for $\log 2$. (Note that the power series defines a continuous function on $(-1, 1]$ while $\log(1 + x)$ is continuous for all $x > -1$.)

(2) We have $(1 + x^2)^{-1} = \sum_{n=0}^{\infty}(-1)^n x^{2n}$ on $(-1, 1)$ and convergence is uniform on $[-x, x]$, for all $x \in [0, 1)$. Applying Theorem 4.7.4, we have for $x \in (-1, 1)$,

$$\tan^{-1}(x) = \sum_{n=0}^{\infty}(-1)^n \frac{x^{2n+1}}{2n + 1}.$$

It follows from Abel's theorem that convergence of $\sum_{n=0}^{\infty}(-1)^n \frac{x^{2n+1}}{2n+1}$ is uniform on $[0, 1]$ and so, as in the previous example, we may take $x = 1$ to get

$$\frac{\pi}{4} = \tan^{-1}(1) = \sum_{n=0}^{\infty}(-1)^n \frac{1}{2n + 1}. \qquad \spadesuit$$

Definition 4.7.6 Let I be an interval. A function $u : I \to \mathbb{R}$ is C^1, or *once continuously differentiable*, if u is continuous, differentiable and $u' : I \to \mathbb{R}$ is continuous. More generally, if $r \in \mathbb{N}$, u is C^r, or *r-times continuously differentiable*, if the first r derivatives of u all exist and are continuous on I. If u is C^r for all $r \in \mathbb{N}$, we say u is C^∞ or *infinitely differentiable*.

Remarks 4.7.7

(1) We allow the interval I to be open, closed, half-open and unbounded. When I is closed, we interpret continuity and differentiability in the usual way. For example, $u : [a, b] \to \mathbb{R}$ is differentiable at $x = a$ if $\lim_{h \to 0+}(u(a+h)-u(a))/h$ exists and the value of the limit is defined to be $u'(a)$.

(2) If $r \in \mathbb{N}$, denote the rth derivative map of u by $u^{(r)}$. If $r = 1, 2$, we write u', u''. Note that if u is C^r, $r > 1$, we require $u^{(r-1)}$ to exist and be continuous and define $u^{(r)}$ to be the derivative of $u^{(r-1)}$.   ✠

Theorem 4.7.8 *Given a sequence (u_n) of C^1 functions defined on the interval $I \subset \mathbb{R}$, suppose that*

(a) *$\sum_{n=1}^{\infty} u'_n$ is uniformly convergent on all closed and bounded subintervals of I.*
(b) *There exists an $a \in I$ such that $\sum_{n=1}^{\infty} u_n(a)$ is convergent.*

Then $\sum_{n=1}^{\infty} u_n$ converges pointwise on I to a C^1 function $U : I \to \mathbb{R}$ and

$$U' = \sum_{n=1}^{\infty} u'_n.$$

The convergence of $\sum_{n=1}^{\infty} u_n$ is uniform on all closed and bounded subintervals of I.

Remark 4.7.9 Condition (b) is clearly necessary. For example, take $u_n \equiv 1$ for all $n \in \mathbb{N}$.   ✠

Proof of Theorem 4.7.8 Since u_n is C^1, u'_n is continuous for all $n \in \mathbb{N}$. Therefore $V = \sum_{n=1}^{\infty} u'_n$ defines a continuous function on I. Apply Theorem 4.7.4 to get for all $x \in I$

$$\int_a^x V(t)\,dt = \int_a^x \sum_{n=1}^{\infty} u'_n(t)\,dt$$

$$= \sum_{n=1}^{\infty} \int_a^x u'_n(t)\,dt$$

$$= \sum_{n=1}^{\infty} (u_n(x) - u_n(a))$$

$$= \sum_{n=1}^{\infty} u_n(x) - \sum_{n=1}^{\infty} u_n(a),$$

where the final step follows by condition (b) and the convergence of $\sum_{n=1}^{\infty}(u_n(x) - u_n(a))$. Hence $\sum_{n=1}^{\infty} u_n$ converges pointwise on I and $\sum_{n=1}^{\infty} u_n(x)$ converges uniformly on all closed and bounded subintervals of I by Theorem 4.7.4. Our argument shows that for all $x \in I$ we have

$$\sum_{n=1}^{\infty} u_n(x) = \sum_{n=1}^{\infty} u_n(a) + \int_a^x V(t)\,dt.$$

Since V is continuous, it follows by the fundamental theorem of calculus that the right-hand side of this equation is differentiable at x with derivative $V(x)$. Hence

$$\left(\sum_{n=1}^{\infty} u_n \right)'(x) = V(x) = \sum_{n=1}^{\infty} u'_n(x),$$

and so we obtain the derivative by term-by-term differentiation. $\qquad \square$

Remark 4.7.10 We leave the C^r version of Theorem 4.7.8 to the exercises. ✲

We end this section with an important application of our results on term-by-integration and differentiation to power series.

Theorem 4.7.11 *Suppose that the power series $U(x) = \sum_{n=0}^{\infty} a_n x^n$ has radius of convergence $R > 0$ (as usual we allow $R = +\infty$). Then U is a C^∞-function on $(-R, R)$ and the derivatives and definite integrals of U may be computed by term-by-term differentiation and integration. Furthermore, the power series giving the derivatives and integrals of U all have radius of convergence R.*

Proof It suffices to show that $\sum_{n=0}^{\infty} a_n x^n$ has radius of convergence R iff $\sum_{n=1}^{\infty} n a_n x^{n-1}$ has radius of convergence R (note that the first series is obtained, up to a constant, by term-by-term integration of the second series). We may prove this either by estimating $|a_n|$ along the lines of the proof of Lemma 4.5.2 or, more simply, by using the root test: $R^{-1} = \limsup |a_n|^{1/n} = \limsup |na_n|^{1/n}$ since $\lim_{n \to \infty} n^{1/n} = 1$.   □

EXERCISES 4.7.12

(1) We gave results on term-by-term differentiation and integration for infinite series. State and prove the corresponding results for sequences of functions.
(2) True or False? If true, explain why; if false give a counterexample.

    (a) Suppose that the sequence (u_n) consists of functions which are C^1 on $\mathbb{R}$. If $\sum_{n=1}^{\infty} u_n'(x)$ is uniformly convergent on $\mathbb{R}$, then $\sum_{n=1}^{\infty} u_n(x)$ converges to a C^1 function on $\mathbb{R}$.

    (b) If $\sum_{n=0}^{\infty} a_n x^n$ has radius of convergence $R = 1$, then $\sum_{n=0}^{\infty} \frac{a_n x^n}{n+1}$ converges when $x = 1$.

(3) Suppose that the sequence (u_n) consists of functions which are C^r on the interval I, where $1 < r < \infty$. Show that if

    (a) $\sum_{n=1}^{\infty} u_n^{(r)}$ is uniformly convergent on all closed and bounded subintervals of I,

    (b) there exists an $a \in I$ such that $\sum_{n=1}^{\infty} u_n^{(s)}(a)$ is convergent for $0 \leq s < r$,

then $\sum_{n=1}^{\infty} u_n$ converges to a C^r-function V on I and

$$V^{(s)} = \sum_{n=1}^{\infty} u_n^{(s)}, \quad 0 < s \leq r.$$

(4) Find an explicit example of a power series $U_0(x) = \sum_{m=0}^{\infty} a_n x^n$ with radius of convergence 1 such that if we define the power series (U_m) inductively by $U_{m+1}(x) = \int_0^x U_m(t)\, dt$, $m \geq 0$, then the power series U_m diverges at $x = 1$ for all $m \in \mathbb{Z}_+$.
(5) Let $(I_n)_{n=1}^{\infty}$ be a sequence of non-empty, mutually disjoint open intervals and (α_n) be a sequence of strictly positive real numbers. Suppose that (f_n) is a sequence of positive continuous functions on $\mathbb{R}$ such that for $n \geq 1$, (a) f_n is non-zero precisely on I_n and (b) the maximum value of f_n is α_n.

    (a) Is $\sum_{n=1}^{\infty} f_n$ always *pointwise* convergent on $\mathbb{R}$?
    (b) Find a necessary and sufficient condition on the sequence (α_n) that allows the M-test to be applied to $\sum_{n=1}^{\infty} f_n$.
    (c) Is it true that if the M-test does not apply, then $\sum_{n=1}^{\infty} f_n$ cannot be uniformly convergent on $\mathbb{R}$? If false, provide an example.
    (d) Denote the length of the interval I_n by ℓ_n. Show that if there exists an $A > 0$ such that $\ell_n \geq A$ for all $n \geq 1$, then $\sum_{n=1}^{\infty} f_n$ is always continuous on $\mathbb{R}$.

(6) Let $u_n(x) = \frac{x^n}{2^n n^3}$, $n \geq 1$.

    (a) Find the radius of convergence R of $\sum_{n=1}^{\infty} u_n$.

    (b) Is the convergence of $\sum_{n=1}^{\infty} u_n$ uniform on $[-R, R]$? Why?/Why not?

    (c) Is $u = \sum_{n=1}^{\infty} u_n$ differentiable on $(-R, R)$? If yes, what is the derivative and does the limit $\lim_{x \to R-} u'(x)$ exist?

(7) Show that the series $U(x) = \sum_{n=1}^{\infty} \frac{1}{n^3 + n^4 x^2}$ defines a C^1 function on $\mathbb{R}$ and find a series representation for $U'(x)$.

(8) Show that $V(x) = \sum_{n=1}^{\infty} \frac{1}{n^2 + n^4 x^2}$ defines a continuous C^1 function on $x \neq 0$ but V is not differentiable at $x = 0$. (Hints for the second part: observe that V is an even function and so if V is differentiable at $x = 0$, we must have $V'(0) = 0$. Now show $\lim_{x \to 0+} (V(x) - V(0))/x$ exists and is non-zero—you might find the estimates of Cauchy's integral test useful.)

(9) For $x > 1$, let $\zeta(x) = \sum_{n=1}^{\infty} \frac{1}{n^x}$. Prove that ζ is C^∞ and find an infinite series for $\zeta^{(n)}(x)$, $n \in \mathbb{N}$.

(10) Define $F(x) = \sum_{n=1}^{\infty} \frac{\sin nx}{n^n}$, $x \in \mathbb{R}$. Prove that F defines a C^∞-function on $\mathbb{R}$. What can we say about $G(x) = \sum_{n=1}^{\infty} \frac{\sin nx}{n^p}$ if p is an integer strictly greater than 1?

(11) Let $(I_n)_{n=1}^{\infty}$ be a sequence of non-empty, mutually disjoint open intervals (for example: $I_n = (n, n+1)$ or $I_n = (\frac{1}{n+1}, \frac{1}{n})$, $n \geq 1$). Suppose that (f_n) is a sequence of positive continuous functions on $\mathbb{R}$ such that for $n \geq 1$, (a) f_n is non-zero precisely on I_n and (b) the maximum value of f_n is 1.

    (a) Show that $\sum_{n=1}^{\infty} f_n$ is *pointwise* convergent on $\mathbb{R}$.

    (b) Show that the M-test can never be applied to $\sum_{n=1}^{\infty} f_n$.

    (c) Show that $\sum_{n=1}^{\infty} f_n$ is never uniformly convergent but there exist choices of (I_n) for which $\sum_{n=1}^{\infty} f_n$ is *always* continuous on $\mathbb{R}$ (you choose the (I_n); the (f_n) satisfy conditions (a,b) listed above).

    (d) Find a choice of (I_n) for which $\sum_{n=1}^{\infty} f_n$ *never* converges to a continuous function on $\mathbb{R}$ (you choose the (I_n); the (f_n) satisfy conditions (a,b) listed above).

4.8   A Continuous Nowhere Differentiable Function

If we consider the infinite series $\sum_{n=1}^{\infty} \frac{\sin nx}{n}$ we can show, using Dirichlet's test, that the series defines a function $U : (-\pi, \pi) \to \mathbb{R}$ which is continuous except at $x = 0$. Term-by-term differentiation of the series for U leads to the infinite series $\sum_{n=1}^{\infty} \cos(nx)$ which, using the partial sum formula, can be shown to diverge at every point of $(-\pi, \pi)$. It is natural to guess that U might not be differentiable on $(-\pi, \pi)$. However, as we see later in Chap. 5, this is *false*. Indeed, U is infinitely differentiable on $(-\pi, \pi)$ except at $x = 0$!

Arising out his work on the zeta-function $\zeta(z) = \sum_{n=1}^{\infty} n^{-z}$, Riemann suggested in 1861 that the continuous function $\sum_{n=1}^{\infty} \sin(n^2 x)/n^2$ might be nowhere differentiable. While this turned out not (quite) to be the case,[1] Riemann's question prompted further work by Weierstrass who investigated the function defined by the series $\sum_{n=0}^{\infty} a^n \sin(b^n \pi x)$, where $0 < a$ and $b > 1$ is an odd positive integer. Since $|a^n \sin(b^n \pi x)| \leq a^n$, for all $x \in \mathbb{R}$, it follows from the M-test that $\sum_{n=0}^{\infty} a^n \sin(b^n \pi x)$ converges uniformly to a continuous function U on $\mathbb{R}$. If we differentiate term-by-term, we obtain the infinite series $\pi \sum_{n=0}^{\infty} (ab)^n \cos(b^n \pi x)$. If $ab > 1$, it again looks unlikely that the series converges pointwise on $\mathbb{R}$, let alone uniformly. This *suggests* that U may not be differentiable anywhere on $\mathbb{R}$, at least if ab is large enough. Weierstrass showed in 1872 that if $ab > 1 + \frac{3}{2}\pi \approx 5.7$, then U is indeed nowhere differentiable on $\mathbb{R}^2$ Although Weierstrass' proof is not hard, we prefer a simpler example, due to van der Waerden (1930), of a nowhere differentiable continuous function. Like Weierstrass' example, van der Waerden's function is defined using a uniformly convergent series of continuous functions.

Define

$$u_0(x) = \begin{cases} x, & \text{if } x \in [0, \frac{1}{2}], \\ 1 - x, & \text{if } x \in [\frac{1}{2}, 1]. \end{cases}$$

Extend u_0 to $\mathbb{R}$ as a 1-periodic function. That is, if $x \in [n, n + 1]$, $n \in \mathbb{Z}$, then $u_0(x) = u_0(x - n)$ (note $x - n \in [0, 1]$). For all $m \in \mathbb{Z}$, $x \in \mathbb{R}$ we have

$$u_0(x + m) = u_0(x).$$

We show the graph of u_0 in Fig. 4.2.

For $n \geq 1$, define

$$u_n(x) = \frac{1}{10^n} u_0(10^n x), \; x \in \mathbb{R}.$$

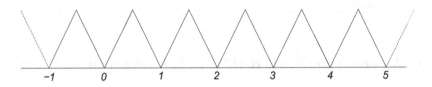

Fig. 4.2 The graph of u_0

[1]More than 100 years later Gerver showed in 1969 that $\sum_{n=1}^{\infty} \sin(n^2 x)/n^2$ is differentiable iff $x = p\pi/q$, where p, q are odd integers.
[2]In 1916, G.H. Hardy improved this result to $ab > 1$.

The function u_n is $\frac{1}{10^n}$-periodic. Indeed, for all $m \in \mathbb{Z}$, $x \in \mathbb{R}$,

$$u_n\left(x + \frac{m}{10^n}\right) = \frac{1}{10^n} u_0\left(10^n\left(x + \frac{m}{10^n}\right)\right)$$

$$= \frac{1}{10^n} u_0(10^n x + m)$$

$$= \frac{1}{10^n} u_0(10^n x)$$

$$= u_n(x).$$

Given $p \in \mathbb{N}$, $p > 1$, define $\frac{1}{p}\mathbb{Z} = \{\frac{m}{p} \mid m \in \mathbb{Z}\}$. For example, $\frac{1}{2}\mathbb{Z} = \{0, \pm\frac{1}{2}, \pm 1, \pm\frac{3}{2}, \cdots\}$. Observe that the set of points where u_0 is not differentiable is precisely $\frac{1}{2}\mathbb{Z}$. Elsewhere the derivative of u_0 is ± 1. The set of points where u_n is not differentiable is $\frac{1}{2\times 10^n}\mathbb{Z} = \{\frac{m}{2\times 10^n} \mid m \in \mathbb{Z}\}$. Elsewhere the derivative of u_n is ± 1.
    Define $U : \mathbb{R} \to \mathbb{R}$ by

$$U(x) = \sum_{n=0}^{\infty} u_n(x).$$

Since $|u_n(x)| \le \frac{1}{10^n}$, it follows by the M-test that $\sum_{n=0}^{\infty} u_n$ is uniformly convergent on $\mathbb{R}$ and U is continuous. We claim that U is nowhere differentiable. We prove the nowhere differentiability of U by showing that for each $x_0 \in \mathbb{R}$, there exists a sequence (x_N) converging to x_0 such that the limit as $N \to \infty$ of $\frac{U(x_N)-U(x_0)}{x_N-x_0}$ does not exist (if U is differentiable at x_0, then $\lim_{N\to\infty} \frac{U(x_N)-U(x_0)}{x_N-x_0} = U'(x_0)$ since $x_N \to x_0$.)
    Let $x_0 \in \mathbb{R}$ and $N \ge 0$. Then there exists a unique $m \in \frac{1}{2}\mathbb{Z}$ such that $x_0 \in [\frac{m}{10^N}, \frac{m+\frac{1}{2}}{10^N})$. The length of the interval $[\frac{m}{10^N}, \frac{m+\frac{1}{2}}{10^N})$ is $\frac{1}{2}\frac{1}{10^N}$. Certainly either $x_0 - \frac{m}{10^N} > 10^{-(N+1)}$ or $\frac{m+\frac{1}{2}}{10^N} - x_0 > 10^{-(N+1)}$ (as $\frac{1}{2}\frac{1}{10^N} > \frac{2}{10^{N+1}}$). Define $x_N = x_0 \pm \frac{1}{10^{N+1}}$ so that $x_N \in [\frac{m}{10^N}, \frac{m+\frac{1}{2}}{10^N}]$. This completes the construction of the sequence (x_N).
    For $n \le N$, we have

$$\frac{u_n(x_N) - u_n(x_0)}{x_N - x_0} = \pm 1.$$

(The set of points where u_n is not differentiable is a proper subset of the set of points where u_N is not differentiable if $n < N$.) On the other hand, if $n > N$ then $u_n(x_N) = u_n(x_0 \pm \frac{1}{10^{N+1}}) = u_n(x_0)$ by the $\frac{1}{10^n}$ periodicity of u_n ($\frac{1}{10^{N+1}}$ is an integer multiple of $\frac{1}{10^n}$ if $n > N$). Hence if $n > N$,

$$\frac{u_n(x_N) - u_n(x_0)}{x_N - x_0} = 0.$$

We have

$$
\frac{U(x_N) - U(x_0)}{x_N - x_0} = \sum_{n=0}^{\infty} \frac{u_n(x_N) - u_n(x_0)}{x_N - x_0}
$$

$$
= \sum_{n=0}^{N} \frac{u_n(x_N) - u_n(x_0)}{x_N - x_0}
$$

$$
= \sum_{n=0}^{N} \pm 1
$$

$$
= Q_N,
$$

where Q_N must be an odd integer if N is even and an even integer if N is odd. Hence the limit of $\frac{U(x_N) - U(x_0)}{x_N - x_0}$ as $N \to \infty$ does not exist and so U cannot be differentiable at x_0.

Remark 4.8.1 Are these examples of nowhere differentiable continuous functions exceptional and pathological? Pathological perhaps, but certainly not exceptional. 'Most', in a sense that can be made precise, continuous functions $f : [a, b] \to \mathbb{R}$ are nowhere differentiable.   ✱

EXERCISES 4.8.2

(1) Let u_0 be the sawtooth function defined in Sect. 4.8. For $n \in \mathbb{Z}_+$, define $v_n(x) = 2^{-2^n} u_0(2^{2^n} x)$. Show that $\sum_{n=0}^{\infty} v_n(x)$ is continuous and nowhere differentiable on $\mathbb{R}$.

(2) In this question, we address the nowhere differentiability of the Weierstrass function. Let $0 < a < b$ and suppose that b is an odd integer. Fix $x_0 \in \mathbb{R}$.

(a) Show that given $m \in \mathbb{N}$, there exists an $N \in \mathbb{Z}$ such that $b^m x_0 - N - 1 \in [-\frac{1}{2}, +\frac{1}{2}]$.

(b) Show that if we set $x_m = N/b^m$ and $n \geq m$, then

$$
\cos(b^n \pi x_m) - \cos(b^n \pi x_0) = (-1)^N (1 + \cos(b^{n-m} \pi (b^m x_0 - N - 1)))
$$
$$
= (-1)^N I(m, n),
$$

where $I(m, m) \geq 1$ and $I(m, n) \geq 0$, all $n > m$.

(c) Show that

$$
\left| \sum_{n=m}^{\infty} a^n \frac{\cos(b^n \pi x_m) - \cos(b^n \pi x_0)}{x_m - x_0} \right| \geq \frac{a^m}{|x_m - x_0|} \geq \frac{2}{3}(ab)^m,
$$

where we have used $|x_m - x_0| \leq 3/(2b^m)$.

(d) Using the mean value theorem, show that for all $n \in \mathbb{Z}_+$,

$$\left| a^n \frac{\cos(b^n \pi x_m) - \cos(b^n \pi x_0)}{x_m - x_0} \right| \le (ab)^n \pi.$$

(e) Show that

$$\left| \sum_{n=0}^{m-1} a^n \frac{\cos(b^n \pi x_m) - \cos(b^n \pi x_0)}{x_m - x_0} \right| < \pi \frac{(ab)^m}{ab - 1}.$$

(f) Show that for all $m \in \mathbb{N}$

$$\left| \sum_{n=0}^{\infty} a^m \frac{\cos(b^n \pi x_m) - \cos(b^n \pi x_0)}{x_m - x_0} \right| > (ab)^m \left(\frac{2}{3} - \frac{\pi}{ab - 1} \right),$$

and hence deduce that $\sum_{n=0}^{\infty} a^n \cos(b^n \pi x)$ cannot be differentiable at x_0 if $ab > 1 + 3\pi/2$.

(3) Let $f : [a, b] \to \mathbb{R}$ be C^{∞}. Show we can construct a sequence $(u_n) \subset C^0([a, b])$ such that (a) (u_n) converges uniformly to f, and (b) for all n, u_n is nowhere differentiable.

(4) Let $f : \mathbb{R} \to \mathbb{R}$ be the continuous 2-periodic function defined by

$$f(x) = \begin{cases} 0, & \text{if } x \in [0, 1/2], \\ 6x - 3, & \text{if } x \in [1/2, 2/3], \\ 1, & \text{if } x \in [2/3, 1], \\ 2 - x, & \text{if } x \in [1, 2]. \end{cases}$$

Define $E = (X, Y) : \mathbb{R} \to \mathbb{R}^2$ by

$$E(t) = \left(\sum_{n=1}^{\infty} 2^{-n} f(3^{2n-1} t), \sum_{n=1}^{\infty} 2^{-n} f(3^{2n} t) \right).$$

(a) Show that E is continuous and E maps the unit interval $[0, 1]$ onto $[0, 1] \times [0, 1]$.
(Hint: Given $(x, y) \in [0, 1] \times [0, 1]$, write x, y in binary form as $x = \sum_{n=1}^{\infty} 2^{-n} a_{2n-1}$, $y = \sum_{n=1}^{\infty} 2^{-n} a_{2n}$, where $a_i \in \{0, 1\}$, $i \ge 1$. Show that if $t = 2 \sum_{i=1}^{\infty} 3^{-1-i} a_i$, then $f(3^k t) = a_k$ and so $E(t) = (x, y)$.)
(b) Show that the result of (a) does not depend on the values of f on $(1, 2)$— subject to $f(1) = 1, f(2) = 0$.
(c) Modifying f on $(1, 2)$ as needed, show that E can be nowhere differentiable.

(This elementary example of a space filling curve was given by I.J. Schoenberg in 1938. No such examples can exist if E is differentiable.)

Chapter 5
Functions

5.1 Introduction

In this chapter we investigate and compare several natural classes of functions that play an important role in analysis. We begin with a general overview and then, in subsequent sections, study specific classes of functions using the tools developed in the previous chapter.

The most regular, and familiar, class of functions on the real line is the space $P(\mathbb{R})$ of *polynomials* on $\mathbb{R}$. Recall that if $p \in P(\mathbb{R})$, then either $p \equiv 0$ or we may write $p(x) = \sum_{j=0}^{n} a_j x^{n-j}$ where $a_0 \neq 0$, n is the *degree* of p, and the expression for p is unique. If $p \in P(\mathbb{R})$ then p is smooth (that is, infinitely differentiable or C^∞) and the derivatives and integrals of p are obtained by term-by-term differentiation and integration of p. At the other extreme we have the space $C^0(\mathbb{R})$ of continuous functions on $\mathbb{R}$. As we indicated at the end of Chap. 4, typical functions in $C^0(\mathbb{R})$ may have unpleasant properties such as nowhere differentiability. We can interpolate between continuous and polynomial functions using spaces of differentiable functions. To this end, if $1 \le r \le \infty$, let $C^r(\mathbb{R})$ denote the space of C^r-functions on $\mathbb{R}$. We have the sequence of strict inclusions

$$C^0(\mathbb{R}) \supset C^1(\mathbb{R}) \supset \cdots \supset C^r(\mathbb{R}) \supset \cdots \supset C^\infty(\mathbb{R}) \supset P(\mathbb{R}).$$

There is another class of functions, intermediate between polynomials and C^∞-functions, that play an important historic role in analysis (especially complex analysis). Recall that if $f \in C^\infty(\mathbb{R})$, then the *Taylor series* Tf_{x_0} of f at $x_0 \in \mathbb{R}$ is defined by

$$Tf_{x_0}(x) = \sum_{n=0}^{\infty} \frac{f^{(n)}(x_0)}{n!} (x - x_0)^n.$$

© Springer International Publishing AG 2017
M. Field, *Essential Real Analysis*, Springer Undergraduate Mathematics Series,
https://doi.org/10.1007/978-3-319-67546-6_5

In general, the Taylor series of f at x_0 may have zero radius of convergence and even if it converges it may not converge to f—we give examples shortly. However, for many classical functions of analysis (such as e^x, $\sin x$, and $\cos x$), the Taylor series at x_0 does converge to f if x is close enough to x_0. We encode this property in a definition.

Definition 5.1.1 A C^∞-function $f : \mathbb{R} \to \mathbb{R}$ is (real) *analytic* if for every $x_0 \in \mathbb{R}$, there exists a $\delta = \delta(x_0) > 0$ such that

$$f(x) = \sum_{n=0}^{\infty} \frac{f^{(n)}(x_0)}{n!}(x - x_0)^n \text{ for all } x \text{ satisfying } |x - x_0| < \delta.$$

Remark 5.1.2 We give the definition of a real analytic function defined on an open interval $(a, b) \subset \mathbb{R}$ in Sect. 5.4. ✠

Let $C^\omega(\mathbb{R})$ denote the space of all real analytic functions on $\mathbb{R}$. Evidently we have

$$C^0(\mathbb{R}) \supset C^\infty(\mathbb{R}) \supset C^\omega(\mathbb{R}) \supset P(\mathbb{R}).$$

We start by developing the theory of C^∞-functions and, in particular, show that a C^∞-function need not be analytic (hence all the inclusions above are strict). Next we show that even though a continuous function f may be nowhere differentiable, we can uniformly approximate f as close as we wish on closed bounded intervals by polynomials (the "Weierstrass approximation theorem"). Next we develop some of the classical theory of analytic functions and show, for example, that e^x, $\sin x$ and $\cos x$ all define analytic functions on $\mathbb{R}$. Finally, we conclude the chapter with two sections on Fourier series—for this we will use many results on pointwise and uniform convergence from the previous chapter as well as a version of the Weierstrass approximation theorem.

5.2 Smooth Functions

We start by constructing a smooth (that is, C^∞) non-analytic function. Specifically, we construct a smooth bounded function $\Phi : \mathbb{R} \to \mathbb{R}$ that is strictly positive on $x > 0$ and zero on $x \leq 0$. Subsequently, we use Φ as a building block for the construction of a wide range of smooth non-analytic functions satisfying various properties and thereby illustrate how to construct a smooth function with specified properties.

The function Φ cannot be built by piecing together simple functions.

Example 5.2.1 Define $F : \mathbb{R} \to \mathbb{R}$ by $F(x) = x^4, x \geq 0$, and $F(x) = 0, x < 0$. The graph of F looks 'smooth' near $x = 0$; see Fig. 5.1. However, although it is easily checked that F is C^3 (we have $F'(0) = F''(0) = F'''(0) = 0$), F is not 4-times differentiable at $x = 0$. Indeed, for $x > 0$, $F'''(x) = 24x$, and if $x < 0$, $F'''(x) = 0$.

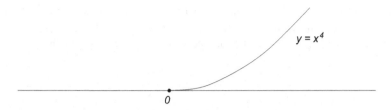

Fig. 5.1 Graph of C^3 but not four times differentiable function

Therefore

$$\lim_{h\to 0+} \frac{Fn'''(h) - F'''(0)}{h} = 24 \neq 0 = \lim_{h\to 0-} \frac{F'''(h) - F'''(0)}{h}$$

and so F''' is not differentiable at $x = 0$. The moral of this example is that if we want to construct a smooth function, we cannot just piece together bits of standard functions like polynomials and trigonometric functions. ♠

Before we construct our example of a smooth non-analytic function we need a technical lemma.

Lemma 5.2.2 *If $q \in P(\mathbb{R})$ and $p \in \mathbb{Z}$, then*

$$\lim_{x\to 0+} \frac{q(x)}{x^p} e^{-\frac{1}{x}} = 0.$$

Proof If q is of degree m, then $q(x) = \sum_{j=0}^{m} b_j x^{m-j}$, and so $x^{-p}q(x) = \sum_{j=0}^{m} b_j x^{m-j-p}$. Since the limit of a finite sum is the sum of the limits of the terms in the sum, it is enough to show that $\lim_{x\to 0+} x^{-k}e^{-\frac{1}{x}} = 0$, for all $k \in \mathbb{Z}$. Setting $y = 1/x$, it suffices to show

$$\lim_{y\to +\infty} y^k e^{-y} = 0,$$

for all $k \in \mathbb{Z}$. A proof of this standard result about the growth of the exponential function is given in Sect. 2.9.3 (Proposition 2.9.9). □

Proposition 5.2.3 *Define $\Phi : \mathbb{R} \to \mathbb{R}$ by*

$$\Phi(x) = \begin{cases} 0, & x \leq 0, \\ e^{-\frac{1}{x}}, & x > 0. \end{cases}$$

Then

(1) $\Phi \in C^\infty(\mathbb{R})$.
(2) $\Phi^{(j)}(0) = 0$, *for all $j \geq 0$.*

Proof It is clear that Φ restricted to either $(-\infty, 0)$ or $(0, \infty)$ is C^∞. We have to show that Φ is infinitely differentiable with all derivatives continuous at $x = 0$. We start by finding expressions for the derivatives of Φ at non-zero points of $\mathbb{R}$. Let $j \geq 1$. We claim that

$$\Phi^{(j)}(x) = \begin{cases} 0, & x < 0, \\ \frac{q_j(x)}{x^{2j}} e^{-\frac{1}{x}}, & x > 0, \end{cases}$$

where q_j is a polynomial in x of degree $j-1$ with constant term $+1$. To see this, note that $\Phi^{(j)}(x) = 0$ if $x < 0$ since Φ vanishes identically on $(-\infty, 0)$. The expression for $x > 0$ is an easy inductive argument that we leave to the reader. We prove that for $j \geq 0$, $\Phi^{(j)}(0)$ exists and is equal to zero and $\Phi^{(j)}$ is continuous at $x = 0$. If $j = 0$, $\Phi^{(0)}(0) = \Phi(0) = 0$ (by definition of Φ) and Φ will be continuous at $x = 0$ since $\lim_{x \to 0+} e^{-\frac{1}{x}} = 0$ by Lemma 5.2.2. Proceeding inductively, suppose that we have shown for $j < n$ that $\Phi^{(j)}(0)$ exists and is equal to zero and that $\Phi^{(j)}$ is continuous at $x = 0$. First we show that $\Phi^{(n-1)}$ is differentiable at $x = 0$ with zero derivative. We have

$$\lim_{x \to 0-} \frac{\Phi^{(n-1)}(x) - \Phi^{(n-1)}(0)}{x} = \frac{0 - 0}{x} = 0.$$

It remains to consider the limit as $x \to 0+$. We have

$$\lim_{x \to 0+} \frac{\Phi^{(n-1)}(x) - \Phi^{(n-1)}(0)}{x} = \lim_{x \to 0+} \frac{\frac{q_{n-1}(x)}{x^{2n-2}} e^{-\frac{1}{x}} - 0}{x}$$

$$= \lim_{x \to 0+} \frac{q_{n-1}(x)}{x^{2n-1}} e^{-\frac{1}{x}}$$

$$= 0,$$

by Lemma 5.2.2, with $m = 2n - 1$. Hence $\Phi^{(n-1)}$ is differentiable at $x = 0$ with zero derivative and so Φ is n times differentiable at $x = 0$ with $\Phi^{(n)}(0) = 0$. To complete the inductive step, we must show that $\Phi^{(n)}$ is continuous at $x = 0$; that is, $\lim_{x \to 0} \Phi^{(n)}(x) = 0$. Obviously, $\lim_{x \to 0-} \Phi^{(n)}(x) = 0$. Since $\Phi^{(n)}(x) = \frac{q_n(x)}{x^{2n}} e^{-\frac{1}{x}}$ if $x > 0$, we have $\lim_{x \to 0+} \Phi^{(n)}(x) = 0$ by Lemma 5.2.2. $\qquad\square$

Example 5.2.4 The C^∞-function $\Phi : \mathbb{R} \to \mathbb{R}$ defined in Proposition 5.2.3 is not analytic. Indeed, Φ is strictly positive on $x > 0$ and so is non-zero on $x > 0$. On the other hand the Taylor series $T\Phi_0$ of Φ at the origin is $\sum_{n=0}^{\infty} \frac{\Phi^{(n)}(0)}{n!} x^n = \sum_{n=0}^{\infty} 0 x^n = 0$. Hence the Taylor series of Φ at the origin does not converge to Φ on any interval $(a, -a), a > 0$, and therefore Φ cannot be analytic. ♠

Remarks 5.2.5

(1) In general, the Taylor series of a smooth function bears little relation to the function. There is a classical result of E. Borel (1895) that shows that given

any sequence $(a_n)_{n\geq 0}$ of real numbers, there exists a C^∞-function $f : \mathbb{R} \to \mathbb{R}$ with Maclaurin series (the Taylor series at zero) given by $\mathcal{T}f_0 = \sum_{n=0}^{\infty} \frac{a_n}{n!}x^n$ (so $f^{(n)}(0) = a_n$). If we choose a rapidly increasing sequence such as $a_n = n^{n^n}$, the radius of convergence of the Maclaurin series will be zero even though f is defined on all of $\mathbb{R}$. See also the exercises at the end of the section.

(2) A necessary condition for $f : \mathbb{R} \to \mathbb{R}$ to be analytic is that $f^{-1}(c)$ must be a countable subset of $\mathbb{R}$ for all $c \in \mathbb{R}$. In particular, if $f^{-1}(c)$ contains an open interval, f cannot be analytic (see Sect. 5.4 for properties of analytic functions). ✠

5.2.1 Constructing Smooth Functions

We use the smooth function Φ constructed in Proposition 5.2.3 as a building block to construct many other smooth non-analytic functions.

Examples 5.2.6

(1) Given $a \in \mathbb{R}$, we construct a smooth function $f : \mathbb{R} \to \mathbb{R}$ such that $f(x) < 0$, for $x < a$ and $f(x) = 0$ for $x \geq a$ by

$$f(x) = -\Phi(a - x).$$

Observe that $f^{(n)}(a) = 0$, $n \in \mathbb{Z}_+$. The obvious variations can be made on this function by considering $\pm\Phi(\pm(x - a))$. More generally, observe that if $g : \mathbb{R} \to \mathbb{R}$ is any smooth function then $f(x) = \Phi(g(x))$ is smooth and

$$f^{-1}(0) = \{x \in \mathbb{R} \mid g(x) \leq 0\}.$$

We have $f^{(n)}(x) = 0$ for all $n \geq 0$ at every point $x \in f^{-1}(0)$. In particular, $f(x) = \Phi(x^2)$ is a smooth positive non-analytic function with zero set $f^{-1}(0) = \{0\}$.

(2) Given $a < b \in \mathbb{R}$, we find a smooth function $\Psi_{a,b} : \mathbb{R} \to \mathbb{R}$ satisfying

$$\Psi_{a,b}(x) = 0, \text{ if } x \notin (a, b),$$

$$\Psi_{a,b}(x) > 0, \text{ if } x \in (a, b).$$

To this end we define

$$\Psi_{a,b}(x) = \Phi(b - x)\Phi(x - a), \ x \in \mathbb{R}.$$

Observe that $\Phi(x - a) = 0$ iff $x \leq a$ and $\Phi(b - x) = 0$ iff $x \geq b$. Hence $\Psi_{a,b}(x) = 0$ iff $x \notin (a, b)$. Since $\Phi(x) > 0$ if $x > 0$, we have $\Psi_{a,b}(x) > 0$ if $x \in (a, b)$. Since $\Psi_{a,b}$ is the product of C^∞-functions, $\Psi_{a,b}$ is C^∞. Note that if

(a)
graph($\Psi_{a,b}$)

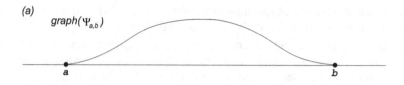

(b)
graph($\Theta_{a,b}$)

Fig. 5.2 Smooth positive bump functions on $\mathbb{R}$. (**a**) Smooth positive bump function which is non-zero on (a, b). (**b**) Tabletop function which is non-zero on $(-b, b)$ and equal to 1 on $[-a, a]$

$z \notin (a, b)$, then $\Psi_{a,b}^{(j)}(z) = 0, j \geq 0$. We show the graph of $\Psi_{a,b}$ in Fig. 5.2a (the graph is symmetric about the mid-point $(a + b)/2$).

(3) Given $a, b \in \mathbb{R}$, with $0 < a < b < \infty$, we construct a smooth function $\Theta_{a,b}$ satisfying

$$\Theta_{a,b}(x) = 0, \quad \text{if } |x| \geq b,$$

$$\Theta_{a,b}(x) = 1, \quad \text{if } |x| \leq a,$$

$$\Theta_{a,b}(x) \in (0, 1), \quad \text{if } |x| \in (a, b).$$

For this we define

$$\Theta_{a,b}(x) = \frac{\Phi(b^2 - x^2)}{\Phi(b^2 - x^2) + \Phi(x^2 - a^2)}, \quad x \in \mathbb{R}.$$

Since $0 < a < b$, the denominator is never zero and so $\Theta_{a,b}$ is well defined and C^∞. If $|x| \geq b$, then the numerator is zero; if $|x| \leq a$, the denominator is equal to the numerator and so $\Theta_{a,b}(x) = 1$. If $|x| \in (a, b)$, then the numerator is strictly less than the denominator and so $\Theta_{a,b}(x) \in (0, 1)$. We remark that all the derivatives of $\Theta_{a,b}$ at x are zero if $|x| \notin (a, b)$. In particular, $\Theta_{a,b}$ is not analytic. We show the graph of $\Theta_{a,b}$ in Fig. 5.2b.   ♠

Remark 5.2.7 The two functions constructed in the previous examples are often called "bump" functions. Granted the map Φ, their construction depends more on simple logic than difficult analysis.   ✠

Examples 5.2.8

(1) We construct a smooth function with zero set equal to $\{0\} \cup \{\pm 1/n \mid n \geq 1\}$. As a first try, we might consider $f(x) = x \sin(\pi/x)$, $x \neq 0$, $f(0) = 0$. This function is continuous and has the specified zero set but it is not differentiable at $x = 0$ as $\lim_{x \to 0}(f(x) - f(0))/x = \lim_{x \to 0} \sin(\pi/x)$, which does not exist. If we instead try $f(x) = x^2 \sin(\pi/x)$, $x \neq 0$, $f(0) = 0$, we find that f is differentiable at $x = 0$ but not C^1. More generally, if we define $f(x) = x^{2n+1} \sin(\pi/x)$, $x \neq 0$, and $f(0) = 0$, then f can be shown to be C^n, $n \geq 1$ (we leave this to the exercises). In order to find a C^∞-function with the correct properties, we try

$$f(x) = \begin{cases} \Phi(x^2) \sin(\pi/x), & x \neq 0, \\ 0, & x = 0. \end{cases}$$

Just as in the proof of Proposition 5.2.3, we may use Lemma 5.2.2 to show that f is C^∞ and all the derivatives of f vanish at zero. In particular, f is not analytic. Notice the way we use Φ to 'smooth' out the irregularities near $x = 0$ of $\sin(\pi/x)$.

(2) We show how to construct a C^∞-function $F : \mathbb{R} \to \mathbb{R}$ satisfying

(a) $F(x) = 2$, $x \leq -2$.
(b) $F(x) \in (0, 2)$ for $x \in (-2, -1)$.
(c) $F(x) = 0$, for $x \in [-1, 0]$.
(d) $F(x) \geq 0$ on $[0, 1]$ and $F(x) = 0$ iff $x = 1/n$ or $1 - 1/n$ for some $n \in \mathbb{N}$.
(e) $F(x) \in (-1, 0)$, for $x \in (1, 5)$.
(f) $F(x) = -1$, for $x \geq 5$.

We express F as a sum of functions $F_1 + F_2 + F_3$, where

$$F_1(x) = 2\frac{\Phi(-x - 1)}{\Phi(-x - 1) + \Phi(x + 2)},$$

$$F_2(x) = \begin{cases} \Phi(x)\Phi(1 - x) \sin^2(\frac{\pi}{x}) \sin^2(\frac{\pi}{1-x}), & x \in (0, 1), \\ 0, & x \notin (0, 1), \end{cases}$$

$$F_3(x) = -\frac{\Phi(x - 1)}{\Phi(5 - x) + \Phi(x - 1)}.$$

The denominator of F_1 is never zero and so F_1 defines a smooth function on $\mathbb{R}$ which satisfies (a,b). Further, $F_1(x) = 0$, for all $x \geq -1$. The function F_2 is zero outside $[0, 1]$ and is positive on $[0, 1]$ with zeros at $1/2, 1/3, 2/3, 1/4, 3/4, \cdots$. The factors $\Phi(x)$, $\Phi(1 - x)$ ensure that F_2 is smooth at $x = 0, 1$. Finally, F_3 vanishes for $x \leq 1$ and satisfies (e,f). Since the denominator of F_3 is never zero, F_3 defines a smooth function on $\mathbb{R}$. The function $F = F_1 + F_2 + F_3$ is a sum of smooth functions and therefore defines a smooth function on $\mathbb{R}$ which satisfies (a–f). ♠

EXERCISES 5.2.9

(1) Define $f(x) = x^3 \sin(\frac{\pi}{x})$, $x \neq 0$, $f(0) = 0$. Show that

    (a) f is continuous on $\mathbb{R}$ (you may assume that f is C^1 on $x \neq 0$).

    (b) f is differentiable at $x = 0$ and $f'(0) = 0$ (you will need to work from the definition of the derivative as a limit).

    (c) f' is continuous on $\mathbb{R}$. (You will need to find $\lim_{x \to 0} f'(x)$.)

    (d) Is f' differentiable at $x = 0$?

More generally, show that if $f(x) = x^{2n} \sin(\frac{\pi}{x})$, $x \neq 0$, and $f(0) = 0$, then f is C^{n-1} and n-times differentiable but not C^n ($f^{(n)}$ is not continuous at $x = 0$). What about if $f(x) = x^{2n+1} \sin(\frac{\pi}{x})$, $x \neq 0$, and $f(0) = 0$?

(2) Define

$$f(x) = \begin{cases} x^2 \sin\left(\frac{\pi}{\sqrt{x}}\right), & x > 0, \\ 0, & x \leq 0. \end{cases}$$

You may assume f is smooth on $x \neq 0$. Show that

    (a) f is continuous on $\mathbb{R}$.

    (b) f is differentiable at $x = 0$ and $f'(0) = 0$.

    (c) f' is continuous on $\mathbb{R}$.

    (d) f is not twice differentiable at $x = 0$.

What is the zero set ($f^{-1}(0)$) of f?

(3) Find (explicit) smooth (C^∞) functions $f, g : \mathbb{R} \to \mathbb{R}$ such that

    (a) $f(0) = 0$ and $f(\frac{1}{n}) = 0$, $n \geq 1$. Elsewhere $f > 0$.

    (b) $g(x) \in (0, 1)$, for all $x \in (0, 1) \cup (2, 3) \cup (3, 4) \cup (5, 6)$, $g = 1$ on $[1, 2] \cup [4, 5]$, elsewhere $g = 0$.

(4) Let $a, b \in \mathbb{R}$, $a < b$. Find a smooth function $f : \mathbb{R} \to \mathbb{R}$ such that

$$f(x) = 0, \quad x \leq a,$$

$$f(x) \in (0, 1), \quad x \in (a, b),$$

$$f(x) = 1, \quad x \geq b.$$

(5) Let $-\infty < a < b < c < d < +\infty$. Using the function Φ find a C^∞ $\rho \in C^\infty(\mathbb{R})$ such that

    (a) $\rho(x) = 0$ if $x \leq a$ or $x \geq b$.

    (b) $\rho(x) = 1$ if $x \in [b, c]$.

    (c) For all other $x \in \mathbb{R}$, $\rho(x) \in (0, 1)$.

Extend the definition of ρ as far as you can so as to remove the strict inequalities in $-\infty < a < b < c < d < +\infty$ (for example, $-\infty \leq a < b \leq c < d < +\infty$).

(6) Using the C^∞-function Φ, find a C^∞-function $G : \mathbb{R} \to \mathbb{R}$ which satisfies all of the following conditions:

   (a) $G(x) = 2$ if $x \le -1$.
   (b) $G(x) \in (0, 2)$ if $x \in (-1, 0)$.
   (c) $G(x) \ge 0$ on $[0, 1]$ and equals zero iff $x = \frac{1}{n^2}$ or $1 - \frac{1}{n}$ for some $n \in \mathbb{N}$.
   (d) $G(x) \in (-3, 0)$ if $x \in (1, 5)$.
   (e) $G(x) = -3$ if $x \ge 5$.

   Indicate briefly why your function G is smooth at $x = 0, 1$.

(7) Using the function Φ

   (a) Find a C^∞-function e such that $e > 0$ on $(-\infty, 0) \cup (1, \infty)$ and $e \equiv 0$ on $[0, 1]$.
   (b) Find a C^∞-function f such that $f(0) = 0$, elsewhere $f < 0$ and $f^{(j)}(0) = 0$, $j \ge 0$.
   (c) Find a C^∞-function g such that the zero set of g is $\{\pm n^3 | n \in \mathbb{Z}\}$, elsewhere $g < 0$.
   (d) Find a C^∞-function h such that $h(x) = 0$, $x \le 0$, and

   (a) $h(x) = n + 1$, if $x \in [2n + 1, 2n + 2]$, $n \ge 0$,
   (b) $h(x) \in (n, n + 1)$, if $x \in (2n, 2n + 1)$, $n \ge 0$.

   (You are advised to draw the graph first. One step at a time.)

(8) Using the function Φ

   (a) Find a C^∞-function e such that (a) $e > 0$ on $(-\infty, 0)$, (b) $e \equiv 0$ on $[0, \infty)$.
   (b) Find a C^∞-function f such that (a) $f > 0$ on $(-\infty, 1)$, (b) $f \equiv 0$ on $[1, \infty)$.
   (c) Find a C^∞-function g such that (b) $g > 0$ on $(0, 1)$, (b) $g(x) = 0$ if $x \notin (0, 1)$, (c) g has a unique maximum value at $x = \frac{1}{2}$. (In particular, g is not a tabletop function—it is simpler). What are $g^{(n)}(1)$, $g^{(n)}(0)$, $n \ge 0$?
   (d) Find a C^∞-function $F(x)$ such that $F(\frac{1}{n}) = F(1 - \frac{1}{n}) = 0$, $n \ge 1$; elsewhere F is strictly positive. What are $F^{(n)}(0)$, $F^{(n)}(1)$, $n \ge 0$? (For this problem it suffices to give a brief indication of why your function F is infinitely differentiable at the points $0, 1$.)

(9) Using the C^∞-function Φ, find a C^∞-function $G : \mathbb{R} \to \mathbb{R}$ which satisfies all of the following conditions:

   (a) $G(x) = 2$ if either $x \le -1$ or $x \ge 2$.
   (b) $G(x) \in (0, 2)$ if either $x \in (-1, 0)$ or $x \in (1, 2)$.
   (c) $G(x) \ge 0$ on $[0, 1]$ and equals zero iff $x = \frac{1}{n}$ or $1 - \frac{1}{n}$ for some $n \in \mathbb{N}$.

   Indicate briefly why your function G is smooth at $x = 0, 1$.

(10) (E. Borel's theorem.) For $b > 0$, define $\Xi_b(x) = \Theta_{1,1/2}(x/b)$, where $\Theta_{1,1/2}$ is the tabletop function defined in Examples 5.2.6(3) with $a = 1, b = 1/2$. Given a sequence $(a_n)_{n \ge 1}$, show that it is possible to choose a sequence $(b_n)_{n \ge 0} \subset$

$\mathbb{R}(> 0)$ such that the series

$$f(x) = \sum_{n=0}^{\infty} \Xi_{b_n}(x) \frac{a_n x^n}{n!}$$

defines a smooth function on $\mathbb{R}$ with Taylor series $Tf_0 = \sum_{n=0}^{\infty} \frac{a_n x^n}{n!}$. (Hints. Choose (b_n) monotone decreasing to zero so that $b_{n+1} \leq b_n/2, n \geq 0$. Observe that $\sum_{n=0}^{N} \Xi_{b_n}(x) \frac{a_n x^n}{n!} = \sum_{n=0}^{N} \frac{a_n x^n}{n!}$ on $[-b_{N+1}, b_{N+1}], N \geq 0$. Show that if (b_n) decreases fast enough, then f is smooth. Note that $\|\Xi_{b_n}^{(j)}\|$ is bounded by $C_j b_n^{-j}$, where C_j depends only on $\|\Theta_{1,1/2}^{(j)}\|$.)

5.3   The Weierstrass Approximation Theorem

In this section we consider uniform approximation of continuous functions by polynomials. For general results we need restrictions on the domain of f. For example, as $x \to \infty$, e^x increases much faster than any polynomial and so it is unreasonable to expect to be able to approximate e^x on $\mathbb{R}$ by a polynomial. Similarly, we cannot expect to approximate $f(x) = x^{-1}$ on $(0, 1)$ by a polynomial (every polynomial is bounded on $(0, 1)$). Instead, we consider approximation of continuous functions on closed and bounded intervals. We prove the *Weierstrass approximation theorem*: every continuous function on a closed and bounded interval can be uniformly approximated by a polynomial. Our proof is relatively elementary and uses Bernstein polynomials.

All the work lies in proving a special case of the theorem that applies to continuous functions on the closed unit interval.

Theorem 5.3.1 *Every continuous function on $I = [0, 1]$ can be uniformly approximated by polynomials. That is, if $f \in C^0(I)$ and $\varepsilon > 0$, then there exists a polynomial p such that*

$$\rho(f, p) = \sup_{x \in I} |f(x) - p(x)| < \varepsilon.$$

Our proof of Theorem 5.3.1 will be constructive: given a continuous function $f : [0, 1] \to \mathbb{R}$, we construct an explicit sequence of polynomials that converge uniformly to f.

5.3.1   *Bernstein Polynomials*

Set $I = [0, 1]$. Let $f \in C^0(I)$ and $n \geq 1$. The *nth Bernstein polynomial* $B_n(f)$ of f is the polynomial of degree at most n defined by

$$B_n(f)(x) = \sum_{p=0}^{n} \binom{n}{p} f(\frac{p}{n}) x^p (1 - x)^{n-p}.$$

Lemma 5.3.2 *We have for $n \geq 1$*

(1) $B_n(cf) = cB_n(f), f \in C^0(I), c \in \mathbb{R}$.
(2) $B_n(f + g) = B_n(f) + B_n(g), f, g \in C^0(I)$.
(3) $B_n(f) > 0$ on I if $f > 0$ on I.
(4) $B_n(1) = 1$.
(5) $B_n(t)(x) = x$ *(here $f(t) = t$).*
(6) $B_n(t^2)(x) = x^2 + \frac{x-x^2}{n}$ *(here $f(t) = t^2$).*

Remarks 5.3.3

(1) Statements (1,2) imply $B_n : C^0(I) \rightarrow C^0(I)$ is linear.
(2) Statement (3) implies that if $f > g$ then $B_n(f) > B_n(g)$ (on I).
(3) When we replace f by an actual function the variable for f will always be t—as in $f(t)$. The Bernstein polynomial will always be a function of $x \in I$. Thus, in statements (5,6) the variable t is a 'dummy' variable which just indicates the functional form of f.   ✤

Proof of Lemma 5.3.2 (1,2,3) are obvious (for (3) observe that $x^p(1-x)^{n-p} > 0$ on $(0,1)$).

(4) $B_n(1)(x) = \sum_{p=0}^{n} \binom{n}{p} 1 x^p (1-x)^{n-p} = (x + (1-x))^n = 1$.
(5) We assume $n \geq 2$—the result is easy if $n = 1$. We have

$$B_n(t)(x) = \sum_{p=0}^{n} \binom{n}{p} \frac{p}{n} x^p (1-x)^{n-p}$$

$$= \sum_{p=1}^{n} \frac{n!}{p!(n-p)!} \frac{p}{n} x^p (1-x)^{n-p}$$

$$= \sum_{p=1}^{n} \frac{(n-1)!}{(p-1)!(n-p)!} x^p (1-x)^{n-p}$$

$$= \sum_{p=1}^{n} \frac{(n-1)!}{(p-1)!((n-1)-(p-1))!} x x^{p-1} (1-x)^{(n-1)-(p-1)}$$

$$= x \sum_{q=0}^{n-1} \binom{n-1}{q} x^q (1-x)^{(n-1)-q}, \quad (q = p-1)$$

$$= x B_{n-1}(1)(x) = x.$$

(6) Again we assume $n \geq 3$—see below for the case $n = 2$. We have

$$B_n(t^2)(x) = \sum_{p=0}^{n} \binom{n}{p} (\frac{p}{n})^2 x^p (1-x)^{n-p}$$

$$= \sum_{p=1}^{n} \frac{(n-1)!}{(p-1)!(n-p)!} \frac{p}{n} x^p (1-x)^{n-p}, \quad \text{as in (5)}$$

$$= \sum_{p=1}^{n} \frac{(n-1)!}{(p-1)!(n-p)!} \left(\frac{p-1}{n} + \frac{1}{n} \right) x^p (1-x)^{n-p}$$

$$= A + B,$$

where

$$A = \sum_{p=1}^{n} \frac{(n-1)!}{(p-1)!(n-p)!} \frac{p-1}{n} x^p (1-x)^{n-p},$$

$$B = \sum_{p=1}^{n} \frac{(n-1)!}{(p-1)!(n-p)!} \frac{1}{n} x^p (1-x)^{n-p}.$$

Checking the proof of (5), we see that $B = \frac{1}{n} B_{n-1}(t)(x) = \frac{x}{n}$. It remains to evaluate A. Cancelling the factor $(p-1)$ and taking out factors $(n-1)/n$ and x^2 we have (just as in the proof of (5))

$$A = x^2 \frac{(n-1)}{n} \sum_{p=2}^{n} \frac{(n-2)!}{(p-2)!((n-2)-(p-2))!} x^{p-2} (1-x)^{(n-2)-(p-2)}$$

$$= x^2 \frac{(n-1)}{n} B_{n-2}(1)(x)$$

$$= x^2 \frac{(n-1)}{n}.$$

(Note that if $n = 2$, $\sum_{p=2}^{2} \frac{(2-2)!}{(p-2)!((2-2)-(p-2))!} x^{p-2} (1-x)^{(2-2)-(p-2)} = 1$.)
Finally,

$$A + B = x^2 \frac{(n-1)}{n} + \frac{x}{n} = x^2 + \frac{x-x^2}{n},$$

and so $B_n(t^2)(x) = x^2 + \frac{x-x^2}{n}$. $\square$

Proof of Theorem 5.3.1 Let $f \in C^0(I)$ and $\varepsilon > 0$. Since I is closed and bounded, $f : I \to \mathbb{R}$ is uniformly continuous (Theorem 2.4.15) and so $\exists \delta > 0$ such that for all

$t, x \in I$ satisfying $|x - t| < \delta$ we have

$$- \varepsilon/2 < f(t) - f(x) < \varepsilon/2 \quad \text{(that is, } |f(t) - f(x)| < \varepsilon/2\text{).} \qquad (5.1)$$

Since f is continuous on I, $M = \sup_{s \in I} |f(s)| < \infty$. The next inequality follows from the triangle inequality

$$- 2M < f(t) - f(x) < 2M, \quad \text{for all } t, x \in I. \qquad (5.2)$$

Observe that the function $\frac{2M}{\delta^2}(t - x)^2$ is greater than or equal to $2M$ provided that $|t - x| \geq \delta$. It follows from (5.1), (5.2) that for all $t, x \in I$ we have

$$- \varepsilon/2 - \frac{2M}{\delta^2}(t - x)^2 < f(t) - f(x) < \frac{2M}{\delta^2}(t - x)^2 + \varepsilon/2. \qquad (5.3)$$

Regard each term in this inequality as a function of t (so x is fixed). Noting property (3) of Bernstein polynomials we have for all $n \geq 1$ the inequality between *fnunctions* (of x)

$$B_n\left(-\varepsilon/2 - \frac{2M}{\delta^2}(t - x)^2\right) < B_n(f) - B_n(f(x)) < B_n\left(\frac{2M}{\delta^2}(t - x)^2 + \varepsilon/2\right).$$

(What are we doing? We fix x, set $t = \frac{p}{n}$ in (5.3), multiply by $\binom{n}{p}x^p(1 - x)^{n-p}$ and sum from $p = 0$ to $p = n$. In particular, $B_n(f(x)) = f(x)B_n(1) = f(x)$, using property (1)).

Using properties (1,2,4), we have

$$B_n\left(\frac{2M}{\delta^2}(t - x)^2 + \varepsilon/2\right) = \frac{2M}{\delta^2}B_n((t - x)^2) + \varepsilon/2,$$

$$B_n\left(-\frac{2M}{\delta^2}(t - x)^2 - \varepsilon/2\right) = -\frac{2M}{\delta^2}B_n((t - x)^2) - \varepsilon/2.$$

Hence for all $x \in I$ we have

$$- \varepsilon/2 - \frac{2M}{\delta^2}B_n((t - x)^2)(x) < B_n(f)(x) - f(x) < \frac{2M}{\delta^2}B_n((t - x)^2)(x) + \varepsilon/2. \qquad (5.4)$$

We claim that $\exists N$ such that for $n \geq N$, $|\frac{2M}{\delta^2}B_n((t - x)^2)(x)| < \varepsilon/2$ for all $x \in I$. It then follows from (5.4) that for $n \geq N$, $x \in I$, $|B_n(f)(x) - f(x)| < \varepsilon/2 + \varepsilon/2 = \varepsilon$ and we are done.

In order to prove the claim we evaluate $B_n((t - x)^2)$. Since $(t - x)^2 = t^2 - 2tx + x^2$, we have

$$B_n((t - x)^2) = B_n(t^2) - 2xB_n(t) + x^2B_n(1).$$

Evaluating at x, this gives us (using (4,5,6))

$$B_n((t-x)^2)(x) = B_n(t^2)(x) - 2xB_n(t)(x) + x^2 B_n(1)(x)$$

$$= \left(x^2 + \frac{x-x^2}{n}\right) - 2xx + x^2 1$$

$$= \frac{x-x^2}{n}.$$

The maximum value of $x - x^2$ on $[0, 1]$ is $1/4$ and so $0 \le B_n((t-x)^2)(x) < 1/4n$. Hence for $x \in I$,

$$0 < \frac{2M}{\delta^2} B_n((t-x)^2) + \varepsilon/2 < \frac{M}{2n\delta^2} + \varepsilon/2.$$

Now choose N so that $\frac{M}{2n\delta^2} < \varepsilon/2, n \ge N$.   □

Remarks 5.3.4

(1) Bernstein polynomials are named after the Russian mathematician Sergei Natanovich Bernstein. They were first used by him to give a constructive proof of the Weierstrass approximation theorem.

(2) The proof of the Weierstrass approximation theorem using Bernstein polynomials may seem slightly magical (especially from Eq. (5.3)). However, there is a simple probabilistic interpretation of the argument that we now briefly describe (we refer to Lamperti [22, pages 38–40] for more details and background). Consider coin tossing where the probability of falling heads is $x \in [0, 1]$ (and therefore the probability of falling tails is $1 - x$). If we toss the coin n times, the probability of there being exactly p tosses that results in heads is $\binom{n}{p} x^p (1-x)^{n-p}$ (this is the *binomial distribution*). Now suppose that X is the random variable defined as the number of times the coin falls heads in n coin tosses. Then X has the binomial distribution defined above. By the *weak law of large numbers* $\lim_{n \to \infty} P(|X/n - x| > \delta) = 0$, for all $\delta > 0$. Moreover, this estimate is uniform in $x \in [0, 1]$. Suppose we are given a continuous function $f : [0, 1] \to \mathbb{R}$. We evaluate f at the points $p/n, p = 0, \cdots, n$. The *expectation* $\mathbb{E}_n(f)$ of $f(X/n)$ is then $B_n(f)(x)$. Because f is uniformly continuous on $[0, 1]$, it follows that $\lim_{n \to \infty} P(|f(X/n) - f(x)| > \delta^*) = 0$, uniformly in x, for all $\delta^* > 0$. From this one can show—as in the proof of Theorem 5.3.1—that $\lim_{n \to \infty} \mathbb{E}_n(f) = f(x)$ uniformly in $x \in [0, 1]$.   ✸

Theorem 5.3.5 (The Weierstrass Approximation Theorem) *Let $[a, b]$ be a closed and bounded interval. Every continuous function on $[a, b]$ can be uniformly approximated by polynomials.*

Proof Let $L : [0, 1] \to [a, b]$ be the linear bijection defined by $L(x) = (b - a)x + a$, $x \in [0, 1]$. We denote the inverse of L by K and note that $K(y) = (y - a)/(b - a)$, $y \in [a, b]$.

Let $f : [a, b] \to \mathbb{R}$ be continuous and set $F = f \circ L : [0, 1] \to \mathbb{R}$. Since F is continuous, Theorem 5.3.1 implies that there exists a sequence (p_n) of polynomials such that $\lim_{n\to\infty} \sup_{x\in[0,1]} |F(x) - p_n(x)| = 0$. Set $P_n(y) = p_n(K(y))$, $y \in [a, b]$, and note that since K is linear, P_n is a polynomial. Now $|F(x) - p_n(x)| = |f(y) - P_n(y)|$, where $L(x) = y$. Since K is 1:1 onto, we have

$$\sup_{x\in[0,1]} |F(x) - p_n(x)| = \sup_{y\in[a,b]} |f(y) - P_n(y)|,$$

and so $\lim_{n\to\infty} \sup_{y\in[a,b]} |f(y) - P_n(y)| = 0$. $\qquad\qquad\square$

5.3.2 An Application of the Weierstrass Approximation Theorem

Proposition 5.3.6 *Let* $f : [a, b] \to \mathbb{R}$ *be continuous and suppose that* $\int_a^b f(x)x^n \, dx = 0$, *for all* $n \geq 0$. *Then* $f \equiv 0$.

Proof Since f is continuous it suffices to prove $\int_a^b f(x)^2 \, dx = 0$.

We start by observing that if $p(x) = a_n x^n + \cdots + a_1 x + a_0$, then

$$\int_a^b f(x)p(x) \, dx = \sum_{j=0}^{n} a_j \int_a^b f(x)x^j \, dx = 0,$$

by our assumption.

Let $M = 1 + \sup_{x\in[a,b]} |f(x)| \geq 1$. By the Weierstrass approximation theorem, we can find a polynomial p such that $\sup_{x\in[a,b]} |f(x) - p(x)| < \varepsilon/(M(b - a))$. We have

$$\int_a^b f(x)^2 \, dx = \int_a^b f(x)(f(x) - p(x)) \, dx + \int_a^b f(x)p(x) \, dx$$

$$= \int_a^b f(x)(f(x) - p(x)) \, dx.$$

Now

$$\left| \int_a^b f(x)(f(x) - p(x)) \, dx \right| \leq \int_a^b |f(x)||f(x) - p(x)| \, dx$$

$$< (b - a)M \frac{\varepsilon}{M(b - a)} = \varepsilon.$$

Our argument shows that for all $\varepsilon > 0$, $\int_a^b f(x)^2 \, dx < \varepsilon$. Hence $\int_a^b f(x)^2 \, dx = 0$. $\qquad\square$

5.3.3 Uniform Approximation of a Family

For our applications to Fourier series we will need a slightly stronger version
of the Weierstrass approximation theorem that applies to continuous families
of continuous functions. For this we need one or two elementary results about
continuous functions defined on rectangles in $\mathbb{R}^2$. We give elementary proofs of
these results in this section but remark that everything we say is an easy consequence
of the general theory we develop later in Chap. 7.

Let $[a, b]$, $[c, d]$ be closed bounded intervals. Suppose that $f : [a, b] \times [c, d] \to \mathbb{R}$.
Given $v \in [c, d]$, define $f_v : [a, b] \to \mathbb{R}$ by

$$f_v(x) = f(x, v), \ x \in [a, b].$$

We may regard $\{f_v \mid v \in [c, d]\}$ as defining a *family* of functions $f_v : [a, b] \to \mathbb{R}$
parameterized by $v \in [c, d]$.

The map $f : [a, b] \times [c, d] \to \mathbb{R}$ is *continuous* if for every $\varepsilon > 0$, there exists a
$\delta > 0$ such that

$$|f(x_1, v_1) - f(x_2, v_2)| < \varepsilon, \ \text{if } |x_1 - x_2|, |v_1 - v_2| < \delta.$$

If this conditions holds, then $f_v : [a, b] \to \mathbb{R}$ is continuous for all $v \in [c, d]$. We refer
to $\{f_v \mid v \in [c, d]\}$ as a *continuous family* of continuous functions on $[a, b]$.

We start by showing that a continuous family satisfies a (weak) version of
uniform continuity.

Lemma 5.3.7 *Let $\{f_v \mid v \in [c, d]\}$ be a continuous family of continuous functions
on $[a, b]$. Given $\varepsilon > 0$, there exists a $\delta > 0$ such that for all $v \in [c, d]$ we have*

$$|f_v(x) - f_v(y)| < \varepsilon, \ |x - y| < \delta.$$

Proof Suppose the contrary. Then there exists an $\varepsilon > 0$ such that for every $n \in \mathbb{N}$
there exist $x_n, y_n \in [a, b]$ and $v_n \in [c, d]$ such that

$$|x_n - y_n| < 1/n \text{ and } |f_{v_n}(x_n) - f_{v_n}(y_n)| \geq \varepsilon.$$

As in the proof of Theorem 2.4.15, the Bolzano–Weierstrass theorem implies that
the bounded sequences $(x_n), (y_n) \subset [a, b]$ have a convergent subsequence. A second
application of the Bolzano–Weierstrass theorem to the corresponding subsequence
of (v_n) yields convergent subsequences of $(x_n), (y_n)$ and (v_n), say $(x_{n_k}), (y_{n_k})$ and
(v_{n_k}), with $\lim_{k \to \infty} x_{n_k} = \lim_{k \to \infty} y_{n_k}$. We derive the required contradiction by
letting $k \to \infty$ in $|f_{v_{n_k}}(x_{n_k}) - f_{v_{n_k}}(y_{n_k})|$.   □

Lemma 5.3.8 *Let $\{f_v \mid v \in [c, d]\}$ be a continuous family of continuous functions
on $[a, b]$. There exists an $M \geq 0$ such that for all $v \in [c, d]$ we have*

$$\sup_{x \in [a,b]} |f_v(x)| \leq M.$$

Proof The function $g : [c, d] \to \mathbb{R}$ defined by $g(v) = f(a, v)$ is continuous and so there exists an $N \geq 0$ such that $|f(a, v)| \leq N$ for all $v \in [c, d]$. Take $\varepsilon = 1$ in Lemma 5.3.7 to obtain $\delta > 0$ such that $|f_v(x) - f_v(y)| < 1$, whenever $|x - y| < \delta$. Observe that $|f_v(x)| \leq |f_v(a)| + (b - a)\delta^{-1} + 1$, $x \in [a, b]$. Hence $|f_v(x)| \leq N + (b - a)\delta^{-1} + 1$ for all $x \in [a, b]$, $v \in [c, d]$. Take $M = N + (b - a)\delta^{-1} + 1$. $\square$

Remarks 5.3.9

(1) Lemma 5.3.8 shows that a continuous function on a bounded closed rectangle is bounded. This is a natural generalization of our earlier result on continuous functions defined on a closed and bounded interval. The proof of Lemma 5.3.7 can easily be extended to prove uniform continuity on a bounded closed rectangle. As we shall see in Chap. 7 we can prove far more general results that apply to functions defined on arbitrary bounded and 'closed' subsets of $\mathbb{R}^n$.

(2) For our continuous families of continuous functions we require *joint* continuity of f in (x, v). Everything we have said breaks down badly if we only assume separate continuity. That is, for fixed x, $f(x, v)$ is continuous on $[c, d]$, and for fixed v, $f(x, v)$ is continuous on $[a, b]$. ✠

Theorem 5.3.10 *Let $\{f_v \mid v \in [c, d]\}$ be a continuous family of continuous functions on $[a, b]$. There exists a sequence (p_v^n) of continuous polynomial families $\{p_v^n : [a, b] \to \mathbb{R} \mid v \in [c, d]\}$ converging uniformly to the family $\{f_v \mid v \in [c, d]\}$. That is, for each $\varepsilon > 0$, there exists an $N \in \mathbb{N}$ such that for each $v \in [c, d]$,*

$$\sup_{x \in [a,b]} |f_v(x) - p_v^n(x)| < \varepsilon, \ n \geq N.$$

Proof Without loss of generality, assume $[a, b] = [0, 1]$. For $v \in [c, d]$, define

$$p_v^n = B_n(f_v), \ n \in \mathbb{N}.$$

We now just repeat the proof of Theorem 5.3.1—using Lemmas 5.3.7, 5.3.8, we choose the constants δ, M that occur in the proof of Theorem 5.3.1 to be independent of $v \in [c, d]$. $\square$

EXERCISES 5.3.11

(1) Let $f(x) = |x - \frac{1}{2}|$, $x \in [0, 1]$. Compute $B_n(f)$, $n = 1, 2, 3$.

   (a) Sketch the graph of f, together with the graphs of the approximations $B_n(f)$, $n = 1, 2, 3$.
   (b) Where is the approximation poor?
   (c) Compute $B_8(f)(1/2)$ and hence show $\rho(f, B_8(f)) > 0.13$.
   (c) Suppose we take $\varepsilon = 1/10$. Find a value of N for which $\rho(f, B_n(f)) < 1/10$, for all $n \geq N$. (Note: Do not strive for the best estimate of N. Just get a value—even if it is quite large. You may want to look back over the proof of the Weierstrass approximation theorem.)

(2) Let $C^r(I)$ denote the space of r-times continuously differentiable functions on $I = [0, 1]$, $0 \le r < \infty$. Show that given $\varepsilon > 0$, there exists a polynomial p such that

$$\rho(f^{(s)}, p^{(s)}) < \varepsilon, \ 0 \le s \le r.$$

(Uniform approximation of a function and its first r-derivatives.) Hint: Start by approximating $f^{(r)}$ and then work back to f.

(3) For $\eta > 0$, define $\phi_\eta = \frac{1}{\sqrt{2\pi\eta}} \exp(-\frac{x^2}{2\eta})$. Show that if $f \in C^0(\mathbb{R})$ is bounded and we define $f_\eta(x) = \int_{-\infty}^{\infty} f(t)\phi_\eta(x - t)\, dt$, then

   (a) f_η is C^∞, $\eta > 0$. (You will need results on differentiation under the integral sign—see Lemma 6.1.6.)
   (b) f_η converges uniformly to f on all closed bounded subintervals of $\mathbb{R}$ (that is, given $[a, b]$ and $\varepsilon > 0$, there exists an $\eta_0 > 0$ such that $\sup_{x \in [a,b]} |f(x) - f_\eta(x)| < \varepsilon$, for all $\eta \in (0, \eta_0)$).

(For part (b) you will need (A) $\int_{-\infty}^{\infty} \phi_\eta(t)\, dt = 1$ for all $\eta > 0$, and (B) if $\delta, \varepsilon > 0$, there exists an $\eta_0 > 0$ such that $\int_{-\delta}^{\delta} \phi_\eta(t)\, dt > 1 - \varepsilon$ for all $\eta \in (0, \eta_0]$.) Show how this result can be used to prove that we can uniformly approximate continuous functions on $[a, b]$ by smooth functions.

(4) Show how to extend the proof of Lemma 5.3.7 to obtain uniform continuity of a continuous family of continuous functions. Show that uniform continuity implies boundedness (we assume the domain is a bounded rectangle).

(5) Show that Lemma 5.3.7 and Lemma 5.3.8 both fail if f is only separately continuous (see Remarks 5.3.9).

5.4 Analytic Functions

Definition 5.4.1 A C^∞-function $f : (a, b) \to \mathbb{R}$ is *(real) analytic* if for every $x_0 \in (a, b)$, there exists an $r > 0$ such that

$$f(x) = \sum_{n=0}^{\infty} \frac{f^{(n)}(x_0)}{n!} (x - x_0)^n, \ x \in (x_0 - r, x_0 + r) \cap (a, b).$$

That is, f is analytic if for every point x_0 in the domain of f, f is equal to the Taylor series of f at x_0 on some open interval containing x_0.

Examples 5.4.2

(1) If $f : (a, b) \to \mathbb{R}$, $g : (c, d) \to \mathbb{R}$ are analytic then $f \pm g$ is an analytic function on $(a, b) \cap (c, d)$.
(2) Every polynomial is an analytic function on $\mathbb{R}$. This requires us to show that if $p(x) = a_0 x^n + \cdots + a_n$, then for every $x_0 \in \mathbb{R}$, we may find

constants $A_0, \cdots, A_n \in \mathbb{R}$ such that $p(x) = \sum_{j=0}^n A_j(x - x_0)^{n-j}$, for all $x \in \mathbb{R}$. We leave this as an easy exercise for the reader (see also the proof of Proposition 5.4.3). ♠

The result given by the next proposition is certainly what one would expect, but the proof requires some work.

Proposition 5.4.3 *Suppose that the power series $\sum_{n=0}^\infty a_n x^n$ has radius of convergence $R > 0$. Then $f(x) = \sum_{n=0}^\infty a_n x^n$ defines an analytic function on $(-R, R)$. More generally, if $c \in \mathbb{R}$, then $\sum_{n=0}^\infty a_n(x - c)^n$ defines an analytic function on $(c - R, c + R)$.*

Proof We are required to show that if $x_0 \in (-R, R)$, then there exists an $r > 0$ such that the Taylor series $Tf_{x_0}(x) = \sum_{n=0}^\infty \frac{f^{(n)}(x_0)}{n!}(x - x_0)^n$ converges to $f(x)$, for all $x \in (x_0 - r, x_0 + r)$.

Since the derivatives of f on $(-R, R)$ are obtained by term-by-term differentiation of the power series $\sum_{n=0}^\infty a_n x^n$, we have

$$\frac{f^{(n)}(x_0)}{n!} = \sum_{m=n}^\infty \binom{m}{n} a_m x_0^{m-n}, \ n \geq 0.$$

We start by noting a special case of the result. If f is a polynomial of degree p, then

$$\sum_{n=0}^p a_n x^n = \sum_{n=0}^p \left(\sum_{m=n}^p \binom{m}{n} a_m x_0^{m-n}\right)(x - x_0)^n,$$

since it is easy to check that both sides of the equation are polynomials of degree at most p and have the same derivatives of order $\leq p$ at $x = x_0$.

The proof of the general case has two parts. First, we estimate $\frac{|f^{(n)}(x_0)|}{n!}$ so as to show that Tf_{x_0} has a non-zero radius of convergence. Then we prove that the partial sums of $\sum_{n=0}^\infty a_n x^n$ converge to $Tf_{x_0}(x)$—this will use the special case together with estimates on remainders.

Fix $b \in (|x_0|, R)$. By Lemma 4.5.1, there exists a $C \geq 0$ such that

$$|a_n| \leq Cb^{-n}, \ n \geq 0.$$

Using this estimate it is easy to show that $\sum_{m=n}^\infty \binom{m}{n} a_m x_0^{m-n}$ is absolutely convergent and that we have the estimate

$$\frac{|f^{(n)}(x_0)|}{n!} \leq Cb^n \left(\sum_{m=n}^\infty \binom{m}{n} \frac{|x_0|^{m-n}}{b}\right)$$

$$= Cb^n \left(1 - \frac{|x_0|}{b}\right)^{-(n+1)},$$

where the last equality follows from the binomial theorem. Choose $r > 0$ so that

$$\frac{br}{1 - \frac{|x_0|}{b}} < 1 \text{ and } [x_0 - r, x_0 + r] \subset (-R, R).$$

We claim that the Taylor series Tf_{x_0} converges on $[x_0 - r, x_0 + r]$. We have

$$\left| \frac{f^{(n)}(x_0)}{n!}(x - x_0)^n \right| \le Cb^n (1 - \frac{|x_0|}{b})^{-(n+1)} |x - x_0|^n$$

$$< Cb^n (1 - \frac{|x_0|}{b})^{-(n+1)} r^n, \text{ if } x \in (x_0 - r, x_0 + r)$$

$$= D \left(\frac{rb}{1 - \frac{|x_0|}{b}} \right)^n,$$

where $D = C/(1 - \frac{|x_0|}{b})$. Since $\sum_{n=0}^{\infty} (\frac{rb}{1 - \frac{|x_0|}{b}})^n$ is convergent (by our choice of r), the Taylor series converges for all $x \in [x_0 - r, x_0 + r]$.

Finally, we need to show that $Tf_{x_0}(x)$ converges to $f(x)$ for all $x \in [x_0 - r, x_0 + r]$. For this it suffices to show that if $\varepsilon > 0$ then there exists an $N \in \mathbb{N}$ such that $|Tf_{x_0}(x) - \sum_{n=0}^{p} a_n x^n| < \varepsilon$, for all $p \ge N$ and $x \in [x_0 - r, x_0 + r]$.

Let $\varepsilon > 0$. Fix $x \in [x_0 - r, x_0 + r]$ and choose $N \in \mathbb{N}$ so that for all $p \ge N$ we have

$$\left| \sum_{n=p+1}^{\infty} \left(\sum_{m=n}^{\infty} \binom{m}{n} |a_m| r^m \right) \right|, \left| \sum_{n=0}^{p} \left(\sum_{m=p+1}^{\infty} \binom{m}{n} |a_m| r^m \right) \right| < \varepsilon/2. \quad (5.5)$$

For $p \ge N$ define

$$I_1 = \sum_{n=0}^{p} \left(\sum_{m=n}^{p} \binom{m}{n} a_m x_0^{m-n} \right) (x - x_0)^n,$$

$$I_2 = \sum_{n=p+1}^{\infty} \left(\sum_{m=n}^{\infty} \binom{m}{n} a_m x_0^{m-n} \right) (x - x_0)^n,$$

$$I_3 = \sum_{n=0}^{p} \left(\sum_{m=p+1}^{\infty} \binom{m}{n} a_m x_0^{m-n} \right) (x - x_0)^n.$$

For $x \in [x_0 - r, x_0 + r]$, we have (by absolute convergence)

$$\sum_{n=0}^{\infty} \frac{f^{(n)}(x_0)}{n!} (x - x_0)^n = I_1 + I_2 + I_3.$$

Now $I_1 = \sum_{n=0}^{p} a_n x^n$ (special case: $\sum_{n=0}^{p} a_n x^n$ is a polynomial of degree p). Since $x \in [x_0 - r, x_0 + r]$, we have by (5.5), $|I_2|, |I_3| < \varepsilon/2$, if $p \geq N$. Hence

$$\left| Tf_{x_0}(x) - \sum_{n=0}^{p} a_n x^n \right| = |I_1 + I_2| < \varepsilon, \quad \text{for all } p \geq N.$$

Hence the sequence $(\sum_{n=0}^{p} a_n x^n)$ of partial sums converges pointwise to $Tf_{x_0}(x)$ on $[x_0 - r, x_0 + r]$. □

Remark 5.4.4 As we show in the next examples, the radius of convergence of Tf_{x_0} may be strictly bigger than R. It is straightforward to show that it is always at least $\min\{R - x_0, x_0 + R\}$. ✽

Examples 5.4.5

(1) The exponential series $\sum_{n=0}^{\infty} \frac{x^n}{n!}$ defines an analytic function $\exp(x)$ on $\mathbb{R}$. We claim that (a) $\exp(0) = 1$, (b) $\exp'(x) = \exp(x)$ for all $x \in \mathbb{R}$, (c) $\exp(x)\exp(-x) = 1$, for all $x \in \mathbb{R}$, and (d) $\exp(x + y) = \exp(x)\exp(y)$, for all $x, y \in \mathbb{R}$. (a) is immediate from the series definition and (b) follows by term-by-term differentiation of the power series defining $\exp(x)$. By the chain rule $\frac{d}{dx}(\exp(-x)) = -\exp(-x)$ and so $\frac{d}{dx}(\exp(x)\exp(-x)) = 0$ for all $x \in \mathbb{R}$. Hence $\exp(x)\exp(-x)$ is constant and, taking $x = 0$, we have $\exp(x)\exp(-x) = 1$ for all $x \in \mathbb{R}$. Finally, using (b) again, we have $\frac{d}{dx}(\exp(x+y)\exp(-x)\exp(-y)) = 0$ and so $\exp(x + y)\exp(-x)\exp(-y)$ is constant as a function of x. Take $x = -y$ and use (a,c) to deduce that $\exp(x + y)\exp(-x)\exp(-y) = 1$ for all $x \in \mathbb{R}$. Hence, applying (c) again, we deduce that $\exp(x + y) = \exp(x)\exp(y)$. If we set $\exp(1) = e \approx 2.718 \cdots$, then (c,d) imply that we may write $\exp(x) = e^x$ where e^x satisfies the exponent laws for a power.

(2) The power series $\sum_{n=0}^{\infty} x^n$ has radius of convergence 1 and converges to $f(x) = (1 - x)^{-1}$ on $(-1, 1)$. Given $a \in (-1, 1)$, we have $f^{(n)}(a) = n!(1 - a)^{-n}$ and so $Tf_a(x) = \sum_{n=0}^{\infty} \frac{f^{(n)}(a)}{n!}(x-a)^n = \sum_{n=0}^{\infty} (\frac{x-a}{1-a})^n$. The radius of convergence of this series is $1 - a$. Observe that $1 - a > 1$ if $a < 0$ and so the radius of convergence of the Taylor series of a power series can be strictly bigger than the radius of convergence of the power series. In this example, the analytic function defined by $\sum_{n=0}^{\infty} x^n$ on $(-1, 1)$ is equal to $(1 - x)^{-1}$ and the latter function is naturally defined on $(-\infty, 1)$ as an analytic function. ♠

Proposition 5.4.6 *Suppose that $f : I \to \mathbb{R}$ and $g : J \to \mathbb{R}$ are analytic.*

(1) *The product $f \times g : I \cap J \to \mathbb{R}$ is analytic.*
(2) *If g is non-zero on $I \cap J$ then the quotient $f/g : I \cap J \to \mathbb{R}$ is analytic.*
(3) *If $f(I) \subset J$ then the composite $g \circ f : I \to \mathbb{R}$ is analytic.*

Proof Statement (1) can be proved using Proposition 5.4.3 and the result on products of power series (Proposition 4.5.10). Similarly (2) follows from Proposition 4.5.11. We omit the proof of (3)—see the remarks below. □

Remark 5.4.7 The easiest way of proving analyticity of the composite of analytic functions is to complexify and use complex analytic methods based on Cauchy's integral theorem. For a proof of analyticity using real power series methods, we refer to Krantz and Parks [20, §1.3]. See also the exercises at the end of Sect. 9.13.1 on Faà di Bruno's formula in Chap. 9.   ✠

Proposition 5.4.8 *Suppose that $f : (a, b) \to \mathbb{R}$ is analytic and not identically zero. Then*

(1) *The zeros of f are* isolated: *if $f(x_0) = 0$, then there exists an $s > 0$ such that the only zero of f on $(x_0 - s, x_0 + s)$ is x_0.*
(2) *If $f(x_0) = 0$ then there exists a unique $p \in \mathbb{N}$ and analytic function g on (a, b) such that $g(x_0) \neq 0$ and*

$$f(x) = (x - x_0)^p g(x).$$

Proof Suppose that $f(x_0) = 0$. Without loss of generality, take $x_0 = 0$. For some $r > 0$ we may write

$$f(x) = \sum_{n=0}^{\infty} a_n x^n, \; x \in (-r, r).$$

Since $f(0) = 0$, we must have $a_0 = 0$. Let p be the smallest integer for which $a_p \neq 0$. Then

$$f(x) = \sum_{n=p}^{\infty} a_n x^n = x^p \sum_{n=0}^{\infty} a_{n+p} x^n = x^p g(x),$$

where $g(x) = \sum_{n=0}^{\infty} a_{n+p} x^n, \; x \in (-r, r)$. Since $a_p \neq 0$, $g(0) \neq 0$. Moreover, the radius of convergence of the power series defining g is at least r and so g is analytic on $(-r, r)$. In particular, g is continuous on $(-r, r)$ and non-zero at $x = 0$. Hence there exists an $s > 0$ such that $g \neq 0$ on $(-s, s)$. Therefore the only zero of f on $(-s, s)$ is at $x = 0$, proving (1). We define $g : (a, b) \to \mathbb{R}$ by $g(0) = a_p$ and $g(x) = x^{-p} g(x), x \neq 0$. We leave it to the exercises for the reader to verify that g is analytic.   □

Remark 5.4.9 It follows from Proposition 5.4.8 that if $f : (a, b) \to \mathbb{R}$ is analytic and not constant, then for all $c \in \mathbb{R}, f^{-1}(c)$ is a countable subset of (a, b) consisting of isolated points.   ✠

5.4.1   Analytic Continuation

The next result is very special to analytic functions—it fails completely for C^∞-functions.

Proposition 5.4.10 *Let $f : I \to \mathbb{R}$ and $g : J \to \mathbb{R}$ be analytic functions defined on the open intervals I, J. If there exists an $x_0 \in I \cap J$ such that*

$$f^{(n)}(x_0) = g^{(n)}(x_0), \text{ for all } n \geq 0,$$

then $f = g$ on $I \cap J$. Otherwise said, if the analytic functions f and g have the same Taylor series at some point then $f = g$ on their common domain.

Proof Let $X = \{x \in I \cap J \mid f^{(n)}(x) = g^{(n)}(x) \text{ for all } n \geq 0\}$. It suffices to prove $X = I \cap J$. Since $x_0 \in X$, $X \neq \emptyset$. Moreover, if $x \in X$, then f and g have the same power series representation on an open interval $K \subset I \cap J$ containing x and therefore $K \subset I \cap J$. Suppose $X \neq I \cap J$. Without loss of generality suppose there exists a $z \in I \cap J$, $z < x_0$. Let $z_0 = \sup\{z < x_0 \mid z \notin X\}$. Clearly, $(z_0, x_0) \subset X$. Choose a sequence $(y_j) \subset (z_0, x_0)$ such that $\lim_{j \to \infty} y_j = z_0$. By sequential continuity of $f^{(n)}, g^{(n)}$, we have $\lim_{j \to \infty} f^{(n)}(y_j) = f^{(n)}(z_0)$, $\lim_{j \to \infty} g^{(n)}(y_j) = g^{(n)}(z_0)$ for all $n \geq 0$. But since $(y_j) \subset X$, we have $f^{(n)}(y_n) = g^{(n)}(y_n)$ for all $n \geq 0$ and so $f^{(n)}(z_0) = g^{(n)}(z_0)$, $n \geq 0$. Hence $z_0 \in X$. But if $z_0 \in X$, then there is an open interval $(z_0 - r, z_0 + r) \subset X \cap (I \cap J)$, contradicting the definition of z_0 as the supremum of points $z < x_0$ not in X. Hence $X = I \cap J$. $\qquad\square$

The next result is an immediate corollary of Proposition 5.4.10.

Corollary 5.4.11 *If $f, g : I \to \mathbb{R}$ are analytic functions which are equal on a nonempty open subinterval of I, then $f = g$ on I.*

Definition 5.4.12 Let I, J be open intervals and $f : I \to \mathbb{R}$, $g : J \to \mathbb{R}$ be analytic functions. We call g an *analytic continuation* of f if (a) $J \supset I$ and (b) $g = f$ on I.

Proposition 5.4.13 *Every analytic function $f : I \to \mathbb{R}$ has a unique maximal analytic continuation $F : J \to \mathbb{R}$.*

Proof Let $\mathcal{A} = \{g_\lambda : J_\lambda \to \mathbb{R} \mid \lambda \in \Lambda\}$ denote the set of all analytic continuations of f. Define $J = \cup_{\lambda \in \Lambda} J_\lambda$. Given $x \in J$, there exists a $\lambda \in \Lambda$ such that $x \in J_\lambda$ and we define $F(x) = g_\lambda(x)$. As an immediate consequence of Corollary 5.4.11, the value $F(x)$ is independent of the choice of $\lambda \in \Lambda$ such that $x \in J_\lambda$ (if $x \in J_\lambda, J_\mu$, then $x \in J_\lambda \cap J_\mu \supset I$). The map F is analytic (since $F = g_\lambda$ on each J_λ) and obviously F is the maximal analytic continuation of f. $\qquad\square$

Example 5.4.14 The analytic function $f(x) = \sum_{n=0}^{\infty} (-1)^n x^n$, $|x| < 1$, has maximal analytic continuation $F(x) = 1/(1 + x)$ defined on $(-1, \infty)$. ♠

5.4.2 Analytic Functions and Ordinary Differential Equations

A natural way of constructing analytic functions is as solutions to linear ordinary differential equations.

Example 5.4.15 Consider the linear differential equation $y' = ay$, where $a \in \mathbb{R}$ and $y' = \frac{dy}{dx}$. We search for a solution $y(x)$ which satisfies the initial condition $y(0) = y_0$ (the analysis is the same if we specify $y(x_0)$, $x_0 \neq 0$).

We start by observing that if $y : \mathbb{R} \to \mathbb{R}$ is a C^∞ solution to $y' = ay$, then all the derivatives $y^{(n)}(0)$ are all uniquely determined by the initial condition. Indeed, since $y' = ay$ we have $y'(0) = ay(0) = ay_0$. Differentiating once, y must satisfy $y'' = ay'$ and so $y''(0) = ay'(0) = a^2 y_0$. Proceeding inductively, it is clear that for $n \geq 0$ we have

$$y^{(n)}(0) = a^n y_0.$$

Assume that y is analytic. Then $y(x) = \sum_{n=0}^{\infty} \frac{y^{(n)}(0)}{n!} x^n$ for $x \in (-r, r)$, where $r > 0$. Using our computed values of $y^{(n)}(0)$ we see that

$$y(x) = y_0 \sum_{n=0}^{\infty} \frac{(ax)^n}{n!}.$$

This power series has radius of convergence $R = \infty$. Using our results on term-by-term differentiation of a power series, we see easily that $y(x) = y_0 \sum_{n=0}^{\infty} \frac{(ax)^n}{n!} = y_0 e^{ax}$ is a solution of $y' = ay$ which is defined for all $x \in \mathbb{R}$ and satisfies the initial condition $y(0) = y_0$. Moreover, the solution is unique. To see this, suppose that $u(x)$ is a differentiable function defined on an open interval I containing $x = 0$ which satisfies $u' = au$ on I and $u(0) = y_0$. Define $v(x) = e^{-ax} u(x)$. For $x \in I$ we have

$$v'(x) = -ae^{-ax} u(x) + e^{-ax} u'(x) = -ae^{-ax} u(x) + ae^{-ax} u(x) = 0.$$

Therefore, v is constant on I. We have $v(0) = u(0) = y_0$ and so $v(x) = y_0$ for all $x \in I$. That is, $u(x) = y_0 e^{ax}$, $x \in I$.   ♠

Remark 5.4.16 It is worth summarizing the method used in the previous example. Given the initial condition, all the higher derivatives of a solution are uniquely determined. As a result the Taylor series of the solution at the origin is uniquely determined. We show that the Taylor series has non-zero radius of convergence and observe, using term-by-term differentiation, that the Taylor series defines a solution to the differential equation with the correct initial condition. Finally, we compare a solution with the right initial condition to the constructed solution and so verify uniqueness. In practice, for higher-order linear constant coefficient differential equations it is usually best to work over the complex numbers, though in some cases it is possible to work using just real numbers—see the exercises at the end of the section.   ✱

The next proposition gives a general result on the existence of analytic solutions to a linear ordinary equation. We omit the proof—which is most easily done using complex variable methods.

Proposition 5.4.17 *Consider the second-order linear differential equation*

$$y'' + a(x)y' + b(x)y = 0, \tag{5.6}$$

where $a, b \in C^{\omega}(\mathbb{R})$. Given $y_0, y_0' \in \mathbb{R}$, there exist solutions $y_1, y_2 \in C^{\omega}(\mathbb{R})$ to (5.6) satisfying

(a) $y_1(0) = y_0$, $y_1'(0) = 0$; $y_2(0) = 0$, $y_2'(0) = y_0'$.
(b) *If $y : I \to \mathbb{R}$ is a solution to (5.6) such that $y(0) = y_0$, $y'(0) = y_0'$ (so $0 \in I$), then $y = y_0 y_1 + y_0' y_2$ on I (in particular, solutions are uniquely specified by their initial conditions).*

EXERCISES 5.4.18

(1) Consider the ordinary differential equation $y'' = -y$. Suppose that $y(x) = \sum_{n=0}^{\infty} a_n x^n$ is a power series solution of the equation (assume a non-zero radius of convergence—you will verify this assumption later). Show that $y(x)$ is uniquely determined by $y(0)$ and $y'(0)$.

(a) If $y(0) = 1$, $y'(0) = 0$ denote the solution by $c(x)$. Verify that the power series you get has radius of convergence $R = \infty$.
(b) If $y(0) = 0$, $y'(0) = 1$ denote the solution by $s(x)$. Verify that the power series you get has radius of convergence $R = \infty$.
(c) Verify that $s' = c$, $c' = -s$ and hence that $s^2 + c^2 \equiv 1$ on $\mathbb{R}$.

(2) Let $\alpha \in \mathbb{R}$. Consider the analytic differential equation

$$y'(x) = \frac{\alpha}{1 + x} y(x)$$

on $(-1, 1)$.

(a) Verify that the unique solution with initial condition $y(0) = 1$ is given by $y(x) = (1 + x)^{\alpha}$. (Assume the uniqueness theorem for solutions of ordinary differential equations—see Chap. 7, Theorem 7.17.12.)
(b) Verify that the binomial series

$$y(x) = \sum_{n=0}^{\infty} \binom{\alpha}{n} x^k$$

has radius of convergence $R = 1$ for all $\alpha \in \mathbb{R}$, $\alpha \notin \mathbb{Z}_+$, and is a solution to the differential equation with $y(0) = 1$.
(c) Deduce the binomial series $(1 + x)^{\alpha} = \sum_{n=0}^{\infty} \binom{\alpha}{n} x^k$, $|x| < 1$.

(For an alternative proof, using Taylor's theorem, see Exercises 4.5.12(4). It can also be shown that the complex binomial series converges to $(1+z)^{\alpha}$ if $\alpha, z \in \mathbb{C}$, $|z| < 1$.)

5.5   Trigonometric and Fourier Series

In this section we consider the problems of approximating periodic functions by trigonometric polynomials and the representation of periodic functions by a trigonometric or *Fourier* series.

We start by giving the definition of a periodic function.

Definition 5.5.1 A function $f : \mathbb{R} \to \mathbb{R}$ is *periodic* with period $\tau > 0$ if

$$f(x + \tau) = f(x), \text{ for all } x \in \mathbb{R}.$$

We say f is τ-periodic.

Remarks 5.5.2

(1) We generally assume that the period τ is the smallest strictly positive real number such that $f(x + \tau) = f(x)$ for all $x \in \mathbb{R}$ (τ is then called the *prime period* of f). Of course, if $f(x + \tau) = f(x)$ for all $x \in \mathbb{R}$ and all $\tau > 0$, then f is constant.
(2) If $f : \mathbb{R} \to \mathbb{R}$ is τ-periodic then $f(x + m\tau) = f(x)$ for all $m \in \mathbb{Z}$.
(3) We may require that the period $\tau = 2\pi$—if not, define $\bar{f}(x) = f(\tau x/2\pi)$ and note that $\bar{f}$ has period 2π. ·   ✺

Example 5.5.3 Let $\omega > 0$. The functions $\sin(\omega x)$, $\cos(\omega x)$ both have period $2\pi/\omega$. The function $\sin^2(\omega x)$ has period π/ω.   ♠

Definition 5.5.4 A function $T : \mathbb{R} \to \mathbb{R}$ is a *trigonometric polynomial* of degree $N \geq 1$ if T can be written in the form

$$T(x) = a_0 + \sum_{n=1}^{N} (a_n \cos nx + b_n \sin nx),$$

where $a_N^2 + b_N^2 \neq 0$ is non-zero. Note that the period of T is 2π.

Example 5.5.5 Let $T(x) = a_0 + \sum_{n=1}^{N} (a_n \cos nx + b_n \sin nx)$ be a trigonometric polynomial of degree N. If we set $z = e^{ix}$, $\bar{z} = e^{-ix}$ then we may write T as a polynomial $p_T(z, \bar{z}) = \sum_{i=0}^{N} (c_i z^i + \bar{c}_i \bar{z}^i)$, where $c_0, \cdots, c_N \in \mathbb{C}$ are uniquely determined by the coefficients a_i, b_i. This observation explains the use of the term 'polynomial' in the definition of trigonometric polynomial.   ♠

We used the Weierstrass approximation theorem to uniformly approximate continuous functions on a closed bounded interval by polynomials. We may also use the Weierstrass approximation theorem to show that continuous periodic functions can be uniformly approximated by trigonometric polynomials. We give the proof of the next result in the appendix to this chapter.

Theorem 5.5.6 (Second Weierstrass Approximation Theorem) *Every continuous 2π-periodic function on $\mathbb{R}$ can be uniformly approximated by trigonometric polynomials.*

In practice, it turns out to be much more interesting to represent periodic functions by trigonometric series.

Definition 5.5.7 A *trigonometric series* is a series of the form

$$a_0 + \sum_{n=1}^{\infty}(a_n \cos nx + b_n \sin nx).$$

We will mainly be interested in the classes of piecewise continuous and piecewise differentiable functions. These are functions which have only jump discontinuities. We give the precise definition we use (see also Sect. 2.5.2).

Definition 5.5.8 A function $f : [a, b] \to \mathbb{R}$ is *piecewise continuous* if there exist a finite subset $\{d_j \mid j = 1, \cdots N\}$ of $[a, b]$ such that

(a) $a \leq d_1 < \cdots < d_N \leq b$.
(b) f is continuous, except at $x = d_1, \cdots, d_N$.
(c) For each j, $\lim_{x\to d_j-} f(x) = f(d_j-)$ and $\lim_{x\to d_j+} f(x) = f(d_j+)$ exist and are finite (we make the obvious variations if either $d_1 = a$ or $d_N = b$).

If f is defined on $\mathbb{R}$, then f is piecewise continuous if it is piecewise continuous restricted to every bounded closed interval $[a, b]$. A function f is piecewise C^1 if both f and f' are piecewise continuous.

Remarks 5.5.9

(1) We refer to the type of discontinuity described in Definition 5.5.8 as a *jump discontinuity*. The jump at a jump discontinuity d of f is defined to be $f(d+) - f(d-)$.
(2) Let $f : [a, b] \to \mathbb{R}$ be piecewise continuous with discontinuity points $a < d_1 < \cdots < d_N < b$. Set $d_0 = a$, $d_{N+1} = b$. It is sometimes useful to regard f as defining a *continuous* function f_j on each subinterval $[d_j, d_{j+1}]$, $j \in \{0, \ldots, N\}$. For this, we define $f_j(d_j) = f(d_j+)$ and $f_j(d_{j+1}) = f(d_{j+1}-)$. ✠

Examples 5.5.10

(1) If $f(x) = \sin(1/x)$, $x \neq 0$, and $f(0) = 0$, then f is *not* piecewise continuous.
(2) If we define $S(x) = 1$, $x \in [2n\pi, (2n + 1)\pi)$ and $S(x) = -1$, $x \in [(2n + 1)\pi, (2n + 2)\pi)$, $n \in \mathbb{Z}$, then S is piecewise continuous (indeed, piecewise smooth) and 2π-periodic. The function S defines a *square wave*.
(3) The function $f(x) = |x|$ is continuous and piecewise C^1. We have $\lim_{x\to 0+} f'(x) = +1$, $\lim_{x\to 0-} f'(x) = -1$. ♠

If the trigonometric series $a_0 + \sum_{n=1}^{\infty}(a_n \cos nx + b_n \sin nx)$ converges for all $x \in \mathbb{R}$, then $U(x) = a_0 + \sum_{n=1}^{\infty}(a_n \cos nx + b_n \sin nx)$ is 2π-periodic: $U(x + 2\pi) = U(x)$ for all $x \in \mathbb{R}$. In this section, we will be interested in representing 2π-

periodic functions as trigonometric series. Initially, we obtain results on pointwise convergence. Later we obtain results on uniform convergence. However, uniform convergence is not the most natural form of convergence to use when studying trigonometric series (unlike power series). Much better is the concept of *mean square convergence*, which we address later in the section.

We start by showing how every continuous (or piecewise continuous) 2π-periodic function $f : \mathbb{R} \rightarrow \mathbb{R}$ naturally determines a trigonometric series which we call the Fourier series of f. The problem will be to relate the Fourier series to the original function f.

Definition 5.5.11 Let f be a 2π-periodic function on $\mathbb{R}$ and assume that f is piecewise continuous (so f has finitely many jump discontinuities on $[0, 2\pi]$). The *Fourier series* $\mathcal{F}(f)$ of f is defined to be the infinite series

$$a_0 + \sum_{n=1}^{\infty} (a_n \cos nx + b_n \sin nx),$$

where

$$a_0 = \frac{1}{2\pi} \int_0^{2\pi} f(x) \, dx = \frac{1}{2\pi} \int_{-\pi}^{\pi} f(x) \, dx,$$

$$a_n = \frac{1}{\pi} \int_0^{2\pi} f(x) \cos nx \, dx = \frac{1}{\pi} \int_{-\pi}^{\pi} f(x) \cos nx \, dx, \ n \geq 1,$$

$$b_n = \frac{1}{\pi} \int_0^{2\pi} f(x) \sin nx \, dx = \frac{1}{\pi} \int_{-\pi}^{\pi} f(x) \sin nx \, dx, \ n \geq 1.$$

We refer to a_n, b_n as the *Fourier coefficients* of f.

Remarks 5.5.12

(1) It is common in the literature to take the first coefficient a_0 in the Fourier series to be $a_0/2$. With this convention, a_0 is half the average of f on $[-\pi, \pi]$, rather than the average as we have defined it. One way or another one has to deal with an anomalous factor or divisor of 2.
(2) If f is an even function ($f(-x) = f(x)$), then $b_n = 0$ for all $n \in \mathbb{N}$ since the integrand $f(x) \sin nx$ will be odd. Similarly if f is an odd function ($f(x) = -f(x)$) then $a_n = 0$ for all $n \in \mathbb{Z}_+$. ✠

Example 5.5.13 Let S be the 2π-periodic square wave function defined in Examples 5.5.10(2). Since S is odd, we have $a_n = 0$, $n \geq 0$ (see the previous remarks). On the other hand

$$b_n = \frac{1}{\pi} \int_{-\pi}^{\pi} S(x) \sin nx \, dx = \frac{2}{\pi} \int_0^{\pi} \sin nx \, dx.$$

It follows easily that for $n \geq 0$ we have

$$b_{2n} = 0,$$

$$b_{2n+1} = \frac{4}{(2n+1)\pi}.$$

Hence the Fourier series is $\mathcal{F}(S) = \frac{4}{\pi} \sum_{n=0}^{\infty} \frac{\sin(2n+1)x}{2n+1}$. ♠

Remark 5.5.14 Although most of the time we assume functions are 2π-periodic, in Chap. 6 we consider Fourier series of 1-periodic functions. For future reference, the Fourier coefficients of a 1-periodic function f are defined by

$$a_0 = \int_0^1 f(x)\, dx,$$

$$a_n = 2 \int_0^1 f(x) \cos 2n\pi x\, dx, \ n \geq 1,$$

$$b_n = 2 \int_0^1 f(x) \sin 2n\pi x\, dx, \ n \geq 1,$$

and the corresponding Fourier series $\mathcal{F}(f)$ is

$$a_0 + \sum_{n=1}^{\infty} (a_n \cos 2\pi nx + b_n \sin 2\pi nx).$$

✠

5.5.1 The Orthogonality Relations

We compute the Fourier coefficients of $\cos px$, $\sin px$, $p \geq 0$,

Lemma 5.5.15

$$\frac{1}{\pi} \int_0^{2\pi} \cos px \cos nx\, dx = \begin{cases} 1 & \text{if } p = n,\ p, n \geq 1, \\ 0 & \text{if } p \neq n, \\ 2 & \text{if } p = n = 0, \end{cases}$$

$$\frac{1}{\pi} \int_0^{2\pi} \sin px \cos nx\, dx = 0, \ \text{if } p \geq 1,\ n \geq 0,$$

$$\frac{1}{\pi} \int_0^{2\pi} \sin px \sin nx\, dx = \begin{cases} 0 & \text{if } p \neq n,\ p, n \geq 1, \\ 1 & \text{if } p = n \geq 1. \end{cases}$$

Proof All of the statements follow using standard trigonometric identities such as $\cos A \cos B = \frac{1}{2}(\cos(A+B)+\cos(A-B))$ and we leave the details to the reader. □

As an important corollary of the second Weierstrass approximation theorem (Theorem 5.5.6) we have

Theorem 5.5.16 *Let $f : \mathbb{R} \to \mathbb{R}$ be continuous and 2π-periodic. If all the Fourier coefficients of f are zero, then $f \equiv 0$. In particular, if continuous and 2π-periodic functions $f, g : \mathbb{R} \to \mathbb{R}$ have the same Fourier series, then $f = g$.*

Proof We leave it to the exercises at the end of the section. □

Remarks 5.5.17

(1) The reader should note the similarity of this result to Proposition 5.3.6.
(2) Theorem 5.5.16 shows that the Fourier coefficients are important invariants of the function f—notwithstanding any issues about convergence of the Fourier series.
(3) Theorem 5.5.16 is true if we only assume piecewise continuity. ✠

We can give additional justification for our definition of the Fourier coefficients a_n, b_n when the Fourier series converges uniformly.

Proposition 5.5.18 *Suppose that $a_0 + \sum_{n=1}^{\infty}(a_n \cos nx + b_n \sin nx)$ converges uniformly on $[0, 2\pi]$ to the function f. Then the a_n and b_n must be the Fourier coefficients of f.*

Proof If $a_0 + \sum_{n=1}^{\infty}(a_n \cos nx + b_n \sin nx)$ converges uniformly on $[0, 2\pi]$ to f then f is continuous on $[0, 2\pi]$ since the partial sums $S_N = a_0 + \sum_{n=1}^{N}(a_n \cos nx + b_n \sin nx)$ are continuous. If (S_N) converges uniformly to f then $S_N(x) \cos nx$ converges uniformly to $f(x) \cos nx$ on $[0, 2\pi]$ for all $n \geq 0$. Hence

$$\frac{1}{\pi} \int_0^{2\pi} S_N(x) \cos nx \, dx \to \frac{1}{\pi} \int_0^{2\pi} f(x) \cos nx \, dx.$$

Using the orthogonality relations (Lemma 5.5.15), we have

$$\frac{1}{\pi} \int_0^{2\pi} S_N(x) \cos nx \, dx = \begin{cases} a_n, & \text{if } N \geq n > 0, \\ 2a_0, & \text{if } n = 0. \end{cases}$$

Hence $a_0 = \frac{1}{2\pi} \int_0^{2\pi} f(x) \, dx$ and $a_n = \frac{1}{\pi} \int_0^{2\pi} f(x) \cos nx \, dx$, $n \geq 1$. A similar analysis applies to the coefficients b_n, $n \geq 0$. □

Remark 5.5.19 Theorem 5.5.16 implies that a continuous 2π-periodic function is uniquely determined by its Fourier coefficients. Proposition 5.5.18 gives conditions under which we can reconstruct f given the Fourier coefficients. A general resolution of this 'inverse' problem motivates much of the more advanced work on Fourier series. For example, a very natural question to ask is which trigonometric series are the Fourier series of a continuous or smooth function. ✠

5.5.2 The Riemann–Lebesgue Lemma

Lemma 5.5.20 (The Riemann–Lebesgue Lemma) *Let* $f : [a, b] \to \mathbb{R}$ *be piecewise continuous. Then*

$$\lim_{\lambda \to \infty} \int_a^b f(x) \cos \lambda x \, dx = \lim_{\lambda \to \infty} \int_a^b f(x) \sin \lambda x \, dx = 0.$$

Proof We prove that $\lim_{\lambda \to \infty} \int_a^b f(x) \cos \lambda x \, dx = 0$, the analysis for the second integral is similar. We start by assuming f is C^1 on $[a, b]$. Integrating by parts, we have

$$\int_a^b f(x) \cos \lambda x \, dx = \frac{1}{\lambda}(f(b) \sin \lambda b - f(a) \sin \lambda a) - \frac{1}{\lambda} \int_a^b f'(x) \sin \lambda x \, dx.$$

Let C be an upper bound for $|f|$ and $|f'|$ on $[a, b]$ (this uses the continuity of f, f'). We have the estimate

$$\left| \int_a^b f(x) \cos \lambda x \, dx \right| \leq 2C/\lambda + C(b - a)/\lambda.$$

Letting $\lambda \to \infty$ we have $\lim_{\lambda \to \infty} \int_a^b f(x) \cos \lambda x \, dx = 0$. Next, assume only that f is continuous. Given $\varepsilon > 0$, it suffices to show that there exists a $\lambda_0 > 0$ such that $|\int_a^b f(x) \cos \lambda x \, dx| < \varepsilon$, for all $\lambda \geq \lambda_0$. By the Weierstrass approximation theorem, we can find a polynomial $p : [a, b] \to \mathbb{R}$ such that

$$\|f - p\| = \sup_{x \in [a,b]} |f(x) - p(x)| < \frac{\varepsilon}{2(b - a)}.$$

Since p is C^1, we can find a $\lambda_0 > 0$ such that

$$\left| \int_a^b p(x) \cos \lambda x \, dx \right| \leq \frac{\varepsilon}{2}, \quad \text{for all } \lambda \geq \lambda_0.$$

We have

$$\left| \int_a^b f(x) \cos \lambda x \, dx \right| = \left| \int_a^b (f - p) \cos \lambda x \, dx + \int_a^b p \cos \lambda x \, dx \right|$$

$$\leq \left| \int_a^b (f - p)(x) \cos \lambda x \, dx \right| + \left| \int_a^b p(x) \cos \lambda x \, dx \right|$$

$$\leq \int_a^b |f(x) - p(x)| \, dx + \left| \int_a^b p(x) \cos \lambda x \, dx \right|$$

$$< (b - a) \frac{\varepsilon}{2(b - a)} + \frac{\varepsilon}{2} = \varepsilon, \quad \text{for all } \lambda \geq \lambda_0.$$

It remains to prove the case when f is only piecewise continuous. Suppose that f has jump discontinuities at $d_1 < d_2 < \cdots < d_{N-1}$, where $a < d_1$ and $d_N < b$. We give two proofs.

Method 1. Set $d_0 = a$, $d_N = b$. We have

$$\int_a^b f(x) \cos \lambda x \, dx = \sum_{n=0}^{N-1} \int_{d_n}^{d_{n+1}} f_n(x) \cos \lambda x \, dx,$$

where f_n is defined as in Remarks 5.5.9(2). Since f_n is continuous on $[d_n, d_{n+1}]$, we have $\lim_{\lambda \to \infty} \int_{d_n}^{d_{n+1}} f(x) \cos \lambda x \, dx = 0$, $n = 0, \cdots, N-1$. The result follows.

Method 2. Given $\varepsilon > 0$, we can approximate f by a continuous function f_ε so that $\int_a^b |f(x) - f_\varepsilon(x)| \, dx < \varepsilon/(b-a)$—see Fig. 5.3. We have

$$\int_a^b f(x) \cos \lambda x \, dx = \int_a^b f_\varepsilon(x) \cos \lambda x \, dx + \int_a^b (f_\varepsilon(x) - f(x)) \cos \lambda x \, dx.$$

Therefore $|\int_a^b f(x) \cos \lambda x \, dx| \le |\int_a^b f_\varepsilon(x) \cos \lambda x \, dx| + \varepsilon$ and

$$\lim_{\lambda \to \infty} \int_a^b f(x) \cos \lambda x \, dx \le \varepsilon.$$

Since this holds for all $\varepsilon > 0$, the result follows.   □

Remarks 5.5.21

(1) It is easy to extend the Riemann–Lebesgue Lemma to bounded functions on $[a, b]$ with finite or countably many discontinuities. The proof follows Method 2 of the proof of Lemma 5.5.20. Stronger versions hold if we use the Lebesgue integral rather than the Riemann integral.

(2) It is useful to have a slightly stronger version of the Riemann–Lebesgue lemma that holds for continuous families of continuous functions. Suppose $f : [a, b] \times [c, d] \to \mathbb{R}$ is continuous and set $f_v(x) = f(x, v)$, $x \in [a, b]$, $v \in [c, d]$. If we define $C(\lambda, v) = \int_a^b f_v(x) \cos \lambda x \, dx$, $S(\lambda, v) = \int_a^b f_v(x) \sin \lambda x \, dx$, then given

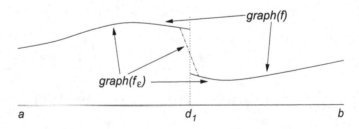

Fig. 5.3 Approximating a piecewise continuous function by a continuous function

$\varepsilon > 0$, there exists a λ_0 such that

$$|C(\lambda, \nu)|, |S(\lambda, \nu)| < \varepsilon,$$

for all $\lambda \geq \lambda_0$ and $\nu \in [c, d]$. The proof is the same as that given above except that we use the Weierstrass approximation theorem for continuous families, Theorem 5.3.10. We can even allow for jump discontinuities at points $d_1, d_2, \cdots, d_{N-1} \in (a, b)$ provided we assume (say) that the discontinuity points do not depend on the parameter ν. ✖

5.5.3 Integral Formula for Partial Sums of a Fourier Series

Definition 5.5.22 Let $n \geq 0$. The nth *Dirichlet kernel* $D_n(x)$ is defined by

$$D_n(x) = 1 + 2 \sum_{j=1}^{n} \cos jx, \ x \in \mathbb{R}.$$

The collection $\{D_n \mid n \geq 1\}$ is called the *Dirichlet kernel*.

The next lemma gives two elementary but useful properties of the Dirichlet kernel.

Lemma 5.5.23 *If $n \geq 0$, then*

(1) $\int_0^{2\pi} D_n(x)\, dx = \int_{-\pi}^{\pi} D_n(x)\, dx = 2\pi$,

(2) $D_n(x) = \frac{\sin((n+\frac{1}{2})x)}{\sin \frac{x}{2}}$, *if x is not an integer multiple of 2π.*

Proof The first statement is an immediate consequence of the orthogonality relations. Next, from the trigonometric identities in the appendix to Chap. 3, we have

$$1 + 2 \sum_{j=1}^{n} \cos jx = 1 + \frac{2 \cos(\frac{n+1}{2}x) \sin(\frac{nx}{2})}{\sin(\frac{x}{2})}.$$

Since $2 \cos A \sin B = \sin(A+B) - \sin(A-B)$, it is easy to verify that the right-hand side is equal to $\frac{\sin((n+\frac{1}{2})x)}{\sin \frac{x}{2}}$. □

Remarks 5.5.24

(1) By Lemma 5.5.23(2), or the definition of $D_n(x)$, we have $\lim_{x \to 0} D_n(x) = 2n + 1$.

(2) As we shall soon see the function $D_n(x)$ plays an important role in the convergence theory of Fourier series. The integral $\int_{-\pi}^{\pi} |D_n(x)|\, dx$ grows like $\log n$ and this lack of convergence is reflected in the fact that the Fourier series of a continuous function may not converge pointwise at every point. It can be

shown that the Fourier series of a continuous function does converge at 'most' points. However, the proof of this result, due to Carleson (1966), is hard. As we shall see, adding a little regularity to the function improves the convergence properties of the Fourier series.   ✠

Suppose $f : \mathbb{R} \to \mathbb{R}$ is a piecewise continuous 2π-periodic function. For $n \geq 0$ define the partial sums

$$S_n(f)(x) = a_0 + \sum_{j=1}^{n}(a_j \cos jx + b_j \sin jx),$$

where a_j, b_j are the Fourier coefficients of f.

Lemma 5.5.25 (Partial Sum Formula) *(Notation as above.) For $n \geq 0$ we have*

$$S_n(f)(x) = \frac{1}{2\pi}\int_{-\pi}^{\pi} f(t)D_n(t - x)\, dt$$

$$= \frac{1}{2\pi}\int_{-\pi}^{\pi} f(x - t)D_n(t)\, dt \qquad (5.7)$$

$$= \frac{1}{2\pi}\int_{-\pi}^{\pi} f(x + t)D_n(t)\, dt.$$

Proof We have

$$S_n(f)(x) = \frac{1}{2\pi}\int_{-\pi}^{\pi} f(x)\, dx + \sum_{j=1}^{n}\frac{1}{\pi}\left[\left(\int_{-\pi}^{\pi} f(t)\cos jt\, dt\right)\cos jx\right]$$

$$+ \sum_{j=1}^{n}\frac{1}{\pi}\left[\left(\int_{-\pi}^{\pi} f(t)\sin jt\, dt\right)\sin jx\right]$$

$$= \frac{1}{2\pi}\int_{-\pi}^{\pi} f(t)[1 + 2\sum_{j=1}^{n}\cos(j(x - t))]\, dt,$$

where we have used the trigonometric identity $\cos(A - B) = \cos A \cos B + \sin A \sin B$. Hence, by definition of D_n, we have

$$S_n(f)(x) = \frac{1}{2\pi}\int_{-\pi}^{\pi} f(t)D_n(x - t)\, dt$$

$$= \frac{1}{2\pi}\int_{-\pi}^{\pi} f(t)D_n(t - x)\, dt,$$

since D_n is even. For the second formula (the only formula we use in the sequel), we make the substitution $u = x - t$ to obtain

$$
\begin{aligned}
S_n(f)(x) &= \frac{1}{2\pi} \int_{x+\pi}^{x-\pi} -f(x-u)D_n(-u)\, du \\
&= -\frac{1}{2\pi} \int_{\pi}^{-\pi} f(x-u)D_n(u)\, du, \quad \text{periodicity, evenness of } D_n, \\
&= \frac{1}{2\pi} \int_{-\pi}^{\pi} f(x-u)D_n(u)\, du.
\end{aligned}
$$

The proof of the third formula is similar and left as an exercise.   □

Theorem 5.5.26 *Let* $f : \mathbb{R} \to \mathbb{R}$ *be a* 2π-*periodic piecewise continuous function and let* $x_0 \in \mathbb{R}$. *Set* $f(x_0+) = \lim_{x \to x_0+} f(x)$, $f(x_0-) = \lim_{x \to x_0-} f(x)$ *and assume that*

$$
D_R = \lim_{t \to 0+} \frac{f(x_0 + t) - f(x_0+)}{t},
$$

$$
D_L = \lim_{t \to 0-} \frac{f(x_0 + t) - f(x_0-)}{t}
$$

exist. Then the Fourier series of f *is convergent at* x_0 *and*

$$
\mathcal{F}(f)(x_0) = \frac{1}{2} [f(x_0-) + f(x_0+)] .
$$

In particular, if f *is continuous and piecewise differentiable on* $\mathbb{R}$ *then* $\mathcal{F}(f)(x)$ *converges to* $f(x)$ *for all* $x \in \mathbb{R}$.

Before we start the proof of the theorem we need a technical lemma that allows us to use the differentiability properties of f at x_0.

Lemma 5.5.27 *Assume* f *satisfies the conditions of Theorem 5.5.26 and define* $g :$ $[-\pi, \pi] \to \mathbb{R}$ *by*

$$
g(t) = \begin{cases}
\frac{f(x_0-t)-f(x_0-)}{\sin \frac{t}{2}}, & t > 0, \\
\frac{f(x_0-t)-f(x_0+)}{\sin \frac{t}{2}}, & t < 0, \\
-(D_L + D_R), & t = 0,
\end{cases}
$$

then $\lim_{t \to 0+} g(t) = -2D_L$, $\lim_{t \to 0-} g(t) = -2D_R$. *In particular,*

(a) g *is piecewise continuous on* $[-\pi, \pi]$ *with a discontinuity at* $t = 0$ *if* $D_L \neq D_R$.

(b) *If* f *is differentiable at* x_0, *then* g *is continuous at zero and* $g(0) = -2f'(x_0)$.

Proof For $t > 0$, we have

$$\frac{f(x_0 - t) - f(x_0-)}{\sin \frac{t}{2}} = -2 \frac{f(x_0 - t) - f(x_0-)}{-t} \frac{\frac{t}{2}}{\sin \frac{t}{2}}$$

$$\to -2D_L, \text{ as } t \to 0+.$$

The same argument shows that $\lim_{t \to 0-} g(t) = -2D_R$. □

Proof of Theorem 5.5.26 We start by observing that since $D_n(t)$ is even we have by Lemma 5.5.23(1)

$$\frac{1}{2}[f(x_0-) + f(x_0+)] = \frac{1}{2\pi} \left(\int_0^\pi D_n(t)f(x_0-) \, dt + \int_{-\pi}^0 D_n(t)f(x_0+) \, dt \right).$$

By the partial sum formula (5.7), we see easily that

$$S_n(f)(x_0) - \frac{1}{2}[f(x_0-) + f(x_0+)] = I_- + I_+,$$

where

$$I_- = \frac{1}{2\pi} \int_0^\pi \frac{f(x_0 - t) - f(x_0-)}{\sin(\frac{t}{2})} \sin\left(\left(n + \frac{1}{2}\right)t\right) dt,$$

$$I_+ = \frac{1}{2\pi} \int_{-\pi}^0 \frac{f(x_0 - t) - f(x_0+)}{\sin(\frac{t}{2})} \sin\left(\left(n + \frac{1}{2}\right)t\right) dt.$$

Hence

$$I_- + I_+ = \int_{-\pi}^\pi g(t) \sin\left(\left(n + \frac{1}{2}\right)t\right) dt,$$

where g is piecewise continuous on $[-\pi, \pi]$ by Lemma 5.5.27.

We have $\lim_{n \to \infty} \int_{-\pi}^\pi g(t) \sin((n + \frac{1}{2})t) \, dt = 0$ (by the Riemann–Lebesgue lemma) and so $\lim_{n \to \infty} S_n(f)(x_0) = (f(x_0-) + f(x_0+))/2$. □

Example 5.5.28 Let S be the 2π-periodic square wave function with Fourier series $\frac{4}{\pi} \sum_{n=1}^\infty \frac{\sin((2n+1)x)}{2n+1}$ (Example 5.5.13). As a result of Theorem 5.5.26, we see that

$$\frac{4}{\pi} \sum_{n=1}^\infty \frac{\sin((2n + 1)x)}{2n + 1} = \begin{cases} 1, & \text{if } x \in \bigcup_{n \in \mathbb{Z}}(2n\pi, (2n + 1)\pi), \\ -1, & \text{if } x \in \bigcup_{n \in \mathbb{Z}}((2n + 1)\pi, (2n + 2)\pi), \\ 0, & \text{if } x \text{ is an integer multiple of } \pi. \end{cases}$$

♠

Theorem 5.5.29 *If $f : \mathbb{R} \to \mathbb{R}$ is continuous, 2π-periodic and piecewise C^1, then the Fourier series of f converges uniformly to f.*

Proof Suppose first that f is C^1. Then, as in the proof of Theorem 5.5.26, we may write $S_n(f)(x) - f(x) = \frac{1}{2\pi} \int_{-\pi}^{\pi} g(x, t) \sin((n + \frac{1}{2})t) \, dt$, where $g(x, t)$ is continuous. We regard $g(x, t) = g_x(t)$ as a continuous family of continuous functions and apply the Riemann–Lebesgue lemma for families (Remark 5.5.21(2)) to get the required estimate for uniform convergence. If we assume that f is continuous and piecewise C^1, then the same argument gives the uniform estimates needed for the Riemann–Lebesgue lemma on any closed interval not containing a discontinuity of f' as an interior point. $\qquad\Box$

5.5.4 Failure of Uniform Convergence: Gibbs Phenomenon

The *Gibbs phenomenon* is the appearance of quite large oscillations in the partial sums $S_n(x)$ to the left and right of a jump discontinuity. The resulting 'overshoot' in the partial sums does not die out as $n \to \infty$. We illustrate the phenomenon with an investigation of the convergence properties of the Fourier series of the square wave function.

The 2π-periodic square wave $S(x)$ defined in Examples 5.5.10(2) does not satisfy the conditions of Theorem 5.5.29 as S has discontinuities at integer multiples of 2π. We showed in Example 5.5.13 that S has Fourier series

$$\mathcal{F}(S) = \frac{4}{\pi} \sum_{n=0}^{\infty} \frac{\sin(2n + 1)x}{2n + 1},$$

and in Examples 5.5.28(1) that the series converges pointwise to $S(x)$ except if x is an integer multiple of π. Since the pointwise limit of $\mathcal{F}(S)$ is not continuous, convergence of $\mathcal{F}(S)$ cannot be uniform. On the other hand, a straightforward application of Dirichlet's test shows that convergence of $\mathcal{F}(S)$ is uniform on every closed interval $[a, b]$ which does not contain an integer multiple of 2π (see the section on Dirichlet and Abel's tests in Chap. 4, especially Examples 4.6.3(2)).

We have $S_n(S)(x) = \frac{4}{\pi} \sum_{j=1}^{n} \frac{1}{2j-1} \sin((2j-1)x)$. Taking $x = \frac{\pi}{2n}$, we compute that

$$S_n(S)\left(\frac{\pi}{2n}\right) = \frac{4}{\pi} \sum_{j=1}^{n} \frac{1}{2j - 1} \sin\left(\frac{(2j - 1)\pi}{2n}\right)$$

$$= 2 \sum_{j=1}^{n} \frac{1}{n} G\left(\frac{2j - 1}{2n}\right),$$

where $G(0) = 1$ and $G(x) = \frac{\sin \pi x}{\pi x}, x \neq 0$. Now $\sum_{j=1}^{n} \frac{1}{n} G(\frac{2j-1}{2n})$ is an approximating Riemann sum to

$$\int_0^1 G(x)\, dx = \int_0^1 \frac{\sin \pi x}{\pi x}\, dx = \frac{1}{\pi} \int_0^\pi \frac{\sin u}{u}\, du.$$

$(\sum_{j=1}^{n} \frac{1}{n} G(\frac{2j-1}{2n})$ is the sum from $j = 0, \cdots, n-1$ of the value of G at the mid-point of $[j/n, (j+1)/n]$ times the length of the interval—$1/n$.) Since G is continuous, we therefore have

$$\lim_{n \to \infty} S_n(S) \left(\frac{\pi}{2n}\right) = \frac{2}{\pi} \int_0^\pi \frac{\sin u}{u}\, du.$$

The integral $\int_0^\pi \frac{\sin u}{u}\, du$ may be computed numerically and has approximate value 1.8519. Hence $\lim_{n \to \infty} S_n(S)(\frac{\pi}{2n}) \approx 3.7038/\pi \approx 1.179$. The jump in S at $x = 0$ is equal to 2 and so we see that

$$\lim_{n \to \infty} S_n(S) \left(\frac{\pi}{2n}\right) \approx 1.179 \approx 1 + 0.0895 * 2.$$

Hence the overshoot is approximately 8.95% of the jump at $x = 0$.

Remarks 5.5.30

(1) The overshoot described above is a universal phenomenon: whenever there is a jump discontinuity in a piecewise C^1-function, there will be an overshoot in the partial sums near the discontinuity and this overshoot in the limit is approximately 8.9490% of the jump at the discontinuity. The phenomenon was originally described (partly incorrectly) by Gibbs in 1848 and later corrected by him in 1898.

(2) It is worth remarking that the rate at which the Fourier coefficients of a function f converge to zero depends on the smoothness of f. If f is C^∞ (or analytic) the coefficients decay very rapidly (see Exercises 5.5.33(12)). If the function is only piecewise continuous then the series of coefficients is never absolutely convergent (else the M-test would imply convergence to a continuous function). ✠

5.5.5 The Infinite Product Formula for sin x

We show how to derive the infinite product formula $\sin x = x \prod_{n=1}^{\infty} (1 - \frac{x^2}{n^2 \pi^2}), x \in \mathbb{R}$, using methods based on Fourier series.

We start by finding the Fourier series of the 2π-periodic continuous piecewise C^1 function f on $\mathbb{R}$ defined by

$$f(x) = \cos \frac{\lambda x}{\pi}, \quad x \in [-\pi, \pi],$$

where we shall assume λ is not an integer multiple of π. Since $\cos x$ is even, f is an even function of x and so all the Fourier sine coefficients $b_n = 0$. We have

$$a_0 = \frac{1}{\pi} \int_0^\pi \cos \frac{\lambda x}{\pi} \, dx = \frac{\sin \lambda}{\lambda}$$

and

$$
\begin{aligned}
a_n &= \frac{2}{\pi} \int_0^\pi \cos \frac{\lambda x}{\pi} \cos nx \, dx \\
&= \frac{1}{\pi} \int_0^\pi \cos \left(\frac{\lambda}{\pi} + n \right) x + \cos \left(\frac{\lambda}{\pi} - n \right) x \, dx \\
&= \frac{1}{\pi} \left[\frac{\sin(\frac{\lambda}{\pi} + n)x}{\frac{\lambda}{\pi} + n} + \frac{\sin(\frac{\lambda}{\pi} - n)x}{\frac{\lambda}{\pi} - n} \right]_{x=0}^{x=\pi} \\
&= \frac{1}{\pi} \left[\frac{\sin(\frac{\lambda}{\pi} + n)\pi}{\frac{\lambda}{\pi} + n} + \frac{\sin(\frac{\lambda}{\pi} - n)\pi}{\frac{\lambda}{\pi} - n} \right] \\
&= \frac{(-1)^n \sin \lambda}{\frac{\lambda}{\pi} + n} + \frac{(-1)^n \sin \lambda}{\frac{\lambda}{\pi} - n} \\
&= (-1)^n \frac{2\lambda \sin \lambda}{\lambda^2 - \pi^2 n^2}.
\end{aligned}
$$

Since f is continuous and piecewise differentiable, the Fourier series of f converges pointwise to f and so for all $x \in \mathbb{R}$ we have

$$\cos \frac{\lambda x}{\pi} = \frac{\sin \lambda}{\lambda} + \sum_{n=1}^\infty (-1)^n \frac{2\lambda \sin \lambda}{\lambda^2 - \pi^2 n^2} \cos nx.$$

If we take $x = \pi$ and divide both sides by $\sin \lambda$ we get the partial fraction expansion for $\cot \lambda$:

$$\cot \lambda = \frac{1}{\lambda} + \sum_{n=1}^\infty \frac{2\lambda}{\lambda^2 - n^2\pi^2}, \quad \text{for all } \lambda \in \mathbb{R} \smallsetminus \pi\mathbb{Z}.$$

Let $0 < \varepsilon \le x < \pi$. We have

$$
\begin{aligned}
\int_\varepsilon^x \left(\cot \lambda - \frac{1}{\lambda} \right) d\lambda &= \int_\varepsilon^x \sum_{n=1}^\infty \frac{2\lambda}{\lambda^2 - n^2\pi^2} \, d\lambda \\
&= -\sum_{n=1}^\infty \int_\varepsilon^x \frac{2\lambda}{n^2\pi^2 - \lambda^2} \, d\lambda,
\end{aligned}
$$

since it follows easily from the M-test that the series $\sum_{n=1}^{\infty} \frac{2\lambda}{\lambda^2 - n^2\pi^2}$ is uniformly convergent on $[\varepsilon, x]$ (this requires $\lambda^2 - n^2\pi^2 \neq 0$ on $[\varepsilon, x]$, which is so since $0 < \varepsilon \le x < \pi$). Integrating, we see that

$$[\log(\sin\lambda) - \log\lambda]_{\lambda=\varepsilon}^{\lambda=x} = \sum_{n=1}^{\infty} [\log(n^2\pi^2 - \lambda^2)]_{\lambda=\varepsilon}^{\lambda=x}.$$

Evaluating, we obtain the identity

$$\log\left(\frac{\sin x}{x}\right) - \log\left(\frac{\sin\varepsilon}{\varepsilon}\right) = \sum_{n=1}^{\infty} \log\left(\frac{n^2\pi^2 - x^2}{n^2\pi^2 - \varepsilon^2}\right).$$

Letting $\varepsilon \to 0+$, we get

$$\log\left(\frac{\sin x}{x}\right) = \sum_{n=1}^{\infty} \log\left(1 - \frac{x^2}{n^2\pi^2}\right).$$

Exponentiating this expression and multiplying both sides by x gives

$$\sin x = x \prod_{n=1}^{\infty}\left(1 - \frac{x^2}{n^2\pi^2}\right), \quad x \in [0, \pi).$$

Since both $\sin x$ and $x \prod_{n=1}^{\infty}\left(1 - \frac{x^2}{n^2\pi^2}\right)$ are odd functions and vanish at $x = \pm\pi$, we have $\sin x = x \prod_{n=1}^{\infty}\left(1 - \frac{x^2}{n^2\pi^2}\right)$ on $[-\pi, \pi]$. Finally, $\sin x$ and $x \prod_{n=1}^{\infty}\left(1 - \frac{x^2}{n^2\pi^2}\right)$ are both 2π-periodic (if $G(x)$ denotes the infinite product, it is enough to show $G(x + \pi) = -G(x)$—see the exercises at the end of the section). Hence $\sin x = x \prod_{n=1}^{\infty}\left(1 - \frac{x^2}{n^2\pi^2}\right)$ for all $x \in \mathbb{R}$.

Remark 5.5.31 This proof only applies when $x \in \mathbb{R}$. The proof we gave in Chap. 3 holds for $x \in \mathbb{C}$. Note that the infinite product converges by Lemmas 3.9.4, 3.9.6 (or Proposition 3.9.10). ✠

Example 5.5.32 (Wallis' Formula for π) Dividing the infinite product formula for $\sin x$ by x and taking $x = \pi/2$ gives the identity

$$\frac{2}{\pi} = \prod_{n=1}^{\infty}\left(1 - \frac{1}{4n^2}\right) = \prod_{n=1}^{\infty}\left(\frac{4n^2 - 1}{4n^2}\right).$$

Taking the reciprocal of both sides and noting that $\frac{4n^2}{4n^2-1} = \frac{(2n)(2n)}{(2n-1)(2n+1)}$ gives Wallis' formula for $\pi/2$:

$$\frac{\pi}{2} = \prod_{n=1}^{\infty} \frac{(2n)(2n)}{(2n-1)(2n+1)} = \frac{2}{1} \cdot \frac{2}{3} \cdot \frac{4}{3} \cdot \frac{4}{5} \cdot \frac{6}{5} \cdot \frac{6}{7} \cdot \frac{8}{7} \cdot \frac{8}{9} \cdots . \qquad \spadesuit$$

EXERCISES 5.5.33

(1) Show that every 2π-periodic polynomial $p : \mathbb{R} \to \mathbb{R}$ is constant.
(2) Let f be continuous and 2π-periodic. Using the second Weierstrass approximation theorem, show that if all the Fourier coefficients of f are zero then $f \equiv 0$ (Theorem 5.5.16).
(3) Show that the Riemann–Lebesgue lemma holds on $[0, b]$ if $f(x) = 1/\sqrt{x}$.
(4) Let f be a piecewise continuous 2π-periodic functions. Show that the Fourier coefficients a_n, b_n of f converge to zero as $n \to \infty$.
(5) Extend Theorem 5.5.16 and Proposition 5.3.6 to piecewise continuous functions. (Hint: use the second method given at the end of the proof of the Riemann–Lebesgue Lemma 5.5.20.) Show that the same method shows that these results also hold if we only assume f is (a) bounded, and (b) has finitely many discontinuities.
(6) Let $f : \mathbb{R} \to \mathbb{R}$ be continuous and 2π-periodic with Fourier series $\mathcal{F}(f) = a_0 + \sum_{n=1}^{\infty}(a_n \cos nx + b_n \sin nx)$. Suppose that (A) $\mathcal{F}(f)$ converges at one point of $\mathbb{R}$, (B) the series $\sum_{n=1}^{\infty}(-na_n \sin nx + nb_n \cos nx)$ is uniformly convergent on $\mathbb{R}$. Show that f is C^1 and that $\mathcal{F}(f)$ converges uniformly to f on $\mathbb{R}$. (Hints: $\mathcal{F}(\mathcal{F}(f)) = \mathcal{F}(f)$ and the result of (2) above).
(7) Define the continuous 2π-periodic function $T : \mathbb{R} \to \mathbb{R}$ by

$$T(x) = \pi - |x|, \ x \in [-\pi, \pi].$$

(a) Sketch the graph of T.
(b) Find the Fourier series of T.
(c) Does the Fourier series converge pointwise to T? uniformly on $\mathbb{R}$ to T? Why/Why not?

(8) Define the piecewise continuous 2π-periodic function $S : \mathbb{R} \to \mathbb{R}$ on $[-\pi, \pi]$ by

$$S(x) = \begin{cases} x, \ x \in (-\pi, \pi), \\ 0, \ x = \pm\pi. \end{cases}$$

(a) Sketch the graph of S.
(b) Find the Fourier series of S.
(c) Does the Fourier series converge pointwise to S? uniformly on $\mathbb{R}$ to S?

If you have access to a program like *Maple* or *Matlab*, plot the graphs of the partial sums $S_{20}(S)$ and $S_{50}(S)$ over the range $[-3\pi, 3\pi]$ and estimate the overshoot as a percentage of the jump 2π.

(9) Show that the Fourier series of the 2π-periodic sawtooth function defined by

$$S(x) = \begin{cases} \frac{\pi-x}{2}, & x \in (0, 2\pi), \\ 0, & x = 0, 2\pi \end{cases}$$

is given by

$$\mathcal{F}(S) = \sum_{n=1}^{\infty} \frac{\sin nx}{n}.$$

(a) Show the partial sums S_n of the Fourier series of S satisfy

$$S_n(x) = \frac{1}{2} \int_0^x D_n(t)\, dt - \frac{x}{2}.$$

(b) Using the approximation $\sin t \approx t$, for t small, deduce that there exists a $C \geq 0$ (independent of n) such that

$$S_n(x) = \int_0^x \frac{\sin(n + \frac{1}{2})t}{t}\, dt + e_n(x),$$

where $|e_n(x)| \leq Cx$, $x \in (0, 2\pi)$.

(c) Take $x = \pi/(n + \frac{1}{2})$ in (b) and deduce that

$$S_n\left(\frac{\pi}{n + \frac{1}{2}}\right) = \int_0^\pi \frac{\sin u}{u}\, du + \alpha_n,$$

where $|\alpha_n| \leq c/n$. Using the approximate value 1.852 for $\int_0^\pi \frac{\sin u}{u}\, du$, deduce that $S_n(\pi/(n + \frac{1}{2})) \approx \pi/2 + 0.09 \times \pi$ for n large. That is, for large n the overshoot is (at least) 9% of the size of the jump at the discontinuity.

(10) Let $F : \mathbb{R} \to \mathbb{R}$ be the piecewise continuous 2π-periodic function defined on $[-\pi, \pi]$ by

$$F(x) = \begin{cases} 0, & \text{if } x \in [-\pi, -\pi/2] \cup [\pi/2, \pi], \\ \pi, & \text{if } x \in (-\pi/2, \pi/2). \end{cases}$$

(a) Find the Fourier series of F.

(b) At what points of $[-\pi, \pi]$ does $\mathcal{F}(F)$ converge pointwise to F? If $\mathcal{F}(F)$ does not converge pointwise to F at x_0, what is $\mathcal{F}(F)(x_0)$? Are your answers consistent with Theorem 5.5.26?

(c) Using (a,b), find $\sum_{n=1}^{\infty}(-1)^{n+1}\frac{1}{2n-1}$.

(11) Suppose that the 2π-periodic continuous function $f : \mathbb{R} \to \mathbb{R}$ has Fourier series $\mathcal{F}(f) = a_0 + \sum_{n=1}^{\infty}(a_n \cos nx + b_n \sin nx)$. Assuming that (a) $\mathcal{F}(f)$ converges at at least one point, and (b) $\sum_{n=1}(-na_n \sin nx + nb_n \cos nx)$ is uniformly convergent, explain why the series $\mathcal{F}(f)$ converges (uniformly) to f on $\mathbb{R}$.

(12) Suppose $f : \mathbb{R} \to \mathbb{R}$ is 2π-periodic and C^{∞}. Show that the Fourier coefficients of f decay faster than any power of $1/n$. Specifically, show that for each $m \geq 1$, there exists a $C_m \geq 0$ such that $|a_n|, |b_n| \leq C_m n^{-m}$, for all $n \geq 1$. Conversely, show that if f is continuous and this condition holds then f is C^{∞}. (The decay is exponentially fast if f is analytic.)

(13) Show that $G(x) = x\prod_{n=1}^{\infty}\left(1 - \frac{x^2}{n^2\pi^2}\right)$ is 2π-periodic. (Hint: We know that $G(x) = \sin(x)$ on $[-\pi, \pi]$ and so it suffices to prove G is 2π-periodic. Show that 2π-periodicity follows from $G(x + \pi) = -G(x)$. Let $G_k(x) = x\prod_{n=1}^{k}\left(1 - \frac{x^2}{n^2\pi^2}\right)$. Find a simple expression for $G_k(x + \pi)/G_k(x)$, $x \neq n\pi$, and let $k \to \infty$.)

(14) Show that Wallis' formula implies that $\lim_{n\to\infty}\frac{4^{2n}(n!)^4}{[(2n)!]^2(2n+1)} = \frac{\pi}{2}$. Deduce that $\sqrt{\pi} = \lim_{n\to\infty}\frac{(n!)^2 2^{2n}}{(2n)!\sqrt{n}}$. (These formulas can also be found by evaluating $I_p = \int_0^{\pi/2}\sin^p dx$ and then finding $\lim_{n\to\infty}I_{2n}/I_{2n+1}$.)

(15) There is an infinite product for cosine: $\cos x = \prod_{n=1}^{\infty}\left(1 - \frac{4x^2}{(2k-1)^2\pi^2}\right)$. Take logs and differentiate to find a fractional series for $\tan x$. Can you derive the cosine product from the sine product using the identity $\cos x = \sin(\frac{\pi}{2} - x)$? (See the previous exercise for hints.)

(16) Derive the infinite product for $\cos x$ using the trigonometric identity $\sin 2x = 2\sin x \cos x$ and the infinite product for $\sin x$.

(17) Assuming that the product formula for $\sin x$ is valid for all $x \in \mathbb{C}$ (it is, see Chap. 3), find an infinite product formula for $\sinh x$.

(18) Let (x_n) be a sequence of points in $[0, 1]$. Given $N \in \mathbb{N}$ and $0 \leq a < b \leq 1$, define $A_N(a, b)$ to be the cardinality of the set $\{j \in [1, N] \mid x_j \in [a, b]\}$. The sequence (x_n) is *uniformly distributed* if for all $0 \leq a < b \leq 1$ we have

$$\lim_{N\to\infty}\frac{A_N(a, b)}{N} = b - a.$$

A sequence $(x_n) \subset \mathbb{R}$ is *uniformly distributed mod 1* if the fractional parts of x_n are uniformly distributed in $[0, 1]$.

(a) Let $f : [0, 1] \to \mathbb{R}$ be continuous and $(x_n) \subset [0, 1]$. Show that (x_n) is uniformly distributed iff

$$\lim_{N\to\infty} \frac{1}{N} \sum_{j=1}^{N} f(x_j) = \int_0^1 f(s) \, ds.$$

(b) (Weyl criterion.) Show that the sequence $(x_n) \subset \mathbb{R}$ is uniformly distributed mod 1 iff

$$\lim_{N\to\infty} \frac{1}{N} \sum_{j=1}^{N} e^{2\pi m \imath x_n} = 0,$$

for all $m \in \mathbb{Z}$.

(Hints for (b): for necessity, use (a). For sufficiency, use the second Weierstrass approximation theorem to show that (a) holds for all continuous $f : \mathbb{R} \to \mathbb{R}$ of period 1.)

(19) Show that

(a) If α is irrational, then $(n\alpha)$ is uniformly distributed mod 1.
(b) $(\log n)$ is not uniformly distributed mod 1.

(Hint for (b): use the Euler–Maclaurin formula. See also [21, Chap. 1].)

5.6   Mean Square Convergence

So far we have focused on the question of whether or not the Fourier series of a continuous function f converges pointwise to f. A more natural notion of convergence for Fourier series is *mean-square* or L^2-convergence: $\lim_{n\to\infty} \int_{-\pi}^{\pi} |f(x) - S_n(f)(x)|^2 \, dx = 0$. Although the full development of this theory depends on using a more sophisticated version of integration such as the Lebesgue integral, we can at least indicate why mean-square convergence is a natural concept for Fourier series.

Definition 5.6.1 Given continuous 2π-periodic functions $f, g : \mathbb{R} \to \mathbb{R}$, the *scalar* or *inner product* of f and g is defined by

$$\langle f, g \rangle = \frac{1}{2\pi} \int_{-\pi}^{\pi} f(x)g(x) \, dx.$$

We leave the proof of the next lemma to the exercises.

Lemma 5.6.2 *Let $f, g, h : \mathbb{R} \to \mathbb{R}$ be continuous and 2π-periodic.*

(1) $\langle af + bg, h \rangle = a\langle f, h \rangle + b\langle g, h \rangle$ *for all $a, b \in \mathbb{R}$.*
(2) $\langle f, g \rangle = \langle g, f \rangle$.
(3) $\langle f, f \rangle \geq 0$ *and* $\langle f, f \rangle = 0$ *iff $f = 0$.*

Definition 5.6.3 Given a continuous 2π-periodic function $f : \mathbb{R} \to \mathbb{R}$, define the L^2-norm of f by

$$|f|_2 = \sqrt{\langle f, f \rangle}.$$

Lemma 5.6.4 *Let $f, g : \mathbb{R} \to \mathbb{R}$ be continuous and 2π-periodic. We have*

(1) $|f|_2 \geq 0$ *and* $|f|_2 = 0$ *iff $f = 0$.*
(2) $|af|_2 = |a| |f|_2$ *for all $a \in \mathbb{R}$.*
(3) $|\langle f, g \rangle| \leq |f|_2 |g|_2$ *(Cauchy–Schwarz inequality).*
(4) $|f + g|_2 \leq |f|_2 + |g|_2$ *(triangle inequality).*

Proof (1,2) are immediate from Lemma 5.6.2. In order to prove (3) we shall use the necessary and sufficient condition $A, C \geq 0$ and $B^2 < AC$ for a quadratic form $Ax^2 + 2Bxy + Cy^2$ to be positive semi-definite. Let $x, y \in \mathbb{R}$. By Lemma 5.6.2,

$$\langle xf + yg, xf + yg \rangle = x^2 \langle f, f \rangle + 2xy \langle f, g \rangle + y^2 \langle g, g \rangle$$
$$= x^2 |f|_2^2 + 2xy \langle f, g \rangle + y^2 |g|_2^2.$$

Since $\langle xf + yg, xf + yg \rangle \geq 0$, the quadratic form $x^2 |f|_2^2 + 2xy \langle f, g \rangle + y^2 |g|_2^2$ is positive for all $x, y \in \mathbb{R}$, and so we must have $\langle f, g \rangle^2 \leq |f|_2^2 |g|_2^2$, proving (3). Finally we have $|f + g|_2^2 = |f|_2^2 + 2\langle f, g \rangle + |g|_2^2 \leq |f|_2^2 + 2|f|_2 |g|_2 + |g|_2^2$ by (3). That is, $|f + g|_2^2 \leq (|f|_2 + |g|_2)^2$, proving (4). ☐

Definition 5.6.5 The continuous non-zero 2π-periodic functions $f, g : \mathbb{R} \to \mathbb{R}$ are *orthogonal* if $\langle f, g \rangle = 0$.

Lemma 5.6.6 *The set*

$$1, \cos x, \cos 2x, \cdots, \sin x, \sin 2x, \cdots$$

of 2π-periodic functions are pairwise orthogonal.

Proof Use the orthogonality relations (Lemma 5.5.15). ☐

Lemma 5.6.7 (Pythagoras' Theorem) *Suppose that $f_1, \cdots, f_n$ are pairwise orthogonal (that is $\langle f_i, f_j \rangle = 0$, $i \neq j$). Then*

$$|f_1 + \cdots + f_n|_2^2 = |f_1|_2^2 + \cdots + |f_n|_2^2.$$

Proof We have

$$\left\langle \sum_{i=1}^{n} f_i, \sum_{j=1}^{n} f_j \right\rangle = \sum_{i,j=1}^{n} \langle f_i, f_j \rangle$$

$$= \sum_{i=1}^{n} \langle f_i, f_i \rangle$$

$$= \sum_{i=1}^{n} |f_i|_2^2.$$

Since $\langle \sum_{i=1}^{n} f_i, \sum_{j=1}^{n} f_j \rangle = |\sum_{i=1}^{n} f_i|_2^2$, the result follows.   □

We define a new distance function $\rho_2(f, g)$ on piecewise continuous 2π-periodic functions by

$$\rho_2(f, g) = |f - g|_2 = \sqrt{\frac{1}{2\pi} \int_{-\pi}^{\pi} |f(x) - g(x)|^2 \, dx}.$$

It follows from Lemma 5.6.4 that $\rho_2(f, g)$ satisfies the usual properties of a distance function; in particular, the triangle inequality: $\rho_2(f, h) \le \rho_2(f, g) + \rho_2(g, h)$.

Recalling the uniform metric $\rho(f, g) = \sup_{x \in [-\pi, \pi]} |f(x) - g(x)|$, we have

$$\rho_2(f, g) = \sqrt{\frac{1}{2\pi} \int_{-\pi}^{\pi} |f(x) - g(x)|^2 \, dx}$$

$$\le \sqrt{\frac{1}{2\pi} \int_{-\pi}^{\pi} \rho(f, g)^2, \, dx} \le \rho(f, g),$$

for all piecewise continuous 2π-periodic functions f, g.

Example 5.6.8 In general, $\rho_2(f, g)$ may be much smaller than $\rho(f, g)$. For example, if we define

$$f_N(x) = \begin{cases} 1, x \in [-1/N, 1/N], \\ 0, x \in (-\pi, \pi] \smallsetminus [-1/N, 1/N], \end{cases}$$

then $\rho(f_N, 0) = 1$ for all $N \ge 0$ but $\rho_2(f_N, 0) = \sqrt{1/N\pi}$.   ♠

Proposition 5.6.9 *Let f be a piecewise continuous 2π-periodic function. For $n \ge 0$, let S_n denote the partial sum $a_0 + \sum_{j=1}^{n} (a_j \cos jx + b_j \sin jx)$ of the Fourier series of f. The infimum of $\rho_2(f, T)$ over all trigonometric polynomials T of degree less than or equal to n is given by $\rho_2(f, S_n)$ and is attained only when $g = S_n$. Moreover,*

$$|f - S_n|_2^2 = |f|_2^2 - (a_0^2 + \frac{1}{2} \sum_{j=1}^{n} (a_j^2 + b_j^2)).$$

Proof Let $T(x) = A_0 + \sum_{j=1}^{n}(A_j \cos jx + B_j \sin jx)$ be any trigonometric polynomial of degree at most n. We have

$$\rho_2(f, T)^2 = |f - T|_2^2 = |(f - S_n) + (S_n - T)|_2^2.$$

Now $f - S_n$ is orthogonal to $\cos jx$, $\sin jx$, $0 \le j \le n$ since, for example, if $j \le n$,

$$\langle f - S_n, \cos jx \rangle = \langle f, \cos jx \rangle - \langle S_n, \cos jx \rangle$$
$$= a_j - a_j = 0,$$

by the orthogonality relations and the definition of a_j. It follows by Pythagoras' theorem (Lemma 5.6.7) that

$$|f - T|^2 = |f - S_n|_2^2 + |S_n - T|_2^2,$$

and so $\rho_2(f, T)^2 = \rho_2(f, S_n)^2 + \rho_2(S_n, T)^2 \ge \rho_2(f, S_n)^2$ with equality iff $T = S_n$. The final statement follows taking $T = 0$. $\square$

Lemma 5.6.10 *Let f be a piecewise continuous 2π-periodic function. Given $\varepsilon > 0$, there exists a trigonometric polynomial T such that*

$$\rho_2(f, T) < \varepsilon.$$

Proof If f is continuous, then by the second Weierstrass approximation theorem we can choose a trigonometric polynomial T such that $\rho(f, T) < \varepsilon$. But $\rho_2(f, T) \le \rho(f, T)$ and so the result is proved if f is continuous. If f is piecewise continuous, we may choose a continuous 2π-periodic function g such that $\rho_2(f, g) < \varepsilon/2$ (we may require $f = g$ outside of small intervals containing the discontinuity points). As we did above, we may choose a trigonometric polynomial T such that $\rho_2(g, T) < \varepsilon/2$. Now $\rho_2(f, T) \le \rho_2(f, g) + \rho_2(g, T) < \varepsilon/2 + \varepsilon/2 = \varepsilon$. $\square$

Theorem 5.6.11 *Let f a piecewise continuous 2π-periodic function with Fourier coefficients a_n, b_n. Then*

(a) $\rho_2(f, S_n) \to 0$ *as* $n \to \infty$.
(b) $|f|_2^2 = \frac{1}{2\pi} \int_{-\pi}^{\pi} |f(x)|^2 \, dx = a_0^2 + \frac{1}{2}\sum_{n=1}^{\infty}(a_n^2 + b_n^2)$.

Proof Immediate from Proposition 5.6.9 and Lemma 5.6.10. $\square$

Remarks 5.6.12

(1) Statement (b) of Theorem 5.6.11 is known as *Parseval's identity*.
(2) Theorem 5.6.11 suggests a natural inverse problem: Given sequences (a_n), (b_n) such that $a_0^2 + \frac{1}{2}\sum_{n=1}^{\infty}(a_n^2 + b_n^2) < \infty$, does there exist a function f with Fourier coefficients a_n, b_n and which satisfies Parseval's identity? In order to give a satisfactory answer to the problem we have to expand the class of functions to allow for functions which may not be continuous anywhere on $\mathbb{R}$ but which are

nevertheless square integrable. For this to make sense we need to work with a more powerful version of the integral that allows for functions which may have no points of continuity. All of this can be, and has been, done but lies beyond the scope of this text.   ✽

Example 5.6.13 We recall that the Fourier series of the square wave function S (Example 5.5.13) is given by

$$S(x) = \frac{4}{\pi} \sum_{n=0}^{\infty} \frac{\sin((2n+1)x)}{2n+1}.$$

Applying Parseval's identity, we see that

$$1 = \frac{1}{2\pi} \int_{-\pi}^{\pi} S(x)^2 \, dx = \frac{1}{2} \left(\frac{4}{\pi} \right)^2 \sum_{n=0}^{\infty} \frac{1}{(2n+1)^2}.$$

Hence $\sum_{n=0}^{\infty} \frac{1}{(2n+1)^2} = \pi^2/8$.   ♠

EXERCISES 5.6.14

(1) Verify the statements of Lemma 5.6.2.
(2) The Fourier sine series of the saw-tooth function $S(x) = (\pi - x)/2, x \in (0, 2\pi)$, $S(0) = S(2\pi) = 0$, is $\sum_{n=1}^{\infty} \sin(nx)/n$. Using Parseval's identity, deduce that $\sum_{n=1}^{\infty} 1/n^2 = \pi^2/6$.
(3) Show that the Fourier sine series of the 2π-periodic odd function defined on $[0, \pi]$ by $f(x) = x(\pi - x)$ is $\frac{8}{\pi} \sum_{n=0}^{\infty} \frac{\sin((2n+1)x)}{(2n+1)^3}$. Hence show that $\sum_{n=0}^{\infty} \frac{1}{(2n+1)^6} = \frac{\pi^6}{960}$.
(4) For $n \geq 0$, define the *Legendre polynomials* by

$$P_n(x) = \frac{1}{2^n n!} \frac{d^n}{dx^n} (x^2 - 1)^n.$$

(a) Show that $P_n(x)$ is a polynomial of degree n and find the coefficient of x^n. Deduce that every polynomial $p(x)$ of degree n can be written as $p(x) = \sum_{k=0}^{n} c_k P_k(x)$, where $c_0, \cdots, c_n \in \mathbb{R}$ are unique.
(b) Show that $\{P_n \mid n \geq 0\}$ define an orthogonal family of polynomials on $[-1, 1]$. Specifically, show that

$$\int_{-1}^{1} P_n(x) P_m(x) \, dx = \begin{cases} 0, & \text{if } n \neq m, \\ \frac{2}{2n+1}, & \text{if } n = m. \end{cases}$$

(c) Show that if $f : [-1, 1] \to \mathbb{R}$ is continuous and $\int_{-1}^{1} f(x) P_n(x) \, dx = 0$ for all $n \geq 0$, then $f = 0$.

5.7 Appendix: Second Weierstrass Approximation Theorem

In this appendix we prove Theorem 5.5.6: every continuous 2π-periodic function on $f : \mathbb{R} \to \mathbb{R}$ can be uniformly approximated by trigonometric polynomials (the second Weierstrass approximation theorem).

Since f is 2π-periodic, it is enough to show that we can uniformly approximate f by trigonometric polynomials on $[-\pi, \pi]$. We break the proof into a number of lemmas.

Lemma 5.7.1 *If $f : \mathbb{R} \to \mathbb{R}$ is even ($f(-x) = f(x)$, for all $x \in \mathbb{R}$), then we can uniformly approximate f by trigonometric polynomials.*

Proof Since f is even the values of f on $[-\pi, \pi]$ are uniquely determined by the values of f on $[0, \pi]$. Therefore it suffices to uniformly approximate f on $[0, \pi]$ by *even* trigonometric polynomials.

Define $g(t) = f(\cos^{-1} t)$, $t \in [-1, 1]$. Since $\cos^{-1} : [-1, 1] \to [0, \pi]$ is continuous, g is continuous on $[-1, 1]$. By the Weierstrass approximation theorem, we may uniformly approximate g on $[-1, 1]$ by polynomials. That is, given $\varepsilon > 0$, there exists a $p \in P(\mathbb{R})$ such that

$$\sup_{t\in[-1,1]} |g(t) - p(t)| < \varepsilon. \tag{5.8}$$

Set $t = \cos x$, $x \in [0, \pi]$. We can rewrite (5.8) as $\sup_{x\in[0,\pi]} |g(\cos x) - p(\cos x)| < \varepsilon$. Since $g(\cos x) = f(\cos^{-1}(\cos x)) = f(x)$, we have

$$\sup_{x\in[0,\pi]} |f(x) - p(\cos x)| < \varepsilon.$$

Using standard trigonometric identities it is well-known (and easy) to show that every power of $\cos x$ can be written as linear combinations of $\cos jx$, $j \in \mathbb{N}$. Hence $p(\cos x)$ can be written as a trigonometric polynomial with no sine terms:

$$p(\cos x) = a_0 + \sum_{j=1}^{n} a_j \cos jx.$$

This function is even and so we have uniformly approximated f on $[0, \pi]$ by an even trigonometric polynomial.   □

Lemma 5.7.2 *If f is even, then $f(x) \sin^2 x$ can be uniformly approximated by trigonometric polynomials.*

Proof Using Lemma 5.7.1, we first uniformly approximate f by trigonometric polynomials then we use standard trigonometric identities to obtain the required uniform approximations of $f(x) \sin^2 x$ by trigonometric polynomials.   □

Lemma 5.7.3 *If f is odd $(f(-x) = -f(x))$ then $f(x)\sin x$ can be uniformly approximated by trigonometric polynomials.*

Proof Since f is odd, $g(x) = f(x)\sin x$ is even and so we may apply Lemma 5.7.1.
□

Lemma 5.7.4 *Every continuous function $f : \mathbb{R} \to \mathbb{R}$ may be written uniquely as a sum $f_e + f_o$ of even and odd continuous functions. If f is 2π-periodic, so are f_e, f_o.*

Proof Define $f_e(x) = \frac{f(x)+f(-x)}{2}, f_o(x) = \frac{f(x)-f(-x)}{2}$.
□

Lemma 5.7.5 *If f is 2π-periodic, then we can uniformly approximate $f(x)\sin^2 x$ by trigonometric polynomials.*

Proof Using Lemmas 5.7.4 and 5.7.2, we reduce to the case when f is odd. Now apply Lemma 5.7.3 to $f(x)\sin x$ and finally multiply the approximating trigonometric polynomials by $\sin x$ and apply the trigonometric identities $\sin x \cos jx = \frac{1}{2}(\sin(j+1)x - \sin(j-1)x)$ to obtain the required uniform approximations to $f(x)\sin^2 x$.
□

Lemma 5.7.6 *If f is 2π-periodic then we can uniformly approximate $f(\frac{\pi}{2}-x)\sin^2 x$ by trigonometric polynomials.*

Proof Apply Lemma 5.7.4 to $\tilde{f}(x) = f(\frac{\pi}{2} - x)$.
□

Proof of Theorem 5.5.6 Taking $y = \frac{\pi}{2} - x$ in Lemma 5.7.6, we see that $f(x)\cos^2 x$ can be uniformly approximated by trigonometric polynomials. Hence, by Lemma 5.7.2, $f(x)\sin^2 x + f(x)\cos^2 x = f(x)$ can be uniformly approximated by trigonometric polynomials.
□

Chapter 6
Topics from Classical Analysis: The Gamma-Function and the Euler–Maclaurin Formula

In this chapter we look at two topics from classical analysis: the Gamma-function and the Euler–Maclaurin formula. Our investigation of the Gamma-function will require many of the ideas we have developed on convergence and involves infinite products, improper integrals and other techniques and results from analysis such as differentiation under the integral sign and multiple integrals. We also need some standard results on multiple integrals (in our situation these results are elementary as we almost always assume rectangular domains and continuous, even smooth, integrands—see Exercises 2.8.10(10)). The Euler–Maclaurin formula is easy to prove but has powerful applications to estimation and asymptotics. For example, using the Euler–Maclaurin formula, we prove Stirling's formula (estimating $n!$) and also estimate Euler's constant γ and the sums of various infinite series.

6.1 The Gamma-Function

The Gamma-function gives an extension of the factorial $n!$ to all positive real numbers. We start by giving the definition (due to Euler) which involves a doubly improper integral. Once we have checked that the integral converges, it is relatively straightforward to derive the basic properties of the Gamma-function. Along the way we encounter a number of fairly standard techniques often seen in applications of analysis: estimates yielding convergence of infinite integrals and conditions that allow us to differentiate under the integral sign (yet another instance of interchanging limits—in this case involving a *triple* limit).

Definition 6.1.1 The Gamma-function is defined for $x > 0$ by

$$\Gamma(x) = \int_0^\infty t^{x-1} e^{-t} \, dt.$$

© Springer International Publishing AG 2017

M. Field, *Essential Real Analysis*, Springer Undergraduate Mathematics Series,
https://doi.org/10.1007/978-3-319-67546-6_6

Since the definition of the Gamma-function involves an infinite integral and t^{x-1} blows-up at $t = 0$ if $x < 1$, we need to take some care with this definition. We start by giving conditions for the convergence of improper integrals.

Lemma 6.1.2 *Let $f : [a, \infty) \to \mathbb{R}$ be a continuous function which is either positive or negative. The integral $\int_a^\infty f(t)\, dt$ converges iff there exists an $M \geq 0$ such that*

$$\left| \int_a^b f(t)\, dt \right| \leq M,$$

for all $b \geq a$.

Proof Without loss of generality suppose $f \geq 0$ (if not, replace f by $-f$). Then $G(b) = \int_a^b f(t)\, dt$ is a monotone increasing function of b. If the condition of the lemma holds, then G is bounded above since $|G(b)| \leq M$ for all $b \geq a$. Since G is increasing, we have $\lim_{b \to \infty} \int_a^b f(t)\, dt = \sup_{b \geq a} G(b) < \infty$, proving convergence. The converse is obvious. □

Lemma 6.1.3 *Let $f : (a, A] \to \mathbb{R}$ be continuous and either positive or negative. The integral $\int_a^A f(t)\, dt$ converges iff there exists an $M \geq 0$ such that*

$$\left| \int_\alpha^A f(t)\, dt \right| \leq M,$$

for all $\alpha \in (a, A]$.

Proof We use the same method of proof as that of Lemma 6.1.2. We leave the details to the reader. □

Remark 6.1.4 We leave to the exercises versions of Lemmas 6.1.2, 6.1.3 that hold without the assumption that f is of constant sign.   ✠

Using these lemmas, it is easy to show the Gamma-function is well defined.

Lemma 6.1.5 *The improper integral $\int_0^\infty t^{x-1} e^{-t}\, dt$ converges for $x > 0$.*

Proof We start by showing that for every $x > 0$, there exists an $M \geq 0$ such that $\int_1^b t^{x-1} e^{-t}\, dt \leq M$, for all $b \geq 1$. Fix $x > 0$. Since $\lim_{t \to \infty} t^{x-1} e^{-t/2} = 0$, there exists a $C \geq 0$ such that $t^{x-1} e^{-t} < C e^{-t/2}$, for all $t \geq 1$. Hence

$$\int_1^b t^{x-1} e^{-t}\, dt \leq C \int_1^b e^{-t/2}\, dt \leq 2C e^{-1/2},$$

and so we satisfy the conditions of Lemma 6.1.2 with $M = 2C e^{-1/2}$. It remains to show that $\int_0^1 t^{x-1} e^{-t}\, dt$ converges if $x \in (0, 1)$. Since $t^{x-1} e^{-t} \leq t^{x-1}$, $t > 0$, we have for all $a \in (0, 1]$,

$$\int_a^1 t^{x-1} e^{-t} \, dt \leq \int_a^1 t^{x-1}, dt = \frac{1}{x}(1 - a^x) \leq 1/x.$$

Hence $\int_0^1 t^{x-1} e^{-t} \, dt$ converges by Lemma 6.1.3. $\qquad \Box$

We need a result on differentiation under the integral sign before we establish the main properties of the Gamma-function.

Lemma 6.1.6 *Let I be an open interval (bounded or unbounded) and $g : I \times (a, \infty) \to \mathbb{R}$ be continuous. Assume that*

(1) $\frac{\partial g}{\partial x}(x, t)$, $\frac{\partial^2 g}{\partial x^2}(x, t)$ *exist and are continuous on $I \times (a, \infty)$.*
(2) *The integrals $\int_a^\infty g(x, t) \, dt$, $\int_a^\infty \frac{\partial g}{\partial x}(x, t) \, dt$, $\int_a^\infty \frac{\partial^2 g}{\partial x^2}(x, t) \, dt$ exist for all $x \in I$.*
(3) *If $x \in I$, $\delta > 0$ and $[x - \delta, x + \delta] \subset I$, then there exists an $M \geq 0$ such that $|\int_\alpha^\beta \frac{\partial g}{\partial x}(y, t) \, dt|, |\int_\alpha^\beta \frac{\partial^2 g}{\partial x^2}(y, t) \, dt| \leq M$ for all $\alpha, \beta \geq a$ and $y \in [x - \delta, x + \delta]$.*

If we define $F(x) = \int_a^\infty g(x, t) \, dt$, $x \in I$, then F is C^1 on I and

$$F'(x) = \int_a^\infty \frac{\partial g}{\partial x}(x, t) \, dt, \ x \in I.$$

Remark 6.1.7 The conditions of Lemma 6.1.6 are not intended to be optimal: indeed they are not! However, not only do the conditions lead to a simple proof of the result on the validity of differentiating under the integral sign but the conditions are easy to verify for our intended application. Note that condition (3) of the lemma is only really needed because we are dealing with an improper integral. If we assume that $g(x, t)$ is continuous on a product of closed and bounded intervals then the estimates (3) follow using uniform continuity arguments. $\qquad \maltese$

Proof of Lemma 6.1.6 Fix $x \in I$ and choose $\delta > 0$ so that $[x - \delta, x + \delta] \subset I$. Let $h \in [-\delta, \delta]$. Applying (the trivial case of) Taylor's theorem with integral remainder—Theorem 2.7.7 with $r = 1$—we get

$$g(x + h, t) = g(x, t) + h \int_0^1 \frac{\partial g}{\partial x}(x + sh, t) \, ds.$$

Integrating from a to ∞ with respect to t, we obtain

$$F(x + h) - F(x) = h \int_a^\infty \int_0^1 \frac{\partial g}{\partial x}(x + sh, t) \, ds \, dt.$$

Interchanging the order of integration (use Exercises 2.8.10(10)(f)) gives

$$F(x + h) - F(x) = h \int_0^1 \left(\int_a^\infty \frac{\partial g}{\partial x}(x + sh, t) \, dt \right) ds.$$

Now by (3), there exists an $M \geq 0$ such that $|\int_a^\infty \frac{\partial g}{\partial x}(x + sh, t)\, dt| \leq M$ for all $h \in [-\delta, \delta]$. Hence $|F(x + h) - F(x)| \leq |h|M$ for all $h \in [-\delta, \delta]$ and so F is continuous at x.

Next we consider the differentiability of F at x. For this we again use Taylor's theorem with integral remainder applied to $g(x, t)$, regarded as a function of x. For $h \in [-\delta, \delta]$, $h \neq 0$, we have

$$g(x + h, t) = g(x, t) + h\frac{\partial g}{\partial x}(x, t) + h^2 \int_0^1 (1 - s)\frac{\partial^2 g}{\partial x^2}(x + sh, t)\, ds.$$

Hence

$$\frac{g(x + h, t) - g(x, t)}{h} - \frac{\partial g}{\partial x}(x, t) = h \int_0^1 (1 - s)\frac{\partial^2 g}{\partial x^2}(x + sh, t)\, ds.$$

Integrating from a to ∞ with respect to t, we obtain

$$\frac{F(x + h) - F(x)}{h} - \int_a^\infty \frac{\partial g}{\partial x}(x, t)\, dt = R(x, t; h),$$

where

$$R(x, t; h) = h \int_a^\infty \int_0^1 (1 - s)\frac{\partial^2 g}{\partial x^2}(x + sh, t)\, ds dt$$

$$= h \int_0^1 (1 - s) \left(\int_a^\infty \frac{\partial^2 g}{\partial x^2}(x + sh, t)\, dt \right) ds,$$

and the interchange of order of integration follows by Fubini's theorem—Exercises 2.8.10(10)(f) again. By condition (3), there exists an $M \geq 0$ such that $|\int_a^\infty \frac{\partial^2 g}{\partial x^2}(x + sh, t)\, dt| \leq M$ for all $h \in [-\delta, \delta]$, $s \in [0, 1]$. Hence

$$\left| h \int_0^1 (1 - s) \left(\int_a^\infty \frac{\partial^2 g}{\partial x^2}(x + sh, t)\, dt \right) ds \right| \leq M|h| \int_0^1 (1 - s)\, ds = M|h|/2.$$

This estimate implies that for $h \in [-\delta, \delta]$, $h \neq 0$, we have

$$\left| \frac{F(x + h) - F(x)}{h} - \int_a^\infty \frac{\partial g}{\partial x}(x, t)\, dt \right| \leq M|h|/2.$$

Now let $h \to 0$ to obtain the differentiability of F at x. Finally, in order to prove that F is C^1 we use the same argument used to prove F is continuous. We omit the details.   □

6.1.1 Properties of $\Gamma(x)$

Theorem 6.1.8

(1) $\Gamma(x+1) = x\Gamma(x)$, *for all* $x > 0$.
(2) $\Gamma(n+1) = n!$, $n \geq 1$.
(3) Γ *is a smooth function on* $(0, \infty)$ *and*

$$\Gamma^{(n)}(x) = \int_0^\infty (\log t)^n t^{x-1} e^{-t} \, dt, \ n \geq 1.$$

Proof Let $0 < a < R < \infty$. Integrating by parts, we have

$$\int_a^R t^x e^{-t} \, dt = [-e^{-t} t^x]_a^R + x \int_a^R t^{x-1} e^{-t} \, dt$$

$$= e^{-a} a^x - e^{-R} R^x + x \int_a^R t^{x-1} e^{-t} \, dt.$$

If $x > 0$, we may take limits as $a \to 0+$ and $R \to \infty$ to get $\Gamma(x+1) = x\Gamma(x)$, proving (1). For (2), observe that by (1) we get $\Gamma(n+1) = n\Gamma(n) = \cdots = n!\Gamma(1)$. We have $\Gamma(1) = \int_0^\infty e^{-t} \, dt = 1$. It remains to prove (3). Set $k(x, t) = t^{x-1} e^{-t}$. Differentiating with respect to x we have

$$\frac{\partial^n k}{\partial x^n} = (\log t)^n k(x, t), \ n \geq 0.$$

For all $n \geq 0$, $(\log t)^n k(x, t)$ is continuous on $(0, \infty) \times (0, \infty)$. Choose $\delta > 0$ so that $[x - \delta, x + \delta] \subset (0, \infty)$. Just as in the proof of Lemma 6.1.5, the integrals $\int_0^\infty (\log t)^n k(x, t) \, dt$ converge for all $n \geq 0$ (see the exercises for details). Similar arguments also show that for each $n \geq 0$, there exists an $M_n \geq 0$ such that $|\int_0^\infty (\log t)^m k(y, t) \, dt| \leq M_n$ for all $y \in [x - \delta, x + \delta]$ and $0 \leq m \leq n$. Now we are in a position to apply Lemma 6.1.6. Since conditions (1,2,3) of Lemma 6.1.6 hold with $g(x, t) = t^{x-1} e^{-t}$, Γ is C^1. Proceeding inductively, suppose we have proved Γ is C^n and that $\Gamma^{(n)}(x) = \int_0^\infty (\log t)^n t^{x-1} e^{-t} \, dt$. We apply Lemma 6.1.6 with $g(x, t) = (\log t)^n t^{x-1} e^{-t}$ to obtain $\Gamma^{(n)}$ is C^1 (and so Γ is C^{n+1}) and $\Gamma^{(n+1)}(x) = (\Gamma^{(n)})'(x) = \int_0^\infty (\log t)^{n+1} t^{x-1} e^{-t} \, dt$. $\quad\square$

6.1.2 Convexity of $\log \Gamma$

Recall that a C^2 function $f : [a, b] \to \mathbb{R}$ is *convex* if $f'' \geq 0$ on $[a, b]$ and *strictly convex* if $f'' > 0$ on (a, b). If f is defined on an open or half-open interval, then f is defined to be convex if f is convex on all closed subintervals.

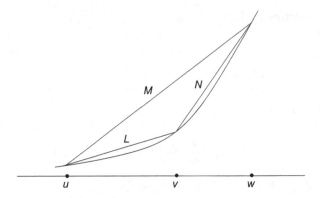

Fig. 6.1 Chord triples for a convex function

Example 6.1.9 The function $-\log : (0, \infty) \to \mathbb{R}$ is strictly convex since $(-\log)''(x) = 1/x^2 > 0$ for all $x \in (0, \infty)$. ♠

Everything we need about convex functions is contained in the next result.

Lemma 6.1.10 *Suppose that the C^2 function $f : [a, b] \to \mathbb{R}$ is convex. Let $a \leq u < v < w \leq b$. Then*

$$\frac{f(v) - f(u)}{v - u} \leq \frac{f(w) - f(u)}{w - u} \leq \frac{f(w) - f(v)}{w - v}.$$

If the inequalities are strict then f is strictly convex.

In Fig. 6.1 we show the geometrically transparent relationship between the slopes of the chords L, M, N given by Lemma 6.1.10.

Proof Since f is convex, f' is increasing on $[a, b]$. For $x \geq v$, define $g(x) = \frac{f(x)-f(u)}{x-u} - \frac{f(v)-f(u)}{v-u}$. We have $g(v) = 0$ and $g'(x) = \frac{1}{(x-u)^2}((x-u)f'(x) - (f(x)-f(u))$. By the mean value theorem, there exists a $\theta \in (u, x)$ such that $f(x) - f(u) = (x - u)f'(\theta)$ and so $g'(x) = \frac{1}{(x-u)}(f'(x) - f'(\theta))$. Since f' is increasing on $[a, b]$, it follows that $g' \geq 0$ on $[v, w]$. Hence, since $g(v) = 0$, $g(w) \geq 0$. This proves the first inequality. The proof of the second inequality is similar. The case of strict inequality follows easily using the same arguments. □

Theorem 6.1.11 $\log \Gamma$ *is strictly convex on* $(0, \infty)$.

Proof We show that $(\log \Gamma)'' > 0$ on $(0, \infty)$. Computing, we find that $(\log \Gamma)'' = (\Gamma \Gamma'' - (\Gamma')^2)/\Gamma^2$. It suffices therefore to show that $\Gamma \Gamma'' > (\Gamma')^2$. We may write

$$\Gamma(x)\Gamma''(x) = \left(\int_0^\infty t^{x-1}e^{-t}\, dt\right)\left(\int_0^\infty (\log s)^2 s^{x-1}e^{-s}\, ds\right)$$

$$= \int_0^\infty \int_0^\infty (\log s)^2 t^{x-1} s^{x-1} e^{-t} e^{-s}\, dtds$$

$$= \int_0^\infty \int_0^\infty (\log s)^2 (ts)^{x-1} e^{-t-s} \, dt ds,$$

$$\Gamma'(x)^2 = \int_0^\infty \int_0^\infty \log t \log s \, (ts)^{x-1} e^{-t-s} \, dt ds.$$

Observe that

$$\Gamma(x)\Gamma''(x) = \int_0^\infty \int_0^\infty (\log s)^2 (ts)^{x-1} e^{-t-s} \, dt ds$$

$$= \int_0^\infty \int_0^\infty (\log t)^2 (ts)^{x-1} e^{-t-s} \, dt ds,$$

using the symmetry of the integrand $(ts)^{x-1} e^{-t-s}$ and the region of integration—positive quadrant—in t, s. Hence

$$2(\Gamma(x)\Gamma''(x) - \Gamma'(x)^2)$$

$$= \int_0^\infty \int_0^\infty [(\log t)^2 + (\log s)^2 - 2\log t \log s](ts)^{x-1} e^{-t-s} \, dt ds$$

$$= \int_0^\infty \int_0^\infty (\log t - \log s)^2 (ts)^{x-1} e^{-t-s} \, dt ds$$

$$> 0.$$

$\square$

Theorem 6.1.12 *Suppose that $f : (0, \infty) \to \mathbb{R}$ is C^2 and satisfies*

(a) $f(x + 1) = x f(x)$, $x > 0$.
(b) $f(1) = 1$.
(c) $\log f$ *is convex.*

Then $f(x) = \lim_{n \to \infty} [\frac{n! n^x}{x(x+1)\cdots(x+n)}]$. In particular, since Γ satisfies (a,b,c),

$$\Gamma(x) = \lim_{n \to \infty} \left[\frac{n! n^x}{x(x+1)\cdots(x+n)} \right].$$

Remarks 6.1.13

(1) It is enough to assume $\phi = \log f$ is convex in the sense that $\phi(\lambda x + (1-\lambda)y) \le \lambda \phi(x) + (1-\lambda)\phi(y)$, $\lambda \in [0, 1]$, without requiring that ϕ is differentiable or C^2 (for a proof, see Rudin [27]).
(2) The limit $\lim_{n \to \infty} [\frac{n! n^x}{x(x+1)\cdots(x+n)}]$ exists provided $x \ne 0, -1, -2, \cdots$ and so the theorem gives an extension of the Gamma-function to all of the real line except the negative integers. ✠

Proof of Theorem 6.1.12 Let $\phi(x) = \log f(x)$, $x > 0$. It follows from (a,b) that

$$\phi(n+1) = \log(n!), \quad n \geq 0. \tag{6.1}$$

In general, we have $\phi(x+1) = \phi(x) + \log x$ and so for $n \geq 0$ we have

$$\phi(n+1+x) = \phi(x) + \log(x+n) + \cdots + \log x \tag{6.2}$$
$$= \phi(x) + \log[x(x+1)\cdots(x+n)].$$

In view of (a,b), it is enough to show that $f(x) = \lim_{n\to\infty}[\frac{n!n^x}{x(x+1)\cdots(x+n)}]$ for $x \in$ (0, 1). Consider the quotients

$$\frac{\phi(n+1) - \phi(n)}{1}, \quad \frac{\phi(n+1+x) - \phi(n+1)}{x}, \quad \frac{\phi(n+2) - \phi(n+1)}{1}.$$

Noting that $\phi(m+1) - \phi(m) = \log m$ and applying Lemma 6.1.10 to the convex function ϕ we have

$$\log n \leq \frac{\phi(n+1+x) - \phi(n+1)}{x} \leq \log(n+1).$$

Substituting for $\phi(n+1+x)$, using (6.1), (6.2) and multiplying through by x gives

$$x \log n \leq \phi(x) + \log[x(x+1)\cdots(x+n)] - \log(n!) \leq x\log(n+1).$$

Hence,

$$0 \leq \phi(x) - \log(n!) - \log(n^x) + \log[x(x+1)\cdots(x+n)] \leq x\log(n+1) - x\log n.$$

That is,

$$0 \leq \phi(x) - \log\left[\frac{n!n^x}{x(x+1)\cdots(x+n)}\right] \leq x\log\left(1 + \frac{1}{n}\right).$$

Since $\lim_{n\to\infty} x\log(1 + \frac{1}{n}) = 0$, it follows from the squeezing lemma that $\phi(x) = \lim_{n\to\infty} \log\left[\frac{n!n^x}{x(x+1)\cdots(x+n)}\right]$ and so

$$f(x) = e^{\phi(x)} = \lim_{n\to\infty}\left[\frac{n!n^x}{x(x+1)\cdots(x+n)}\right]. \qquad \square$$

6.1.3   The Gamma-Product

In this section we obtain an infinite product formula for $\Gamma(x)$, involving Euler's constant, and which easily leads to a relation between the Gamma and sine functions.

Let $n \in \mathbb{N}$. For $x \in \mathbb{R}$, we have

$$n^{-x}\frac{x(x+1)\cdots(x+n)}{n!} = xn^{-x}\prod_{j=1}^{n}\left(1+\frac{x}{j}\right).$$

The infinite product $\prod_{n=1}^{\infty}(1+\frac{x}{n})$ is not convergent (for example, by Lemma 3.9.13). However, there is a powerful trick due to Weierstrass that enables us to manufacture a convergent infinite product from $\prod_{n=1}^{\infty}(1+\frac{x}{n})$.

Lemma 6.1.14 *The infinite product $\prod_{n=1}^{\infty}\left[(1+\frac{x}{n})e^{-\frac{x}{n}}\right]$ converges for all $x \in \mathbb{R}$ and is only zero if $x = 0, -1, -2, \cdots$.*

Proof We give a proof that works for $x \in \mathbb{C}$. For a simpler argument, valid only for $x \in \mathbb{R}$, use Exercises 6.1.23(6). Set $a_n = (1+\frac{x}{n})e^{-\frac{x}{n}} - 1, n \geq 1$. We have

$$a_n = -e^{-\frac{x}{n}}\left(e^{\frac{x}{n}} - 1 - \frac{x}{n}\right)$$

and so

$$|a_n| = |e^{-\frac{x}{n}}|\left|\sum_{j=2}^{\infty}\frac{1}{j!}\left(\frac{x}{n}\right)^j\right|$$

$$\leq e^{\frac{|x|}{n}}\left[\sum_{j=2}^{\infty}\frac{1}{j!}\left(\frac{|x|}{n}\right)^j\right]$$

$$= e^{\frac{|x|}{n}}\frac{|x|^2}{2n^2}\left(1 + \frac{1}{3}\frac{|x|}{n} + \cdots\right)$$

$$\leq e^{\frac{|x|}{n}}\frac{|x|^2}{2n^2}\left[\sum_{j=0}^{\infty}\frac{1}{j!}\left(\frac{|x|}{n}\right)^j\right]$$

$$= e^{2\frac{|x|}{n}}\frac{|x|^2}{2n^2}$$

$$\leq \frac{1}{2n^2}e^{2|x|}|x|^2$$

$$= cn^{-2},$$

where c is independent of n. Therefore $\sum_{n=1}^{\infty} a_n$ is absolutely convergent and so by Lemmas 3.9.4, 3.9.6, $\prod_{n=1}^{\infty}\left[(1+\frac{x}{n})e^{-\frac{x}{n}}\right]$ is convergent for all $x \in \mathbb{R}$ and is zero iff x is a negative integer. □

Remark 6.1.15 Lemma 6.1.14 is valid for $x \in \mathbb{C}$, x not a negative integer. The proof is exactly the same except that absolute value is everywhere replaced by modulus of a complex number and we use Proposition 3.9.10 rather than Lemmas 3.9.4, 3.9.6. ✠

Theorem 6.1.16 *For $x > 0$, we have*

$$\Gamma(x) = \frac{1}{xe^{\gamma x}\prod_{n=1}^{\infty}\left[(1+\frac{x}{n})e^{-\frac{x}{n}}\right]},$$

where γ denotes Euler's constant. Moreover, the expression on the right is defined and finite for all $x \in \mathbb{C}$ provided only that x is not a negative integer.

Proof We have

$$\frac{n!n^x}{x(x+1)\cdots(x+n)} = \frac{e^{x\log n}e^{-x\sum_{j=1}^{n}\frac{1}{j}}}{x\prod_{j=1}^{n}\left[(1+\frac{x}{j})e^{-\frac{x}{j}}\right]}.$$

By Theorem 6.1.12, $\lim_{n\to\infty}\frac{n!n^x}{x(x+1)\cdots(x+n)} = \Gamma(x)$. On the other hand, by Lemma 6.1.14, $\prod_{j=1}^{n}\left[(1+\frac{x}{j})e^{-\frac{x}{j}}\right]$ is convergent for all $x \in \mathbb{R}$. Finally, since

$$\lim_{n\to\infty} e^{x\log n}e^{-x\sum_{j=1}^{n}\frac{1}{j}} = \lim_{n\to\infty} e^{x(\log n-\sum_{j=1}^{n}\frac{1}{j})} = e^{-\gamma x},$$

we have

$$\lim_{n\to\infty}\frac{e^{x\log n}e^{-x\sum_{j=1}^{n}\frac{1}{j}}}{x\Pi_{j=1}^{n}\left[(1+\frac{x}{j})e^{-\frac{x}{j}}\right]} = \frac{1}{xe^{\gamma x}\Pi_{n=1}^{\infty}\left[(1+\frac{x}{n})e^{-\frac{x}{n}}\right]}.$$

□

As a result of Theorem 6.1.16, we may regard the Gamma-function as defined on all of $\mathbb{R}$ (or $\mathbb{C}$), except for the non-positive integers, by the formula

$$\Gamma(x) = \frac{1}{xe^{\gamma x}\prod_{n=1}^{\infty}\left[(1+\frac{x}{n})e^{-\frac{x}{n}}\right]}.$$

Lemma 6.1.17 *With the extended definition of Γ, we have*

$$\Gamma(x+1) = x\Gamma(x),\ x \notin \{0,-1,-2,\cdots\}.$$

Proof We leave this to the exercises. □

Theorem 6.1.18

$$\Gamma(x)\Gamma(1-x) = \frac{\pi}{\sin(\pi x)}, \quad x \notin \mathbb{Z}.$$

Proof By Lemma 6.1.17, we have $\Gamma(1-x) = -x\Gamma(-x)$ and so, provided $x \notin \mathbb{Z}$, we have

$$\Gamma(x)\Gamma(1-x) = -x\Gamma(x)\Gamma(-x)$$

$$= \lim_{n\to\infty} \frac{-x}{xe^{\gamma x}\prod_{j=1}^{n}\left[(1+\frac{x}{j})e^{-\frac{x}{j}}\right] \times -xe^{-\gamma x}\prod_{j=1}^{n}\left[(1-\frac{x}{j})e^{\frac{x}{j}}\right]}$$

$$= \lim_{n\to\infty} \frac{1}{x\prod_{j=1}^{n}\left(1-\frac{x^2}{j^2}\right)}$$

$$= \frac{\pi}{\sin(\pi x)},$$

where the last statement follows by the infinite product formula for $\sin x$ (Sect. 5.5.5). □

Remark 6.1.19 Theorem 6.1.18 holds for $x \in \mathbb{C}$, $x \notin \mathbb{Z}$—use the infinite product formula for $\sin z$, $z \in \mathbb{C}$, Proposition 3.9.17. ✱

Example 6.1.20 Taking $x = \frac{1}{2}$ in Theorem 6.1.18, we obtain $\Gamma(\frac{1}{2}) = \sqrt{\pi}$. An alternative proof of this result, which does not use the product formula, can be based on Exercises 6.1.23(9).

6.1.4 An Integral Formula for the Beta Function

We conclude this section on the Gamma function with a useful integral formula.

Theorem 6.1.21 *For $x, y > 0$ we have*

$$\int_0^1 (1-t)^{x-1}t^{y-1}\,dt = \frac{\Gamma(x)\Gamma(y)}{\Gamma(x+y)}.$$

Proof We have

$$\Gamma(x)\Gamma(y) = \left(\int_0^\infty t^{x-1}e^{-t}\,dt\right)\left(\int_0^\infty s^{y-1}e^{-s}\,ds\right)$$

$$= \int_0^\infty \int_0^\infty t^{x-1}s^{y-1}e^{-(t+s)}\,dt\,ds.$$

Making the change of variables $u = s + t$, $v = s$, we find

$$\int_0^\infty \int_0^\infty t^{x-1} s^{y-1} e^{-(t+s)} \, dt \, ds = \int \int_R (u-v)^{x-1} v^{y-1} e^{-u} \, du \, dv,$$

where $R = \{(u, v) \mid 0 \le v \le u\}$. Now

$$\int \int_R (u-v)^{x-1} v^{y-1} e^{-u} \, du \, dv$$

$$= \int_0^\infty \left[\int_0^u (u-v)^{x-1} v^{y-1} \, dv \right] e^{-u} \, du$$

$$= \int_0^\infty \left[\int_0^1 u^{x-1} (1-t)^{x-1} u^{y-1} t^{y-1} u \, dt \right] e^{-u} \, du,$$

where the second integral is obtained from the first by the change of variable $t = v/u$. But now the second integral is equal to

$$\left(\int_0^\infty e^{-u} u^{x+y-1} \, du \right) \left(\int_0^1 (1-t)^{x-1} t^{y-1} \, dt \right),$$

which is the product of $\Gamma(x+y)$ with the integral we want. $\square$

Remark 6.1.22 The integral $\int_0^1 (1-t)^{x-1} t^{y-1} \, dt$ is usually called the *beta function* of x, y and denoted by $B(x, y)$.   ✱

EXERCISES 6.1.23

(1) Let $f : [a, \infty) \to \mathbb{R}$ be a continuous function. The integral $\int_a^\infty f(t) \, dt$ converges iff for every $\varepsilon > 0$, there exists an $N \in [a, \infty)$ such that $|\int_\alpha^\beta f(t) \, dt| < \varepsilon$, for all $\alpha < \beta \in [N, \infty)$.

(2) Let $f : (a, A] \to \mathbb{R}$ be continuous on $(a, A]$. The integral $\int_a^A f(t) \, dt$ converges iff for every $\varepsilon > 0$, there exists a $b \in (a, A]$ such that $|\int_\alpha^\beta f(t) \, dt| < \varepsilon$. for all $\alpha < \beta \in (a, b]$.

(3) Complete the details of the proof of Lemma 6.1.10.

(4) Suppose that $f : [a, b] \to \mathbb{R}$ is C^2. Show that f is convex iff $f(\lambda x + (1 - \lambda) y) \le \lambda f(x) + (1 - \lambda) f(y)$, for all $x < y \in [a, b]$ and $\lambda \in [0, 1]$. (Hint: use Lemma 6.1.10.) What about strict convexity?

(5) Verify Lemma 6.1.17.

(6) Show that for all $x \in \mathbb{R}$, $e^x \ge 1 + x$. Deduce that for all $x \ge -1$, $1 \ge (1 + x)e^{-x} \ge 1 - x^2$. Using this estimate, obtain a simple proof of Lemma 6.1.14, valid only for real values of x.

(7) Verify the *duplication formula* of Legendre:

$$\Gamma(x)\Gamma\left(x+\frac{1}{2}\right) = \sqrt{\pi}\,2^{1-2x}\Gamma(2x).$$

(Hint: Use the product representation of $\Gamma(x)$ given in Theorem 6.1.12—note that $\Gamma(x) = \lim_{n\to\infty}\left[\frac{n!n^{x-1}}{x(x+1)\cdots(x+n-1)}\right]$.)

(8) Show that

$$(1-x)\left(1+\frac{x}{2}\right)\left(1-\frac{x}{3}\right)\left(1+\frac{x}{4}\right)\cdots = \frac{\sqrt{\pi}}{\Gamma(1+\frac{x}{2})\Gamma(\frac{1}{2}-\frac{x}{2})}.$$

Show also that $(1-x)(1-\frac{x}{3})(1+\frac{x}{2})(1-\frac{x}{5})(-)(+)(-)(-)(+)\cdots$ converges and find the limit. This provides an example of rearrangement to a different limit for a conditionally convergent infinite product.

(9) Show that the substitution $t = s^2$ in the defining integral for $\Gamma(x)$ leads to the formula

$$\Gamma(x) = 2\int_0^\infty s^{2x-1}e^{-s^2}\,ds,\ x > 0.$$

Deduce that $\int_{-\infty}^\infty e^{-s^2}\,ds = \Gamma(\frac{1}{2}) = \sqrt{\pi}$.

(10) Show that

$$\int_0^{\pi/2}(\sin\theta)^{2x-1}(\cos\theta)^{2y-1}\,d\theta = \frac{\Gamma(x)\Gamma(y)}{2\Gamma(x+y)},$$

where $x, y > 0$. (Hint: use Theorem 6.1.21.)

(11) Show that

$$\int_0^{\pi/2}\frac{d\theta}{\sqrt{1+\sin^2\theta}}\,d\theta = \frac{\Gamma(\frac{1}{4})^2}{4\sqrt{2\pi}}.$$

6.2   Bernoulli Numbers and Bernoulli Polynomials

We define the Bernoulli numbers and Bernoulli polynomials and establish some of their basic properties. We will make much use of these ideas in our subsequent development and applications of the Euler–Maclaurin formula.

Proposition 6.2.1   *There is a unique sequence* $(B_n)_{n\geq 0}$ *of real numbers characterized by*

(a) $B_0 = 1$.
(b) $B_n = \sum_{k=0}^n \binom{n}{k}B_k,\ n > 1$.

Proof Condition (b) implies that if $n > 1$, then

$$B_n = B_0 + \binom{n}{1} B_1 + \cdots + nB_{n-1} + B_n,$$

and so for $n \geq 2$,

$$B_{n-1} = -\frac{1}{n}\left(B_0 + \binom{n}{1}B_1 + \cdots + \binom{n}{n-2}B_{n-2}\right). \tag{6.3}$$

If $B_0 = 1$, we can use (6.3) to inductively define B_n for all $n \geq 1$.   □

Definition 6.2.2 The numbers $B_0, B_1, \cdots$ given by Proposition 6.2.1 are called the *Bernoulli numbers*.

We may use (6.3) to compute the first few Bernoulli numbers. We find

$$B_0 = 1, \ B_1 = -\frac{1}{2}, \ B_2 = \frac{1}{6}, \ B_3 = 0, \ B_4 = -\frac{1}{30}, \ B_5 = 0, \ B_6 = \frac{1}{42}.$$

Appearances to the contrary, the sequence (B_n) is unbounded since $|B_{2n}| \to \infty$ as $n \to \infty$. Sometimes, B_1 is taken to be zero. However, for us it is convenient to take $B_1 = -\frac{1}{2}$.

In the next lemma we give a rather devious proof that $B_{2n+1} = 0, n \geq 1$. We give an alternative proof later in the section.

Lemma 6.2.3 $B_{2n+1} = 0, n \geq 1$.

Proof Define $g(x) = \frac{x}{e^x - 1}$, $x \neq 0$, and $g(0) = 1$. Since $e^x - 1 = xf(x)$, where $f(0) = 1$ and $f(x)$ is analytic, it follows from the results of Chap. 4 that $g(x)$ is an analytic function on some interval $(-\delta, \delta)$ containing the origin and so we may write

$$g(x) = \sum_{n=0}^{\infty} b_n \frac{x^n}{n!}, \ x \in (-\delta, \delta).$$

Since $f(0) = 1$, $b_0 = 1$. Multiplying both sides by $(e^x - 1)$ gives $(e^x - 1)(\sum_{n=0}^{\infty} b_n \frac{x^n}{n!}) = x$. That is $(\sum_{m=1}^{\infty} \frac{x^m}{m!})(\sum_{n=0}^{\infty} \frac{b_n}{n!}x^n) = x$. Computing the coefficient of x^n we find that

$$\sum_{p=0}^{n-1} b_p \frac{1}{p!(n-p)!} = 0, \ n > 1.$$

Multiplying by $n!$ and adding b_n to both sides yields

$$b_n = \sum_{p=0}^{n} \binom{n}{p} b_p.$$

Hence, since $b_0 = 1$, it follows by Proposition 6.2.1 that $b_n = B_n$, all $n \geq 0$. Now $B_1 = b_1 = -1/2$. We prove that $B_{2n+1} = 0, n \geq 1$, by showing that $g(x) + x/2$ is an even function of x. We have $\frac{x}{e^x - 1} + \frac{x}{2} = \frac{-x}{e^{-x} - 1} - \frac{x}{2}$ iff

$$x(e^{-x} - 1) + \frac{x}{2}(e^x - 1)(e^{-x} - 1) = -x(e^x - 1) - \frac{x}{2}(e^x - 1)(e^{-x} - 1).$$

Computing, we find that both sides are equal to $\frac{x}{2}(e^{-x} - e^x)$. Hence $g(x) + x/2$ is even and $B_{2n+1} = b_{2n+1} = 0, n \geq 1$. $\square$

The *Bernoulli polynomials* $B_n(x)$ (not to be confused with the Bernstein polynomials) are defined for $n \geq 0$ by

$$B_n(x) = \sum_{k=0}^{n} \binom{n}{k} B_{n-k} x^k.$$

Computing the first few polynomials, we find that

$$B_0(x) = 1, \; B_1(x) = x - \frac{1}{2}, \; B_2(x) = x^2 - x + \frac{1}{6}, \; B_3(x) = x^3 - \frac{3}{2}x^2 + \frac{1}{2}x.$$

Lemma 6.2.4 *For* $n \neq 1$,

$$B_n(0) = B_n(1) = B_n.$$

Proof Left to the exercises. $\square$

Remark 6.2.5 Note that $B_1(0) = -\frac{1}{2} \neq B_1(1) = \frac{1}{2}$.   ✱

Lemma 6.2.6

$$B_n'(x) = nB_{n-1}(x), \; n \geq 1.$$

Proof Since $B_n(x) = \sum_{k=0}^{n} \binom{n}{k} B_{n-k} x^k$, we have

$$B_n'(x) = \sum_{k=1}^{n} \binom{n}{k} k B_{n-k} x^{k-1}$$

$$= \sum_{k=1}^{n} n \binom{n-1}{k-1} B_{n-k} x^{k-1}$$

$$= n \sum_{k-1=0}^{n-1} \binom{n-1}{k-1} B_{(n-1)-(k-1)} x^{k-1}$$

$$= n \sum_{j=0}^{n-1} \binom{n-1}{j} B_{(n-1)-j} x^j$$

$$= n B_{n-1}(x).$$

□

As a useful corollary to Lemmas 6.2.4 and 6.2.6 we have

Proposition 6.2.7 $B_n(x) = n \int_0^x B_{n-1}(t)\, dt + B_n$, $n \geq 1$.

Remark 6.2.8 We can use Proposition 6.2.7 to recursively compute the Bernoulli polynomials. ✠

6.2.1 The 1-Periodic Functions $\widetilde{B}_n$

Let $\widetilde{B}_n$ denote the 1-periodic extension of B_n restricted to $[0, 1]$ to $\mathbb{R}$. That is, if $x \in \mathbb{R}$, choose $p \in \mathbb{Z}$ such that $x - p \in [0, 1]$ and define

$$\widetilde{B}_n(x) = B_n(x - p).$$

Since $B_n(0) = B_n(1)$ when $n \neq 1$, it is immediate that $\widetilde{B}_n$ is uniquely determined and continuous provided $n \neq 1$. When $n = 1$, we need to be careful as $B_1(0) \neq B_1(1)$. What we do is take $\widetilde{B}_1(x) = B_1(x - p)$ if $x \notin \mathbb{Z}$ and define $\widetilde{B}_1(x) = 0$ if x is an integer. The resulting function will then have a jump discontinuity at integer points. See Fig. 6.2.

For all $x \in \mathbb{R}$, $p \in \mathbb{Z}$, and $n \geq 0$, we have $\widetilde{B}_n(x+p) = \widetilde{B}_n(x)$ (that is, the functions $\widetilde{B}_n$ are all *1-periodic*). If $x \in (j, j+1)$, then

$$\widetilde{B}_1(x) = x - j - \frac{1}{2}.$$

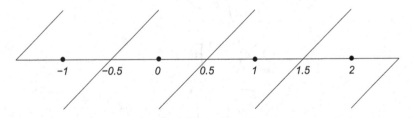

Fig. 6.2 Graph of $\widetilde{B}_1$

Lemma 6.2.9 *For all $A \geq 0$,*

$$\left| \int_0^A \widetilde{B}_1(x)\, dx \right| \leq \frac{1}{8}.$$

Proof We have $\int_j^{j+1} \widetilde{B}_1(x)\, dx = 0$, for all $j \in \mathbb{Z}$. It is clear from Fig. 6.2 that if $j + y \in [j, j+1]$, then we maximize $|\int_j^{j+y} \widetilde{B}_1(x)\, dx|$ when $y = \frac{1}{2}$. Obviously, $\int_j^{j+\frac{1}{2}} \widetilde{B}_1(x)\, dx = \frac{1}{8}$. The result follows since we can write A uniquely as $j + y$, where $y \in [0,1), j \in \mathbb{Z}$.   □

We now compute the Fourier series of $\widetilde{B}_n$. This will give new and remarkable expressions for the original Bernoulli polynomials $B_n(x)$.

Theorem 6.2.10

(1) *If $n \geq 2$ is even,*

$$\widetilde{B}_n(x) = \frac{2(-1)^{\frac{n}{2}+1} n!}{(2\pi)^n} \sum_{k=1}^{\infty} \frac{\cos(2\pi k x)}{k^n}, \quad x \in \mathbb{R}.$$

(2) *If $n \geq 1$ is odd,*

$$\widetilde{B}_n(x) = \frac{2(-1)^{\frac{n+1}{2}} n!}{(2\pi)^n} \sum_{k=1}^{\infty} \frac{\sin(2\pi k x)}{k^n}, \quad x \in \mathbb{R}.$$

Remark 6.2.11 Theorem 6.2.10 gives expressions for the Bernoulli polynomials $B_n(x)$—restrict $\widetilde{B}_n(x)$ to $[0, 1]$. There is one proviso: the infinite series formula we get for $B_1(x)$ is only valid for $x \in (0, 1)$.   ✠

Before we prove Theorem 6.2.10, we give several corollaries.

Corollary 6.2.12

$$\sum_{k=1}^{\infty} \frac{1}{k^n} = \frac{(-1)^{\frac{n}{2}+1} B_n (2\pi)^n}{2(n!)}, \quad n = 2, 4, 6, \cdots$$

Proof Take $x = 0$ in (1) of Theorem 6.2.10.   □

Example 6.2.13 $\sum_{k=1}^{\infty} \frac{1}{k^2} = \frac{\pi^2}{6}, \sum_{k=1}^{\infty} \frac{1}{k^4} = \frac{\pi^4}{90}$.   ♠

Remark 6.2.14 The problem of computing $\sum_{k=1}^{\infty} \frac{1}{k^n}$ when n is odd is much less well understood. It was only in 1978 that it was shown that $\sum_{k=1}^{\infty} \frac{1}{k^3} \notin \mathbb{Q}$.   ✠

Corollary 6.2.15 *The Bernoulli number B_{2n} is strictly positive iff n is odd, else B_{2n} is strictly negative.*

Proof Left to the exercises.   □

Corollary 6.2.16 $B_{2n+1} = 0$, $n \geq 1$.

Proof Take $x = 0$ in (2) of Theorem 6.2.10 and use $B_n(0) = B_n$, Lemma 6.2.4. □

Corollary 6.2.17 $B_{2n} \sim \frac{2(-1)^{n+1}(2n)!}{(2\pi)^{2n}}$. *That is,*

$$\lim_{n \to \infty} B_{2n} \Big/ \left(\frac{2(-1)^{n+1}(2n)!}{(2\pi)^{2n}} \right) = 1.$$

Proof $B_{2n} \Big/ \left(\frac{2(-1)^{n+1}(2n)!}{(2\pi)^{2n}} \right) = \sum_{k=1}^{\infty} \frac{1}{k^{2n}}$. We leave it to the exercises to show $\lim_{n \to \infty} \sum_{k=1}^{\infty} \frac{1}{k^{2n}} = 1$. □

Corollary 6.2.18 $\lim_{n \to \infty} |B_{2n}| = \infty$.

Proof This is a simple consequence of the previous corollary—it suffices to show that $\lim_{n \to \infty} |B_{2n+2}/B_{2n}| = \infty$. We leave the details to the exercises. □

Remark 6.2.19 For a more precise estimate on the growth of B_{2n}, see Exercises 6.3.9(4). ✸

Corollary 6.2.20

$$|\widetilde{B}_{2n}(x)| \leq |B_{2n}|, \ n \in \mathbb{N}, \ x \in \mathbb{R}.$$

Proof By (1) of Theorem 6.2.10 the maximum value of $|B_{2n}(x)|$ is attained at $x = 0$. But $|B_{2n}(0)| = |B_{2n}|$ (Lemma 6.2.4). □

Lemma 6.2.21 *For all $n \in \mathbb{N}$,*

(1) $\int_{j}^{j+1} \widetilde{B}_n(x)\, dx = 0$, $j \in \mathbb{Z}$.

(2) $|\int_{1}^{A} \widetilde{B}_{2n}(x)\, dx| \leq |B_{2n}|$, *all $A \geq 1$, $n \in \mathbb{N}$.*

Proof We have

$$\int_{j}^{j+1} \widetilde{B}_n(x)\, dx = \int_{j}^{j+1} \frac{d}{dx} \frac{\widetilde{B}_{n+1}(x)}{n+1}\, dx$$

$$= (\widetilde{B}_{n+1}(j+1) - \widetilde{B}_{n+1}(j))/(n+1)$$

$$= 0,$$

by 1-periodicity of $\widetilde{B}_{n+1}$. The second statement follows from Corollary 6.2.20 and (1). □

Proof of Theorem 6.2.10 In order to compute the Fourier coefficients, we use integration by parts together with Lemmas 6.2.4, 6.2.6. Suppose that $n \geq 1$ and

$$a_0 + \sum_{k=1}^{\infty} (a_k \cos(2\pi kx) + b_k \sin(2\pi kx))$$

is the Fourier series of $\tilde{B}_n$. Since $\tilde{B}_n$ is piecewise smooth for $n > 1$, the Fourier series of $\tilde{B}_n$ converges uniformly to $\tilde{B}_n$ if $n > 1$, by Theorem 5.5.29. In case $n = 1$, our definition of the value of $\tilde{B}_1$ at integer points guarantees by Theorem 5.5.26 that the Fourier series converges pointwise to $\tilde{B}_1$.

We can combine the computation of the cosine and sine coefficients by making use of complex numbers. More precisely, given $n \geq 1$, we define $c_k^{(n)} \in \mathbb{C}$ by

$$c_0^{(n)} = a_0, \ 2c_k^{(n)} = a_k - \imath b_k, \ k \geq 1.$$

The choice of the factor 2 and the minus sign is purely to optimize the computations. With these conventions we have

$$a_k = 2\mathrm{Re}(c_k^{(n)}), \ b_k = -2\mathrm{Im}(c_k^{(n)}), \ k \geq 1.$$

Using Remark 5.5.14, we have

$$c_k^{(n)} = \int_0^1 B_n(t)e^{-\imath 2\pi kt} \, dt, \ k \geq 0.$$

Taking $k = 0$, we have

$$
\begin{aligned}
c_0^{(n)} &= \int_0^1 B_n(t) \, dt \\
&= \int_0^1 \frac{B_{n+1}'(t)}{n+1} \, dt, \ \text{by Lemma 6.2.6} \\
&= \frac{B_{n+1}(1) - B_{n+1}(0)}{n+1} \\
&= 0, \ \text{by Lemma 6.2.4.}
\end{aligned}
$$

Next suppose $k \geq 1$, $n \geq 2$. We have

$$
\begin{aligned}
c_k^{(n)} &= \int_0^1 B_n(t)e^{-\imath 2\pi kt} \, dt \\
&= \left[\frac{-1}{\imath 2\pi k} B_n(t)e^{-\imath 2\pi kt} \right]_0^1 + \frac{1}{\imath 2\pi k} \int_0^1 B_n'(t)e^{-\imath 2\pi kt} \, dt \\
&= \frac{n}{\imath 2\pi k} \int_0^1 B_{n-1}(t)e^{-\imath 2\pi kt} \, dt, \ \text{by Lemmas 6.2.4, 6.2.6} \\
&= \frac{n}{\imath 2\pi k} c_k^{(n-1)}.
\end{aligned}
$$

Iterating, we obtain

$$c_k^{(n)} = \frac{n!}{(\imath 2\pi k)^{n-1}} c_k^{(1)}, \quad n \geq 2.$$

It remains to compute $c_k^{(1)}$

$$c_k^{(1)} = \int_0^1 B_1(t) e^{-\imath 2\pi kt}\, dt$$

$$= \int_0^1 \left(t - \frac{1}{2}\right) e^{-\imath 2\pi kt}\, dt, \quad \text{definition of } B_1$$

$$= \left[\frac{-1}{\imath 2\pi k}\left(t - \frac{1}{2}\right) e^{-\imath 2\pi kt}\right]_0^1 + \frac{1}{\imath 2\pi k}\int_0^1 e^{-\imath 2\pi kt}\, dt$$

$$= -\frac{1}{\imath 2\pi k}\left(\frac{1}{2} - \left(-\frac{1}{2}\right)\right)$$

$$= -\frac{1}{\imath 2\pi k}.$$

Therefore,

$$c_k^{(n)} = \frac{-n!}{(\imath 2\pi k)^n}, \quad n \geq 1,\ k \geq 1.$$

Since $a_k = 2\mathrm{Re}(c_k^{(n)})$ and $b_k = 2\mathrm{Im}(c_k^{(n)})$, we have

$$a_k = \frac{-2(n!)(-1)^{\frac{n}{2}}}{(2\pi k)^n}, \quad b_k = \qquad 0, \qquad n \text{ even,}$$

$$a_k = \qquad 0, \qquad b_k = \frac{2(n!)(-1)^{\frac{n+1}{2}}}{(2\pi k)^n}, \quad n \text{ odd.}$$

This completes the proof of Theorem 6.2.10.   □

EXERCISES 6.2.22

(1) Prove Lemma 6.2.4.
(2) Verify that $B_3(x) = x^3 - \frac{3}{2}x^2 + \frac{1}{2}x$ and $B_4(x) = x^4 - 2x^3 + x^2 - \frac{1}{30}$.
(3) Prove Corollary 6.2.15.
(4) Show that $\lim_{n\to\infty}\sum_{k=1}^{\infty}\frac{1}{k^n} = 1$.
(5) Show that $\sum_{n=0}^{\infty}\frac{(-1)^n}{(2n+1)^3} = \frac{\pi^3}{32}$. More generally, for all $p \in \mathbb{N}$ find a formula for $\sum_{n=0}^{\infty}\frac{(-1)^n}{(2n+1)^{2p-1}}$.
(6) Complete the proof of Corollary 6.2.18.

(7) Prove that for all $n \in \mathbb{N}$, $\int_n^\infty \frac{\widetilde{B_3}(x)}{x^3}\, dx \geq 0$ (we use this result in the next section).

(8) (Generating function for Bernoulli polynomials.) Prove that $\frac{xe^{xy}}{e^x-1} = \sum_{n=0}^\infty B_n(y)\frac{x^n}{n!}$ (for this note the proof of Lemma 6.2.3 and that $\frac{x}{e^x-1}$ is the generating function for the Bernoulli numbers).

6.3 The Euler–Maclaurin Formula

Let $f : [1, \infty) \to \mathbb{R}$ be a smooth or analytic function. In this section we develop a formula, known as the Euler–Maclaurin (summation) formula, that allows us to estimate a sum $\sum_{k=1}^n f(k)$ in terms of $\int_1^n f(x)\, dx$ and various expressions, involving Bernoulli numbers, together with a remainder term. More precisely, given an integer $r \geq 0$, we have

$$
\sum_{k=1}^n f(k) = \int_1^n f(x)\, dx + \frac{f(1) + f(n)}{2}
$$

$$
+ \sum_{k=1}^r \frac{B_{2k}}{(2k)!} \left(f^{(2k-1)}(1) - f^{(2k-1)}(n) \right) + R(r, n).
$$

Typically, the infinite series $\sum_{k=1}^\infty \frac{B_{2k}}{(2k)!} \left(f^{(2k-1)}(1) - f^{(2k-1)}(n) \right)$ does not converge and so the remainder term $R(r, n)$ may diverge as $r \to \infty$. However, $R(r, n)$ is given explicitly as an integral and, with a careful choice of r (preferably not too large but typically depending on n), we can often make $R(r, n)$ very small and thereby get a good estimate on $\sum_{k=1}^n f(k)$ (see also below where we discuss the strategy for applying the Euler–Maclaurin formula).

The formula was discovered independently by Euler and Maclaurin in about 1735. Euler applied the formula to compute $\sum_{n=1}^\infty n^{-2}$ to 20 decimal places and likely used the result to conjecture that the sum was $\pi^2/6$—a result he proved later that year (1735). As an indication of the power of the Euler–Maclaurin formula, a direct computation of $\sum_{n=1}^\infty n^{-2}$ requires about 10^{20} terms to get 20 decimal places of accuracy (over three trillion years work at one calculated term per second).

There are many applications of the Euler–Maclaurin formula including Stirling's formula (estimating $n!$) as well as good estimates for Euler's constant γ and sums of infinite series such as $\sum_{n=1}^\infty n^{-3}$.

We start by proving a simple special case of the formula and then, after giving an application to Stirling's formula, we proceed to state and prove the general case.

6.3.1 The Euler–Maclaurin Formula for r = 0

Proposition 6.3.1 *Let $f : [1, \infty) \to \mathbb{R}$ be C^1. For $n \in \mathbb{N}$ we have*

$$\sum_{k=1}^{n} f(k) = \int_{1}^{n} f(x)\,dx + \frac{f(1) + f(n)}{2} + \int_{1}^{n} \widetilde{B}_1(x) f'(x)\,dx.$$

Proof We have $\int_{1}^{n} f(x)\,dx = \sum_{k=1}^{n-1} \int_{k}^{k+1} f(x)\,dx$. Now

$$\int_{k}^{k+1} f(x)\,dx = \int_{k}^{k+1} \frac{d}{dx}\left(x - k - \frac{1}{2}\right) f(x)\,dx$$

$$= \left[\left(x - k - \frac{1}{2}\right) f(x)\right]_{k}^{k+1} - \int_{k}^{k+1} \left(x - k - \frac{1}{2}\right) f'(x)\,dx$$

$$= \frac{f(k+1) + f(k)}{2} - \int_{k}^{k+1} f'(x) \widetilde{B}_1(x)\,dx,$$

where we have used the observation that the 1-periodic extension of $B_1(x) = x - \frac{1}{2}$ to $\mathbb{R}$ is equal to $x - k - \frac{1}{2}$ on $(k, k+1)$. Summing from $k = 1$ to $n - 1$, we get

$$\int_{1}^{n} f(x)\,dx = \frac{2\sum_{k=1}^{n} f(k) - f(1) - f(n)}{2} - \int_{1}^{n} \widetilde{B}_1(x) f'(x)\,dx,$$

and rearranging we obtain the required result. □

Example 6.3.2 (Stirling's Formula—Version 1) We show that $n! = \sqrt{2\pi n}\left(\frac{n}{e}\right)^n$ $e^{-\delta_n}$, where $\lim_{n \to \infty} \delta_n = 0$.
Taking $f(x) = \log x$ and applying Proposition 6.3.1, we find that

$$\log(n!) = \sum_{k=1}^{n} \log k = n \log n - n + 1 + \frac{\log n}{2} + \int_{1}^{n} \frac{\widetilde{B}_1(x)}{x}\,dx,$$

where we have used $\int_{1}^{n} \log x\,dx = n \log n - n + 1$. Noting that $n \log n - n + \frac{\log n}{2} = \log(n^{n+\frac{1}{2}} e^{-n})$, we have

$$\log(n!) = \log(n^{n+\frac{1}{2}} e^{-n}) + 1 + \int_{1}^{\infty} \frac{\widetilde{B}_1(x)}{x}\,dx - \delta_n,$$

where $\delta_n = \int_{n}^{\infty} \frac{\widetilde{B}_1(x)}{x}\,dx$. Set $C = 1 + \int_{1}^{\infty} \frac{\widetilde{B}_1(x)}{x}\,dx$ so that $\log n! = \log(n^{n+\frac{1}{2}} e^{-n}) + C - \delta_n$. Exponentiating, we obtain

$$n! = e^C n^{n+\frac{1}{2}} e^{-n} e^{-\delta_n}.$$

It remains to prove that (a) $\delta_n \to 0$ as $n \to \infty$, and (b) $e^C = \sqrt{2\pi}$.

(a) Let $A \geq n$. By Lemma 6.2.9 and the 1-periodicity of $\widetilde{B}_1$, we have

$$\left| \int_n^A \widetilde{B}_1(x)\, dx \right| = \left| \int_0^{A-n} \widetilde{B}_1(x)\, dx \right| \leq 1/8.$$

Set $F(x) = \int_n^x \widetilde{B}_1(t)\, dt$. Integrating by parts, we have

$$\int_n^A \frac{\widetilde{B}_1(x)}{x}\, dx = F(x)/x\big|_{x=n}^A + \int_n^A F(x)/x^2\, dx$$

$$= F(A)/A + \int_n^A F(x)/x^2\, dx.$$

Therefore

$$\left| \int_n^A \frac{\widetilde{B}_1(x)}{x}\, dx \right| \leq \frac{1}{8A} + \int_n^A |F(x)|/x^2\, dx$$

$$\leq \frac{1}{8A} + \frac{1}{8} \int_n^A \frac{1}{x^2}\, dx$$

$$= \frac{1}{8A} + \frac{1}{8} \left[-\frac{1}{x} \right]_{x=n}^A$$

$$= \frac{1}{8n}.$$

Since this estimate holds for all $A \geq n$, we have shown that $|\delta_n| \leq \frac{1}{8n}$ and so $\lim_{n\to\infty} \delta_n = 0$, proving (a).

(b) We recall Wallis's formula from Example 5.5.32: $\lim_{n\to\infty} \frac{4^{2n}(n!)^4}{[(2n)!]^2(2n+1)} = \frac{\pi}{2}$. Taking square roots gives

$$\lim_{n\to\infty} \frac{4^n(n!)^2}{(2n)!\sqrt{2n+1}} = \sqrt{\frac{\pi}{2}}.$$

Substituting our expressions for $n!$ and $(2n)!$ in $\frac{4^n(n!)^2}{(2n)!\sqrt{2n+1}}$, we have

$$\frac{4^n(n!)^2}{(2n)!\sqrt{2n+1}} = \frac{2^{2n}[e^C n^{n+\frac{1}{2}} e^{-n} e^{-\delta_n}]^2}{e^C (2n)^{2n+\frac{1}{2}} e^{-2n} e^{-\delta_{2n}} \sqrt{2n+1}}$$

$$= e^C \frac{\sqrt{n}}{\sqrt{2(2n+1)}} e^{\delta_{2n}-2\delta_n}.$$

Since $\lim_{n\to\infty} \delta_n = 0$,

$$\sqrt{\frac{\pi}{2}} = \lim_{n\to\infty} e^C \frac{\sqrt{n}}{\sqrt{2(2n+1)}} e^{\delta_{2n}-2\delta_n} = e^C/2.$$

Hence $e^C = \sqrt{2\pi}$. ♦

Remark 6.3.3 Using the explicit form $\tilde{B}_1(x) = x - j - \frac{1}{2}$, $x \in [j, j+1]$, together with the strict monotonicity of $1/x$, we may show easily that $\delta_n < 0$. Consequently, $e^{-\delta_n} > 1$ and, since $|\delta_n| \le \frac{1}{8n}$, we have the estimate

$$n! \in \left[\sqrt{2\pi n} \left(\frac{n}{e}\right)^n, \sqrt{2\pi n} \left(\frac{n}{e}\right)^n e^{1/8n} \right], \quad n \in \mathbb{N}.$$

We improve on this estimate shortly. ✱

6.3.2 General Version of the Euler–Maclaurin Formula

Theorem 6.3.4 *Let n, r be positive integers with $n > 0$ and let $f : [1, \infty) \to \mathbb{R}$ be at least C^{2r+1}. Then*

$$\int_1^n f(x)\,dx = \sum_{k=1}^n f(k) - \left(\frac{f(1)+f(n)}{2}\right)$$

$$- \sum_{j=1}^r \frac{B_{2j}}{(2j)!} [f^{(2j-1)}(n) - f^{(2j-1)}(1)]$$

$$+ \frac{1}{(2r)!} \int_1^n \tilde{B}_{2r}(x) f^{(2r)}(x)\,dx.$$

Moreover,

$$\frac{1}{(2r)!} \int_1^n \tilde{B}_{2r}(x) f^{(2r)}(x)\,dx = -\frac{1}{(2r+1)!} \int_1^n \tilde{B}_{2r+1}(x) f^{(2r+1)}(x)\,dx.$$

Remarks 6.3.5

(1) The utility of this result depends on being able to show that the remainder or error term $\frac{1}{(2r)!} \int_1^n \tilde{B}_{2r}(x) f^{(2r)}(x)\,dx$ converges rapidly as $n \to \infty$. This is typically the case provided that the higher derivatives $f^{(2r)}(x)$ or $f^{(2r+1)}(x)$ converge rapidly to zero as $x \to \infty$.

(2) Observe that the error term will vanish if f is a polynomial of degree less than or equal to $2r$. Hence the theorem gives an explicit formula for $\sum_{k=1}^n p(k)$, when p is a polynomial.

(3) The *proof* of the Euler–Maclaurin formula is rather easy, quite formal and similar to the proofs of Taylor's theorem with integral remainder (see Sect. 2.7 reviewing results from the differential calculus). Matters get more interesting when one starts to estimate. ✠

Proof of Theorem 6.3.4 We proceed by induction on r. We have already proved the result for $r = 0$—Proposition 6.3.1. Suppose the theorem is proved for $r \leq R$. We prove it for $R + 1$. By Lemma 6.2.6, we have

$$\frac{1}{(2R)!} \int_1^n \widetilde{B}_{2R}(x) f^{(2R)}(x) \, dx \tag{6.4}$$

$$= \frac{1}{(2R)!} \int_1^n \frac{1}{2R+1} \widetilde{B}'_{2R+1}(x) f^{(2R)}(x) \, dx.$$

Integrating by parts,

$$\frac{1}{(2R)!} \int_1^n \frac{1}{2R+1} \widetilde{B}'_{2R+1}(x) f^{(2R)}(x) \, dx$$

$$= \left[\frac{1}{(2R+1)!} \widetilde{B}_{2R+1}(x) f^{(2R)}(x) \right]_{x=1}^{x=n}$$

$$\quad - \frac{1}{(2R+1)!} \int_1^n \widetilde{B}_{2R+1}(x) f^{(2R+1)}(x) \, dx$$

$$= - \frac{1}{(2R+1)!} \int_1^n \widetilde{B}_{2R+1}(x) f^{(2R+1)}(x) \, dx,$$

since $B_{2R+1} = 0$, if $R > 0$ by Corollary 6.2.16.

Now

$$- \frac{1}{(2R+1)!} \int_1^n \widetilde{B}_{2R+1}(x) f^{(2R+1)}(x) \, dx$$

$$= - \frac{1}{(2R+2)!} \int_1^n \widetilde{B}'_{2R+2}(x) f^{(2R+1)}(x) \, dx$$

$$= - \left[\frac{1}{(2R+2)!} \widetilde{B}_{2R+2}(x) f^{(2R+1)}(x) \right]_{x=1}^{x=n}$$

$$\quad + \frac{1}{(2R+2)!} \int_1^n \widetilde{B}_{2R+2}(x) f^{(2R+2)}(x) \, dx$$

$$= - \frac{B_{2R+2}}{(2R+2)!} [f^{(2R+1)}(n) - f^{(2R+1)}(1)]$$

$$\quad + \frac{1}{(2R+2)!} \int_1^n \widetilde{B}_{2R+2}(x) f^{(2R+2)}(x) \, dx.$$

This proves the Euler–Maclaurin formula for $r = R + 1$ and completes the inductive step. The final statement of the theorem is explicit in the second step of our computation of $\frac{1}{(2R)!} \int_1^n \widetilde{B}_{2R}(x) f^{(2R)}(x)\, dx$.   □

6.3.3   The Strategy

The Euler–Maclaurin formula states that

$$\sum_{k=1}^n f(k) = \int_1^n f(x)\, dx + \left(\frac{f(1) + f(n)}{2} \right)$$

$$+ \sum_{j=1}^r \frac{B_{2j}}{(2j)!} [f^{(2j-1)}(n) - f^{(2j-1)}(1)]$$

$$- \frac{1}{(2r)!} \int_1^n \widetilde{B}_{2r}(x) f^{(2r)}(x)\, dx.$$

One way we can use the result is to fix n and estimate the remainder $\frac{1}{(2r)!} \int_1^n \widetilde{B}_{2r}(x) f^{(2r)}(x)\, dx$. This is exactly what we did in Example 6.3.2. However, suppose that $\lim_{n \to \infty} f^{(s)}(n) = 0$ for all $s \le 2r$, then letting $n \to \infty$ we find that

$$\sum_{k=1}^\infty f(k) = \int_1^\infty f(x)\, dx + \frac{f(1)}{2}$$

$$- \sum_{j=1}^r \frac{B_{2j}}{(2j)!} f^{(2j-1)}(1)$$

$$- \frac{1}{(2r)!} \int_1^\infty \widetilde{B}_{2r}(x) f^{(2r)}(x)\, dx.$$

Not only does this give an expression for $\sum_{k=1}^\infty f(k)$ but we can subtract the original finite sum formula to get

$$\sum_{k=1}^\infty f(k) = \sum_{k=1}^n f(k) - \frac{f(n)}{2} - \sum_{j=1}^r \frac{B_{2j}}{(2j)!} f^{(2j-1)}(n)$$

$$- \frac{1}{(2r)!} \int_n^\infty \widetilde{B}_{2r}(x) f^{(2r)}(x)\, dx + \int_n^\infty f(x)\, dx.$$

Now choose a small value of n, say $n = 10$. Provided that we can integrate f, we can often easily compute all the terms on the right-hand side except the integral involving $\widetilde{B}_{2r}(x)f^{(2r)}(x)$. This we have to estimate. In many cases a judicious choice of n and r will make this term very small—we give one or two examples shortly.

6.3.4 Application to Stirling's Formula

If we apply the Euler–Maclaurin formula to $f(x) = \log x$ with $r > 0$, we can obtain better estimates of $n!$

Theorem 6.3.6 (Stirling's Formula, Version 2)

$$\sqrt{2\pi n}\left(\frac{n}{e}\right)^n \le n! \le \sqrt{2\pi n}\left(\frac{n}{e}\right)^n e^{\frac{1}{12n}}, \quad n \ge 1.$$

Proof It follows from the Euler–Maclaurin formula with $f(x) = \log x$ that

$$\log\left(\frac{n!}{n^{n+\frac{1}{2}}e^{-n}}\right) = \sum_{j=1}^{r}\frac{B_{2j}}{2j(2j-1)}\left[\frac{1}{n^{2j-1}}-1\right] + \frac{1}{2r}\int_1^n \frac{\widetilde{B}_{2r}(x)}{x^{2r}}\,dx.$$

Let $n \to \infty$ and we get, using the version of Stirling's formula proved in Example 6.3.2

$$\log(\sqrt{2\pi}) = -\sum_{j=1}^{r}\frac{B_{2j}}{2j(2j-1)} + \frac{1}{2r}\int_1^\infty \frac{\widetilde{B}_{2r}(x)}{x^{2r}}\,dx.$$

Subtract this from our expression for $\log\left(\frac{n!}{n^{n+\frac{1}{2}}e^{-n}}\right)$ to get

$$\log\left(\frac{n!}{\sqrt{2\pi}n^{n+\frac{1}{2}}e^{-n}}\right)$$

$$= \sum_{j=1}^{r}\frac{B_{2j}}{2j(2j-1)}\frac{1}{n^{2j-1}} - \frac{1}{2r}\int_n^\infty \frac{\widetilde{B}_{2r}(x)}{x^{2r}}\,dx. \tag{6.5}$$

Take $r = 1$ in (6.5). The right-hand side equals $\frac{1}{12n} - \frac{1}{2}\int_n^\infty \frac{\widetilde{B}_2(x)}{x^2}\,dx$. We have

$$\frac{1}{2}\int_n^\infty \frac{\widetilde{B}_2(x)}{x^2}\,dx = \frac{1}{2}\int_n^\infty \frac{1}{x^2}\frac{d}{dx}\left(\frac{\widetilde{B}_3(x)}{3}\right)\,dx$$

$$= -\frac{1}{6}\int_n^\infty \widetilde{B}_3(x)\frac{d}{dx}\left(\frac{1}{x^2}\right)\,dx$$

$$= \frac{1}{3}\int_n^\infty \frac{\widetilde{B}_3(x)}{x^3}\,dx.$$

Using Exercises 6.2.22(3), we have $\int_n^\infty \frac{\widetilde{B_3}(x)}{x^3}\,dx \geq 0$. Using the explicit formula for $B_3(x)$ (see Exercises 6.2.22(2)), we find that $|\widetilde{B_3}(x)| \leq 1/25$ and so $|\frac{1}{3}\int_n^\infty \frac{\widetilde{B_3}(x)}{x^3}\,dx| \leq \frac{1}{75}\int_n^\infty \frac{1}{x^3}\,dx = \frac{2}{75n^2} \leq \frac{1}{12n}$, $n \geq 1$. Hence $0 \leq \frac{1}{12n} - \frac{1}{2}\int_n^\infty \frac{\widetilde{B_2}(x)}{x^2}\,dx \leq \frac{1}{12n}$ and

$$0 < \log\left(\frac{n!}{\sqrt{2\pi}n^{n+\frac{1}{2}}e^{-n}}\right) \leq \frac{1}{12n}, \quad n \geq 1.$$

Exponentiating, we get the result.   ☐

Remark 6.3.7 If we take $r > 1$, we can find sharper estimates on the error. Indeed the result we proved above gives the second term in *Stirling's series*:

$$n! = \sqrt{2\pi n}\left(\frac{n}{e}\right)^n\left(1 + \frac{1}{12n} + \frac{1}{288n^2} - \frac{139}{51840n^3} - \frac{571}{2488320n^4} + \cdots\right).$$

This series does *not* converge. It is an example of an *asymptotic expansion*. For any given n there are only so many initial terms that give a good approximation. Taking more terms makes the approximation worse.   ✠

6.3.5 Computing Euler's Constant

Take $f(x) = \frac{1}{x}$ and $r = 2$ in the Euler–Maclaurin formula. We have $f'(x) = -\frac{1}{x^2}$, $f''(x) = \frac{2}{x^3}, f^{(3)}(x) = -\frac{6}{x^4}, f^{(4)}(x) = -\frac{24}{x^5}$, and $\int_1^n \frac{dx}{x} = \log n$. Substituting, we get

$$\log n = \sum_{k=1}^n \frac{1}{k} - \frac{1+\frac{1}{n}}{2} - \frac{\frac{1}{6}}{2}\left[\left(-\frac{1}{n^2}\right) - \left(-\frac{1}{1^2}\right)\right]$$

$$- \frac{(-\frac{1}{30})}{24}\left[\left(\frac{-6}{n^4}\right) - \left(\frac{-6}{1^4}\right)\right]$$

$$+ \frac{1}{4!}\int_1^n \widetilde{B_4}(x)\frac{4!}{x^5}\,dx.$$

After some simplifying, this gives

$$\log n = \sum_{k=1}^n \frac{1}{k} - \frac{1}{2} - \frac{1}{12} + \frac{1}{120} - \frac{1}{2n} + \frac{1}{12n^2} - \frac{1}{120n^4} + \int_1^n \frac{\widetilde{B_4}(x)}{x^5}\,dx,$$

and so

$$\sum_{k=1}^{n} \frac{1}{k} - \log n = \frac{1}{2} + \frac{1}{12} - \frac{1}{120} + \frac{1}{2n} - \frac{1}{12n^2} + \frac{1}{120n^4} - \int_1^n \frac{\widetilde{B}_4(x)}{x^5} \, dx.$$

Letting $n \to \infty$, we get

$$\gamma = \frac{1}{2} + \frac{1}{12} - \frac{1}{120} - \int_1^{\infty} \frac{\widetilde{B}_4(x)}{x^5} \, dx.$$

Since $\int_1^n = \int_1^{\infty} - \int_n^{\infty}$, this gives us

$$\sum_{k=1}^{n} \frac{1}{k} - \log n = \gamma + \frac{1}{2n} - \frac{1}{12n^2} + \frac{1}{120n^4} + \int_n^{\infty} \frac{\widetilde{B}_4(x)}{x^5} \, dx,$$

and so we obtain an asymptotic formula for Euler's constant:

$$\gamma = \sum_{k=1}^{n} \frac{1}{k} - \log n - \frac{1}{2n} + \frac{1}{12n^2} - \frac{1}{120n^4} - E_n,$$

where the error term $E_n = \int_n^{\infty} \frac{\widetilde{B}_4(x)}{x^5} \, dx$. Ignoring the error term for the moment, we find that if we take n=10 then

$$\log 10 = 2.302585092994$$
$$1/20 = 0.05$$
$$1/1200 = 0.0008\dot{3}$$
$$1/1200000 = 0.0000008\dot{3}$$

From this, we get

$$\sum_{k=1}^{10} \frac{1}{k} - \log 10 - \frac{1}{20} + \frac{1}{1200} - \frac{1}{12000000} = 0.577215660974 \cdots$$

The true value of γ is $\gamma = 0.577215664901\ldots$ and so our estimate is accurate to 8 decimal places. We can verify this by estimating the error term $E_{10} = \int_{10}^{\infty} \frac{\widetilde{B}_4(x)}{x^5} \, dx$.

6.3.6 Estimating E_{10}

We have $|\widetilde{B}_4(x)| \leq |B_4|$, $x \in \mathbb{R}$, by Corollary 6.2.20. Therefore

$$\left| \int_{10}^{\infty} \frac{\widetilde{B}_4(x)}{x^5}\, dx \right| \leq \int_{10}^{\infty} \frac{|B_4|}{x^5}$$

$$= \frac{1}{30} \int_{10}^{\infty} \frac{1}{x^5}\, dx$$

$$= \frac{1}{30} \left[\frac{x^{-4}}{-4} \right]_{10}^{\infty}$$

$$= \frac{1}{120} \times 10^{-4}$$

$$\leq 10^{-6}.$$

Hence $|E_{10}| \leq 10^{-6}$. We can do better by using the second form of the error term in the Euler–Maclaurin formula. We have

$$\int_{10}^{\infty} \frac{\widetilde{B}_4(x)}{x^5}\, dx = \int_{10}^{\infty} \frac{\widetilde{B}_5(x)}{x^6}\, dx.$$

This time we have to deal with an odd Bernoulli polynomial and we can no longer use Corollary 6.2.20. What we shall do is use an estimate on improper integrals—really an integral version of Dirichlet's test—and then reduce to a problem that involves estimating $\int_{10}^{\infty} \widetilde{B}_6(x)\, dx$, where we can again use Corollary 6.2.20.

First we prove a powerful lemma on improper integrals.

Lemma 6.3.8 *Suppose that $g, h : [a, \infty) \to \mathbb{R}$ and g is continuous and h is C^1. Assume further that*

(a) *$h(x)$ is decreasing and converges to zero as $x \to \infty$,*
(b) *there exists an $M \geq 0$ such that $|\int_a^A g(x)\, dx| \leq M$ for all $A \geq a$.*

Then $\int_a^{\infty} g(x)h(x)\, dx$ exists and

$$\left| \int_a^{\infty} g(x)h(x)\, dx \right| \leq Mh(a).$$

Proof We verify the estimate and leave the proof of convergence to the exercises (the proof uses Exercises 6.1.23(1), but the crucial step is done below). Let $A \geq a$.

Set $G(x) = \int_a^x g(t)\,dt$. Integrating by parts we have

$$\int_a^A g(x)h(x)\,dx = [G(x)h(x)]_{x=a}^{x=A} - \int_a^A G(x)h'(x)\,dx$$

$$= G(A)h(A) - \int_a^A G(x)h'(x)\,dx.$$

We have to estimate both terms in this equation. Obviously,

$$|G(A)h(A)| \le Mh(A).$$

Since $h'(x) \le 0$, we have

$$\left| \int_a^A G(x)h'(x)\,dx \right| = \left| \int_a^A G(x)(-h'(x))\,dx \right|$$

$$\le \int_a^A |G(x)|(-h'(x))\,dx|$$

$$\le M \int_a^A -h'(x)\,dx$$

$$= M(h(a) - h(A)).$$

Therefore

$$\left| \int_a^A g(x)h(x)\,dx \right| \le Mh(A) + M(h(a) - h(A)) = Mh(a).$$

Letting $A \to \infty$ the result follows. $\square$

If we take $g(x) = \widetilde{B}_5(x)$ then g satisfies (b) of Lemma 6.3.8 by Lemma 6.2.21 and the 1-periodicity of $\widetilde{B}_5$. Since $h(x) = x^{-6}$ obviously satisfies (a) of Lemma 6.3.8, we have

$$\left| \int_{10}^{\infty} \frac{\widetilde{B}_5(x)}{x^6}\,dx \right| \le 10^{-6} \times \sup_{A \ge 1} \left| \int_1^A \widetilde{B}_5(x)\,dx \right|.$$

But

$$\left| \int_1^A \widetilde{B}_5(x)\,dx \right| = \left| \int_1^A \frac{d}{dx} \frac{\widetilde{B}_6(x)}{6}\,dx \right|$$

$$= \left| \frac{\widetilde{B}_6(A) - B_6}{6} \right|$$

$$\leq \frac{2B_6}{6}, \quad \text{Corollary 6.2.20}$$

$$= \frac{1}{3 \times 42} \quad \left(B_6 = \frac{1}{42}\right).$$

Hence $|\int_{10}^{\infty} \frac{\widetilde{B}_4(x)}{x^5} dx| \leq \frac{1}{126} \times 10^{-6} < 10^{-8}$. Hence the error $|E_{10}|$ in our computation of Euler's constant is less than 10^{-8}.

6.3.7   Estimating $\sum_{k=1}^{\infty} \frac{1}{k^2}$

This time we apply Euler–Maclaurin to $f(x) = 1/x^2$ and take $r = 1$. We have $f'(x) = -\frac{2}{x^3}, f''(x) = \frac{6}{x^4}$ and $\int_1^n \frac{dx}{x^2} = 1 - \frac{1}{n}$. Substituting, we get

$$1 - \frac{1}{n} = \sum_{k=1}^{n} \frac{1}{k^2} - \frac{1 + \frac{1}{n^2}}{2} - B_2 \left[\left(-\frac{2}{n^3}\right) - \left(-\frac{2}{1^3}\right)\right] + 3 \int_1^n \frac{\widetilde{B}_2(x)}{x^4} dx.$$

Taking $B_2 = 1/6$, we get

$$\sum_{k=1}^{n} \frac{1}{k^2} = \frac{11}{6} - \frac{1}{n} + \frac{1}{2n^2} - \frac{1}{3n^3} - 3 \int_1^n \frac{\widetilde{B}_2(x)}{x^4} dx.$$

Letting $n \to \infty$ gives

$$\sum_{k=1}^{\infty} \frac{1}{k^2} = \frac{11}{6} - 3 \int_1^{\infty} \frac{\widetilde{B}_2(x)}{x^4} dx.$$

Writing $\int_1^n = \int_1^{\infty} - \int_n^{\infty}$ and substituting, we get an asymptotic formula for $\sum_{k=1}^{\infty} \frac{1}{k^2}$:

$$\sum_{k=1}^{\infty} \frac{1}{k^2} = \sum_{k=1}^{n} \frac{1}{k^2} + \frac{1}{n} - \frac{1}{2n^2} + \frac{1}{3n^3} - 3 \int_n^{\infty} \frac{\widetilde{B}_2(x)}{x^4} dx.$$

Using the known value $\sum_{k=1}^{\infty} \frac{1}{k^2} = \pi^2/6$, this gives an estimate accurate to 4 decimal places if we take $n = 10$ and ignore the remainder. We can do much better if we take $r = 2$.

EXERCISES 6.3.9

(1) Using the Euler–Maclaurin formula with $r = 1$, show that

$$\sum_{k=1}^{n} \frac{1}{k^3} = \frac{5}{4} - \frac{1}{2n^2} + \frac{1}{2n^3} - \frac{1}{4n^4} - 6 \int_1^n \frac{\widetilde{B}_2(x)}{x^5} dx.$$

Hence find a formula for $\sum_{k=1}^{\infty} \frac{1}{k^3}$ in terms of $\int_1^{\infty} \frac{\widetilde{B_2}(x)}{x^5} dx$ and deduce the asymptotic formula

$$\sum_{k=1}^{\infty} \frac{1}{k^3} = \sum_{k=1}^{n} \frac{1}{k^3} + \frac{1}{2n^2} - \frac{1}{2n^3} + \frac{1}{4n^4} - 6 \int_n^{\infty} \frac{\widetilde{B_2}(x)}{x^5} dx.$$

Verify that $|6 \int_n^{\infty} \frac{\widetilde{B_2}(x)}{x^5} dx| \leq \frac{1}{n^5}$ and hence find the smallest value of n that you can take in the formula above to estimate $\sum_{k=1}^{\infty} \frac{1}{k^3}$ to within 10^{-3}.

(2) Show that the estimate used in Q1 can be improved to $|6 \int_n^{\infty} \frac{\widetilde{B_2}(x)}{x^5} dx| \leq \frac{1}{n^6}$.
(3) Using Stirling's formula, show that

$$\lim_{n \to \infty} \frac{\sqrt{3}\,[3^n (n!)]^3}{n(3n)!}$$

exists and equals 2π. What is $\lim_{n \to \infty} \frac{\sqrt{5}\,[5^n(n!)]^5}{n^2 (5n)!}$?
(4) Show that if $\alpha = 1 + \frac{p}{n}$, where $0 \leq p < n$, then B_{2n} grows at least as fast as $(n\alpha)!$ (that is, there exists a $C = C(p) > 0$ such that for all sufficiently large n, $B_{2n} \geq C(n\alpha)!$). Hint: Stirling's formula and Corollary 6.2.17.
(5) Prove the existence of the infinite integral in Lemma 6.3.8 (you will need Exercises 6.1.23(1)).
(6) Prove that the integral $\int_1^{\infty} \frac{\sin x}{x^{\alpha}} dx$ converges provided that $\alpha > 0$. What can you say about $\int_0^1 \frac{\sin x}{x^{\alpha}} dx$?
(7) Define $F(\alpha) = \int_0^{\infty} \frac{\sin x}{x^{\alpha}} dx$. Verify that F is a C^{∞}-function on $(0, 2)$. (You may assume or easily show that given $p \in \mathbb{N}$, $\alpha > 0$, then $(\log x)^p x^{-\alpha}$ is decreasing for all sufficiently large x and if $\alpha > 0$, there exists $0 < \beta < \alpha$, $C > 0$, such that $|(\log x)^p x^{-\alpha}| \leq Cx^{-\beta}$ on $(0, 1]$).

Chapter 7
Metric Spaces

In this chapter we develop the theory of *metric spaces*. The idea of a metric or distance on a set is simple, intuitive and powerful. The concept is natural for many important mathematical structures including vector spaces, geometric objects such as surfaces or manifolds, and spaces of continuous and differentiable functions. Although metric spaces need not have any vector space structure, they provide an ideal abstract framework for studying sequences, convergence and continuous functions. In the first half of the chapter we focus on foundations and examples. In particular, we define open and closed sets and show that a metric space has a natural *topology* of open sets. Using this idea we will be able to give a natural 'preservation of structure' definition of continuity that avoids the surfeit of quantifiers in the ε, δ-definition. In the remainder of the chapter we develop theory and give results, such as the Arzelà–Ascoli theorem, that generalize the Bolzano–Weierstrass theorem to spaces of continuous functions. We also prove the simple, yet very powerful, *contraction mapping lemma*. We use this result in Chap. 8 to prove a fundamental result on the existence of fractals and in the final chapter to prove results including the implicit and inverse function theorems and the existence and uniqueness theorem for ordinary differential equations.

7.1 Basic Definitions and Examples

Definition 7.1.1 A metric space (X, d) consists of a (non-empty) set X together with a real-valued function $d(x, y)$ on X^2 which satisfies

(1) $d(x, y) \geq 0$ for all $x, y \in X$.
(2) $d(x, y) = 0$ iff $x = y$.
(3) $d(x, y) = d(y, x)$ for all $x, y \in X$.
(4) $d(x, z) \leq d(x, y) + d(y, z)$ for all $x, y, z \in X$.

© Springer International Publishing AG 2017
M. Field, *Essential Real Analysis*, Springer Undergraduate Mathematics Series,
https://doi.org/10.1007/978-3-319-67546-6_7

We call d a *metric* (or distance) on X and often refer to the *metric space* X if the metric d is clear from the context.

Remark 7.1.2 We refer to (4) of Definition 7.1.1 as the *triangle inequality*—see examples (1,2) below for justification.   ✸

Examples 7.1.3

(1) $(\mathbb{R}, |\cdot|)$, where $|\cdot|$ denotes absolute value; that is $d(x,y) = |x-y|$. Properties (1,2,3) are immediate. The triangle inequality $|x+y| \le |x| + |y|$ implies (4) if we replace x by $x-y$ and y by $y-z$.

(2) The *Euclidean metric d* (or d_2) on $\mathbb{R}^n$ is defined by

$$d(\mathbf{x}, \mathbf{y}) = \sqrt{\sum_{i=1}^{n}(x_i - y_i)^2},$$

where $\mathbf{x} = (x_1, \cdots, x_n), \mathbf{y} = (y_1, \cdots, y_n) \in \mathbb{R}^n$ (if $n = 1$, $d(x,y) = |x-y|$, the metric defined in (1)). Properties (1,2,3) are easily verified. It remains to verify the triangle inequality. If we let $(\mathbf{u}, \mathbf{v})$ denote the inner (or dot) product of vectors $\mathbf{u}, \mathbf{v}$ and $\|\mathbf{u}\| = (\mathbf{u}, \mathbf{u})^{\frac{1}{2}}$ denote the corresponding Euclidean norm, then it suffices to show that $\|\mathbf{u} + \mathbf{v}\| \le \|\mathbf{u}\| + \|\mathbf{v}\|$ for all $\mathbf{u}, \mathbf{v} \in \mathbb{R}^n$. Squaring, it is easy to see that this inequality follows from the Cauchy–Schwarz inequality $|(\mathbf{u}, \mathbf{v})| \le \|\mathbf{u}\|\|\mathbf{v}\|$. We prove the Cauchy–Schwarz inequality by observing that the quadratic form

$$(x\mathbf{u} + y\mathbf{v}, x\mathbf{u} + y\mathbf{v}) = x^2(\mathbf{u}, \mathbf{u}) + 2xy(\mathbf{u}, \mathbf{v}) + y^2(\mathbf{v}, \mathbf{v})$$

is positive for all $x, y \in \mathbb{R}$. The result follows since a quadratic form $Ax^2 + 2Bxy + Cy^2$ is positive for all $x, y \in \mathbb{R}$ iff $A, B, AC - B^2 \ge 0$.

(3) Let X be any non-empty set. We define the *discrete metric* on X by

$$d(x,y) = \begin{cases} 1, & \text{if } x \ne y, \\ 0, & \text{if } x = y. \end{cases}$$

We leave it to the exercises for the reader to verify that d satisfies the conditions for a metric.

(4) Let $C^0([a,b])$ denote the vector space of real-valued continuous functions on the closed and bounded interval $[a,b]$. Define the *uniform metric ρ* on $C^0([a,b])$ by

$$\rho(f,g) = \sup_{x\in[a,b]} \{|f(x) - g(x)| \mid x \in [a,b]\}.$$

Obviously $\rho(f,g) \ge 0$ for all $f, g \in C^0([a,b])$. If $\rho(f,g) = 0$, then $|f(x) - g(x)| = 0$ for all $x \in [a,b]$ and so $f = g$, proving (2). Since $|f(x) - g(x)| = |g(x) - f(x)|$, we obviously have $\rho(f,g) = \rho(g,f)$. Finally, if $f, g, h \in C^0([a,b])$

and $x \in [a, b]$, we have by the triangle inequality

$$|f(x) - g(x)| + |g(x) - h(x)| \geq |f(x) - h(x)|.$$

Since $|f(x) - g(x)| \leq \rho(f, g)$, $|g(x) - h(x)| \leq \rho(g, h)$ for all $x \in [a, b]$ (by definition of the supremum), we have $\rho(f, g) + \rho(g, h) \geq |f(x) - h(x)|$. Since this estimate holds for all $x \in [a, b]$, $\rho(f, g) + \rho(g, h)$ is an upper bound for $\{|f(x) - g(x)| \mid x \in [a, b]\}$ and so $\rho(f, g) + \rho(g, h) \geq \rho(f, h)$. Not surprisingly, the uniform metric is particularly well adapted for the study of uniform convergence. In our final example (6), we define another metric on $C^0([a, b])$ that is particularly appropriate for the study of Fourier series.

(5) Let $B([a, b])$ denote the vector space of real-valued *bounded functions* on the interval $[a, b]$. We define the uniform metric ρ on $B([a, b])$ exactly as we did for $C^0([a, b])$ and the proof that ρ defines a metric on $B([a, b])$ is unchanged. Note that we can replace $[a, b]$ by any non-empty subset of $\mathbb{R}$.

(6) Recall from Sect. 5.6 that the L^2-metric on $C^0([a, b])$ is defined by

$$\rho_2(f, g) = \left(\int_a^b |f(x) - g(x)|^2 \, dx \right)^{\frac{1}{2}}, \ f, g \in C^0([a, b]).$$

It follows from the results of Sect. 5.6, notably Lemma 5.6.4, that ρ_2 is a metric and that $\rho_2(f, g) \leq \rho(f, g)$, for all $f, g \in C^0([a, b])$. ♠

The following variant of the triangle inequality is very useful.

Lemma 7.1.4 *Let (X, d) be a metric space. Then for all $x, y, z \in X$ we have*

$$d(x, z) \geq |d(x, y) - d(z, y)|.$$

Proof By the triangle inequality, we have $d(x, z) + d(z, y) \geq d(x, y)$ and so

$$d(x, z) \geq d(x, y) - d(z, y).$$

Interchanging x and z in this inequality we obtain

$$d(x, z) = d(z, x) \geq d(z, y) - d(x, y).$$

Hence $d(x, z) \geq |d(x, y) - d(z, y)|$. □

Definition 7.1.5 Let Y be a non-empty subset of the metric space (X, d). The *induced metric* d_Y on Y is defined by

$$d_Y(y_1, y_2) = d(y_1, y_2), \ y_1, y_2 \in Y,$$

Lemma 7.1.6 *The induced metric is a metric on Y.*

Fig. 7.1 Metrics on the unit circle

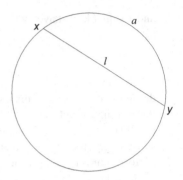

Proof Immediate, since d_Y inherits all the properties of d.   □

Remark 7.1.7 If we take the induced metric on a subset Y of (X, d), we often drop the subscript Y and generally refer to Y as a subspace (rather than subset) of X. ✠

Examples 7.1.8

(1) The metric induced on the subset $\mathbb{Z}$ of $\mathbb{R}$ by the standard metric is given by $d_{\mathbb{Z}}(m, n) = |m - n|$, $m, n \in \mathbb{Z}$.
(2) Take the standard Euclidean metric d on $\mathbb{R}^2$ and let S^1 denote the unit circle, centre the origin, in $\mathbb{R}^2$. Referring to Fig. 7.1, $d_{S^1}(x, y)$ equals the length of the chord $\overline{xy}$. The induced metric d_{S^1} is obviously different from the more natural metric on S^1 defined by arc length (referring to Fig. 7.1, $\ell < a$ if $x \neq y$). ♠

EXERCISES 7.1.9

(1) Verify that the discrete metric (Examples 7.1.3(3)) is a metric.
(2) Regard the unit circle $S^1 \subset \mathbb{R}^2$ as parametrized by angle $\theta \in [0, 2\pi)$. Define $d(\theta, \psi) = \min\{|\theta - \psi|, 2\pi - |\theta - \psi|\}$ (arc length) and verify that d defines a metric on S^1.
(3) Let $\mathbf{x} = (x_1, x_2)$, $\mathbf{y} = (y_1, y_2) \in \mathbb{R}^2$ and define

$$d_1(\mathbf{x}, \mathbf{y}) = |x_1 - y_1| + |x_2 - y_2|,$$

$$d_\infty(\mathbf{x}, \mathbf{y}) = \max\{|x_1 - y_1|, |x_2 - y_2|\},$$

$$d_{\frac{1}{2}}(\mathbf{x}, \mathbf{y}) = \left(\sqrt{|x_1 - y_1|} + \sqrt{|x_2 - y_2|}\right)^2.$$

Show that d_1, d_∞ define metrics on $\mathbb{R}^2$ but that $d_{\frac{1}{2}}$ is not a metric on $\mathbb{R}^2$. (For $d_{\frac{1}{2}}$ you will need to find a triple of points in $\mathbb{R}^2$ for which the triangle inequality fails.)
(4) Define the appropriate extensions of d_1 and d_∞ to $\mathbb{R}^n$, $n > 2$, and verify that they are metrics on $\mathbb{R}^n$.
(5) (Product metric.) Let (X, d_X), (Y, d_Y) be metric spaces. Let $X \times Y = \{(x, y) \mid x \in X, y \in Y\}$. Define the *product metric* d on $X \times Y$ by

$d((x_1, y_1), (x_2, y_2)) = \max\{d_X(x_1, x_2), d_Y(y_1, y_2)\}$. Verify that d is a metric on $X \times Y$.

(6) Let X be a set and suppose that $d(x, y)$ satisfies $d(x, y) = 0$ iff $x = y$, and $d(x, z) \leq d(x, y) + d(z, y)$, for all $x, y, z \in X$. Show that d is a metric on X (that is, d satisfies (1–4) of Definition 7.1.1).

(7) Let (X, d) be a metric space and suppose that $x_1, \cdots, x_n \in X, n \geq 3$. Show that $d(x_1, x_n) \leq \sum_{i=1}^{n-1} d(x_i, x_{i+1})$.

(8) Let X be a non-empty set and $B(X, \mathbb{R})$ denote the set of all *bounded* functions $f : X \to \mathbb{R}$. That is, $f \in B(X, \mathbb{R})$ iff there exists an $M \geq 0$ such that $|f(x)| \leq M$ for all $x \in X$. Show that

  (a) $B(X, \mathbb{R})$ is a vector space (if $f, g \in B(X, \mathbb{R})$, $\lambda \in \mathbb{R}$, then $f + \lambda g \in B(X, \mathbb{R})$).
  (b) $\rho(f, g) = \sup_{x \in X} |f(x) - g(x)| < \infty$ for all $f, g \in B(X, \mathbb{R})$.
  (c) ρ defines a metric on $B(X, \mathbb{R})$.

  (Notes: X is a general set and $f : X \to \mathbb{R}$ is not assumed to be continuous. ρ is called the *uniform metric* on $B(X, \mathbb{R})$.)

(9) Let (X, d) be a metric space and let $D = \sup_{x, y \in X} d(x, y)$. We refer to D as the *diameter* of X. Show by means of examples that we can have (a) $D < \infty$, (b) $D = \infty$. Show that if we define $\rho(x, y) = \min\{1, d(x, y)\}$, then ρ defines a metric on X and that that the diameter of X with respect to the metric ρ is at most one.

(10) Let $\langle \cdot, \cdot \rangle$ be an inner product on the vector space V. Define $\|X\| = \sqrt{\langle X, X \rangle}$, $X \in V$. Show that the Cauchy–Schwarz inequality holds:

$$|\langle X, Y \rangle| \leq \|X\| \|Y\|, \text{ for all } X, Y \in V.$$

(Use the method of Examples 7.1.3(2).) Deduce that if we define $d(X, Y) = \|X - Y\|$, then (V, d) is a metric space.

(11) Let X be a set. Metrics d, ρ on X are said to be *equivalent* if there exist constants $C, c > 0$ such that for all $x, y \in X$, $cd(x, y) \leq \rho(x, y) \leq Cd(x, y)$.

  (a) Show that if d, ρ are equivalent, then we can find constants $c', C' > 0$ such that for all $x, y \in X$, $c'\rho(x, y) \leq d(x, y) \leq C'\rho(x, y)$.
  (b) Verify that if d, ρ are equivalent and ρ, η are equivalent, then d, η are equivalent.
  (c) Show that every metric on a finite set X is equivalent to the discrete metric on X.
  (d) Show that the induced metric on $\mathbb{Z}$ (Examples 7.1.8(1)) is not equivalent to the discrete metric on $\mathbb{Z}$.
  (e) Show that the induced metric on the unit circle S^1 (Examples 7.1.8(2)) is equivalent to the metric defined by arc length (see exercise (2)).
  (f) Show that the metrics d_1, d_∞ on $\mathbb{R}^2$ are equivalent to the Euclidean metric on $\mathbb{R}^2$. Generalize to $\mathbb{R}^n$.

(h) Show that the uniform and L^2 metrics on $C^0([a, b])$ are not equivalent. (Hint: for $n \in \mathbb{N}$, construct $f_n \in C^0([a, b])$ such that $\rho(f, 0) = 1$, $\rho_2(f, 0) = 1/n$. See also Sect. 5.6.)

7.2 Distance from a Subset

Suppose that A is a non-empty subset of the metric space (X, d). Given $x \in X$, it is natural to define the distance $d(x, A)$ from x to A. Roughly speaking, this should be the shortest distance from x to A. More formally, we define

$$d(x, A) = \inf_{a \in A} d(x, a).$$

Notice that $0 \leq d(x, A) \leq d(x, a)$ for all $a \in A$. As we shall see shortly, it may not be possible to pick a point $a' \in A$ such that that $d(x, A)$ is equal to $d(x, a')$. The main properties of distance to a subset are given in the next proposition.

Proposition 7.2.1 *Let A be a non-empty subset of the metric space (X, d).*

(1) $d(x, A) \geq 0$ *for all* $x \in X$.
(2) *If* $x \in A$, *then* $d(x, A) = 0$.
(3) *If* $A \subset B \subset X$, *then* $d(x, A) \geq d(x, B)$ *for all* $x \in X$.
(4) *If* $x, x' \in X$, *then*

$$|d(x, A) - d(x', A)| \leq d(x, x').$$

Proof Statements (1) and (2) are obvious. For statement (3), observe that for every $a \in A$, we can choose $b = a \in B$ such that $d(x, a) = d(x, b)$. Hence $\inf_{b \in B} d(x, b) \leq \inf_{a \in A} d(x, a)$. It remains to prove (4). Given $a \in A$, we have

$$d(x, x') + d(x', a) \geq d(x, a) \geq d(x, A).$$

Hence $d(x, A)$ is a lower bound for $d(x, x') + d(x', a)$ and so

$$d(x, A) \leq \inf_{a \in A}(d(x, x') + d(x', a)) = d(x, x') + d(x', A).$$

Therefore, $d(x, x') \geq d(x, A) - d(x', A)$. Interchanging x and x', we have $d(x, x') = d(x', x) \geq d(x', A) - d(x, A)$. Combining the two inequalities gives $|d(x, A) - d(x', A)| \leq d(x, x')$. □

Remark 7.2.2 Proposition 7.2.1(4) generalizes Lemma 7.1.4. ✠

Examples 7.2.3

(1) Let $X = \mathbb{R}$ (standard metric) and $A = (a, b)$, $0 < a < b < \infty$. Observe that $d(a, A) = 0$ but $a \notin A$ and so the converse of Proposition 7.2.1(2) is false.
(2) If $A \subsetneq B \subset X$, it is possible to have $d(x, A) = d(x, B)$ for all $x \in X$. For example, take $A = (a, b)$, $B = [a, b]$, $X = \mathbb{R}$. ♠

Remarks 7.2.4

(1) It is useful to extend the definition of $d(x, A)$ to allow for A to be the empty set. We define $d(x, \emptyset) = +\infty$, where the symbol $+\infty$ satisfies $+\infty > x$ for all $x \in \mathbb{R}$. This convention is compatible with Proposition 7.2.1(1,2,3).
(2) Later, in Chap. 8, we define the distance between non-empty subsets A and B of a metric space. The distance we define will be a metric—though for this we need to restrict the class of subsets we work with. The definition of the distance between a point and a subset will suffice for our discussion of open and closed subsets of a metric space—the topic of the next few sections. ✠

EXERCISES 7.2.5

(1) Suppose that $\{A_i \mid i \in I\}$ is a family of subsets of X. Let $x \in X$ and define $a_i = d(x, A_i)$, $i \in I$. Show that $d(x, \cup_{i \in I} A_i) = \inf_{i \in I} a_i$. Verify this result remains true if we allow some of the sets A_i to be empty. (See Remarks 7.2.4(1).)
(2) Verify that (1–3) of Proposition 7.2.1 are compatible with our definition of $d(x, A)$ if $A = \emptyset$.

7.3 Open and Closed Subsets of a Metric Space: Intuition

Suppose that A is a non-empty subset of the metric space X—see Fig. 7.2.

We discuss some simple features of A that relate to the metric structure. First of all the *outside* of A is defined as the complement $X \smallsetminus A$. It is natural to define the *boundary* of A—usually denoted by ∂A—as the set of points in X which are of zero distance from both A and $X \smallsetminus A$. This leads naturally to the question of

Fig. 7.2 Subset A of metric space X

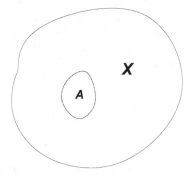

characterizing the points of A which do not lie on the boundary of A. It turns out that this set—the *interior* of A—is an example of an *open set*. As we shall see, open sets play a central role in the theory of metric spaces and our investigations will start by giving a careful, but very simple and natural, definition of an open set. Later we shall see that continuity can be formulated entirely in terms of open sets. We might also consider all the points in X which are at zero distance from A. This set of points—called the *closure* of A—gives an example of a *closed set*. As we shall see, a set $F \subset X$ is closed iff $X \smallsetminus F$ is open. Closed sets are important because, for example, a subset of X is closed iff it can be represented as the solution set of a continuous real-valued function on X. After we have established the basic properties of open and closed sets, we then develop the theory of sequences and continuous functions in metric spaces. Much of what we do is formally very similar to what we have previously done on the real line. In many ways it will be simpler as we will not get distracted by any extraneous structure of the real line (such as arithmetical properties). Basically, we work with a set X and metric d. A very simple yet, as we shall see, very rich structure.

7.4   Open and Closed Sets

We start with the definition of an open set.

Definition 7.4.1 A subset U of the metric space X is *open* if

$$d(u, X \smallsetminus U) > 0, \quad \text{for all } u \in U.$$

Examples 7.4.2

(1) An open interval $(a, b) \subset \mathbb{R}$ is an open set: if $x \in (a, b)$, then $d(x, (-\infty, a] \cup [b, \infty)) = \min\{x - a, b - x\} > 0$.
(2) If we give the set X the discrete metric, then every subset of X is open: if $Z \subsetneq X$, we have $d(u, X \smallsetminus Z) = 1 > 0$ for all $u \in Z$.   ♠

Lemma 7.4.3 *The empty set $\emptyset$ and X are open subsets of X.*

Proof If we take the negation of the definition of an open set, we see that a subset Z is not open if there exists a $u \in Z$ such that $d(u, X \smallsetminus Z) = 0$. Since $\emptyset$ contains no points, it cannot satisfy the 'not open' condition and so $\emptyset$ must be open. The set X is an open subset of X since $d(x, \emptyset) > 0$ by our convention on distance (Remarks 7.2.4(1)).   □

We can find more examples of open sets by generalizing the definition of an open interval to a general metric space.

If (X, d) is a metric space, $x_0 \in X$ and $r > 0$, we define

$$D_r(x_0) = \{x \in X \mid d(x, x_0) < r\}.$$

We call $D_r(x_0)$ the *open disk* or *open ball* of centre x_0 and radius r.

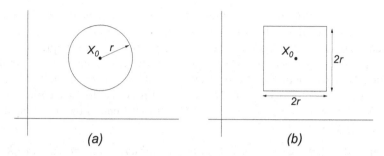

Fig. 7.3 Round (Euclidean) and square disks in $\mathbb{R}^2$. **(a)** A round Euclidean disk in R^2, and **(b)** A square disk in R^2 for the d_∞-metric

Examples 7.4.4

(1) If $X = \mathbb{R}$, standard metric, then $D_r(x_0) = (x_0 - r, x_0 + r)$ (the open interval, centre x_0, length $2r$).
(2) If $X = \mathbb{R}^2$ and we take the Euclidean metric, then $D_r(x_0)$ is the Euclidean disk of radius r and centre x_0, see Fig. 7.3a. On the other hand, if we use the metric $d_\infty(x, y) = \max\{|x_1 - y_1|, |x_2 - y_2|\}$, $D_r(x_0)$ will be the square centred at x_0 with side-length $2r$, see Fig. 7.3b.
(3) If d is the discrete metric on the set X, then $D_r(x_0) = \{x_0\}$, if $r \leq 1$, and $D_r(x_0) = X$ if $r > 1$. ♠

Lemma 7.4.5 *Let X be a metric space. An open disk $D_r(x_0) \subset X$ is an open set.*

Proof We may assume that $D_r(x_0) \neq X$. We have to show that for all $u \in D_r(x_0)$, $d(u, X \smallsetminus D_r(x_0)) > 0$. Let $d(u, x_0) = s < r$. For every $v \in X \smallsetminus D_r(x_0)$ we have by Lemma 7.1.4

$$d(u, v) \geq |d(u, x_0) - d(v, x_0)| \geq r - s > 0,$$

since $d(v, x_0) \geq r$. □
   We make frequent use of the next result.

Lemma 7.4.6 *If $A \subset X$, then $D_r(a) \subset A$ iff $d(a, X \smallsetminus A) \geq r$.*

Proof Suppose that $D_r(a) \subset A$. If $v \in X \smallsetminus A$, then, since $v \notin D_r(a)$, $d(a, v) \geq r$ and so $d(a, X \smallsetminus A) \geq r$. We prove the converse by contradiction. Suppose that $d(a, X \smallsetminus A) = r > 0$. If $D_r(a) \not\subset A$, there exists a $u \in D_r(a) \cap (X \smallsetminus A)$. Therefore $d(a, X \smallsetminus A) \leq d(a, u) < r$, contradicting our assumption that $d(a, X \smallsetminus A) \geq r$. □

Proposition 7.4.7 *Let (X, d) be a metric space. A subset U of X is open iff for every $u \in U$, there exists an $r = r(u) > 0$ such that $D_r(u) \subset U$.*

Proof Suppose first that U is open. Let $u \in U$ and set $d(u, X \smallsetminus U) = r(u) = r > 0$. By Lemma 7.4.6, $D_r(u) \subset U$. Conversely, suppose that $u \in U$ and there exists an $r = r(u) > 0$ such that $D_r(u) \subset U$. By Lemma 7.4.6, $d(u, X \smallsetminus U) \geq r$. □

Remark 7.4.8 It is common in the literature to define an open subset of a metric space (X, d) by requiring that for every $u \in U$, there exists an $r = r(u) > 0$ such that $D_r(u) \subset U$. We prefer our definition because it is (a) simpler and more natural, (b) uses only one quantifier rather than the two demanded by the disk definition. However, notice that the disk definition of an open set automatically gives X as an open subset of X. In particular, we do not need the distance convention in Remarks 7.2.4(1). We give a number of exercises at the end of the section that use the disk definition. ✠

Theorem 7.4.9 *Let $\mathcal{U}$ denote the set of all open subsets of the metric space X. We have*

(1) $X \in \mathcal{U}$.
(2) $\emptyset \in \mathcal{U}$.
(3) *If $\{U_i \mid i \in I\}$ is a family of open subsets of X (not necessarily countable), then $\cup_{i\in I} U_i \in \mathcal{U}$. (Arbitrary unions of open sets are open.)*
(4) *If $U_1, \cdots, U_n \in \mathcal{U}$, then $\cap_{i=1}^n U_i \in \mathcal{U}$. (Finite intersections of open sets are open.)*

Proof (1,2) are just Lemma 7.4.3. (3) We may assume that at least one of the sets U_i is non-empty (else the result follows from (2)). It suffices to show that if $u \in \cup_{i\in I} U_i$, then $d(u, X \smallsetminus \cup_{i\in I} U_i) > 0$. So suppose $u \in U_{i_0}$. We have $X \smallsetminus \cup_{i\in I} U_i \subset X \smallsetminus U_{i_0}$ and so by Lemma 7.2.1(3)

$$d(u, X \smallsetminus \cup_{i\in I} U_i) \geq d(u, X \smallsetminus U_{i_0}).$$

Since $u \in U_{i_0}$ and U_{i_0} is open, $d(u, X \smallsetminus U_{i_0}) > 0$. (4) Assume $\cap_{i=1}^n U_i \neq \emptyset$. If $u \in \cap_{i=1}^n U_i$, then $u \in U_i, i \in \{1, \cdots, n\}$ and so $d_i = d(u, X \smallsetminus U_i) > 0, i \in \{1, \cdots, n\}$. If we set $d = \min_i\{d_i\}$, then $d > 0$ and

$$d(u, X \smallsetminus \cap_{i=1}^n U_i) = d(u, \cup_{i=1}^n (X \smallsetminus U_i)) = \min_i\{d(u, X \smallsetminus U_i)\} = d > 0.$$

Hence $\cap_{i=1}^n U_i$ is open. □

Remark 7.4.10 Given a set X, a collection $\mathcal{U}$ of subsets of X satisfying (1–4) of Theorem 7.4.9 is said to define a *topology* on X. Members of $\mathcal{U}$ are called *open sets* and X, together with the topology $\mathcal{U}$, is called a *topological space*. For a metric space X, we call the associated topology the *metric topology* of X. ✠

Examples 7.4.11

(1) Infinite intersections of open sets will generally not be open. As a simple example, take $U_i = (-1/i, 1 + 1/i) \subset \mathbb{R}, i \geq 1$. Then $\cap_{i=1}^\infty U_i = [0, 1]$.
(2) Let the set X be given the discrete metric. Then the topology of X consists of all subsets of X. This topology is the largest topology one can define on X.

(3) Given a non-empty set X, define $\mathcal{U} = \{\emptyset, X\}$. Then $\mathcal{U}$ is a topology on X and is the smallest topology one can define on a set X. If X has more than one element, this topology cannot be defined by a metric on X. (For more examples, see the exercises at the end of the section.) ♠

Proposition 7.4.12 *Let X be a metric space. Every non-empty open subset of X can be written as a union of open disks.*

Proof Let U be a non-empty open subset of X. Given $x \in U$, we may choose $r(x) > 0$ so that $D_{r(x)}(x) \subset U$ (Proposition 7.4.7). We have $\cup_{x \in U} D_{r(x)}(x) = U$. □

Remark 7.4.13 Every open subset of $(\mathbb{R}, |\cdot|)$ is a *countable* union of *disjoint* open intervals. We leave the proof to the exercises at the end of the section. Later in the chapter we extend the countable union part of this result to a large and important class of metric spaces. ✱

For the remainder of this section, we consider closed subsets of a metric space. Again, the definition is most simply given in terms of the distance function to a subset.

Definition 7.4.14 A subset F of the metric space X is *closed* if $d(u, F) = 0$ implies $u \in F$.

Examples 7.4.15

(1) The sets $\emptyset$ and X are closed subsets of X. Indeed, $\emptyset$ is closed since $d(u, \emptyset)$ is never zero (distance convention, Remarks 7.2.4(1)). On the other hand, X is closed since $d(x, X) = 0$ for all $x \in X$.
(2) A closed interval $[a, b] \subset \mathbb{R}$ is a closed set: if $x \notin [a, b]$, then $d(x, [a, b]) = \max\{\max\{0, a - x\}, \max\{0, x - b\}\} > 0$.
(3) If $x \in X$, then $\{x\}$ is a closed subset of X: $d(y, \{x\}) = d(x, y) = 0$ iff $x = y$.
(4) If we give the set X the discrete metric, then every subset Z of X is closed: $d(u, Z) = 1$ iff $u \notin Z$. ♠

Proposition 7.4.16 *A subset F of the metric space X is closed iff $X \smallsetminus F$ is open.*

Proof Suppose F is a closed subset of X. If $u \in X \smallsetminus F$, then $d(u, X \smallsetminus (X \smallsetminus F)) = d(u, F) > 0$, since otherwise $d(u, F) = 0$, contradicting our assumption that F is closed. For the converse, reverse the previous argument. □

Remark 7.4.17 Proposition 7.4.16 does not say that a subset of X must be either open or closed. Although every subset of a metric space with the discrete metric is open *and* closed, it is usually the case that most subsets of a metric space are neither open nor closed. For a simple example of a subset A of $\mathbb{R}$ which is neither open nor closed, take $A = (0, 1]$. ✱

Theorem 7.4.18 *Let $\mathcal{F}$ denote the set of all closed subsets of the metric space X. Then*

(1) $X \in \mathcal{F}$.
(2) $\emptyset \in \mathcal{F}$.

(3) If $\{F_i \mid i \in I\}$ is a family of closed subsets of $\mathcal{F}$, then $\cap_{i \in I} F_i \in \mathcal{F}$ (an arbitrary intersection of closed sets is closed).

(4) If $F_1, \cdots, F_n \in \mathcal{F}$, then $\cup_{i=1}^n F_n \in \mathcal{F}$ (a finite union of closed sets is closed).

Proof We can prove this in two ways. Either use Theorem 7.4.9 and Proposition 7.4.16 or work directly from the definition. We use the direct approach and prove (3) and leave the remaining cases to the exercises. Suppose then that $d(x, \cap_{i \in I} F_i) = 0$. Since $\cap_{i \in I} F_i \subset F_j$ for all $j \in I$ we have, by Lemma 7.2.1(3), $d(x, F_j) \leq d(x, \cap_{i \in I} F_i) = 0$ for all $j \in I$. Hence $d(x, F_j) = 0$ and so $x \in F_j$ since F_j is closed. Since this holds for all $j \in I$, $x \in \cap_{i \in I} F_i$.   □

Examples 7.4.19

(1) An infinite union of closed sets need not be closed. For example, $\cup_{i \geq 2} [\frac{1}{i}, 1 - \frac{1}{i}] = (0, 1)$, which is not closed (since $d(0, (0, 1)) = 0$).

(2) Let $f : \mathbb{R} \to \mathbb{R}$ be continuous and set $F = f^{-1}(0) = \{x \in \mathbb{R} \mid f(x) = 0\}$ (F is the solution set of $f(x) = 0$). Then F is a closed set. Suppose that $d(x, F) = 0$. Choose a sequence $(x_n) \subset F$ converging to x. We have $\lim_{n \to \infty} d(x_n, x) = 0$. By the sequential continuity of f, we have $0 = \lim_{n \to \infty} f(x_n) = f(x)$. Therefore $x \in F$. Alternatively, we can use Proposition 7.4.16 and prove that $\mathbb{R} \smallsetminus F$ is open. To do this, observe that $\mathbb{R} \smallsetminus F = \{x \in \mathbb{R} \mid f(x) \neq 0\}$. Let $z \in \mathbb{R} \smallsetminus F$ and set $r = |f(z)| \neq 0$. By the continuity of f, and therefore $|f|$, there exists a $\delta > 0$ such that $|f(x)| > r/2$ for all $x \in (z - \delta, z + \delta)$. Hence $(z - \delta, z + \delta) \subset \mathbb{R} \smallsetminus F$ and so $\mathbb{R} \smallsetminus F$ must be open and F closed. As we shall see later, this result holds in great generality. Moreover, F is closed iff F is the zero set of a continuous function: closed sets are precisely the zero sets of continuous functions. ♠

Let $\overline{D}_r(x_0)$ denote the *closed disk* of radius $r > 0$, centre x_0, in the metric space (X, d). That is

$$\overline{D}_r(x_0) = \{x \in X \mid d(x_0, x) \leq r\}.$$

Lemma 7.4.20 *A closed disk is closed.*

Proof If $x \notin \overline{D}_r(x_0)$, then $d(x, x_0) > r$. Hence, $d(x, \overline{D}_r(x_0)) = d(x, x_0) - r \neq 0$. □

Example 7.4.21 If $D_r(x) = \overline{D}_r(x)$ then $D_r(x)$ is open *and* closed. As a simple example where this can happen, let Y be the metric space defined to be the union of the open intervals $(1, 2)$ and $(3, 4)$. Take the induced metric on Y (that is, the metric induced on Y by the standard metric on $\mathbb{R}$). Both $(1, 2)$, $(3, 4)$ are open subsets of Y and therefore, taking complements (in Y!), are also closed subsets of Y. Of course, it is easy to check directly that $d_Y(x, (a, b)) = 0$ iff $x \in (a, b)$. ♠

We give a simple proposition that relates open and closed sets. This result is quite useful when we look at open and closed sets in the induced metric on a subset.

Proposition 7.4.22 *Let $U \subset X$ be open and $F \subset X$ be closed. Then*

(1) $U \smallsetminus F$ *is open.*

(2) $F \smallsetminus U$ *is closed.*

Proof Let A, B be subsets of X. We claim that $X \smallsetminus (A \smallsetminus B) = (X \smallsetminus A) \cup B$. Indeed, $x \in X \smallsetminus (A \smallsetminus B)$ iff $x \notin A$ or $x \in B$. But $x \notin A$ or $x \in B$ iff $x \in (X \smallsetminus A) \cup B$. Apply this result with $A = U, B = F$ to obtain (1). Interchange U and F to get (2). □

Example 7.4.23 Take $X = \mathbb{R}$. Then $(a, b) \smallsetminus [c, d]$ is always an open interval (possibly empty) and $[a, b] \smallsetminus (c, d)$ is always a closed interval (possibly empty). ♠ We conclude with the definition of an isolated point.

Definition 7.4.24 A point x in the metric space X is *isolated* if $d(x, X \smallsetminus \{x\}) > 0$.

Lemma 7.4.25 *The following conditions are equivalent.*

(1) *The point x is an isolated point of X.*
(2) *$\{x\}$ is an open subset of X.*
(3) *$\{x\}$ is an open & closed subset of X.*
(4) *$X \smallsetminus \{x\}$ is closed.*

Proof (1) $\Longleftrightarrow$ (2) by the definition of open set. (2) $\Longleftrightarrow$ (3) since $\{x\}$ is always a closed subset. (3) $\Longleftrightarrow$ (4) by Proposition 7.4.16. □

Examples 7.4.26

(1) The metric space $(\mathbb{R}, |\cdot|)$ contains no isolated points.
(2) The metric space $(\mathbb{Z}, |\cdot|)$ consists of isolated points.
(3) If we give the set X the discrete metric, then X consists of isolated points. ♠

EXERCISES 7.4.27

(1) Describe (draw a figure) the open disk, centre x_0, radius r, for the metric d_1 on $\mathbb{R}^2$ (see Exercises 7.1.9(3) for the definition of d_1).
(2) Let U be a non-empty open subset of $\mathbb{R}$. Show that U can be written as a countable disjoint union of open intervals. (Hint: if $x \in U$, let I_x denote the union of all open intervals $I \subset U$ which contain x. Verify that I_x is an open interval.)
(3) Prove Theorem 7.4.9 using the disk definition of open set.
(4) Prove Proposition 7.4.16 using the disk definition of open set.
(5) Complete the proof of Theorem 7.4.18.
(6) Suppose that h_1 and h_2 are equivalent metrics on X (see Exercises 7.1.9(11) for the definition of equivalent metric). Show that (X, h_1), (X, h_2) have the same open sets. By looking at $\mathbb{Z}$ with the discrete metric and metric induced from $(\mathbb{R}, |\cdot|)$, show that the converse is false—same open sets does not imply equivalent metrics.
(7) Show that x_0 is an isolated point of X iff $\{x_0\}$ and $X \smallsetminus \{x_0\}$ are both open and closed.
(8) Show that the diagonal $\Delta(X) = \{(x, x) \mid x \in X\}$ is a closed subset of X^2 if we take the product metric on $X^2 = X \times X$ (Exercises 7.1.9(5)).
(9) Let (X, d) be a metric space. Suppose that arbitrary intersections of open subsets of X are open. Show that every point of X is isolated (so X has the topology given by the discrete metric).

(10) Let $\mathcal{U}$ consist of the empty set together with all subsets of $\mathbb{R}$ which are the form $\mathbb{R} \smallsetminus F$, where F is a *finite* subset of $\mathbb{R}$. Verify that $\mathcal{U}$ defines a topology on $\mathbb{R}$ (this topology is known as the *Zariski topology* on $\mathbb{R}$ and is used in algebraic geometry for the study of zero sets of polynomials).

7.5   Interior and Closure

In this section we show that there is a natural way of associating an open set and a closed set to every subset of a metric space.

Definition 7.5.1 Let (X, d) be a metric space with topology of open sets $\mathcal{U}$ and closed sets $\mathcal{F}$. Let $A \subset X$.

(1) The *interior* $\mathring{A}$ of A is the largest open subset of A:

$$\mathring{A} = \bigcup_{U \in \mathcal{U}, U \subset A} U.$$

(2) The *closure* $\overline{A}$ of A is the smallest closed superset of A:

$$\overline{A} = \bigcap_{F \in \mathcal{F}, F \supset A} F.$$

Remarks 7.5.2

(1) Since a union of open sets is open, by Theorem 7.4.9(3), $\bigcup_{U \in \mathcal{U}, U \subset A} U$ is an open subset set of A. Since the union contains all open subsets of A, it is the largest open subset of A. Similarly, using Theorem 7.4.18(3), $\bigcap_{F \in \mathcal{F}, F \supset A} F$ is the smallest closed set containing A.

(2) Observe that the definition of interior and closure only uses properties of the topology and does not directly use the metric structure on X. Hence the definition extends to general topological spaces.   ✠

Examples 7.5.3

(1) Take $X = \mathbb{R}$ and suppose $-\infty < c < d < +\infty$. We have $\mathring{[c, d]} = (c, d)$ and $\overline{(c, d)} = [c, d]$.

(2) For all metric spaces (X, d) we have $\mathring{X} = X$, $\mathring{\emptyset} = \emptyset$, $\overline{X} = X$, $\overline{\emptyset} = \emptyset$.

(3) If X has the discrete metric, then $\mathring{A} = \overline{A} = A$ for all subsets A of X.   ♠

Proposition 7.5.4 *If $A \subset X$, then*

(a) $\mathring{A} \subset A \subset \overline{A}$.

(b) *A is open iff $\mathring{A} = A$.*

(c) *A is closed iff* $\overline{A} = A$.

(d) *If* $A \subset B \subset X$, *then* $\overset{\circ}{A} \subset \overset{\circ}{B}$ *and* $\overline{A} \subset \overline{B}$.

Proof (a) is immediate by the definition of interior and closure. (b,c) If A is open, then $\overset{\circ}{A} \supset A$, since A contains the open set A. The proof of (c) is similar. Finally, if $A \subset B$, then every open subset of A is an open subset of B. Hence $\overset{\circ}{A} \subset \overset{\circ}{B}$. The proof that $\overline{A} \subset \overline{B}$ is similar.   □

The next result characterizes the interior and closure of a set using metric properties.

Proposition 7.5.5 *Let* $A \subset X$. *Then*

(1) $\overset{\circ}{A} = \{x \in A \mid d(x, X \smallsetminus A) > 0.$
(2) $\overline{A} = \{x \in X \mid d(x, A) = 0\}.$

Proof We give the proof of (1), the proof of (2) is similar. Suppose first that $a \in A$ and $d(a, X \smallsetminus A) = r$. By Lemma 7.4.6, $D_r(a) \subset A$. But $D_r(a)$ is an open subset of A and so, by definition of the interior, $D_r(a) \subset \overset{\circ}{A}$. Hence $a \in \overset{\circ}{A}$. Conversely, suppose that $a \in \overset{\circ}{A}$. Since $\overset{\circ}{A}$ is open, there exists an $r > 0$ such that $D_r(a) \subset \overset{\circ}{A} \subset A$. Hence, again by Lemma 7.4.6, $d(a, X \smallsetminus A) \geq r$.   □

Remark 7.5.6 We say $a \in A$ is an *interior point* of A if $d(a, X \smallsetminus A) > 0$ and that $x \in X$ is a *closure point* of A if $d(x, A) = 0$. Proposition 7.5.5 shows that the interior (respectively, closure) of A is the set of all interior (respectively, closure) points of A.   ✱

We may easily give a characterization of the interior and closure of a subset in terms of open and closed disks.

Lemma 7.5.7 *Let* A *be a subset of* X.

(1) $x \in \overset{\circ}{A}$ *iff there exists an* $r > 0$ *such that* $D_r(x) \subset A$.
(2) $x \in \overline{A}$ *iff* $D_r(x) \cap A \neq \emptyset$ *for all* $r > 0$.

Proof Left to the exercises.   □

EXERCISES 7.5.8

(1) Take the standard metric on $\mathbb{R}$. Find (a) $\overset{\circ}{\mathbb{Q}}$, (b) $\overline{\mathbb{Q}}$. How would your answer change if we took the discrete metric on $\mathbb{R}$?
(2) Let $D_r(x)$ be the open r-disk in $\mathbb{R}^2$, Euclidean metric. Show that $\overline{D_r(x)} = \overline{D}_r(x)$, for all $r > 0$, $x \in \mathbb{R}^2$. Find an example of a metric space (X, d) for which $\overline{D_r(x)} \subsetneq \overline{D}_r(x)$. Show that we always have $\overline{D_r(x)} \subseteq \overline{D}_r(x)$. Similarly, investigate the relation between $D_r(x)$ and the interior of $\overline{D}_r(x)$ and show that, in general, the interior of a closed disk of radius r is not equal to the open disk of radius r.
(3) If $f \in C^0([a, b])$ and $F = \{g \in C^0([a, b]) \mid \rho(f, g) < r\}$, show that $\overline{F} = \{g \in C^0([a, b]) \mid \rho(f, g) \leq r\}$.
(4) Provide the proof of Proposition 7.5.5(2).

(5) (Proof of Lemma 7.5.7.) Let $A \subset X$. Show that

    (a) $x \in \overset{\circ}{A}$ iff there exists an $r > 0$ such that $D_r(x) \subset A$.
    (b) $x \in \overline{A}$ iff for all $r > 0$, $D_r(x) \cap A \neq \emptyset$.

    ((a) and (b) are commonly used to define the interior and closure of a set.)
(6) If $E_1, \cdots, E_n$ is a finite collection of subsets of the metric space X, show that
the interior of $\cap_{i=1}^{n} E_i$ equals $\cap_{i=1}^{n} \overset{\circ}{E_i}$. Show that the result is false if we allow
arbitrary intersections. What, if anything, can be said relating the interior of a
union of sets to the union of the interiors?

7.6   Open and Closed Subsets of a Subspace

Let Y be a non-empty subset of the metric space (X, d) and let d_Y denote the induced
metric on Y (see Definition 7.1.5). There is a simple relationship between the open
and closed sets of (Y, d_Y) and (X, d).

Proposition 7.6.1 *Let Y be a subset of the metric space (X, d).*

(1) *A subset U of Y is open in (Y, d_Y) iff there exists an open set V of (X, d) such
that $U = Y \cap V$.*
(2) *A subset F of Y is closed in (Y, d_Y) iff there exists a closed set Z of (X, d) such
that $F = Y \cap Z$.*

*In particular, if we denote the topology of (X, d) by $\mathcal{U}$ and that of (Y, d_Y) by $\mathcal{U}_Y$, we
have $\mathcal{U}_Y = \mathcal{U} \cap Y \overset{\text{def}}{=} \{U \cap Y \mid U \in \mathcal{U}\}$.*

Proof Suppose first that U is an open subset of X. We show $U \cap Y$ is an open subset
of Y (relative to the induced metric). Let $y \in U \cap Y$. Since U is an open subset of X,
we have $d(y, X \smallsetminus U) > 0$. But $Y \smallsetminus U \subset X \smallsetminus U$ and so $d(y, Y \smallsetminus U) \geq d(y, X \smallsetminus U) > 0$.
Hence $U \cap Y$ is an open subset of Y. If $F \subset X$ is closed, then $Y \cap (X \smallsetminus F)$ is an
open subset of Y. But $Y \cap F = Y \smallsetminus (X \smallsetminus F)$ and so $Y \cap F$ is a closed subset of Y.
Now suppose that F is a closed subset of Y (induced metric) and let $\overline{F}$ denote the
closure of F in (X, d). We have $F = Y \cap \overline{F}$ since $\overline{F} = \{x \in X \mid d(x, F) = 0\}$ and
so $Y \cap \overline{F} = \{x \in Y \mid d(x, F) = 0\} = F$, completing the proof of (2). Finally, the
converse to (1) follows from (2) by taking complements. That is, if $U \subset Y$ is open
then $U = Y \cap (X \smallsetminus \overline{Y \smallsetminus U})$ (closure taken in X).   □

EXERCISES 7.6.2

(1) Provide an alternative proof of Proposition 7.6.1(1) that uses the disk definition
of open set together with the result that a union of open disks is open.
(2) Suppose that $Y \subset X$ is an open set (relative to the metric d on X). Show that
$U \subset Y$ is open (in the induced metric) iff U is an open subset of X and that
$F \subset Y$ is closed iff there exists an open subset W of X such that $F = Y \smallsetminus W$
(note Proposition 7.4.16). Formulate and prove the corresponding results when
Y is a closed subset of X.

7.7   Dense Subsets and the Boundary of a Set

In this section we give some useful definitions based on closure.

7.7.1   Dense Subsets and Separable Metric Spaces

Definition 7.7.1 A subset A of the metric space X is *dense* in X if $\overline{A} = X$.

Lemma 7.7.2 *Let A be a subset of X. The following conditions are equivalent*

(1) *A is a dense subset of X.*

(2) *$X \smallsetminus A$ has no interior points: $(X \overset{\circ}{\smallsetminus} A) = \emptyset$.*

(3) *$d(x, A) = 0$ for all $x \in X$.*

(4) *For every $x \in X$, and every $r > 0$, $D_r(x) \cap A \neq \emptyset$.*

Proof We leave the proof to the exercises.   □

Definition 7.7.3 A metric space (X, d) is *separable* if X has a countable dense subset.

Examples 7.7.4

(1) $(\mathbb{R}, |\cdot|)$ is separable: the rational numbers $\mathbb{Q}$ are a dense subset of $\mathbb{R}$. More generally, $\mathbb{R}^n$ is separable since $\mathbb{Q}^n$ is a dense subset of $\mathbb{R}^n$, $n \geq 1$ (here we may take the Euclidean metric or either of the metrics d_1, d_∞ on $\mathbb{R}^n$). The simplest proof of density uses (3) of Lemma 7.7.2 and the metric d_∞. Since $\mathbb{Q}^n$ is countable, $\mathbb{R}^n$ is separable, $n \geq 1$.

(2) Let $[a, b] \subset \mathbb{R}$ be a bounded closed interval and let $P \subset C^0([a, b])$ be the set of all polynomial maps $p : [a, b] \to \mathbb{R}$. Then P is a dense subset of $(C^0([a, b]), \rho)$ (uniform metric). This is precisely the Weierstrass approximation theorem. Indeed, the Weierstrass approximation theorem states that for all $f \in C^0([a, b])$ and all $r > 0$ we have $D_r(f) \cap P \neq \emptyset$. Hence $\overline{P} = C^0([a, b])$ by Lemma 7.7.2(3). If we let $P_\mathbb{Q}$ denote the space of polynomial maps $p : [a, b] \to \mathbb{R}$ with rational coefficients, then $P_\mathbb{Q}$ is countable ($P_\mathbb{Q}$ can be written as a countable union of countable sets) and so since $\overline{P_\mathbb{Q}} = \overline{P} = C^0([a, b])$, we see that $(C^0([a, b]), \rho)$ is separable. Similar results are true if we use the L^2-metric on $C^0([a, b])$ (see Theorem 5.6.11).

(3) If we give X the discrete metric then the only subset of X which is dense is X itself (use (2) of Lemma 7.7.2). In particular, X is separable iff X is countable.   ♠

Proposition 7.7.5 *Suppose that the metric space (X, d) is separable. Then there exists a countable family $\mathcal{B}$ of open subsets of X such that every open subset of X is a union of sets from $\mathcal{B}$.*

Proof If X is separable then there exists a countable dense subset $\{q_n \mid n \in \mathbb{N}\}$ of X. Associated to each $n \in \mathbb{N}$, we define $\mathcal{B}_n$ to be the set of all open disks centred at q_n and with radius $r \in \mathbb{Q}$. Since $\mathbb{Q}$ is countable, $\mathcal{B}_n$ is countable and so $\mathcal{B} = \cup_{n\in\mathbb{N}}\mathcal{B}_n$ is a countable set of open subsets of X. Given any open subset U of X, we may write U as a union of open sets from $\mathcal{B}$. Indeed, if $x \in U$, choose $m \in \mathbb{N}$ so that $D_{2/m}(x) \subset U$. Now choose $q_n \in D_{1/m}(x)$. Then $U_x = D_{1/m}(q_n) \subset D_{2/m}(x) \subset U$. We have $U = \cup_{x\in U}U_x$ and so we have expressed U as a union of open sets from $\mathcal{B}$. □

Remark 7.7.6 Any metric space (X, d) with the property that there exists a countable collection $\mathcal{B}$ of open sets such that every open set can be written as a union of sets from $\mathcal{B}$ is called *second countable* and $\mathcal{B}$ is called a *basis* for the topology of (X, d). ✠

Example 7.7.7 If $X = \mathbb{R}$, then every open subset of $\mathbb{R}$ is a countable union of disjoint open intervals (see Exercises 7.4.27(2)). Each of these open intervals can be written as a countable union of open intervals with rational endpoints. However, we cannot generally write U as a *disjoint* union of open intervals with rational endpoints. If U is an open subset of $\mathbb{R}^n$, $n \geq 1$, then U is a countable union of (generally non-disjoint) open disks (with rational radius and rational centre). ♠

7.7.2 Boundary of a Subset

Definition 7.7.8 The *boundary* (also called *frontier*) ∂A of a subset A of the metric space X is defined by

$$\partial A = \overline{A} \cap \overline{(X \setminus A)}.$$

Lemma 7.7.9 *Let A be a subset of the metric space X. Then*

(1) *∂A is a closed subset of X.*
(2) $\partial A = \{x \in X \mid d(x, A) = d(x, X \setminus A) = 0\}$.
(3) $\partial A = \overline{A} \setminus \mathring{A}$.
(4) *$x \in \partial A$ iff $D_r(x) \cap A, D_r(x) \cap (X \setminus A) \neq \emptyset$ for all $r > 0$.*

Proof (1,2) are immediate from the definitions. For (3) observe that if $x \notin \mathring{A}$ then $x \in \overline{X \setminus A}$. For (4) use Exercises 7.5.8(5b). □

Examples 7.7.10

(1) If $[a, b] \subset \mathbb{R}$, $\partial[a, b] = \{a, b\}$. Similarly $\partial(a, b) = \{a, b\}$.
(2) $\partial X = \partial\emptyset = \emptyset$.

(3) $\partial D_r(x), \partial \overline{D}_r(x)$ are subsets of $S_r(x) \overset{\text{def}}{=} \{y \in X \mid d(x, y) = r\}$ ($S_r(x)$ is the 'sphere' of radius r centred at x). In general, $\partial D_r(x) \neq \partial \overline{D}_r(x)$ and $\partial D_r(x), \partial \overline{D}_r(x)$ may be proper subsets of $S_r(x)$.

(4) If X has the discrete metric, then $\partial Y = \emptyset$ for all subsets Y of X.

(5) If A is dense in X then $\partial A = X$ if $X \smallsetminus A$ is dense in X.   ♠

EXERCISES 7.7.11

(1) Prove Lemma 7.7.2.
(2) Let $D_r(x)$ be a disk in the metric space (X, d). Show that $\partial D_r(x)$ may be empty, even if $D_r(x) \neq X$.
(3) Let A be a subset of X. Show that $\partial A \cap \partial(X \smallsetminus A) = X$ iff A and $X \smallsetminus A$ are dense in X.

7.8 Neighbourhoods

Definition 7.8.1 A subset N of (X, d) is a *neighbourhood* of $x \in X$ if $x \in \overset{\circ}{N}$. If N is open, we say N is an *open neighbourhood* of x.

Lemma 7.8.2 *A subset N of (X, d) is a neighbourhood of x iff there exists an $r > 0$ such that $D_r(x) \subset N$.*

Proof By Lemma 7.4.6, if $d(x, X \smallsetminus N) = r > 0$, then $D_r(x) \subset N$ and conversely.   □

Examples 7.8.3

(1) Let $X = \mathbb{R}$, standard metric. The closed interval $[a, b]$ is a neighbourhood of every point $x \in (a, b)$. It is not a neighbourhood of a or b. The open interval (a, b) is an open neighbourhood of every point in (a, b).

(2) If X is a metric space and $N \subset X$, then N is a neighbourhood of x iff $x \in \overset{\circ}{N}$.

(3) The open disk $D_r(x)$ and closed disk $\overline{D}_r(x)$ are neighbourhoods of x.   ♠

Remark 7.8.4 If N is an open neighbourhood of x, then N is an open neighbourhood of every point in N.   ✸

We may characterize the interior and closure of a set using neighbourhoods.

Lemma 7.8.5 *Let A be a subset of X.*

(1) $x \in \overset{\circ}{A}$ *iff there exists a neighbourhood N of x such that $N \subset A$.*
(2) $x \in \overline{A}$ *iff $N \cap A \neq \emptyset$ for all neighbourhoods N of x.*

Proof The result is immediate from Lemmas 7.5.7, 7.8.2.   □

EXERCISES 7.8.6

(1) Show that distinct points of a metric space have disjoint open neighbourhoods.
(2) Let (X, d) be a metric space and suppose that $x \in X$. Show that if x has a neighbourhood containing finitely many points, then x is isolated. Conversely, show that if every neighbourhood of x contains infinitely many points, then x is not isolated.
(3) Let A be a subset of the metric space (X, d). Show that $a \in \overline{A}$ iff $N \cap A \neq \emptyset$ for every neighbourhood N of a. Reformulate the definition of the interior of a set in terms of neighbourhoods and verify that your definition does define the interior.

7.9   Summary and Discussion

Let A be any subset of the metric space X. We have shown there is a maximal open set $\overset{\circ}{A}$ and a minimal closed set $\overline{A}$ such that

$$\overset{\circ}{A} \subset A \subset \overline{A}.$$

Moreover,

(1) $x \in \overset{\circ}{A}$ iff $d(x, X \smallsetminus A) > 0$.
(2) $x \in \overline{A}$ iff $d(x, A) = 0$.

(3) A is open iff $A = \overset{\circ}{A}$.
(4) A is closed iff $A = \overline{A}$.

Notwithstanding that a closed set is just the complement of an open set, open and closed sets have rather different properties. For example, every open subset of $\mathbb{R}$ is a countable union of disjoint open intervals. However, it is not true that a closed subset of $\mathbb{R}$ can be expressed in such a simple way; for example, as a countable union of disjoint closed intervals. Indeed, closed subsets of $\mathbb{R}$ can be extremely complex and pathological (matters are worse in $\mathbb{R}^n$, $n > 1$). If we can write an open set U as the disjoint union $\bigcup_{i=1}^{\infty}(a_n, b_n)$, where $b_n < a_{n+1}$, $n \leq 1$, then the closed set $\mathbb{R} \smallsetminus U$ is the countable union of the disjoint closed intervals $[b_n, a_{n+1}]$. However, although we can write U as the disjoint union of open intervals (a_n, b_n), we *cannot* usually require that $b_n < a_{n+1}$, for all n. The situation is similar to that of the rational numbers: although the rationals are countable, we cannot write them as a sequence (r_n) satisfying $r_n < r_{n+1}$, $n \geq 1$. In Exercises 2.2.14(9), a construction was given of an open subset I of $\mathbb{R}$ which contained every rational number and was such that the total length $|I|$ of the open intervals comprising I was less than some preassigned number $\varepsilon > 0$. Let A denote the complement of I in $\mathbb{R}$. Even though $|I| < \varepsilon$, A can have no interior points since arbitrarily close to every interior point is a rational interior point. Consequently the structure of A is hard to visualize—arbitrarily close to every point of A is a hole where we have removed an interval containing a rational number. At non-isolated points of $a \in A$, there is a sequence of holes converging to a. Granted this complexity, it is perhaps surprising that every closed subset of $\mathbb{R}$ can be represented as the zero set of a continuous (indeed C^∞) function $f : \mathbb{R} \to \mathbb{R}$. Needless to say, the construction of f depends on defining f to be non-zero on the complement of the closed set (see the exercises for an example).

EXERCISES 7.9.1

(1) Let (X, d) be a metric space. Prove that for all $x \in X$, $r > 0$, $S_r(x) = \{y \in X \mid d(x, y) = r\}$ is a closed set.

(2) Let A be a subset of the metric space X. Prove that the diameter of A equals the diameter of the closure of A. Does the same result hold if instead of the closure we take the interior of A?

(3) Let $A_1, \cdots, A_n$ be subsets of the metric space X. Prove that $\overline{\bigcup_{i=1}^{n} A_i} = \bigcup_{i=1}^{n} \overline{A_i}$. Find an example to show this result is generally false for infinite unions. Investigate what happens for intersections.

(4) Let A be a non-empty subset of the metric space X. Show that $\overset{\circ}{A} = X \smallsetminus \overline{X \smallsetminus A}$ and $\overline{A} = X \smallsetminus (X \smallsetminus A)^{\circ}$.

(5) Show that in general $\partial D_r(x) \neq \partial \overline{D}_r(x)$. Also find examples where we have $\partial D_r(x), \partial \overline{D}_r(x) \neq S_r(x)$ ($S_r(x)$ is the sphere of radius r and centre x—see also (1)).

(6) True or false: in each case either prove it or provide a counterexample.

(a) $\overset{\circ}{\overline{E}} = \overline{\overset{\circ}{E}}$?

(b) $\overline{\overset{\circ}{E}} = \overset{\circ}{\overline{E}}$?

(E is a subset of the metric space X.)

(7) If A is a subset of the metric space X show that $\overset{\circ}{A} \cup \partial A \cup (X \smallsetminus A)^{\circ} = X$.

(8) Let $B([a, b])$ denote the space of bounded functions $f : [a, b] \to \mathbb{R}$ with uniform metric $\rho(f, g) = \sup_{x \in X} |f(x) - g(x)|$, $f, g \in B([a, b])$ (see Exercises 7.1.9). Show (a) if $f \in B([a, b])$ is not continuous, then there exists an $r > 0$ such that every $g \in D_r(f)$ is not continuous, (b) the space $C^0([a, b])$ is a closed subset of $B([a, b])$, and (c) the space P of polynomials $p : [a, b] \to \mathbb{R}$ is not dense in $B([a, b])$. (For (c) you should prove it in two ways: either use (b) or construct a bounded function which cannot be approximated by polynomials in the metric ρ.)

(9) Suppose that every subset of the metric space X is either open or closed. Show that at most one point of X is not isolated. What about the converse?

(10) A subset A of the metric space X is *nowhere dense* if the interior of $\overline{A}$ is empty. Show that every finite subset of $\mathbb{R}$ is nowhere dense and construct an example of a countable subset of $[0, 1]$ which is nowhere dense. (Later we give an example of a non-countable subset of $[0, 1]$ which is nowhere dense.)

(11) Let (q_n) be the set of all rational numbers, indexed by the positive integers, and let $\varepsilon > 0$. For $n \geq 1$, set $I_n = (q_n - 2^{-(n+1)}\varepsilon, q_n + 2^{-(n+1)}\varepsilon)$ and define $I = \cup_{n \geq 1} I_n$ (see Exercises 2.2.14(9) and the discussion above).

(a) Show that $\bar{I} = \mathbb{R}$.

(b) Set $A = \mathbb{R} \smallsetminus I$. Show that $\overset{\circ}{A} = \emptyset$.

(c) For $n \geq 1$, construct a continuous function $\phi_n : \mathbb{R} \to \mathbb{R}$ which is non-zero precisely on I_n and has maximum value $2^{-(n+1)}\varepsilon$. Using the M-test show that $\sum_{n=1}^{\infty} \phi_n$ converges to a continuous function on $\mathbb{R}$ which is zero precisely on the set A.

(d) Using the bump function $\Psi_{a,b}$ of Examples 5.2.6(2), construct a smooth function $\psi_n : \mathbb{R} \to \mathbb{R}$ which is non-zero precisely on the interval I_n. For each $n \geq 1$, choose $\alpha_n > 0$ so that

$$\alpha_n(\max\{\sup_{x \in I_n}|\psi_n(x)|, \sup_{x \in I_n}|\psi'_n(x)|, \cdots, \sup_{x \in I_n}|\psi_n^{(n)}(x)|\}) \leq \varepsilon 2^{-(n+1)}.$$

Show that $\sum_{n=1}^{\infty} \alpha_n \psi_n$ converges to a C^∞-function $\psi : \mathbb{R} \to \mathbb{R}$ which is zero precisely on the set A. (You will need both the M-test and results of Chap. 5 that give conditions for an infinite series of functions to converge to a differentiable function.)

(The methods used in this example to construct ϕ and ψ extend to general closed subsets of $\mathbb{R}$. See also Sect. 7.12.)

7.10 Sequences and Limit Points

In this section we develop the theory of convergent sequences in metric spaces. We show how results about convergence are related to closed sets and prove a very useful characterization of a closed subset of a metric space: a subset A of X is closed iff the limit of every convergent sequence $(x_n) \subset A$ lies in A.

Definition 7.10.1 Let (X, d) be a metric space. A *sequence* of points in X consists of an ordered subset (x_n) of X indexed by the positive or strictly positive integers.

Examples 7.10.2

(1) A *constant sequence* (x_n) in the metric space X has the property that for all n, $x_n = x_0$ for some fixed $x_0 \in X$.

(2) If we define $x_n = (\frac{1}{n}, \frac{1}{n})$, $n \geq 1$, then (x_n) is a sequence in $\mathbb{R}^2$.

(3) Let $T : X \to X$. Given $x_0 \in X$, define the sequence (x_n) recursively by $x_{n+1} = T(x_n)$, $n \geq 0$. For example, suppose $X = C^0([a, b])$, $F : \mathbb{R} \to \mathbb{R}$ is continuous

and $C \in \mathbb{R}$. Define $T : X \to X$ by

$$T(f)(x) = C + \int_a^x F(f(t)) \, dt, \ f \in C^0([a, b]), \ x \in [a, b].$$

If we choose an initial function $f_0 \in C^0([a, b])$, then we obtain the sequence (f_n) in $C^0([a, b])$ by the rule $f_{n+1} = T(f_n)$. As we shall see, this iteration turns out to be useful in constructing the solution $y = y(x)$ to the differential equation $\frac{dy}{dx} = F(y)$ with initial condition $y(a) = C$.   ♠

Definition 7.10.3 The sequence (x_n) of points in (X, d) is *convergent* if there exists an $x \in X$ such that $\lim_{n \to \infty} d(x_n, x) = 0$. We call x the *limit* of the sequence and write $\lim_{n \to \infty} x_n = x$.

Lemma 7.10.4 *Let (x_n) be a sequence of points in (X, d). The following statements are equivalent.*

(1) *(x_n) converges to x.*
(2) *For all $r > 0$, there exists an $m = m(r) \in \mathbb{N}$ such that $x_n \in D_r(x)$, for all $n \geq m$.*
(3) *For all $r > 0$, there exists an $m = m(r) \in \mathbb{N}$ such that $x_n \in \overline{D}_r(x)$, for all $n \geq m$.*
(4) *For every neighbourhood N of x, there exists an $m = m(N) \in \mathbb{N}$ such that $x_n \in N$, for all $n \geq m$.*

Proof (1) $\Longrightarrow$ (2) Assume (1) holds. Since $\lim_{n \to \infty} d(x_n, x) = 0$, given $r > 0$, there exists $m \in \mathbb{N}$ such that $d(x, x_n) < r$, $n \geq m$. That is, $x_n \in D_r(x)$, $n \geq m$.

(2) $\Longrightarrow$ (4) Assume (2) holds. Given a neighbourhood N of x, there exists an $r > 0$ such that $D_r(x) \subset N$. Now apply (2).

(4) $\Longrightarrow$ (3) $\overline{D}_r(x)$ is a neighbourhood of x, $r > 0$.

(3) $\Longrightarrow$ (1) Assume (3) holds. Given $\varepsilon > 0$, there exists an $m \in \mathbb{N}$ such that $x_n \in \overline{D}_{\varepsilon/2}(x) \subset D_\varepsilon(x)$, $n \geq m$. That is, $d(x, x_n) < \varepsilon$, $n \geq m$. Hence (x_n) converges to x.   □

Remarks 7.10.5

(1) Note formulation (4) of convergence—framed in terms of neighbourhoods. This is part of our move away from the ε, δ style of definitions to more general and natural definitions given in terms of open sets (or neighbourhoods) and one less quantifier.
(2) Just as for sequences of real numbers, we need a criterion for convergence that does not depend on knowing the limit. However, before we develop that aspect of the theory we introduce some new ideas that relate limits of sequences to closed sets.   ✠

7.10.1   Limit Points of a Set

We recall that a point $x \in X$ is *isolated* if $\{x\}$ is a neighbourhood of x. More generally, if A is a non-empty subset of X, a point $a \in A$ is isolated (in A) if there exists a neighbourhood N of a such that $N \cap (A \setminus \{a\}) = \emptyset$. In terms of the induced topology, a is isolated in A iff a is an isolated point of (A, d_A) (we leave the formal verification to the exercises).

Example 7.10.6 Let $X = \mathbb{R}$, standard metric, and take $A = \{0\} \cup [1, 2]$. Then A has one isolated point: $\{0\}$. As a less trivial example, take $A = \{0\} \cup \{1/n \mid n \geq 1\}$. Then every point of A is isolated *except* $\{0\}$.   ♠

In our next definition we aim to capture points which are not isolated relative to a subset A of X.

Definition 7.10.7 If A is a non-empty subset of the metric space X, then a point $x \in X$ is called a *limit point* (or *accumulation point*) of A if

(1) $d(x, A) = 0$ (equivalently, $x \in \overline{A}$) and
(2) x is not an isolated point of A.

We denote the set of limit points of A by A'.

Examples 7.10.8

(1) Let $A = (a, b) \subset \mathbb{R}$. Every point of $[a, b]$ is a limit point of A and so $[a, b] \subset A'$. Since the closure of (a, b) is $[a, b]$, we have $A' = [a, b]$.
(2) Let $X = \mathbb{R}$ and define $A = \{0\} \cup \{\frac{1}{n} \mid n \geq 1\}$. The subset A has exactly one limit point: 0 (0 is the only point in A which is not isolated). Hence $A' = \{0\}$.
(3) Let $X = \mathbb{R}$ and define $A = \{\frac{1}{n} \mid n \geq 1\}$. The subset A has exactly one limit point: 0 and so $A' = \{0\}$. In this case the limit point does not lie in A.
(4) Let (X, d) be a metric space and suppose x_0 is an isolated point of X. If A is any non-empty subset of X, then $x_0 \notin A'$. In particular, if X has the discrete metric and $A \subset X$, then $A' = \emptyset$.
(5) Take $\mathbb{Q} \subset \mathbb{R}$. Then $\mathbb{Q}' = \mathbb{R}$: we have $\overline{\mathbb{Q}} = \mathbb{R}$ and since $\mathbb{Q}$ has no isolated points, $\mathbb{Q}' = \mathbb{R}$.
(6) Let $P \subset C^0([a, b])$ denote the set of polynomial maps and take the uniform metric on $C^0([a, b])$. Then $P' = C^0([a, b])$ (since P contains no isolated points and $\overline{P} = C^0([a, b])$ by the Weierstrass approximation theorem).
(7) Let $B([a, b])$ denote the space of bounded real-valued functions on $[a, b]$ and take the uniform metric on $B([a, b])$. Suppose that (x_n) is a sequence of *distinct* points of $[a, b]$. For $n \geq 1$ define $\phi_n \in B([a, b])$ by

$$\phi_n(x) = \begin{cases} 1, & \text{if } x = x_n, \\ 0, & \text{if } x \neq x_n. \end{cases}$$

We have $\rho(\phi_n, \phi_m) = 1$ for all $n \neq m$ and so the set $\{\phi_n \mid n \in \mathbb{N}\}$ consists of isolated points and has empty limit point set. Note that $\{\phi_n \mid n \in \mathbb{N}\}$ has no convergent subsequences and is a bounded subset of $B([a, b])$ (it is a subset of $\overline{D}_2(0)$). Consequently, the Bolzano–Weierstrass theorem does not generalize to $B([a, b])$ (a similar remark holds for $(C^0([a, b]), \rho)$, see the exercises at the end of the section).   ♠

We give equivalent formulations of the definition of a limit point.

Lemma 7.10.9 *Let A be a non-empty subset of the metric space (X, d) and suppose $x \in X$. The following statements are equivalent.*

(1) $x \in A'$.
(2) *For all $r > 0$, $D_r(x) \cap (A \smallsetminus \{x\}) \neq \emptyset$.*
(3) *For every neighbourhood N of x, $N \cap (A \smallsetminus \{x\}) \neq \emptyset$.*
(4) *There exists a sequence $(x_n) \subset A \smallsetminus \{x\}$ which converges to x.*

Proof (1) $\Longrightarrow$ (2) Suppose that $x \in A'$. Given $r > 0$, $D_r(x) \cap (A \smallsetminus \{x\}) \neq \emptyset$ (else either x is an isolated point of A—if $x \in A$—or $x \notin A$ and $d(x, A) > 0$).

(2) $\Longleftrightarrow$ (3). For this observe that if N is a neighbourhood of x, then there exists an $n \in \mathbb{N}$ such that $D_{1/n}(x) \subset N$ and so $N \cap (A \smallsetminus \{x\}) \neq \emptyset$. The converse implication is trivial.

(2) $\Longrightarrow$ (4). For all $n \in \mathbb{N}$, $D_{1/n}(x) \cap (A \smallsetminus \{x\}) \neq \emptyset$. Choose $x_n \in D_{1/n}(x) \cap (A \smallsetminus \{x\})$, $n \in \mathbb{N}$. Then $d(x, x_n) < 1/n$, for all $n \in \mathbb{N}$, and so $\lim_{n\to\infty} x_n = x$.

(4) $\Longrightarrow$ (1). Since $\lim_{n\to\infty} d(x_n, x) = 0$ and $d(x, A) \leq d(x, x_n)$, we have $d(x, A) = 0$. Since $(x_n) \subset A \smallsetminus \{x\}$, x cannot be an isolated point of A. Hence $x \in A'$. $\qquad\qquad\qquad\qquad\qquad\qquad\qquad\qquad\qquad\qquad\qquad\qquad\qquad\quad$ $\square$

We collect together some properties of the limit point set in the next result.

Proposition 7.10.10 *Let A be a non-empty subset of the metric space (X, d).*

(1) $x \in A'$ *iff* $x \in \overline{A \smallsetminus \{x\}}$.
(2) A' *is a closed subset of X.*
(3) $\overline{A} = A \cup A'$.
(4) *A is closed iff $A' \subset A$ ("A is closed iff A contains all its limit points").*

Proof

(1) If $x \in \overline{A \smallsetminus \{x\}}$ then $d(x, A \smallsetminus \{x\}) = 0$ and so $x \in \overline{A}$ and x is not an isolated point of A. The converse is equally simple.
(2) It suffices to show that if $x \notin A'$, then $d(x, A') > 0$. If $x \notin A'$, there exists an $r > 0$ such that $D_r(x) \cap (A \smallsetminus \{x\}) = \emptyset$. Hence $D_r(x) \subset X \smallsetminus A'$ and $d(x, A') \geq r > 0$.
(3) Suppose $x \in \overline{A}$. Either $x \in A$ or not. If not, then $d(x, A) = d(x, A \smallsetminus \{x\}) = 0$ and so $x \in A'$ by (1). Hence $\overline{A} \subset A \cup A'$. Conversely, if $x \in A \cup A'$, then $d(x, A) = 0$ (using (1) again) and so $x \in \overline{A}$.

Finally, (4) is immediate from (3) since A is closed iff $A = \overline{A}$. $\qquad\qquad\qquad$ $\square$

7.10.2 Limit Points of a Sequence

We next investigate limit points of sequences—our definition will need to take account of the *order* implicit in the definition of a sequence.

Definition 7.10.11 A *limit point* (or *cluster point*) of the sequence $(x_n) \subset X$ is a point $x \in X$ such that there exists a subsequence (x_{n_k}) of (x_n) converging to x.

Remark 7.10.12 A sequence $(x_n) \subset X$ defines a subset $\{x_n \mid n \in \mathbb{N}\}$. A limit point of the *sequence* (x_n) may not be a limit point of the *subset* $\{x_n \mid n \in \mathbb{N}\}$. For example, a constant sequence $(x_n = x_0)$ has the limit point x_0 but the set $\{x_0\}$ has no limit points.   ✠

Proposition 7.10.13 *Let A be a subset of the metric space X and suppose that* $(x_n) \subset A$. *Every limit point of the sequence* (x_n) *lies in* $\overline{A}$. *In particular, A is closed iff A contains the limit of every convergent sequence of points of A.*

Proof Suppose that (x_{n_k}) is a subsequence of (x_n) converging to $x^\star$. We have $d(x^\star, A) \le d(x^\star, x_{n_k})$, $k \in \mathbb{N}$, and so letting $k \to \infty$, we see $d(x^\star, A) = 0$ and $x^\star \in \overline{A}$. Alternatively, one can base the proof on disk neighbourhoods of $x^\star$ and use Lemma 7.5.7. For the final statement, suppose first that A is closed. If $(x_n) \subset A$ converges to $x^\star$, then $x^\star \in \overline{A} = A$. Conversely, if there exists $(x_n) \subset A$ converging to $x^\star \notin A$, then $x^\star \in A'$ and so A cannot be closed by Proposition 7.10.10(4).   □

Remark 7.10.14 The last part of Proposition 7.10.13 will be very useful when we investigate properties of continuous functions and compact sets. It is worth giving a self-contained proof that uses only the definition of a closed set. Let A be closed and suppose $(x_n) \subset A$ converges to $x^\star$. Since $\lim_{n \to \infty} d(x_n, x^\star) = 0$ and $d(x^\star, A) \le d(x_n, x^\star)$, for all $n \in \mathbb{N}$, $d(x^\star, A) = 0$. Hence $x^\star \in A$. Conversely, suppose A is not closed. Then there exists an $x^\star \in X \smallsetminus A$ such that $d(x^\star, A) = 0$. For $n \in \mathbb{N}$, choose $x_n \in D_{1/n}(x^\star) \cap A$. Then the sequence $(x_n) \subset A$ converges to $x^\star \notin A$.   ✠

We now give a far-reaching generalization of Proposition 7.10.13.

Theorem 7.10.15 *Let* (x_n) *be a sequence in* (X, d). *Define*

$$S = \cap_{n \ge 1} \overline{\{x_m \mid m \ge n\}}.$$

Then $x \in S$ *iff there exists a subsequence* (x_{n_k}) *of* (x_n) *converging to x. In particular,* (x_n) *has a convergent subsequence iff* $S \ne \emptyset$.

This theorem allows us to capture all possible limits of convergent subsequences of a sequence by a process of intersection and closure. We remark that if $(x_n) \subset A$, then $\overline{\{x_m \mid m \ge n\}} \subset \overline{A}$ so Theorem 7.10.15 implies Proposition 7.10.13.

Before we prove the theorem, we give a number of examples to illustrate the ideas.

Examples 7.10.16

(1) Let $(x_n) \subset \mathbb{R}$ be defined by $x_n = n$, $n \ge 1$. For $n \ge 1$ we have

$$\overline{\{x_m \mid m \ge n\}} = \overline{\{m \mid m \ge n\}}$$
$$= \{m \mid m \ge n\}.$$

Hence $\cap_{n \ge 1} \overline{\{x_m \mid m \ge n\}} = \cap_{n \ge 1} \{m \mid m \ge n\} = \emptyset$. In this case, (x_n) has no convergent subsequences.

(2) Let $(q_n) \subset \mathbb{Q} \subset \mathbb{R}$ be a sequence which contains every rational number. Then for $n \geq 1$,

$$\{q_m \mid m \geq n\} = \mathbb{Q} \smallsetminus \text{finite set},$$

and so $\overline{\{q_m \mid m \geq n\}} = \mathbb{R}$. Therefore, $\cap_{n \geq 1} \overline{\{q_m \mid m \geq n\}} = \mathbb{R}$—every real number is the limit of a sequence of distinct rational numbers.

(3) Let $x_n = (-1)^{n+1}$, $n \in \mathbb{N}$. We have $\{x_m \mid m \geq n\} = \{-1, +1\}$ for all $n \in \mathbb{N}$. Hence $\cap_{n \geq 1} \overline{\{x_m \mid m \geq n\}} = \{-1, +1\}$, reflecting the fact that a convergent subsequence of (x_n) must converge to ± 1.

(4) Let $(x_n) \subset \mathbb{R}$ be a bounded sequence of real numbers. We know by Proposition 2.4.3 (corollary to the Bolzano–Weierstrass theorem) that (x_n) has at least one convergent subsequence. Consequently, it follows from Theorem 7.10.15 that $\cap_{n \geq 1} \overline{\{x_m \mid m \geq n\}} \neq \emptyset$. This property does not hold for bounded sequences in a general metric space. Rather than working with bounded sequences, we instead require that sequences are subsets of *compact* subsets of the metric space. We then always have at least one convergent subsequence (this is essentially our definition of the term "compact"). A compact set can be thought of as a far reaching generalization of a closed and bounded interval. The main problem will be to find good characterizations of compactness. ♠

Proof of Theorem 7.10.15 If $x^\star \in S = \cap_{n \geq 1} \overline{\{x_m \mid m \geq n\}}$, then $x^\star \in \overline{\{x_m \mid m \geq n\}}$ for all $n \in \mathbb{N}$. Hence, for every $r > 0$,

$$D_r(x^\star) \cap \{x_m \mid m \geq n\} \neq \emptyset. \tag{7.1}$$

Using (7.1), we construct inductively a subsequence (x_{n_k}) of (x_n) converging to $x^\star$. Taking $r = 1$, there exists an $n_1 \in \mathbb{N}$ such that $x_{n_1} \in D_1(x^\star)$. Suppose we have constructed $x_{n_1}, \cdots, x_{n_k}$ such that $n_1 < n_2 < \cdots < n_k$ and $x_{n_j} \in D_{1/j}(x^\star)$, $j = 1, \cdots, k$. We claim we can pick $n_{k+1} > n_k$ so that $x_{n_{k+1}} \in D_{1/(k+1)}(x^\star)$. If not, this would imply $D_{1/(k+1)}(x^\star) \cap \{x_m \mid m \geq n\} = \emptyset$, for all $n > n_k$, contradicting (7.1). This completes our construction of (x_{n_k}). Since $d(x^\star, x_{n_k}) < 1/k$, we have $\lim_{k \to \infty} x_{n_k} = x^\star$.

Conversely, suppose that (x_{n_k}) is a subsequence of (x_n) converging to $x^\star$. We claim that $x^\star \in S$. It suffices to show that for every $r > 0$, $D_r(x^\star) \cap \{x_m; \mid m \geq n\} \neq \emptyset$ for all $n \in \mathbb{N}$. Since $\lim_{k \to \infty} d(x_{n_k}, x^\star) = 0$, there exists a $k(r) \in \mathbb{N}$ such that $d(x_{n_k}, x^\star) < r$ for all $k \geq k(r)$. Pick $n_k \geq n$ with $k \geq k(r)$. Then $x_{n_k} \in D_r(x^\star) \cap \{x_m \mid m \geq n\}$ and so $D_r(x^\star) \cap \{x_m \mid m \geq n\} \neq \emptyset$. □

Example 7.10.17 Let $(x_n) \subset \mathbb{R}$ be a bounded sequence and set $S = \cap_{n \geq 1} \overline{\{x_m \mid m \geq n\}}$. It follows by the definition of lim inf, lim sup and Lemma 2.5.1(3) that $\inf S = \lim \inf x_n$, $\sup S = \lim \sup x_n$. Therefore the smallest closed interval containing S is $[\lim \inf x_n, \lim \sup x_n]$. Viewed this way, we may think of Theorem 7.10.15 as a generalization of lim sup, lim inf to general metric spaces. ♠

EXERCISES 7.10.18

(1) Suppose that $A \subset X$. Show that $a \in A$ is isolated in A iff a is an isolated point of (A, d_A).

(2) Find *countable* infinite subsets A of $\mathbb{R}$ such that

   (a) A has no limit points.
   (b) A has exactly three limit points.
   (c) A is bounded and $A' = \{0\} \cup \{1/n \mid n \geq 1\}$ (so A' is countable).
   (d) A has non-countably many limit points.

   In which of the cases (a–d) *must* (1) A have isolated points? (2) A have infinitely many isolated points?

(3) Let (X, d) be a metric space with the property that every convergent sequence is eventually constant. Prove that every point in X is isolated. (The topology on X is therefore the same as the topology given by the discrete metric.)

(4) Find an example of a non-empty open subset A of $\mathbb{R}$ for which $(A')^\circ \neq A$.

(5) Let $A \subset X$. Show that if a is an isolated point of A then $d(a, A') > 0$ and deduce that A' is a closed subset of X.

(6) Show by means of an example that in general $(A')' \neq A'$. Is $(A')'$ a subset or superset of A'?

(7) Find an example of a subset A of $\mathbb{R}$ such that A' is countably infinite but $(A')' = \emptyset$.

(8) Let A, B be subsets of the metric space (X, d). Prove that $(A \cup B)' = A' \cup B'$. Is it true that $(A \cap B)'$ is equal to the intersection of A' and B'?

(9) Show that $A' = (\overline{A})'$ for all subsets A of a metric space X.

(10) Prove that $A \subset X$ is closed iff for every convergent sequence $(x_n) \subset A$, the limit of the sequence lies in A.

(11) Show that a sequence $(x_n) \subset X$ is convergent iff $\cap_{n \geq 1} \overline{\{x_m \mid m \geq n\}}$ consists of a single point.

(12) Construct a bounded countable subset (ϕ_n) of $(C^0([a, b]), \rho)$ with no limit points. (Hint: choose ϕ_n so that $\rho(\phi_n, \phi_m) \geq 1$ for all $n > m \geq 1$, see Examples 7.10.8(7)).

(13) Show that every closed nonempty subset F of a closed interval can be written as a union $F = E \cup P$, where E is a countable subset of F and P is closed and contains no isolated points (P is an example of a *perfect* set). Is this result true for general metric spaces? Proof or counterexample. (Hint for the first part: Suppose F is uncountable. Let $P \subset F$ be the set of all points $x \in F$ such that every neighbourhood of x contains uncountably many points of F. Note that taking $P = F'$ does *not* work in general.)

(14) Suppose that F is a closed nonempty subset of the interval $[a, b]$ and that F is uncountable. Show that we can find a subset H of F such that (a) $F \smallsetminus H$ is countable, (b) for every point $z \in H$, we can find sequences $(x_n), (y_n) \subset H$ converging to z such that $x_n < z < y_n$ for all $n \in \mathbb{N}$. (Hint: Use the previous exercise and construct H as a subset of P.)

7.11 Continuous Functions

We start with a general definition of continuity (that works for all topological spaces). We show later that our definition is equivalent to the familiar ε, δ-definition.

Definition 7.11.1 Let (X, d) and (Y, ρ) be metric spaces and $f : X \to Y$. The map f is *continuous at the point* $x_0 \in X$ if for every neighbourhood N of $f(x_0), f^{-1}(N)$ is a neighbourhood of x_0. If f is continuous at every point of X, we say f is *continuous*.

We give a simple application of our definition that shows an advantage of framing continuity in terms of neighbourhoods.

Lemma 7.11.2 *Let* X, Y, Z *be metric spaces and suppose* $f : X \to Y$ *is continuous at* $x_0 \in X$, $g : Y \to Z$ *is continuous at* $f(x_0) = y_0 \in Y$. *Then the composite* $g \circ f : X \to Z$ *is continuous at* x_0.

Proof It suffices to show that if Q is a neighbourhood of $g(y_0)$, then $(g \circ f)^{-1}(Q)$ is a neighbourhood of x_0. Since Q is a neighbourhood of $g(y_0)$ and g is continuous at $y_0, g^{-1}(Q)$ is a neighbourhood of y_0. Since f is continuous at $x_0, f^{-1}(g^{-1}(Q)) = (g \circ f)^{-1}(Q)$ is a neighbourhood of x_0. □

If we work with continuous maps from X to Y, we can give an elegant characterization of continuity in terms of open or closed sets.

Theorem 7.11.3 *Let* X *and* Y *be metric spaces and* $f : X \to Y$. *The following statements are equivalent.*

(1) f *is continuous.*
(2) *For every open subset* U *of* $Y, f^{-1}(U)$ *is an open subset of* X.
(3) *For every closed subset* F *of* $Y, f^{-1}(F)$ *is a closed subset of* X.

Proof We start by noting that (2) and (3) are equivalent since

$$f^{-1}(Y \smallsetminus U) = f^{-1}(Y) \smallsetminus f^{-1}(U) = X \smallsetminus f^{-1}(U), \ f^{-1}(Y \smallsetminus F) = X \smallsetminus f^{-1}(F).$$

It suffices to prove (1) and (2) are equivalent. Suppose (2) holds. Let $x_0 \in X$ and N be a neighbourhood of $f(x_0)$. It suffices to show $f^{-1}(N)$ is a neighbourhood of x_0. Certainly, $\overset{\circ}{N}$ is an open neighbourhood of $f(x_0)$ and so $f^{-1}(\overset{\circ}{N})$ is an open subset of X. Since $x_0 \in f^{-1}(\overset{\circ}{N}) \subset f^{-1}(N), f^{-1}(N)$ is a neighbourhood of x_0. Conversely, suppose (1) holds. Let U be an open subset of Y. Then U is a neighbourhood of every point $y \in U$. Since f is continuous, $f^{-1}(U)$ will be a neighbourhood of every point $x \in X$ such that $f(x) \in U$. In other words, the interior of $f^{-1}(U)$ is precisely $f^{-1}(U)$ and so $f^{-1}(U)$ is open. □

Examples 7.11.4

(1) The identity map $I : X \to X$ of a metric space X is continuous: for every open subset of U of $X, I^{-1}(U) = U$, which is open.

(2) Let $f : X \to \mathbb{R}$ be continuous. Then $f^{-1}(0)$ is a closed subset of X: solutions
   sets of continuous functions are closed. Generally, if $f : X \to Y$ is continuous
   and $y_0 \in Y$, then $f^{-1}(y_0) = \{x \in X \mid f(x) = y_0\}$ is a closed subset of X. If
   we work with strict inequality, we obtain open sets. For example, if $a \in \mathbb{R}$ and
   $f : X \to \mathbb{R}$ is continuous, then $\{x \mid f(x) > a\}$ is an open subset of X. ♠

In the next lemma, we show that continuity as we have defined it is equivalent to
the usual ε, δ definition.

Lemma 7.11.5 *Let (X, d) and (Y, ρ) be metric spaces, $f : X \to Y$ and $x_0 \in X$. The
following statements are equivalent.*

(1) *f is continuous at x_0.*
(2) *For every $\varepsilon > 0$, there exists a $\delta > 0$ such that $f^{-1}(D_\varepsilon(f(x_0))) \supset D_\delta(x_0)$.*
(3) *For every $\varepsilon > 0$, there exists a $\delta > 0$ such that $\rho(f(x), f(x_0)) < \varepsilon$ if $d(x, x_0) < \delta$.*

Proof (1) $\Longrightarrow$ (2) Taking $N = D_\varepsilon(f(x_0))$, we see that $f^{-1}(N)$ is a neighbourhood
of x_0. Hence there exists a $\delta > 0$ such that $D_\delta(x_0) \subset f^{-1}(N)$ and $f^{-1}(D_\varepsilon(f(x_0))) \supset$
$D_\delta(x_0)$. (2) $\Longleftrightarrow$ (3) If $f^{-1}(D_\varepsilon(f(x_0))) \supset D_\delta(x_0)$ then $D_\varepsilon(f(x_0)) \supset f(D_\delta(x_0))$.
Obviously this implies the equivalence of (2) and (3). Finally, we show (2) $\Longrightarrow$
(1). Let N be a neighbourhood of $f(x_0)$. Choose $\varepsilon > 0$ such that $D_\varepsilon(f(x_0)) \subset N$.
Now there exists a $\delta > 0$ such that $f^{-1}(D_\varepsilon(f(x_0))) \supset D_\delta(x_0)$ and so $f^{-1}(D_\varepsilon(f(x_0)))$
is a neighbourhood of x_0. Therefore, $f^{-1}(N) \supset f^{-1}(D_\varepsilon(f(x_0)))$ is a neighbourhood
of x_0.   □

Remarks 7.11.6

(1) Our definition of continuity simply says that continuous functions are exactly
   those functions that preserve open sets. That is, $f^{-1}(U)$ is open for every open
   set U. The disadvantages of the ε, δ definition are firstly that it requires three
   quantifiers ('for all $x_0 \in X$, 'for all $\varepsilon > 0$', 'there exists a $\delta > 0$') and secondly
   that it uses metrically defined disks which are not preserved by continuous
   functions ($f^{-1}(D_r(x))$ is usually not a disk). In this sense the definition is not at
   all natural.
(2) Note that the continuity definition uses the *inverse* image of sets, not the *forward*
   images. This is characteristic of many definition in mathematics. It is also often
   the case that the properties are not preserved under forward images. However,
   it is usually interesting when they are; we encounter two important examples
   shortly (compactness and connectedness).   ✱

Example 7.11.7 If $f : X \to Y$ is continuous, then $f(F)$ is generally not a closed
subset of Y if F is closed in X. Similarly, $f(U)$ will generally not be open in Y if U
is an open subset of X. For example, suppose $f : \mathbb{R} \to \mathbb{R}$ is given by $f(x) = x^2$. Take
$U = (-1, 1)$. Then $f(U) = [0, 1)$, which is not an open subset of $\mathbb{R}$. For an example
where f does not map closed sets to closed sets, let $F \subset \mathbb{R}^2$ be the graph of the
continuous strictly positive function $g(x) = (1 + x^2)^{-1}$. Since g is continuous, F is a
closed subset of $\mathbb{R}^2$: F is the zero set of the continuous function $G(x, y) = y - f(x)$.

Let $f : \mathbb{R}^2 \to \mathbb{R}$ be the projection on the y-axis: $f(x, y) = y$. Then $f(F) = (0, 1]$, which is not closed. ♦

For future reference, we give the definition of uniform continuity in metric spaces.

Definition 7.11.8 Let $(X, d), (Y, \rho)$ be metric spaces. The map $f : X \to Y$ is *uniformly continuous* if for each $\varepsilon > 0$, there exists a $\delta > 0$ such that

$$\rho(f(x), f(x')) < \varepsilon, \text{ for all } x, x' \in X \text{ satisfying } d(x, x') < \delta.$$

Remark 7.11.9 Unlike continuity, the definition of uniform continuity requires structure beyond that of open and closed sets. ✱

EXERCISES 7.11.10

(1) Suppose X, Y, Z are metric space and $f : X \to Y$, $g : Y \to Z$ are continuous. Prove that the composite $g \circ f : X \to Z$ is continuous. Show that if f, g are uniformly continuous, then so is $g \circ f$.

(2) Suppose that $f : X \to \mathbb{R}$ is continuous. Prove that the maps $f_+(x) = \max(0, f(x))$ and $f_-(x) = \min(0, f(x))$ are continuous.

(3) Let (X, d), (Y, ρ) be metric spaces. An *isometry* of X and Y is an onto map $f : X \to Y$ such that $\rho(f(x_1), f(x_2)) = d(x_1, x_2)$ for all $x_1, x_2 \in X$. Prove that every isometry is 1:1 and continuous (even uniformly continuous). Show that if f is an isometry then the inverse map $f^{-1} : Y \to X$ is also an isometry.

(4) Suppose that $f : X_1 \to Y_1$ and $g : X_2 \to Y_2$ are continuous maps of metric spaces. Define metrics d^i on $X_i \times Y_i$ by $d_i((u, v), (a, b)) = \max\{d^{X_i}(u, a), d^{Y_i}(v, b)\}$, $i = 1, 2$. Show that $f \times g : X_1 \times Y_1 \to X_2 \times Y_2$ is continuous.

(5) Let A be a non-empty subset of the metric space (X, d). Show that the distance function $d(x, A) = \inf_{a \in A} d(x, a)$ is uniformly continuous.

(6) Let X be a metric space and suppose that every function $f : X \to \mathbb{R}$ is continuous. Show that every subset of X is open and closed.

(7) Suppose that the metric space X is written as a union $\cup_{i \in I} U_i$ of open subsets of X. Given $f : X \to \mathbb{R}$ show that if $f : U_i \to \mathbb{R}$ is continuous for all $i \in I$ then f is continuous. What about if we write X as a finite or infinite union of closed sets F_i and we assume $f : F_i \to \mathbb{R}$ is continuous?

(8) We showed that a continuous map $f : X \to Y$ need not map open sets to open sets. Find examples of maps $f : X \to Y$ which map open sets to open sets but which are not continuous. (Hints: (a) Let Y have the discrete topology; (b) take $X = Y = C^0([0, 1])$, f the identity map of X but inequivalent metrics on X and Y.)

(9) Find an example of a map $f : \mathbb{R} \to \mathbb{R}$ which maps closed sets to closed sets but which is not continuous.

(10) Let $f : X \to Y$, where (X, d), (Y, ρ) are metric spaces. Show if $f(x_0)$ is an interior point of $f(D_\delta(x_0))$ for all $\delta > 0$ then it does *not* necessarily follow that f is continuous at x_0. (Hint: Take $X = Y = \mathbb{R}$, $x_0 = f(x_0) = 0$. Choose f so that $f(-\delta, \delta) = [-1, 1]$, for all $\delta > 0$! Why do we need something like this?)

(11) Take the Zariski topology on $\mathbb{R}$ (see Exercises 7.5.8). Show that if $p : \mathbb{R} \to \mathbb{R}$ is a polynomial then $p^{-1}(U)$ is Zariski open for every Zariski open subset U of $\mathbb{R}$. Would this result be true if $p : \mathbb{R} \to \mathbb{R}$ was continuous or smooth but not a polynomial? Why?

(12) A map $f : X \to Y$ between metric spaces is a *homeomorphism* if f is 1:1 onto and both f and f^{-1} are continuous. Show that if $f : X \to X$ is 1:1 and onto then f is a homeomorphism iff $f(U), f^{-1}(U)$ are open subsets of X for all open subsets U of X. Show, by means of examples, that a homeomorphism need not be uniformly continuous.

(13) Show that every metric space is homeomorphic to a metric space of finite diameter.

(14) Extend the definitions of $\bar{f}(x\pm)$, $\underline{f}(x\pm)$, $\omega_f(x)$ given in Sect. 2.5.2 to maps $f : X \to \mathbb{R}$, where X is a metric space.

(15) Let $f : [a, b] \to \mathbb{R}$ be bounded and not necessarily continuous. Given $\ell > 0$, define $\overline{F}_\ell = \{x \in [a, b] \mid \bar{f}(x-) = \bar{f}(x+)\}$ and $\underline{F}_\ell = \{x \in [a, b] \mid \underline{f}(x-) = \underline{f}(x+)\}$. Show that $\overline{F}_\ell$ and $\underline{F}_\ell$ are closed subsets of $[a, b]$. (For notation and terminology, see Sect. 2.5.2.)

(16) Prove Young's theorem: Suppose $f : [a, b] \to \mathbb{R}$ and let $\overline{F} = \{x \mid \bar{f}(x+) \neq \bar{f}(x-)\}$. Then $\overline{F}$ is countable (see also Remarks 2.5.8(2)). (Hints. Let $\ell, k > 0$. Following the previous exercise, show that $\overline{F}_{\ell,k} = \{x \mid \underline{f}(x+) - \bar{f}(x-) \geq \ell, \bar{f}(x-) \geq k\}$ is a closed subset of $[a, b]$. If $\overline{F}_{\ell,k}$ is not countable, then $\overline{F}_{\ell,k}$ contains an uncountable subset H such that every point of H is a limit from the left and right of points of H (Exercises 7.10.18(14)). This implies that $\bar{f}(x-) \geq k$, for all $x \in H$ and so $\bar{f}(x+) \geq k + \ell$, for all $x \in H$. Proceeding inductively, deduce that $\bar{f}(x+), \bar{f}(x-) = +\infty$, for all $x \in H$, contradicting our definition of $\overline{F}_{\ell,k}$. Hence $\overline{F}_{\ell,k}$ is countable.)

(17) Improve the previous result to show that outside of a countable subset of $[a, b]$ we have

$$\bar{f}(x+) = \bar{f}(x-) \geq f(x) \geq \underline{f}(x+) = \underline{f}(x-).$$

(18) Show, by means of examples, that Young's theorem generally fails for maps $f : X \to \mathbb{R}$, X a metric space.

7.12   Construction and Extension of Continuous Functions

In the last section we defined and gave various characterizations of continuous functions on a metric space. However, we avoided the issue of the existence of non-trivial continuous functions on a general metric space. It is time to address this question. We consider the simplest case of constructing real-valued functions on a metric space. Suppose then that (X, d) is a metric space and let $C^0(X)$ denote the set of continuous functions $f : X \to \mathbb{R}$. Obviously, $C^0(X)$ contains the constant functions. Is it possible to construct non-constant continuous functions? We are assuming nothing about the set X except the presence of a metric. At this level of abstraction, the only way forward appears to be to use the metric to construct continuous functions on (X, d).

Lemma 7.12.1 *Let $a \in X$ and define $d_a : X \to \mathbb{R}$ by $d_a(x) = d(a, x)$. Then d_a is continuous. Consequently, $\{d_a \mid a \in X\} \subset C^0(X)$.*

Proof By Lemma 7.1.4, we have

$$|d_a(x) - d_a(y)| = |d(a, x) - d(a, y)| \le d(x, y), \ x, y \in X.$$

Hence d_a is continuous at x for all $x \in X$ and so $d_a \in C^0(X)$. $\qquad\qquad\square$

Remark 7.12.2 The function d_a is never constant if X contains more than one point. $\qquad\qquad\qquad\qquad\qquad\qquad\qquad\qquad\qquad\qquad\qquad\qquad\qquad\qquad$ ✖
For our purposes we need a slight generalization of Lemma 7.12.1.

Proposition 7.12.3 *Let A be a non-empty subset of the metric space (X, d) and define $d_A : X \to \mathbb{R}$ by*

$$d_A(x) = d(x, A).$$

Then $d_A \in C^0(X)$.

Proof The result follows from Proposition 7.2.1(4) by exactly the same argument used to prove Lemma 7.12.1. $\qquad\qquad\qquad\qquad\qquad\qquad\qquad\qquad\qquad\qquad\square$
    It turns out that the set $\{d_A \mid A \subset X, A \ne \emptyset\}$ is rich enough to allow us to represent the closed sets of a metric space as the zero sets of continuous functions.

Theorem 7.12.4 (Urysohn's Lemma) *Let (X, d) be a metric space and A, B be disjoint closed subsets of X. There exists a continuous function $f : X \to \mathbb{R}$ such that*

(1) $f^{-1}(0) = A$.
(2) $f^{-1}(1) = B$.
(3) $f(X) \subset [0, 1]$.

Proof In the spirit of our constructions of C^∞-functions given in Chap. 5, define

$$f(x) = \frac{d(x, A)}{(1 + d(x, B))(d(x, B) + d(x, A))}, \ x \in X.$$

Since A, B are closed and disjoint $d(x, A) + d(x, B) > 0$ for all $x \in X$ and so f is well defined. Since $d(x, A), d(x, B)$ are continuous by Proposition 7.12.3, f is continuous. We leave it to the reader to complete the simple verification that f satisfies (1–3). $\qquad\qquad\qquad\qquad\qquad\qquad\qquad\qquad\qquad\qquad\qquad\qquad\qquad\qquad\square$

Remarks 7.12.5

(1) We may allow B to be the empty set in Theorem 7.12.4: define $f(x) = d(x, A)/(1 + d(x, A))$ and note that $f(X) \subset [0, 1)$ and $f^{-1}(0) = A$.
(2) Urysohn's lemma holds for *normal* topological spaces which need not be metric spaces. However, at this level of generality, the best that can be claimed is $f^{-1}(1) \supset A, f^{-1}(0) \supset B$. The metric space proof of the Urysohn lemma often uses the Tietze extension theorem (see below). The proof we give is elementary and constructs f so that A, B are level sets of f. ✠

Theorem 7.12.6 (Tietze Extension Theorem) *Let A be a closed subset of the metric space (X, d) and suppose $f : A \to \mathbb{R}$ is continuous and bounded. There exists a continuous map $F : X \to \mathbb{R}$ such that $F(x) = f(x)$, for all $x \in A$. Moreover, we may construct F so that F is bounded and*

$$\inf_{s \in A} f(s) \leq F(x) \leq \sup_{s \in A} f(s), \quad x \in X.$$

Proof It suffices to prove the result under the assumption $f \geq 0$ since we can write f as a difference $\max(0, f) - \max(0, -f)$ of positive continuous functions (note Exercises 7.11.10(2)). Replacing f by $f + 1$, we can further assume $f \geq 1$. Set $M = \sup_{x \in A} f(x)$. We may assume $M > 1$ (else f is constant and the result is trivial). We define the extension F by

$$F(x) = \begin{cases} f(x), & x \in A, \\ (\inf_{y \in A} f(y)d(x, y))/d(x, A), & x \in X \smallsetminus A. \end{cases}$$

Since $X = \overset{\circ}{A} \cup (X \smallsetminus A) \cup \partial A$, it suffices to prove that F is continuous at points of $\overset{\circ}{A}$, $X \smallsetminus A$, and ∂A. Since $F = f$ on $\overset{\circ}{A}$, the continuity of F at points of $\overset{\circ}{A}$ is immediate.
Continuity of F at points of $X \smallsetminus A$.

Let $x \in X \smallsetminus A$. Since $d(x, A) > 0$, it suffices to show that the function $g : X \to \mathbb{R}$ defined by $g(z) = \inf_{y \in A} f(y)d(z, y)$ is continuous at $z = x$. Let $\varepsilon > 0$ and set $\delta = \varepsilon/M$. If $x' \in D_\delta(x)$, we have $d(x, y) \geq d(x', y) - d(x', x)$. Choose $y \in A$ such that $g(x) > f(y)d(x, y) - \varepsilon$. We have $f(y)d(x, y) \geq f(y)d(x', y) - f(y)d(x', x) > g(x') - \varepsilon$ and so $g(x) \geq g(x') - 2\varepsilon$. Similarly, $g(x') \geq g(x) - 2\varepsilon$. Hence $|g(x) - g(x')| < 2\varepsilon$ for all $x' \in D_\delta(x)$ proving the continuity of g at x.
Continuity of F at points of ∂A.

Let $x \in \partial A$ and choose $\varepsilon > 0$. Since f is continuous at x, there exists a $\delta > 0$ such that $|f(x) - f(y)| < \varepsilon$ for all $y \in D_\delta(x) \cap A$. Set $\bar{\delta} = \delta/(M + 1)$. Suppose $x' \in X \smallsetminus A$ and $d(x, x') < \bar{\delta}$. If $y \in A \smallsetminus D_\delta(x)$, we have $d(x', y) \geq d(x, y) - d(x, x') > \delta M/(M + 1) = M\bar{\delta}$ and so, since $f \geq 1$,

$$\inf_{y \in A \smallsetminus D_\delta(x)} f(y)d(x', y) > M\bar{\delta}.$$

Since $f(x) \le M, f(x)d(x',x) \le M\bar{\delta}$ and so

$$\inf_{y \in A} f(y)d(x',y) = \inf_{y \in D_\delta(x) \cap A} f(y)d(x',y). \tag{7.2}$$

If $y \in D_\delta(x) \cap A, f(x) - \varepsilon < f(y) < f(x) + \varepsilon$. Since $\inf_{y \in D_\delta(x) \cap A} d(x',y) = d(x',A)$, it follows from (7.2) that

$$(f(x) - \varepsilon)d(x',A) < \inf_{y \in A}(f(y)d(x',y)) < (f(x) + \varepsilon)d(x',A)$$

and so $|F(x') - f(x)| < \varepsilon$, for $x' \in D_{\bar{\delta}}(x) \cap (X \smallsetminus A)$. Since $F = f$ on A and $\bar{\delta} < \delta$, this gives $|F(x') - f(x)| < \varepsilon$, for $x' \in D_{\bar{\delta}}(x)$ proving the continuity of F at x.

Finally, it is immediate from the definition of F on $X \smallsetminus A$ that $1 \le F(x) \le M$, $x \in X$.   □

Remark 7.12.7 The boundedness assumption cannot be avoided in our argument for the continuity of F on $X \smallsetminus A$ (it is not essential for the continuity on ∂A). In the exercises we indicate the generalization of the Tietze extension theorem to unbounded functions.   ✠

EXERCISES 7.12.8

(1) Show that if A is a closed subset of $\mathbb{R}^n$, then every continuous function $f : A \to \mathbb{R}$ extends to a continuous function $F : \mathbb{R}^n \to \mathbb{R}$. (Hint: Construct a sequence (F_n) of continuous functions $F_n : \mathbb{R}^n \to \mathbb{R}$ such that $F_n = f$ on $D_n(0) \cap A$ and $F_{n+1} = F_n$ on $D_n(0), n \ge 1$.)
(2) Show that the Tietze extension theorem holds if f is unbounded. (Hint: Suppose $f : A \to \mathbb{R}$ is unbounded. Apply Theorem 7.12.6 to $\tilde{f} = \alpha \circ f : A \to \mathbb{R}$, where $\alpha(x) = \tan^{-1}(x), x \in \mathbb{R}$.)

7.12.1 Sequential Continuity

Just as for functions on $\mathbb{R}$, there is a very useful characterization of continuity of functions on a metric space given in terms of convergent sequences. First, a definition.

Definition 7.12.9 Let (X, d), (Y, ρ) be metric spaces. A map $f : X \to Y$ is *sequentially continuous* if given any convergent sequence (x_n) in X, $(f(x_n))$ is a convergent sequence of points in Y and

$$\lim_{n \to \infty} f(x_n) = f(\lim_{n \to \infty} x_n).$$

Remark 7.12.10 We define sequential continuity of f at a point $x_0 \in X$ by restricting to sequences which converge to x_0.   ✠

Examples 7.12.11

(1) Let (X, d) be a metric space and fix $a \in X$. Then $f(x) = d(x, a)$ is sequentially continuous. Indeed, let $\lim_{n\to\infty} x_n = x^\star$. Then $|d(x^\star, a) - d(x_n, a)| \leq d(x^\star, x_n)$ by Lemma 7.1.4. The result follows.

(2) If we take the product metric $D((x_1, y_1), (x_2, y_2)) = \max_i d(x_i, y_i)$ on $X \times X$, then $d : X \times X \to \mathbb{R}$ is sequentially continuous. To see this, observe that if $(X_n = (x_n, y_n))$ is a sequence in $X \times X$, then (X_n) converges to $(x^\star, y^\star)$ in the product metric iff (x_n) converges to $x^\star$ and (y_n) converges to $y^\star$. We claim that if (x_n, y_n) converges to $(x^\star, y^\star)$, then $\lim_{n\to\infty} d(x_n, y_n) = d(x^\star, y^\star)$. We have

$$\begin{aligned} |d(x_n, y_n) - d(x^\star, y^\star)| &\leq |d(x_n, y_n) - d(x_n, y^\star)| + |d(x_n, y^\star) - d(x^\star, y^\star)|, \\ &\leq d(y_n, y^\star) + d(x_n, x^\star), \end{aligned}$$

where the last line follows by Lemma 7.1.4. Now let $n \to \infty$. ♠

Theorem 7.12.12 (Notation as Above) *The function $f : X \to Y$ is continuous iff f is sequentially continuous.*

Proof The proof is formally identical that of the proof of Theorem 2.4.9 in Chap. 2 that applied to real-valued functions on $\mathbb{R}$. In detail, suppose first that f is continuous. Let $(x_n) \subset X$ be a convergent sequence with limit x_0. Since f is continuous at x_0, given $\varepsilon > 0$, there exists an $r > 0$ such that $d(f(x), f(x_0)) < \varepsilon$, if $x \in D_r(x_0)$. Since (x_n) converges to x_0, there exists an $m \in \mathbb{N}$ such that $x_n \in D_r(x_0)$, $n \geq m$ and so $d(f(x_n), f(x_0)) < \varepsilon$, for $n \geq m$. Therefore, $(f(x_n))$ converges to $f(x_0)$. Conversely, suppose that $(f(x_n))$ converges to $f(x_0)$ for every sequence (x_n) converging to x_0. We claim f is continuous at x_0. Suppose the contrary. If f is not continuous at x_0, there exists an $\varepsilon > 0$, such that for every $n \in \mathbb{N}$, there exists an $x_n \in X$ such that $x_n \in D_{1/n}(x_0)$ and $f(x_n) \notin D_\varepsilon(f(x_0))$. Obviously, (x_n) converges to x_0. Since $f(x_n) \notin D_\varepsilon(f(x_0))$, $d(f(x_n), f(x_0)) \geq \varepsilon$ for all $n \in \mathbb{N}$ and so $(f(x_n))$ cannot converge to $f(x_0)$, contradicting the assumption that f is sequentially continuous. Hence f must be continuous at x_0. □

Remark 7.12.13 Theorem 7.12.12 is very much a *metric space* theorem. It does not extend to general topological spaces. It is, however, a powerful result and, as in Chap. 2, leads to simple and transparent proofs of many foundational results for continuous functions. ✱

Example 7.12.14 Let X, Y_1, Y_2 be metric spaces and $f_i : X \to Y_i$, $i = 1, 2$ be continuous. Then $(f_1, f_2) : X \to Y_1 \times Y_2$ is continuous, where we take the product metric on $Y_1 \times Y_2$. By Theorem 7.12.12, it suffices to show (f_1, f_2) is sequentially continuous. Let $(x_n) \subset X$ converge to $x^\star$. By sequential continuity we have $\lim_{n\to\infty} f_i(x_n) = f_i(x^\star)$, $i = 1, 2$. Hence $\lim_{n\to\infty}(f_1(x_n), f_2(x_n)) = (f_1(x^\star), f_2(x^\star))$ and so (f_1, f_2) is sequentially continuous. ♠

We conclude with an application of Theorem 7.12.12.

Proposition 7.12.15 *Let $f, g : X \to Y$ be continuous. Then $S = \{x \in X \mid f(x) = g(x)\}$ is a closed subset of X.*

Proof We give a proof based on sequential continuity. In order to prove that S is closed, it suffices to show that if $(x_n) \subset S$ converges to x_0, then $x_0 \in S$. By sequential continuity of f and g, $\lim_{n \to \infty} f(x_n) = f(x_0)$, $\lim_{n \to \infty} g(x_n) = g(x_0)$. Since $f(x_n) = g(x_n)$ for all n, we have $f(x_0) = g(x_0)$ and so $x_0 \in S$.   □

Remark 7.12.16 Here is a sketch of an alternative proof of the previous proposition which uses Theorem 7.11.3 and Example 7.12.14. Let $(f, g) : X \to Y \times Y$ be the map defined by $(f, g)(x) = (f(x), g(x))$. By Example 7.12.14, (f, g) is continuous. If we define the diagonal $\Delta = \{(y, y) \mid y \in Y\}$, then Δ is a closed subset of $Y \times Y$ (see Exercises 7.4.27(8)). Now use $S = (f, g)^{-1}(\Delta)$ and Theorem 7.11.3. It is worth noting that even though there is a natural way of defining the product topology on $Y \times Y$, once we move away from the setting of metric spaces the diagonal Δ may not be closed in $Y \times Y$.   ✴

EXERCISES 7.12.17

(1) Let $f : X \to Y$ be continuous and e be a limit point of the set $E \subset X$. Show that if f is 1:1 then $f(e)$ is a limit point of $f(E) \subset Y$. True or false if f is not 1:1?
(2) Let $f : X \to Y$ be continuous. Show that if $E \subset X$, then $f(\overline{E}) \subset \overline{f(E)}$. What about the reverse inclusion: $f(\overline{E}) \supset \overline{f(E)}$? (Prove or give a counterexample.)
(3) Construct a function $f : \mathbb{R} \to \mathbb{R}$ such that f is discontinuous at all points of a dense subset Q of $\mathbb{R}$ but is such that the restriction of f to Q is continuous.
(4) Let $f = (f_1, \cdots, f_n) : X \to \mathbb{R}^n$. Prove that f is continuous iff every component function $f_i : X \to \mathbb{R}$ is continuous. (Do this in two ways: an ε, δ-proof and a proof based on neighbourhoods or closed sets.)
(5) Suppose that $f, g : X \to Y$ are continuous functions and that $f = g$ on a dense subset E of X. Show that $f = g$. (Hint: Proposition 7.12.15.)

7.13 Sequential Compactness

In this section our aim is to generalize to metric spaces the result that every continuous real-valued function on a closed and bounded interval is bounded and attains its bounds. More specifically, we want to characterize those subsets of a metric space for which every continuous function defined on the subset is bounded and attains its bounds. We do this by focusing on one property of a closed and bounded interval that follows from the Bolzano–Weierstrass theorem: every sequence contained in a closed and bounded interval has a subsequence converging to a point of the interval. We call sets that satisfy this condition (sequentially) *compact*. We provide some interesting classes of sets which are compact and finally show that continuous functions preserve compactness.

Definition 7.13.1 Let (X, d) be a metric space. A subset A of X is *sequentially compact* if every sequence $(x_n) \subset A$ has a subsequence converging to a point of A.

Remark 7.13.2 To avoid discussion of uninteresting special cases, we generally assume that the set A of Definition 7.13.1 is not empty.   ✠

Example 7.13.3 The closed and bounded interval $[a, b]$ is sequentially compact. Indeed, if $(x_n) \subset [a, b]$ is a sequence, then by Proposition 2.4.3, (x_n) has a convergent subsequence which must converge to a point of $[a, b]$ since $[a, b]$ is closed.   ♠

Proposition 7.13.4 *Let A be a sequentially compact subset of the metric space X. Then A is a closed and bounded subset of X.*

Proof Let $x \in A'$. There exists a sequence $(x_n) \subset A \smallsetminus \{x\}$ which converges to x. Therefore $x \in A$ (since every convergent subsequence of (x_n) converges to x). It remains to prove that A is bounded. That is, there exists an $M \geq 0$ such that $d(x, y) \leq M$ for all $x, y \in A$. Fix $a \in A$ and observe that A is bounded if and only if there exists an $M' \geq 0$ such that $d(x, a) \leq M'$ for $x \in A$. ($d(x, a) \leq M$ for all $x \in A$ implies $d(x, y) \leq d(x, a) + y(y, a) \leq 2M'$ for all $x, y \in A$. The converse is obvious taking $M' = M$.)

Suppose A is not bounded. Then for every $n \in \mathbb{N}$, there exist $x_n \in A$ such that

$$d(x_n, a) > n.$$

Since A is sequentially compact, we can find a subsequence (x_{n_k}) of (x_n) converging to a point $x^\star \in A$. We have $d(x_{n_k}, a) > n_k$, for all $k \geq 1$. Since $f(x) = d(x, a)$ is sequentially continuous,

$$\lim_{k \to \infty} d(x_{n_k}, a) = d(x^\star, a) < \infty.$$

This is a contradiction since $d(x_{n_k}, a) > n_k$ and so $(d(x_{n_k}, a))$ diverges to $+\infty$. □

Example 7.13.5 Although a necessary condition for sequential compactness is boundedness, it is not a sufficient condition. For example, every set X with the discrete metric is bounded (with $M = 1$) but a general sequence $(x_n) \subset X$ is only assured of having a convergent subsequence if X is finite. In particular, if (x_n) consists of distinct points then (x_n) has no convergent subsequence. Somewhat less trivially if (X, d) is any metric space, we can define a new metric D on X by $D(x, y) = \min\{1, d(x, y)\}$. Every subset of X is bounded with respect to the metric D. For example, if we replace the Euclidean metric on $\mathbb{R}^n$ by the metric D, then every closed subset of $\mathbb{R}^n$ is bounded. Obviously, the closed sets $\mathbb{Z}$ or $\mathbb{R}$ are not sequentially compact.   ♠

Notwithstanding the previous examples, there is one important case where sequential compactness is equivalent to being closed and bounded.

Theorem 7.13.6 *Let $m \in \mathbb{N}$. A subset A of $\mathbb{R}^m$ is sequentially compact iff A is closed and bounded. (The metric may be the Euclidean metric, d_1 or d_∞ or any metric equivalent to these metrics.)*

Proof We know by Proposition 7.13.4 that every compact subset of $\mathbb{R}^m$ is closed and bounded. It remains to prove the converse. The proof is by induction on m. Suppose $m = 1$. If (x_n) is a sequence of points in A, then there exists a convergent subsequence by Proposition 2.4.3 and the limit must lie in A since A is closed.

Assume the result has been proved for $m - 1$, $m > 1$. Observe that the product metric d_∞ on $\mathbb{R}^m$ restricts to d_∞ on $\mathbb{R}^p$ where we identify $\mathbb{R}^p$ with the subspace $\{(x_1, \cdots, x_p, 0, \cdots, 0) \mid x_1, \cdots, x_p \in \mathbb{R}\}$ of $\mathbb{R}^m$, $1 \leq p < m$. The same is true for the metrics d_2 and d_∞. We make a choice of one of these metrics and denote it by d. Suppose $(x_n) \subset A$. Write $x_n = (y_n, z_n)$ where $y_n \in \mathbb{R}^{m-1}$, $z_n \in \mathbb{R}$. Since A is bounded, (y_n) is a bounded sequence in $\mathbb{R}^{m-1}$, (z_n) is a bounded sequence in $\mathbb{R}$ (since $d((y, z), 0) \geq d(y, 0), d(z, 0)$). Since (y_n) is a bounded sequence in $\mathbb{R}^{m-1}$ it follows by the inductive hypothesis that there is a convergent subsequence, say (y_{n_k}). Let $\lim_{k \to \infty} y_{n_k} = y^\star$. Now (z_{n_k}) is a bounded sequence in $\mathbb{R}$ and so by the result for $n = 1$, there is a convergent subsequence, which we may denote by (z_{m_k}) (where $m_1 < m_2 < \cdots$ and $\{m_i \mid i \geq 1\} \subset \{n_k \mid k \geq 1\}$). Set $\lim_{k \to \infty} z_{m_k} = z^\star$. Since (y_{m_k}) is a subsequence of the convergent sequence (y_{n_k}), (y_{m_k}) is convergent and $\lim_{k \to \infty} y_{m_k} = y^\star$. Hence (y_{m_k}, z_{m_k}) is convergent in $\mathbb{R}^m$ with limit $x^\star = (y^\star, z^\star)$. Since A is closed, $x^\star \in A$. $\square$

Corollary 7.13.7 *Every bounded sequence in $\mathbb{R}^m$ has a convergent subsequence.*

Proof Let $(x_n) \subset \mathbb{R}^m$ be bounded. Then $A = \overline{\{x_n \mid n \geq 1\}}$ is a closed and bounded subset of $\mathbb{R}^m$. By Theorem 7.13.6, A is sequentially compact and so $(x_n) \subset A$ has a convergent subsequence. $\square$

Theorem 7.13.8 *Let (X, d), (Y, ρ) be metric spaces. If $f : X \to Y$ is continuous and A is a sequentially compact subset of X, then*

(1) *$f(A)$ is a sequentially compact subset of Y,*
(2) *$f : A \to Y$ is uniformly continuous (Definition 7.11.8).*

Proof

(1) We have to show that if $(y_n) \subset f(A)$ is a sequence, then there exists a convergent subsequence with limit in $f(A)$. Since $(y_n) \subset f(A)$, we can find a sequence $(x_n) \subset A$ such that $f(x_n) = y_n$, $n \in \mathbb{N}$. Since A is sequentially compact, there exists a convergent subsequence (x_{n_k}) of (x_n) with limit $x^\star \in A$. By sequential continuity, $\lim_{k \to \infty} f(x_{n_k}) = f(x^\star)$. Therefore (y_{n_k}) is a convergent subsequence of (y_n) with limit equal to $f(x^\star) \in f(A)$.
(2) The proof is formally identical to that of Theorem 2.4.15 and we leave the details to the exercises. $\square$

Theorem 7.13.9 *Let (X, d) be a metric space, A be a sequentially compact subset of X and $f : X \to \mathbb{R}$ be continuous. Then $f : A \to \mathbb{R}$ is bounded and attains its bounds: there exist $a_m, a_M \in A$ such that*

$$-\infty < \inf f(A) = f(a_m) \leq f(x) \leq f(a_M) = \sup f(A) < +\infty,$$

for all $x \in A$.

Proof By Theorem 7.13.8, $f(A)$ is a compact subset of $\mathbb{R}$. Therefore, by Proposition 7.13.4, $f(A)$ is a closed and bounded subset of $\mathbb{R}$. Hence $\sup f(A), \inf f(A) \in f(A)$. Pick $a_m, a_M \in A$ such that $f(a_m) = \inf f(A), f(a_M) = \sup f(A)$.   □

7.13.1   Additional Properties of Compactness

Proposition 7.13.10 *If A is a sequentially compact subset of the metric space X, then every closed subset of A is sequentially compact.*

Proof Let Z be a closed subset of A. It suffices to prove that every sequence (x_n) of points of Z has a convergent subsequence converging to a point of Z. Since A is sequentially compact and $(x_n) \subset Z \subset A$, there exists a convergent subsequence of (x_{n_k}) of (x_n). Since Z is closed and (x_{n_k}) is a convergent sequence of points of Z, $\lim_{k \to \infty} x_{n_k} \in Z$.   □

Proposition 7.13.11 *Let A be a subset of the metric space (X, d). Then A is a sequentially compact subset of X iff (A, d_A) (A with the induced metric) is a sequentially compact metric space.*

Proof Suppose (x_n) is a sequence of points of A. By sequential compactness of A as a subset of X, there exists a convergent subsequence (x_{n_k}) of A such that $\lim_{k \to \infty} x_{n_k} = x^* \in A$. Now $d(x_{n_k}, x^*) = d_A(x_{n_k}, x^*)$ and so clearly (x_{n_k}) is a convergent subsequence in (A, d_A). This argument shows that if A is a sequentially compact subset of X then (A, d_A) is a sequentially compact metric space. The converse is obtained by reversing the argument.   □

Remark 7.13.12 Proposition 7.13.11 shows that sequential compactness is an *absolute* or *intrinsic* property of a set. By contrast, properties like open and closed are *relative* properties. For example, if A is a proper open subset of X which is not closed (in X), then A will always be a closed subset of the metric space (A, d_A). If Z is a subset of A which does not contain all its limit points in X (and so is not closed in X), then Z may contain all of its limit points if viewed as a subset of (A, d_A). For example, if $Z = A \cap F$, where F is a closed subset of X and A is open.   ✻
We now work towards giving some more topological properties of compactness. With the exception of the relatively elementary Theorem 7.13.21 (used in the proof of the Arzelà–Ascoli theorem), no use is made of these results in the remainder of the book.

Theorem 7.13.13 *Let $F_1 \supset F_2 \supset \cdots$ be a decreasing sequence of non-empty sequentially compact subsets of X. Then $\bigcap_{n=1}^{\infty} F_n \neq \emptyset$. Conversely, if it is true that the intersection of every decreasing sequence of closed subsets is non-empty, then X is sequentially compact.*

Proof For each $n \in \mathbb{N}$, pick $x_n \in F_n$. Then (x_n) is a sequence of points in F_1 and so has a convergent subsequence (x_{n_k}) with limit $x^* \in F_1$. We claim $x^* \in \bigcap_{n=1}^{\infty} F_i$. It suffices to show $x^* \in F_m$ for all $m \geq 1$. But $x_{n_k} \in F_m$ for $k \geq m$ ($n_k \geq k$) and so, since F_m is sequentially compact and therefore closed, $x^* \in F_m$.

Conversely, suppose that the intersection of every decreasing sequence (F_n) of closed subsets of X is non-empty. Let (x_n) be a sequence of points in X. Set $F_n = \overline{\{x_m \mid m \geq n\}}$. Then (F_n) is a decreasing sequence of closed subsets of X. Since $\cap_{n \geq 1} F_n \neq \emptyset$ it follows by Theorem 7.10.15 that (x_n) has a convergent subsequence. Hence X is sequentially compact. □

Remark 7.13.14 The property described in Theorem 7.13.13 is exactly the property we used to prove the Bolzano–Weierstrass theorem (Theorem 2.4.1). In that case we looked at a decreasing sequence of closed and bounded intervals. ✙

Corollary 7.13.15 *Suppose that $\mathcal{U} = \{U_i \mid i \in \mathbb{N}\}$ is a countable collection of open subsets of a sequentially compact metric space X such that $\cup_{n=1}^{\infty} U_n = X$, then there exists a finite subset $\{U_{i_1}, \cdots, U_{i_k}\}$ of $\mathcal{U}$ such that $\cup_{j=1}^{k} U_{i_j} = X$.*

Proof Suppose the contrary. Then $V_n = \cup_{i=1}^{n} U_i \neq X$, for all $n \in \mathbb{N}$. For $n \geq 1$, define $F_n = X \setminus V_n$. Then (F_n) is a decreasing sequence of closed subsets of X. Since each F_n is a closed subset of a sequentially compact space, F_n is compact (Proposition 7.13.10). By our hypothesis, $F_n \neq \emptyset$ for all $n \geq 1$. Therefore, by Theorem 7.13.13, $\cap_{n=1}^{\infty} F_n \neq \emptyset$. This contradicts our assumption that $\cup_{n=1}^{\infty} U_n = X$ since $X \setminus \cup_{n=1}^{\infty} U_n = X \setminus \cup_{n=1}^{\infty} V_n = \cap_{n=1}^{\infty} (X \setminus V_n) = \cap_{n=1}^{\infty} F_n$. □

Definition 7.13.16 Let A be a subset of the metric space X. If $\mathcal{U} = \{U_i \mid i \in I\}$ is a collection of open subsets of X, we say $\mathcal{U}$ is an *open cover* of A if

$$A \subset \cup_{i \in I} U_i.$$

If I is finite (respectively, countable), $\mathcal{U}$ is a finite (respectively, countable) open cover of A.

If $\mathcal{V} \subset \mathcal{U}$ is also an open cover of A, then $\mathcal{V}$ is a *subcover* of A.

Remark 7.13.17 Corollary 7.13.15 states that every countable open cover of a sequentially compact metric space has a finite subcover. ✙
A much stronger version of Corollary 7.13.15 is true and the result—stated below— is used to *define* compactness for general topological spaces.

Theorem 7.13.18 *Let A be a compact subset of the metric space X and suppose that $\mathcal{U} = \{U_i \mid i \in I\}$ is an open cover of A. Then there exists a finite subcover of A. That is, there exist $U_{i_1}, \cdots, U_{i_k} \in \mathcal{U}$ such that*

$$\cup_{j=1}^{k} U_{i_j} \supset A.$$

We break the proof of Theorem 7.13.18 into a number of steps, each interesting in its own right. First, we remark that it follows from Proposition 7.13.11 that there is no loss of generality in assuming $A = X$ (else, replace (X, d) by (A, d_A) and then $\mathcal{U}_A = \{U_i \cap A \mid i \in I\}$ will be an open cover of A).

We recall that a metric space is *separable* if it has a countable dense subset. We showed earlier (Proposition 7.7.5) that if X is a separable metric space, then X is second countable: there exists a countable collection $\mathcal{B}$ of open subsets of X such that every open subset of X can be written as a union of open sets from B.

Remark 7.13.19 It is not hard to show that X is a separable metric space iff X is second countable. See the exercises. ✤

Proposition 7.13.20 *Every open cover of a separable metric space has a countable subcover.*

Proof Let $\mathcal{B} = \{B_n \mid n \in \mathbb{N}\}$ be the countable collection of open sets given by Proposition 7.7.5. Let $\mathcal{U} = \{U_i \mid i \in I\}$ be an open cover of X. Let $B_n \in \mathcal{B}$. If there exists a $U_i \in \mathcal{U}$, such that $B_n \subset U_i$, then choose one such U_i and label it as $U_{i(n)}$. In this way, we choose a countable collection $\{U_{i(n)} \mid n \in Q\}$, where Q will be a subset of $\mathbb{N}$ (if there is no U_i such that $B_n \subset U_i$, we make no choice). We claim that $\{U_{i(n)} \mid n \in Q\}$ is an open cover of X. Pick $x \in X$. Then x lies in some U_k and U_k is a union of B_n's. The point x lies in at least one of these B_n's, say B_m. Since $B_m \subset U_k$, one of the U_i's containing B_m must equal $U_{i(m)}$. But $x \in U_{i(m)}$. Therefore, $\{U_{i(n)} \mid n \in Q\}$ is an open cover of X. □

Theorem 7.13.21 *A sequentially compact metric space is separable.*

Proof For each $n \in \mathbb{N}$, we construct a finite subset E_n of X such that $d(x, E_n) < 1/n$ for all $x \in X$. Let $n \in \mathbb{N}$. Suppose we have chosen $z_1, \cdots, z_m \in X$ such that $d(z_i, z_j) \geq 1/n$, $i \neq j$. If $\min_{1 \leq i \leq m} d(x, z_i) < 1/n$ for all $x \in X$, take $E_n = \{z_1, \cdots, z_m\}$. Else, pick $z_{m+1} \in X$ such that $d(z_{m+1}, z_i) \geq 1/n$, $1 \leq i \leq m$. The process eventually terminates since otherwise we construct an infinite sequence $(z_n) \subset X$ such that $d(z_i, z_j) \geq 1/n$ for all $i \neq j$. Such a sequence can have no convergent subsequence, contradicting the assumption that X is sequentially compact. If we define $E = \cup_{n=1}^{\infty} E_n$, E is a countable dense subset of X and so X is separable. □

Proof of Theorem 7.13.18 As indicated previously, we may assume $A = X$. Since X is sequentially compact, X is separable by Theorem 7.13.21. Therefore, by Proposition 7.13.20, an open cover of X has a countable subcover. The result follows from Corollary 7.13.15. □

Remark 7.13.22 It follows from Theorem 7.13.18 and the Bolzano–Weierstrass theorem that every open cover of a closed and bounded subset of $\mathbb{R}^n$ has a finite subcover. This result is known as the *Heine–Borel theorem*. It is possible to use this result to give alternative proofs of many of our results on continuous functions on closed and bounded sets or sequentially compact sets. We give some illustrations in the exercises. However, there is no application presented in this book where a proof using the Heine–Borel theorem is simpler than a proof based on sequential compactness. For this reason, we have preferred to use sequence-based arguments in most of our proofs. ✤

Using Theorem 7.13.18, we may prove a generalization of Theorem 7.13.13 that is important for the study of compactness in general topological spaces.

Theorem 7.13.23 *Let $\mathcal{F} = \{F_i \mid i \in I\}$ be a collection of non-empty closed subsets of the sequentially compact metric space X. Suppose that every finite intersection $\cap_{j=1}^{k} F_{i_j}$ of sets from $\mathcal{F}$ is non-empty, then $\cap_{i \in I} F_i \neq \emptyset$.*

Proof We leave the proof to the exercises. □

EXERCISES 7.13.24

(1) Complete the proof of Theorem 7.13.8 by showing that every continuous function on a sequentially compact set is uniformly continuous.

(2) Suppose that $f : X \to Y$ is a continuous 1:1 onto map and that X is sequentially compact. Prove that f is a homeomorphism. (Hints: see Exercises 7.11.10(12) for the definition of homeomorphism and use Theorem 7.13.8, Proposition 7.13.10 and Theorem 7.11.3.)

(3) Let $f : X \to Y$ be continuous. Show that if $E \subset X$ and $\overline{E}$ is sequentially compact, then $f(\overline{E}) \supset \overline{f(E)}$. Do we have equality?

(4) Prove Theorem 7.13.23.

(5) Provide an alternative proof of Corollary 7.13.15 along the following lines: Let $\{U_i\}$ be a countable open cover of the compact space X. If there is no finite open subcover, then for each $n \in \mathbb{N}$, there exists an $x_n \in X \setminus \cup_{i=1}^{n} U_i$. Complete the proof by obtaining a contradiction.

(6) Let $E_1, \cdots, E_n$ be sequentially compact subsets of the metric space (X, d). Prove that $\cup_{i=1}^{n} E_i$ is sequentially compact.

(7) Suppose that $(X_1, d_1), (X_2, d_2)$ are sequentially compact metric spaces. Show that $X_1 \times X_2$ is sequentially compact if we take the product metric on $X_1 \times X_2$. Generalize to the product of n sequentially compact metric spaces.

(8) Suppose that (X, d) is sequentially compact and let X_∞ denote the space of all sequences $(x_n) \subset X$. Define

$$d_\infty((x_n), (x'_n)) = \sum_{n=1}^{\infty} 2^{-n} d(x_n, y_n).$$

Show that

(a) d_∞ is a metric on X_∞.
(b) (X_∞, d_∞) is sequentially compact.

We remark that it can be shown that an arbitrary product of compact topological spaces is compact—Tychonoff's theorem. We refer to books on general topology (for example, [18, 30]) for the definition of the product topology and the proof of Tychonoff's theorem, which depends on the Axiom of Choice from set theory.

(9) Let $X = \{0, 1\}$ and take the discrete metric on X. Define (X_∞, d_∞) as in the previous question. Show that X_∞ is homeomorphic to the middle-thirds Cantor set $\mathbf{C}$. (Hints: use the ternary expansion for points in $\mathbf{C}$ to define a continuous bijection $h : X_\infty \to \mathbf{C}$. Use exercise (2) above.)

(10) Let $f : X \to Y$ be a continuous map between metric spaces. Suppose that A is a compact subset of Y. Find an example to show that $f^{-1}(A)$ need not be compact.

(11) Let $f : X \to Y$ be a continuous map between metric spaces. Suppose that

(a) For all $y \in Y, f^{-1}(y)$ is either empty or a compact subset of X.
(b) f is closed: f maps closed subsets of X to closed subsets of Y.

Show that if (a,b) hold, then $f^{-1}(A)$ is compact for all compact subsets A of Y. Show, by means of examples, that conditions (a) and (b) are both necessary. (Maps for which inverse images of compact sets are compact are called *proper* maps.)

(12) Let A, B be non-empty subsets of the metric space (X, d). Define $D(A, B) = \inf_{a \in A, b \in B} d(a, b)$.

(a) Show that if A and B are sequentially compact, then there exist $a_0 \in A$, $b_0 \in B$ such that $D(A, B) = d(a_0, b_0)$.
(b) Show that if A is sequentially compact and B is a closed subset of X then $D(A, B) > 0$ iff $A \cap B = \emptyset$.
(c) Show that if A and B are subsets of $\mathbb{R}^n$ (standard metric) and A is sequentially compact, B is closed, then there exist $a_0 \in A$, $b_0 \in B$ such that $D(A, B) = d(a_0, b_0)$. Show that this result does not hold for subsets of general metric spaces. (Hints for second part: One approach can be based on Examples 7.10.8(3). Take $X = \mathbb{R}^2 \smallsetminus \{(0, 0)\}$ and observe that $\{(0, \frac{1}{n}) \mid n \in \mathbb{N}\}$ is a closed subset of X. Alternatively, an example can be constructed based on Examples 7.10.8(7)—suppose $x_n \to x_0 \notin \{x_n\}$.)
(d) Find an example of disjoint closed subsets A, B with $D(A, B) = 0$.

(13) Suppose (X, d), $(Y, \bar{d})$ are metric spaces and (X, d) is sequentially compact. Given continuous functions $f, g : X \to Y$ define

$$\rho(f, g) = \sup\{\bar{d}(f(x), g(x)) \mid x \in X\} \quad \text{(the uniform metric).}$$

Verify

(a) ρ is well defined (that is, $\rho(f, g) < \infty$).
(b) $\exists x_0 \in X$ such that $\rho(f, g) = \bar{d}(f(x_0), g(x_0))$.
(c) If $C^0(X, Y)$ denotes the space of all continuous functions from X to Y, then ρ defines a metric on $C^0(X, Y)$.

Suppose we allow X to be non-compact and let $\mathcal{B}(X, Y)$ denote the space of all continuous functions f from X to Y such that f is bounded (that is, $f(X)$ is a bounded subset of Y: $\exists R = R_f > 0$ such that $f(X) \subset D_R(y)$ for some $y \in Y$). Show that ρ defines a metric on $\mathcal{B}(X, Y)$. Is statement (b) above still valid? (Prove or give a counterexample.) (Hint: Define $G(x) = \bar{d}(f(x), g(x))$ and use Lemma 7.1.4 to prove the estimate $|G(x) - G(x')| \leq \bar{d}(f(x), f(x')) + \bar{d}(g(x), g(x'))$.)

(14) Show that if there exists a countable collection $\mathcal{B}$ of open subsets of X such that every open subset of X can be written as a union of open sets from $\mathcal{B}$, then X is separable. (Hint: Proposition 7.13.20 and cover by open disks.)

(15) Let (X, d) be a metric space and $f : \mathbb{R} \to X$ be continuous. Define $\Omega(f) = \cap_{T \geq 0}\overline{\{f(t) \mid t \geq T\}}$. Show that $x \in \Omega(f)$ if and only if there exists a monotone increasing sequence (t_n), $\lim_{n \to \infty} t_n = +\infty$, such that $\lim_{n \to \infty} f(t_n) = x$. If X is compact (or $\overline{f(\mathbb{R})}$ is compact) show that $\Omega(f) \neq \emptyset$. Show, by means of an example, that if these conditions are not satisfied, $\Omega(f)$ may be empty.

7.14 Compact Subsets of $\mathbb{R}$: The Middle Thirds Cantor Set

The structure of open subsets of $\mathbb{R}$ is relatively simple: every open subset of $\mathbb{R}$ can be written as a countable union of disjoint open intervals (Exercises 7.4.27(2)). Closed sets, even of the real line, can have a highly complex structure. In this section we describe the construction and properties of the (middle-thirds or ternary) Cantor set. The Cantor set is a compact subset of the unit interval $[0, 1]$ which (a) is uncountable, (b) has no interior points, and (c) has no isolated points. It is obtained by removing a countable set of open disjoint intervals from $[0, 1]$ of total length equal to one. A very interesting feature of the middle-thirds Cantor set is that it looks the same at all scales: self-similarity. Cantor-like sets play a very important role in the modern theory of dynamics and we briefly investigate that aspect in the exercises. At the end of the section we give a general definition of a Cantor set. However, when we say *the* Cantor set, we always mean the middle thirds (or ternary) Cantor set.

7.14.1 Construction of the Cantor Set

We give a construction which is based on ideas from dynamics. We define the continuous map $T : \mathbb{R} \to \mathbb{R}$ by

$$T(x) = \begin{cases} 3x, & \text{if } x \leq \frac{1}{2}, \\ 3 - 3x, & \text{if } x \geq \frac{1}{2}. \end{cases}$$

Observe that $T : (-\infty, \frac{1}{2}] \to (-\infty, \frac{3}{2}]$ and $T : [\frac{1}{2}, \infty) \to (-\infty, \frac{3}{2}]$ are 1:1 onto linear maps and that

$$T((-\infty, 0) \cup (1, +\infty)) \subset (-\infty, 0). \tag{7.3}$$

Given $x_0 \in \mathbb{R}$, we define the sequence $(x_n) \subset \mathbb{R}$ inductively by

$$x_{n+1} = T(x_n), \quad n \geq 0.$$

We usually write $x_n = T^n(x_0)$. If $x_0 < 0$, then $x_1 = T(x_0) = 3x_0 < x_0 < 0$ and clearly $x_n = 3^n x_0 < 0$, $n \geq 1$. Hence, using (7.3), we see that

$$\lim_{n \to \infty} x_n = -\infty, \quad \text{if } x_0 \in (-\infty, 0) \cup (1, +\infty).$$

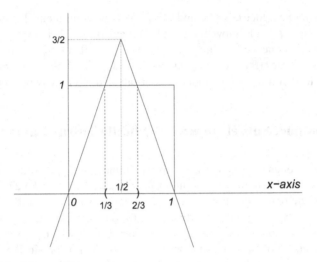

Fig. 7.4 Graph of the map T

If $T(x) \in [0, 1]$, then $x \in [0, 1]$. Consequently, if $x_0 \in [0, 1]$, then one of two things happen, either there exists an $n \geq 0$ such that $x_0, \cdots, x_n \in [0, 1]$ but $x_{n+1} = T(x_n) = T^{n+1}(x_0) > 1$ (see Fig. 7.4) or $x_n = T^n(x_0) \in [0, 1]$ for all $n \geq 0$, In the first case $\lim_{n \to \infty} x_n = -\infty$. In the second case $(x_n) \subset [0, 1]$. Certainly there exist points $x_0 \in [0, 1]$ for which $(x_n) \not\subset [0, 1]$. For example, every point in $(\frac{1}{3}, \frac{2}{3})$ exits $[0, 1]$ under just one application of T. On the other hand there exist points $x_0 \in [0, 1]$ for which $(x_n) \subset [0, 1]$. For example, if we take $x_0 = 0$, then $x_n = 0$, for all $n \geq 0$. Another example is given by taking $x_0 = \frac{1}{3}$. We have $x_1 = 1, x_n = 0, n \geq 2$. We define *the Cantor set* to be the subset $\mathbf{C}$ of $[0, 1]$ consisting of all points x such that $T^n(x) \in [0, 1]$ for all $n \geq 0$:

$$\mathbf{C} = \{x \in [0, 1] \mid T^n(x) \in [0, 1], \text{ for all } n \geq 0\}.$$

7.14.2   Properties of the Cantor Set

We are going to give a precise geometric description of the Cantor set. In order to do this, we need some new notation. Denote the unit interval $[0, 1]$ by I_0 and for $n > 0$ define

$$I_n = \{x \in I_0 \mid T^n(x) \in I_0\}.$$

Note that $I_n = \{x \in I_0 \mid T^j(x) \in I_0, \ 0 \leq j \leq n\}$, since once a point has exited I_0 it never returns, and

$$I_0 \supset I_1 \supset \cdots \supset I_n \supset I_{n+1} \supset \cdots \tag{7.4}$$

We have

$$\mathbf{C} = \bigcap_{n \geq 0} I_n = \bigcap_{n \geq m} I_n, \text{ for all } m \in \mathbb{N}, \tag{7.5}$$

where the last equality follows from (7.4).

Lemma 7.14.1 *For $n \geq m \geq 0$, we have $T^m(I_n) = I_{n-m}$. In particular,*

(1) *for $n \geq m \geq 0$, $(T^n)^{-1}(I_m) = I_{m+n}$,*
(2) $T(\mathbf{C}) = \mathbf{C}$.

Proof We claim that for $k \geq 0$ we have $T(I_{k+1}) = I_k$. Granted the claim, a simple induction verifies that for $n \geq m \geq 0$, we have $T^m(I_n) = I_{n-m}$. In order to verify the claim, observe that if $x \in I_{k+1}$, then $T(x) \in I_k$ and so $T(I_{k+1}) \subset I_k$. Conversely, let $x \in I_k$. Since $T(I_0) \supset I_k$, $k \geq 0$, there exists a $y \in I_0$ such that $T(y) = x$. Since $T^k(x) \in I_0$, $T^{k+1}(y) \in I_0$ and so $y \in I_{k+1}$ and $x \in T(I_{k+1})$. Hence $T(I_{k+1}) \supset I_k$. It remains to prove (1,2). For (1), observe that if $x \in (T^n)^{-1}(I_m)$, then $T^n(x) \in I_m$. Since $T^n(x) \in I_m$ implies that $x \in I_{m+n}$, we have $(T^n)^{-1}(I_m) = I_{m+n}$. (2) Since $\cap_{n \geq 0} I_n = \cap_{n \geq 1} I_n$, we have

$$T(\mathbf{C}) = T(\cap_{n \geq 1} I_n) = \cap_{n \geq 1} T(I_n)$$
$$= \cap_{n \geq 1} I_{n-1} = \cap_{n \geq 0} I_n = \mathbf{C},$$

where the last line follows since $T(I_n) = I_{n-1}$, $n \geq 1$. □

Remark 7.14.2 Although T maps $\mathbf{C}$ onto $\mathbf{C}$, T is not 1:1 (for example, $T(0) = T(1) = 0$ and $0, 1 \in \mathbf{C}$). ✴

Lemma 7.14.3 $\mathbf{C}$ *is a compact subset of I_0.*

Proof Since T^n is continuous and $I_n = (T^n)^{-1}(I_0)$, I_n is a closed subset of I_0. Hence $\mathbf{C} = \cap_{n \geq 0} I_n$ is a closed subset of $[0, 1]$ and therefore $\mathbf{C}$ is compact. □

Example 7.14.4 We have $I_1 = I_0 \smallsetminus (\frac{1}{3}, \frac{2}{3}) = [0, \frac{1}{3}] \cup [\frac{2}{3}, 1]$. Now $T : [0, \frac{1}{3}] \to [0, 1]$ is given by $T(x) = 3x$ and so

$$I_2 \cap \left[0, \frac{1}{3}\right] = \left[0, \frac{1}{3}\right] \cap T^{-1}\left(\left(\frac{1}{3}, \frac{2}{3}\right)\right)$$
$$= \left[0, \frac{1}{3}\right] \smallsetminus \left(\frac{1}{3^2}, \frac{2}{3^2}\right)$$
$$= \left[0, \frac{1}{3^2}\right] \cup \left[\frac{2}{3^2}, \frac{1}{3}\right].$$

Similarly $I_2 \cap [\frac{2}{3}, 1] = [\frac{2}{3}, \frac{7}{3^2}] \cup [\frac{8}{3^2}, 1]$. In other words, we obtain I_1 by removing the middle third of I_0 and we obtain I_2 by removing the middle thirds of the two closed intervals that comprise I_1. ♠

The next lemma, although elementary, will prove useful in unravelling the structure of the sets I_n.

Lemma 7.14.5 *Let $f(x) = mx + c$, where $m, c \in \mathbb{R}$ and $m \neq 0$. Suppose that $f([\alpha, \beta]) = [0, 1]$, where $\alpha < \beta$. We have*

(1) *f maps $[\alpha, \beta]$ 1:1 onto $[0, 1]$.*
(2) *If f preserves orientation (that is, $m > 0$) then $f(\alpha) = 0, f(\beta) = 1$. If f reverses orientation, then $f(\alpha) = 1, f(\beta) = 0$.*
(3) *$f^{-1}((\frac{1}{3}, \frac{2}{3})) = (\alpha + \frac{\beta - \alpha}{3}, \beta + \frac{\alpha - \beta}{3})$. ($f^{-1}$ maps the middle third open interval of $[0, 1]$ to the middle third open interval of $[\alpha, \beta]$.)*

Proof The result is geometrically obvious—see Fig. 7.5—but for completeness we provide an analytic/algebraic proof.
(1) Since $m \neq 0, f : \mathbb{R} \to \mathbb{R}$ is 1:1 onto. Given that $f([\alpha, \beta]) = [0, 1]$, it is immediate that f restricts to a 1:1 map of $[\alpha, \beta]$ onto $[0, 1]$. (2) Suppose that $m > 0$. Then f is an increasing function of x. If $f(\alpha) > 0$, then $f(x) \geq f(\alpha) > 0$ for all $x \in [\alpha, \beta]$ and so $0 \notin f([\alpha, \beta])$, contradicting the assumption that $f([\alpha, \beta]) = [0, 1]$. Hence $f(\alpha) = 0$. Similarly, $f(\beta) = 1$. If $m < 0$, then f is decreasing and we apply the same arguments to show that $f(\alpha) = 1, f(\beta) = 0$. (3) Suppose that $m > 0$ (the argument is similar if $m < 0$). We have $f(\alpha) = 0, f(\beta) = 1$ and so, by linearity,

$$f\left(\alpha + \frac{\beta - \alpha}{3}\right) = m\alpha + m\frac{\beta - \alpha}{3} + c = (m\alpha + c) + \frac{1}{3}(m\beta + c - (m\alpha + c)) = \frac{1}{3}.$$

The same argument shows that $f(\alpha + \frac{\beta - \alpha}{3}) = \frac{2}{3}$. $\square$

Proposition 7.14.6 *For $n \geq 0$ we have*

(1) *I_n is the disjoint union of 2^n closed intervals $I_{nj}, j = 1, \cdots, 2^n$, each of length 3^{-n}.*

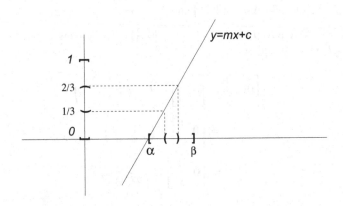

Fig. 7.5 Removing middle thirds, case $m > 0$

(2) *For $1 \leq j \leq 2^n$, $T^n : I_{nj} \to I_0$ is a linear 1:1 onto map and there exists a $b_{nj} \in 3\mathbb{Z}$ such that $T^n(x) = \pm 3^n x + b_{nj}$, for all $x \in I_{nj}$. ($3\mathbb{Z}$ is the set of all integers divisible by 3.)*

(3) *$I_{n+1} \cap I_{nj} = I_{nj} \smallsetminus T^{-n}((\frac{1}{3}, \frac{2}{3}))$. That is, we obtain I_{n+1} from I_n by removing the middle third open interval from each closed interval I_{nj} comprising I_n.*

Proof The proof is by induction on n. In the previous example, we verified the result in case $n = 1$. So suppose the result has been shown for $n = 0, \cdots, m$. We prove it for $n = m + 1$. Let $J = I_{mj}$ be one of the closed intervals comprising I_m. By the inductive hypothesis, $T^m : I_{mj} \to I_0$ is 1:1 onto and we may write $T^m(x) = \pm 3^m x + b$, where $b \in 3\mathbb{Z}$. Suppose that $T^m(x) = 3^m x + b$ (the argument when T^m reverses orientation is similar). Then by Lemma 7.14.5, T^m maps the open middle-thirds interval of J onto $(\frac{1}{3}, \frac{2}{3})$. Hence $I_{m+1} \cap J = J \smallsetminus (T^m)^{-1}(\frac{1}{3}, \frac{2}{3})$. Therefore, $I_{m+1} \cap J$ consists of two closed intervals J_1, J_2, each of length one third the length of J, that is $3^{-(m+1)}$. Now $T^m : J_1 \to [0, \frac{1}{3}]$ and $T^m : J_2 \to [\frac{2}{3}, 1]$ (we assumed T^m preserved orientation). Hence $T^{m+1} : J_1 \to [0, 1]$, $T^{m+1} : J_2 \to [0, 1]$ are 1:1 onto maps. For $x \in J_1$, $T^{m+1}(x) = 3(3^m x + b) = 3^{m+1} x + 3b$ and if $x \in J_2$, $T^{m+1}(x) = 3 - 3(3^m x + b) = -3^{m+1} x + 3(1 - b)$. In both cases, the constant term lies in $3\mathbb{Z}$. Applying this argument to each of the closed subintervals comprising I_m, we see that I_{m+1} is the disjoint union of $2 \times 2^m = 2^{m+1}$ closed intervals each of length $3^{-(m+1)}$. This completes the inductive step. □

We now give a number of corollaries of Proposition 7.14.6.

Corollary 7.14.7 *The total length of all the middle thirds intervals removed in the construction of $\mathbf{C}$ is 1.*

Proof At step one, we remove one interval of length $1/3$. At step two we remove two intervals, each of length $1/3^2$. At the nth step, we remove 2^n intervals each of length $3^{-(n+1)}$. Hence the total length of the intervals removed is

$$\sum_{n=0}^{\infty} 2^n 3^{-(n+1)} = \frac{1}{3} \sum_{n=0}^{\infty} \left(\frac{2}{3}\right)^n.$$

Since $\sum_{n=0}^{\infty} \left(\frac{2}{3}\right)^n = 1/(1 - \frac{2}{3}) = 3$, the result follows. □

Corollary 7.14.8 *If we let $\mathbf{E} = \cup_{n \geq 0} \cup_{1 \leq j \leq 2^n} \partial I_{nj}$ denote the set of end-points of all the closed intervals comprising I_n, $n \geq 0$, then $\mathbf{E}$ is a countable subset of $\mathbf{C}$.*

Proof Since each set $\cup_{1 \leq j \leq 2^n} \partial I_{nj}$ is finite, $\mathbf{E}$ is countable (a countable union of finite sets is finite). Since each I_n is obtained from I_{n-1} by removing middle third intervals, we never remove end-points of the intervals I_{nj}. Hence $\mathbf{E} \subset \cap_{n \geq 0} I_n = \mathbf{C}$. □

Remark 7.14.9 It is natural to guess that the Cantor set $\mathbf{C}$ is equal to $\mathbf{E}$. However, as we shall soon see, this is false. Indeed, $\mathbf{C}$ is an *uncountable* subset of I_0. ✱

Corollary 7.14.10 *Let $n \in \mathbb{N}$ and chose j, $1 \leq j \leq 2^n$. Then T^n maps $\mathbf{C} \cap I_{nj}$ 1:1 onto $\mathbf{C}$.*

Proof We have $T^n(\mathbf{C} \cap I_{nj}) = T^n(\mathbf{C}) \cap T^n(I_{nj}) = \mathbf{C} \cap I_0$, by Lemma 7.14.1(2) and Proposition 7.14.6(2). By Proposition 7.14.6(2), $T^n : I_{nj} \to I_0$ is 1:1.   □

Remark 7.14.11 The property of $\mathbf{C}$ described by the previous corollary implies that the Cantor set is 'self-similar' on all scales. That is, given any of the closed intervals I_{nj}, we find a copy of the Cantor set within I_{nj}. Sets of this type are examples of fractals and we give more examples and constructions in the next chapter.   ✠

We have already remarked that the Cantor set $\mathbf{C}$ is a compact set. We now verify some other metric and topological properties of $\mathbf{C}$.

Definition 7.14.12 A non-empty subset E of the metric space X is *perfect* if $E = E'$.

Remark 7.14.13 A set is perfect iff it is closed and has no isolated points.   ✠

Lemma 7.14.14 *The Cantor set is perfect:* $\mathbf{C} = \mathbf{C}'$.

Proof We already know that $\mathbf{C}$ is a closed subset of $\mathbb{R}$ and so $\mathbf{C}' \subset \mathbf{C}$. It suffices to show that $\mathbf{C}$ has no isolated points. Suppose the contrary and let $x \in \mathbf{C}$ be isolated. Then there exists a $\delta > 0$ such that $(x - \delta, x + \delta) \cap \mathbf{C} = \{x\}$. Since $\mathbf{C} = \cap_{n \geq 0} I_n$, $\mathbf{C} \subset I_n$ for all $n \geq 0$. Consequently, $x \in I_n$, for all $n \geq 0$. Each closed interval I_{nj} comprising I_n has length 3^{-n}. Choose n so that $3^{-n} < \delta$ and suppose that $x \in I_{nj}$. Then $(x - \delta, x + \delta) \cap I_{nj} \supset \partial I_{nj}$. Since $\partial I_{nj} \subset \mathbf{C}$ (Corollary 7.14.8), we see that $(x - \delta, x + \delta) \cap \mathbf{C}$ contains at least two points. Contradiction. Hence x cannot be an isolated point.   □

Definition 7.14.15 A non-empty subset E of $\mathbb{R}$ is *totally disconnected* if E contains no (non-empty) open intervals.

Remark 7.14.16 Later we will define totally disconnected for general metric spaces.   ✠

Example 7.14.17 If E is a subset of $\mathbb{R}$ then E is totally disconnected iff E has no interior points. To see this, observe that x is an interior point of E iff there exists a non-empty open interval $I \subset E$ which contains x.   ♠

Proposition 7.14.18 *The Cantor set is compact, perfect and totally disconnected.*

Proof We have already shown that $\mathbf{C}$ is compact and perfect. It remains to prove that $\mathbf{C}$ is totally disconnected. We give two proofs. The first proof makes essential use of the structure of open subsets of the real line; the second proof uses arguments from dynamics and extends to more general spaces. In what follows $|I|$ denotes the length of the interval I.

Method I. Suppose that $I \subset \mathbf{C}$ is a closed interval. It suffices to show that $|I| = 0$. Since $\mathbf{C} = \cap_{n \geq 0} I_n$, we have $I \subset I_n$ for all $n \geq 0$. Therefore for each n, there exists a j such that $I \subset I_{nj}$. Hence $|I| < 3^{-n}$ for all $n \geq 0$ and so $|I| = 0$.

Method II. Let $I = [\alpha, \beta] \subset \mathbf{C}$, where $\alpha \leq \beta$. Since $I \subset \mathbf{C}$ and $T(\mathbf{C}) = \mathbf{C}$, we have $T^n(I) \subset \mathbf{C}$ for all $n \geq 0$. Since $\mathbf{C} \subset [0, 1]$, it follows that the closed interval $T^n(I)$ must be a subset of $[0, 1]$ for all $n \geq 0$. But $|T^n(I)| = 3^n |I|$, for all $n \geq 0$. If $|I| > 0$, we eventually get $|T^n(I)| > 1$, contradicting $\mathbf{C} \subset I_0$. Hence $|I| = 0$.   □

Definition 7.14.19 A compact metric space is a *Cantor set* if it is perfect and totally disconnected.

Remark 7.14.20 It can be shown [30, Theorem 30.7] that every Cantor set is homeomorphic to the middle thirds Cantor set **C** (see Exercises 7.11.10(12) for the definition of a homeomorphism).   ✵

7.14.3 *Ternary Expansions and the Uncountability of* C

The *ternary expansion* of $x \in \mathbb{R}$ is the expansion of x to base 3. That is, $x = \pm x_0.x_1 \cdots$ will be the ternary expansion of x if $x_i \in \{0, 1, 2\}$ for all $i \geq 1$ and

$$x = \mathrm{sign}(x) \left(x_0 + \sum_{n=1}^{\infty} \frac{x_n}{3^n} \right),$$

where $\mathrm{sign}(x) = +1$ if $x \geq 0$ and $\mathrm{sign}(x) = -1$ if $x < 0$.

Example 7.14.21 Just as for decimal expansions, rational numbers may have more than one ternary expansion. For example, $1 = 0.\bar{2} = 1.\bar{0}$ and $\frac{1}{3} = 0.1\bar{0} = 0.0\bar{2}$. ♠

Let $\Sigma \subset [0, 1]$ denote the set of points which have a ternary expansion $x = 0.x_1x_2 \cdots$ such that $x_n \in \{0, 2\}$ for all n. If $x \in \Sigma$, we always regard the ternary expansion as infinite. That is, we write $0.2\bar{0}$ rather than 0.2.

Example 7.14.22 $1 \in \Sigma$ (since $1 = 0.\bar{2}$) and $\frac{1}{3} \in \Sigma$ (since $\frac{1}{3} = 0.0\bar{2}$). On the other hand, $\frac{1}{2} \notin \Sigma$ as the (unique) ternary expansion of $\frac{1}{2}$ is $0.\bar{1}$. ♠
If $x \in \{0, 1, 2\}$, let $\bar{x} = 2 - x$.

Lemma 7.14.23 *If* $x = 0.x_1x_2 \cdots x_n \cdots \in \mathbf{C}$ *and* $x_1 \neq 1$, *then*

$$T(x) = \begin{cases} 0.x_2x_3 \cdots x_n \cdots, & \text{if } x_1 = 0, \\ 0.\bar{x}_2\bar{x}_3 \cdots \bar{x}_n \cdots, & \text{if } x_1 = 2. \end{cases}$$

In particular, we have $T(\Sigma) = \Sigma$. *Conversely, if* $x \in [0, 1]$ *does not have a ternary expansion in* Σ, *then there exists an* $N \in \mathbb{N}$ *such that* $T^N(x) \notin [0, 1]$.

Proof If $x_1 = 0$, then $x \in [0, \frac{1}{3}]$ and $T(x) = 3x = 0.x_2x_3 \cdots x_n \cdots$. If $x_1 = 2$, then $x \in [\frac{2}{3}, 1]$ and so

$$T(x) = 3 - 2.x_2x_3 \cdots x_n \cdots$$

$$= 1 - 0.x_2x_3 \cdots x_n \cdots$$

$$= 0.22 \cdots 2 \cdots - 0.x_2x_3 \cdots x_n \cdots$$

$$= 0.\bar{x}_2\bar{x}_3 \cdots \bar{x}_n \cdots.$$

Since $\bar{x} \in \{0, 2\}$ if $x \in \{0, 2\}$, we see that if $x \in \Sigma$, then $T(x) \in \Sigma$. Hence $T(\Sigma) \subset \Sigma$. On the other hand, if $x = 0.x_1x_2 \cdots \in \Sigma$, then $T(0.0x_1x_2 \cdots) = x$. Hence $T(\Sigma) = \Sigma$. Finally, suppose that $x \in [0, 1]$ does not have a ternary expansion consisting of 0's and 2's. Let $x = 0.x_1x_2 \cdots$. If $x_1 = 1$, then $x \neq 0.x_1\bar{a}$, $a \in \{0, 2\}$ (else $x \in \Sigma$). Hence $x \in (\frac{1}{3}, \frac{2}{3})$ and $T(x) \notin I_0$. More generally, if x_j, $j > 1$, is the first term in the ternary expansion of x which is equal to 1, then $T^{j-1}(x) = 0.x_j\hat{x}_{j+1} \cdots \hat{x}_n \cdots$, where $\hat{x}_n \in \{x_n, \bar{x}_n\}$. It follows just as before that $T^{j-1}(x) \in (\frac{1}{3}, \frac{2}{3})$ and so $T^j(x) \notin I_0$. □

Theorem 7.14.24 *We have*

(1) $\mathbf{C} = \Sigma$,
(2) $\mathbf{C}$ *is uncountable.*

Proof

(1) Since $T(\Sigma) = \Sigma$, points in Σ never leave I_0 under iteration by T. Hence $\Sigma \subset \mathbf{C}$. On the other hand, if $x \notin \Sigma$, then there exists an $N \in \mathbb{N}$ such that $T^N(x) \notin [0, 1]$ and so $x \notin \mathbf{C}$. Therefore, $\mathbf{C} = \Sigma$.
(2) It suffices to prove Σ is uncountable. Define $B : \Sigma \to [0, 1]$ by $B(0.x_1 \cdots x_n \cdots) = 0.y_1 \cdots y_n \cdots$, where $y_n = 0$ if $x_n = 0$ and $y_n = 1$ if $x_n = 2$, $n \geq 1$. Observe that $B(\Sigma)$ is the set of all binary expansions $0.b_1 \cdots b_n \cdots$ of points in $[0, 1]$ and so B is certainly onto. Since $[0, 1]$ is uncountable so therefore is Σ. □

Example 7.14.25 Theorem 7.14.24 shows that $\mathbf{C}$ contains many more points than those in the (countable) interval end point set $\mathbf{E}$. For example, $\frac{1}{4} \notin \mathbf{E}$ is a point of the Cantor set. To see this observe that $T(\frac{1}{4}) = \frac{3}{4}$ and $T(\frac{3}{4}) = 3 - 3\frac{3}{4} = \frac{3}{4}$. Since $\frac{3}{4}$ is fixed by T, $T^n(\frac{1}{4}) = \frac{3}{4} \in I_0$ for all $n \geq 1$ and so $\frac{1}{4} \in \mathbf{C}$. ♠

EXERCISES 7.14.26

(1) Show that $\mathbb{Q}$ and $\mathbb{R} \setminus \mathbb{Q}$ are totally disconnected subsets of $\mathbb{R}$.
(2) Find examples of subsets of $\mathbb{R}$ which are (a) compact, perfect, not totally disconnected, (b) compact, not perfect, totally disconnected, (c) not compact, perfect, totally disconnected.
(3) Show that a perfect subset E of $\mathbb{R}^n$ is uncountable. (Hint: suppose the contrary and set $E = \{x_n \mid n \in \mathbb{N}\}$. Construct a decreasing sequence $\overline{D}_k$ of closed disks such that (a) $x_1 \in \overline{D}_1$, (b) $\overline{D}_k \cap E \neq \emptyset$, $k \in \mathbb{N}$, (c) $x_k \notin \overline{D}_{k+1}$, $k \geq 1$. Now use Theorem 7.13.13 applied to $\overline{F}_n = D_n \cap E$ to obtain a contradiction.)
(4) A metric space X is *locally compact* if every point in X has a compact neighbourhood. Show that every perfect set in a locally compact metric space is uncountable.
(5) Find all the points $x \in \mathbf{C}$ such that $T(x) = x$ (x is a *fixed point* of T). Find a point $x \in \mathbf{C}$ such that $T^2(x) = x$, but $T(x) \neq x$ (we call x a point of prime period two for T). Can you find a point $x \in \mathbf{C}$ which is of prime period three for T? (Hint: use ternary expansions; alternatively, graph T^3 and find the points of intersection of the graph of T^3 with the diagonal $y = x$.)

(6) Show that there exists a continuous map $f : \mathbf{C} \to [0, 1]$ which maps the Cantor set *onto* $[0, 1]$. (Hint: use the ternary expansion of points in $\mathbf{C}$. It can be shown that every compact metric space is the continuous image of the Cantor set.)

(7) Let $b : [0, 1] \to \mathbb{R}$ be the asymmetric "Baker's" transformation defined by

$$b(x) = \begin{cases} 3x, & 0 \le x < 2/3, \\ 3x - 2, & 2/3 \le x \le 1. \end{cases}$$

Verify that the set of points X in $[0, 1]$ that never exit $[0, 1]$ under iteration by b is the middle thirds Cantor set.

(8) If you construct a middle fifths Cantor set $\subset [0, 1]$—the middle fifth interval of each closed subinterval is removed—what is the total length of all middle fifth intervals that are removed? Prove that the resulting set is compact, perfect and totally disconnected. More generally, define a Cantor set by removing middle $x$ths, starting with $[0, 1]$, $x \in (0, 1)$. Show that the total length of all the intervals removed is one. (Hint: calculate what is removed and left at each step rather than counting the number and lengths of intervals created at each step.)

(9) Can you construct a 'fat' Cantor subset of $[0, 1]$ such that the total length of intervals removed is *less* than one? (Hint: follow the counting strategy of the previous example. You will have to vary the proportion $x_n \in (0, 1)$ removed at the nth step. Exercises 3.9.19(4) will be useful in showing that you get a fat Cantor set iff $\sum_{n=1}^{\infty} x_n < \infty$.)

(10) Let $f : \mathbb{R} \to \mathbb{R}$ be defined by $f(x) = \lambda x(1 - x)$. Show that if $\lambda > 2 + \sqrt{5}$ then $X = \{x \in [0, 1] \mid f^n(x) \in [0, 1], \text{ for all } n \in \mathbb{N}\}$ is a compact, perfect, totally disconnected subset of $[0, 1]$. (Hint: the condition on λ implies that there exists an $a > 1$ such that $|f'(x)| \ge a$ for all $x \in [0, 1]$ such that $f(x) \in [0, 1]$. Use the second method of proof of Proposition 7.14.18 to show that X is totally disconnected.)

7.15 Complete Metric Spaces

One of the most important ideas in our study of convergence of sequences and series of real numbers was that of a *Cauchy sequence*. The definition of a Cauchy sequence naturally generalizes to metric spaces. In this section we develop the theory of Cauchy sequences in metric spaces and show, for example, how results on uniform convergence in Chap. 4 can be naturally reformulated in metric space terms.

Definition 7.15.1 A sequence (x_n) in the metric space (X, d) is a *Cauchy sequence* if $\lim_{m,n \to \infty} d(x_n, x_m) = 0$. That is, if for every $\varepsilon > 0$, there exists an $N \in \mathbb{N}$ such that

$$d(x_n, x_m) < \varepsilon, \text{ for all } m, n \ge N.$$

The next lemma is a metric space version of Lemma 2.4.20.

Lemma 7.15.2 *Let* (x_n) *be a sequence in the metric space* (X, d).

(1) *If* (x_n) *is Cauchy, then* $\{x_n \mid n \in \mathbb{N}\}$ *is a bounded subset of* X.
(2) *If* (x_n) *is convergent, then* (x_n) *is Cauchy.*
(3) *If* (x_n) *is Cauchy and* (x_n) *has a convergent subsequence, then* (x_n) *is convergent.*

Proof The proof, modulo changes of notation, is formally identical to that of Lemma 2.4.20. We prove (3) and leave (1,2) to the exercises. Suppose that the subsequence (x_{n_k}) of (x_n) is convergent with limit $x^\star$. Given $\varepsilon > 0$, choose $N \in \mathbb{N}$ such that $d(x_n, x_m) < \varepsilon/2$, for all $n, m \geq N$, and $d(x_{n_k}, x^\star) < \varepsilon/2$, for all $n_k \geq N$. Choose $n_k \geq N$. Then for all $n \geq N$ we have

$$d(x_n, x^\star) \leq d(x_n, x_{n_k}) + d(x_{n_k}, x^\star)$$
$$< \varepsilon/2 + \varepsilon/2 = \varepsilon.$$

Hence (x_n) is convergent with limit $x^\star$. □

Definition 7.15.3 A metric space X is *complete* if every Cauchy sequence in X converges.

Examples 7.15.4

(1) $\mathbb{R}$ is complete (in the standard metric).
(2) $\mathbb{R}^m$ is complete in the Euclidean metric d_2 (or in either of the metrics d_1 or d_∞). If $\mathbf{x}, \mathbf{y} \in \mathbb{R}^m$, then $d_2(\mathbf{x}, \mathbf{y}) = \sqrt{\sum_{i=1}^{m}(x_i - y_i)^2} \geq |x_i - y_i|$, $1 \leq i \leq m$. Hence, if $(\mathbf{x}^n)$ is a Cauchy sequence in $\mathbb{R}^m$, then (x_i^n) is a Cauchy sequence in $\mathbb{R}$, $1 \leq i \leq m$. Let $\lim_{n\to\infty} x_i^n = x_i^\star$, $1 \leq i \leq m$. We claim that $(\mathbf{x}^n)$ converges with limit $\mathbf{x}^\star = (x_1^\star, \cdots, x_m^\star)$. This follows easily from the estimate $d_2(\mathbf{x}^\star, \mathbf{x}^n) \leq \sqrt{m}\max_{1 \leq i \leq m} |x_i^\star - x_i^n|$. Similar arguments apply for the metrics d_1, d_∞. ♠

The next proposition gives more examples of complete metric spaces.

Proposition 7.15.5 *Let* (X, d) *be a metric space.*

(1) *If* X *is sequentially compact, then* X *is complete.*
(2) *If* X *is complete then every closed subset of* X *is complete (in the induced metric).*

Proof (1) If X is sequentially compact, then every sequence (x_n) in X has a convergent subsequence. Now apply Lemma 7.15.2(3). (2) Suppose that E is a closed subset of the complete metric space X. If (x_n) is a Cauchy sequence of points of (E, d_E), then certainly (x_n) is Cauchy in (X, d). Hence (x_n) converges to a point $x^\star \in X$. Since E is closed, $x^\star \in E$. □

Example 7.15.6 Every closed and bounded subset of $(\mathbb{R}^n, d_2)$ is complete in the induced metric. ♠

7.15.1 Completeness of Spaces of Functions

Let (X, d) be a metric space. Let $C^0(X, \mathbb{R})$, $B(X, \mathbb{R})$ and $B^0(X, \mathbb{R})$ respectively denote the spaces of continuous, bounded and bounded continuous real-valued functions on X. Obviously,

$$C^0(X, \mathbb{R}) \supset B^0(X, \mathbb{R}) \subset B(X, \mathbb{R}).$$

If X is compact then $C^0(X, \mathbb{R}) = B^0(X, \mathbb{R})$. Given $f, g \in B(X, \mathbb{R})$, we define the *uniform metric* ρ on $B(X, \mathbb{R})$ by

$$\rho(f, g) = \sup_{x \in X} |f(x) - g(x)|.$$

(See also Exercises 7.1.9(8).) Let ρ also denote the induced metrics $\rho_{B^0(X, \mathbb{R})}$ on $B^0(X, \mathbb{R})$ and $\rho_{C^0(X, \mathbb{R})}$ on $C^0(X, \mathbb{R})$ (when X is compact).

Theorem 7.15.7 *Let (X, d) be a metric space.*

(1) $(B(X, \mathbb{R}), \rho)$ *and* $(B^0(X, \mathbb{R}), \rho)$ *are complete.*
(2) *If X is compact, then* $(C^0(X, \mathbb{R}), \rho)$ *is complete.*

Proof The proof of this result is similar to that of Theorem 4.3.16. We prove that $(B(X, \mathbb{R}), \rho)$ and $(B^0(X, \mathbb{R}), \rho)$ are complete (which also accounts for (2) since $(C^0(X, \mathbb{R}), \rho) = (B^0(X, \mathbb{R}), \rho)$ if X is compact). Suppose then that (f_n) is a Cauchy sequence of functions in $B(X, \mathbb{R})$. Given $x \in X$, we have $\rho(f_n, f_m) \geq |f_n(x) - f_m(x)|$ for all $n, m \in \mathbb{N}$. Hence $(f_n(x))$ is a Cauchy sequence in $\mathbb{R}$. Since $(\mathbb{R}, |\cdot|)$ is complete, $(f_n(x))$ is convergent. Set $\lim_{n \to \infty} f_n(x) = \hat{f}(x)$. Since x was an arbitrary point of X, this defines the function $\hat{f} : X \to \mathbb{R}$. Let $\varepsilon > 0$. Since (f_n) is Cauchy, there exists an $N \in \mathbb{N}$ such that $\rho(f_n, f_m) \leq \varepsilon$ for all $n, m \geq N$. Letting $m \to \infty$, this gives $\rho(f_n, \hat{f}) \leq \varepsilon$ for all $n \geq N$. Hence (f_n) converges to $\hat{f}$ in $(B(X, \mathbb{R}), \rho)$ (since $\hat{f} - f_n \in (B(X, \mathbb{R}), \rho)$ for all n, $\hat{f}$ is bounded). Now suppose $(f_n) \subset B^0(X, \mathbb{R})$. We prove that $\hat{f}$ is continuous. Given $x \in X$ and $\varepsilon > 0$, it suffices to find $r > 0$ such that $|\hat{f}(x) - \hat{f}(y)| < \varepsilon$, for all $y \in D_r(x)$. Since (f_n) is Cauchy, there exists an $N \in \mathbb{N}$ such that $\rho(f_n, f_m) < \varepsilon/3$ for all $n, m \geq N$. Since f_N is continuous, there exists an $r > 0$ such that $|f_N(x) - f_N(y)| < \varepsilon/3$ for all $y \in D_r(x)$. We have

$$|\hat{f}(x) - \hat{f}(y)| = |\hat{f}(x) - f_N(x) + f_N(x) - f_N(y) + f_N(y) - \hat{f}(y)|$$
$$\leq |\hat{f}(x) - f_N(x)| + |f_N(x) - f_N(y)| + |f_N(y) - \hat{f}(y)|$$
$$< \varepsilon/3 + \varepsilon/3 + \varepsilon/3 = \varepsilon, \text{ if } y \in D_r(x).$$

Hence $\hat{f}$ is continuous at x. □

Examples 7.15.8

(1) If I is a closed interval in $\mathbb{R}$ or more generally a closed and bounded subset of $\mathbb{R}$, Theorem 7.15.7 implies that a Cauchy sequence of continuous real-valued functions on I, uniform metric, is uniformly convergent (general principle of convergence, Theorem 4.3.16).

(2) If we choose a different metric on $B^0(X, \mathbb{R})$ or $C^0(X, \mathbb{R})$, then the corresponding function space may not be complete. As an example, recall that the L^2-metric ρ_2 on $C^0([0, 1], \mathbb{R})$ is defined by

$$\rho_2(f, g) = \left(\int_0^1 |f(x) - g(x)|^2 \, dx \right)^{\frac{1}{2}}, \ f, g \in C^0([0, 1]).$$

The metric space $(C^0([0, 1], \mathbb{R}), \rho_2)$ is not complete. It suffices to find a non-convergent Cauchy sequence in $(C^0([0, 1], \mathbb{R}), \rho_2)$. If we define the sequence (f_n) by

$$f_n(x) = \begin{cases} 0, & \text{if } 0 \le x \le \frac{1}{2} - \frac{1}{n}, \\ \frac{1}{2}(1 + n(x - \frac{1}{2})), & \text{if } \frac{1}{2} - \frac{1}{n} \le x \le \frac{1}{2} + \frac{1}{n}, \\ 1, & \text{if } \frac{1}{2} + \frac{1}{n} \le x \le 1, \end{cases}$$

then it is straightforward to check that (f_n) is a Cauchy sequence in $(C^0([0, 1], \mathbb{R}), \rho_2)$ that does not converge in the L^2-metric to a continuous function. Note that (f_n) is not Cauchy in $(C^0([0, 1], \mathbb{R}), \rho)$.   ♠

It is easy to extend Theorem 7.15.7 to spaces of vector-valued functions. Given $p \in \mathbb{N}$, let $B(X, \mathbb{R}^p)$ denote the space of bounded $\mathbb{R}^p$-valued functions on X, where bounded is relative to one of the standard (equivalent) metrics on $\mathbb{R}^p$: d_2, d_1 or d_∞ (because the metrics are equivalent, $B(X, \mathbb{R}^p)$ does not depend on which particular metric from d_2, d_1, d_∞ we choose). We may similarly define $B^0(X, \mathbb{R}^p)$ and $C^0(X, \mathbb{R}^p)$. Define the uniform metric ρ in the usual way by $\rho(f, g) = \sup_{x \in X} d(f(x), g(x))$, where d is the metric on $\mathbb{R}^p$ and if $f, g \in C^0(X, \mathbb{R}^p)$, we assume X is compact. We have the following useful corollary of Theorem 7.15.7.

Theorem 7.15.9 *Let (X, d) be a metric space and $p \in \mathbb{N}$.*

(1) *$(B(X, \mathbb{R}^p), \rho)$ and $(B^0(X, \mathbb{R}^p), \rho)$ are complete.*
(2) *If X is compact, then $(C^0(X, \mathbb{R}^p), \rho)$ is complete.*

Proof Choose the metric d_∞ on $\mathbb{R}^p$. Suppose $(f^n) \subset B(X, \mathbb{R}^p)$ is a Cauchy sequence. If we write f^n in component form as $(f_1^n, \cdots, f_p^n)$, then $\rho(f^n, f^m) = \sup_{x \in X} \max_{1 \le i \le p} |f_i^n(x) - f_i^m(x)|$ and so $(f_i^n) \subset B(X, \mathbb{R})$ is Cauchy, $1 \le i \le p$. Now apply Theorem 7.15.7 to deduce that there exists an $f^\star = (f_1^\star, \cdots, f_p^\star)$ such that $\lim_{n \to \infty} f_i^n = f_i^\star$, $1 \le i \le p$. Since each component of $f^\star$ is bounded, it is immediate that $f^\star$ is bounded in the d_∞ metric (the bound is the maximum of the bounds of the components) and that (f^n) converges to f^* in $(B(X, \mathbb{R}^p), \rho)$. The continuity statements are immediate since f^* is continuous iff each component of f^* is continuous. $\qquad\qquad\qquad\qquad\qquad\qquad\qquad\qquad\qquad\qquad\qquad\qquad\qquad\square$

7.15.2   Completion of a Metric Space

Definition 7.15.10 Let (X, d) be a metric space. A *completion* of (X, d) consists of a metric space $(\hat{X}, \hat{d})$ such that

(1) $(\hat{X}, \hat{d})$ is a complete metric space.
(2) X is a dense subset of $\hat{X}$ (closure is relative to the metric $\hat{d}$).
(3) $\hat{d}_X = d$ (the metric $\hat{d}_X$ induced by $\hat{d}$ on X equals d).

Example 7.15.11 A completion of $(\mathbb{Q}, |\cdot|)$ is $(\mathbb{R}, |\cdot|)$. ♦

First we prove that if (X, d) has a completion then the completion is essentially unique. Then we shall show that every metric space has a completion. Our proof will depend on the completeness of $(\mathbb{R}, |\cdot|)$.

Before we prove the uniqueness of a completion, we need to review the definition of an isometry. Let (X, d), $(Y, \bar{d})$ be metric spaces. Recall (see Exercises 7.11.10(3)) that an *isometry* of X and Y is a 1:1 onto map $F : X \to Y$ such that

$$\bar{d}(F(x), F(x')) = d(x, x'), \text{ for all } x, x' \in X.$$

We remark that if $F : X \to Y$ is an isometry then so is $F^{-1} : Y \to X$. Both F and F^{-1} are obviously continuous. If $F : X \to Y$ is an isometry, then X and Y are indistinguishable as metric spaces and we say X and Y are *isometric*. We show that any two completions of a metric space are isometric. More precisely, we prove

Proposition 7.15.12 *Let* $(\hat{X}_1, \hat{d}_1)$, $(\hat{X}_2, \hat{d}_2)$ *be completions of the metric space* (X, d). *Then there exists a unique isometry* $F : \hat{X}_1 \to \hat{X}_2$ *such that* F *restricts to the identity map on the subspace* X *of* $\hat{X}_1$.

Proof Let $x^* \in \hat{X}_1$. Since X is dense in $\hat{X}_1$, there exists a sequence $(x_n) \subset X$ which converges to x^* in $(\hat{X}_1, \hat{d}_1)$. Necessarily (x_n) is a Cauchy sequence (with respect to d) and so (x_n) is a Cauchy sequence in $(\hat{X}_2, \hat{d}_2)$ (since $\hat{d}_2$ induces the metric d on X). Since $(\hat{X}_2, \hat{d}_2)$ is complete, (x_n) converges in $(\hat{X}_2, \hat{d}_2)$, say to y^*. We claim that y^* depends only on x^* and not on the particular choice of sequence $(x_n) \subset X$ converging to x^*. This is clear since if $(x'_n) \subset X$ converges to x^* then $d(x_n, x'_n) \to 0$ as $n \to \infty$ and so $\lim_{n \to \infty} \hat{d}_2(x_n, x'_n) = 0$. Hence (x_n) and (x'_n) have the same limit in $(\hat{X}_2, \hat{d}_2)$. We define $F : \hat{X}_1 \to \hat{X}_2$ by $F(x^*) = y^*$. If $x^* \in X$, then $x^* = y^*$ and so F restricts to the identity map on X. If we reverse the construction, to define a map $G : \hat{X}_2 \to \hat{X}_1$, then it is easy to check that $G \circ F$ is the identity on $\hat{X}_1$ and $F \circ G$ is the identity on $\hat{X}_2$. Therefore F is 1:1 onto. We must check that F is an isometry. Suppose $(x_n), (z_n) \subset X$ are convergent sequences with respective limits $x^*, z^* \in \hat{X}_1$.

We have

$$\hat{d}_2(F(x^*), F(z^*)) = \lim_{n\to\infty} \hat{d}_2(F(x_n), F(z_n))$$

$$= \lim_{n\to\infty} d(x_n, y_n)$$

$$= \lim_{n\to\infty} \hat{d}_1(x_n, y_n)$$

$$= \hat{d}_1(x^*, z^*).$$

Finally, we must show that F is unique. Suppose that $F, F' : \hat{X}_1 \to \hat{X}_2$ satisfy the conditions of the proposition. Then $F = F'$ on X. Since X is a dense subset of $\hat{X}_1$ and F, F' are continuous, $F = F'$ by Proposition 7.12.15. $\square$

Corollary 7.15.13 *If (X, d) is a complete metric space and $(\hat{X}, \hat{d})$ is a completion of X, then $\hat{X} = X$, $\hat{d} = d$.*

Theorem 7.15.14 *Let (X, d) be a metric space. Then (X, d) has a completion $(\hat{X}, \hat{d})$.*

Proof The metric space $(B^0(X, \mathbb{R}), \rho)$ is complete (Theorem 7.15.7; as usual ρ denotes the uniform metric). We construct a completion of X by defining an isometry Θ of X onto a subspace Z of $B^0(X, \mathbb{R})$. Identifying X with Z, we define the completion of X to be $(\overline{Z}, \rho)$. In order to define Θ, fix $a \in X$. For $x \in X$, define $\Theta(x)$ to be the map $f_x : X \to \mathbb{R}$ where

$$f_x(y) = d(x, y) - d(y, a), \ y \in X.$$

Lemma 7.1.4 implies that $|f_x(y)| = |d(x, y) - d(y, a)| \leq d(x, a)$ and so f_x is bounded. Since f_x is continuous, $f_x \in B^0(X, \mathbb{R})$ and so $\Theta : X \to B^0(X, \mathbb{R})$. It remains to prove that Θ is an isometry onto $\Theta(X) \subset (B^0(X, \mathbb{R}), \rho)$. For $x, x' \in X$, we have

$$\rho(f_x, f_{x'}) = \sup_{y\in X} |f_x(y) - f_{x'}(y)|$$

$$= \sup_{y\in X} |d(x, y) - d(y, a) - (d(x', y) - d(y, a))|$$

$$= \sup_{y\in X} |d(x, y) - d(x', y)|$$

$$= d(x, x'),$$

where the last equality follows taking $y = x'$ and using Lemma 7.1.4. $\square$

Remark 7.15.15 Theorem 7.15.14 highlights the pivotal role of the completeness of the real numbers—the completion of $\mathbb{Q}$.   ✠

7.15.3 Category

Recall that a subset E of a metric space X is dense if $\overline{E} = X$. At the opposite extreme we can formalize the idea of a subset which is 'nowhere dense'.

Definition 7.15.16 A subset E of the metric space X is *nowhere dense* if $\overset{\circ}{\overline{E}} = \emptyset$.

Examples 7.15.17

(1) Suppose that X contains no isolated points. Any finite subset of X is nowhere dense.
(2) If (x_n) is a convergent sequence in $\mathbb{R}$, then $E = \{x_n \mid n \in \mathbb{N}\}$ is nowhere dense.
(3) The middle thirds Cantor set $\mathbf{C}$ is nowhere dense (even though $\mathbf{C}$ is compact and uncountable). ♠

Definition 7.15.18 A subset E of the metric space X is of the *first category* if E can be written as a countable union of nowhere dense sets; if E is not of the first category, it is of the *second category*.

Lemma 7.15.19 *Let E be a subset of the metric space X which is of the first category. Then if $U \subset X$ is open and non-empty, then there exists a non-empty open subset V of U such that $V \cap E = \emptyset$.*

Proof Suppose the contrary. Then given $x \in U$, every open neighbourhood of x meets E and so $x \in \overline{E}$. Since this holds for all $x \in U$, $U \subset \overline{E}$ and so $\overset{\circ}{\overline{E}} \neq \emptyset$, contradicting the assumption that E is nowhere dense. $\square$

Theorem 7.15.20 (Baire Category Theorem) *A complete metric space is of the second category.*

Proof Suppose the contrary and that the metric space X can be written as $X = \cup_{n=1}^{\infty} A_n$, where the A_n are nowhere dense subsets of X. Apply Lemma 7.15.19 to choose a closed disk $\overline{D}_{r_1}(x_1)$ such that $\overline{D}_{r_1}(x_1) \cap A_1 = \emptyset$. Apply Lemma 7.15.19 again with $U = D_{r_1}(x_1)$ and $E = A_2$, to find a closed disk $\overline{D}_{r_2}(x_2) \subset D_1$ with $r_2 \leq 2^{-1} r_1$. Proceeding inductively, we obtain a decreasing sequence $D_n = \overline{D}_{r_n}(x_n)$ of closed disks such that $D_n \cap A_n = \emptyset$ and $r_n \leq 2^{-n} r_1$. We claim that $\cap_{n \geq 1} D_n \neq \emptyset$. This follows since (x_n) is a Cauchy sequence in X (as $r_n \to 0$) and so, by the completeness of X, $\lim_{n \to x_n} x_n = x^{\star}$ exists. Since each D_n is closed and $D_n \supset \{x_m \mid m \geq n\}$, $x^{\star} \in D_n$, for all $n \geq 1$ and so $x^{\star} \in \cap_{n \geq 1} D_n$. But by construction $x^{\star} \notin A_n$, for all n, contradicting the assumption that $X = \cup_{n=1}^{\infty} A_n$. $\square$

Corollary 7.15.21 *Let Z be a complete metric space. If V_n is an open and dense subset of Z, $n \in \mathbb{N}$, then $\cap_{n \geq 1} V_n \neq \emptyset$.*

Proof If V_n is an open and dense subset of Z, then the closed set $F_n = Z \smallsetminus V_n$ is nowhere dense in Z. Now $Z \smallsetminus \cap_{n \geq 1} V_n = \cup_{n \geq 1} F_n$ and since Z is second category, $\cup_{n \geq 1} F_n \neq Z$. Hence $\cap_{n \geq 1} V_n \neq \emptyset$. $\square$

Corollary 7.15.22 *Let X be a complete metric space. Suppose that for $n \geq 1$, U_n is an open and dense subset of X. Then $\cap_{n\geq 1} U_n$ is a dense subset of X.*

Proof It suffices to show that

$$(\cap_{n\geq 1} U_n) \cap \overline{D}_r(x) = \cap_{n\geq 1} (U_n \cap \overline{D}_r(x)) \neq \emptyset$$

for all $x \in X$, $r > 0$. Take $Z = \overline{D}_r(x)$ and $V_n = U_n \cap \overline{D}_r(x)$. Since $Z = \overline{D}_r(x)$ is a closed subset of a complete metric space, Z is complete in the induced metric and so Corollary 7.15.21 applies. $\qquad\square$

EXERCISES 7.15.23

(1) Prove parts (1) and (2) of Lemma 7.15.2.
(2) Show that $(B^0(X, \mathbb{Q}), \rho)$ is not complete (ρ as usual is the uniform metric. What is the completion of $(B^0(X, \mathbb{Q}), \rho)$)? (Hint for the last part: does $B^0(X, \mathbb{Q}) = C^0(X, \mathbb{Q})$?)
(3) Let $C^1([a, b])$ denote the space of C^1 functions on $[a, b]$. Show that $(C^1([a, b]), \rho)$ is not complete (ρ denotes the uniform metric). Show that if we define $\rho_1(f, g) = \rho(f, g) + \rho(f', g')$, then $(C^1([a, b]), \rho_1)$ is complete.
(4) For $x, y \in \mathbb{R}$, define $\kappa(x, y) = \left| \frac{e^x}{1+e^x} - \frac{e^y}{1+e^y} \right|$. Verify that κ is a metric on $\mathbb{R}$ that defines the same open sets as the standard metric on $\mathbb{R}$. Show that $(\mathbb{R}, \kappa)$ is not complete. What is the completion of $(\mathbb{R}, \kappa)$? (Specifically, what is $\widehat{\mathbb{R}} \smallsetminus \mathbb{R}$?)
(5) Let (X, d), $(Y, \bar{d})$ be metric spaces. Define the uniform metric ρ on $B^0(X, Y)$ by $\rho(f, g) = \sup_{x\in X} \bar{d}(f(x), g(x))$ (see Exercises 7.13.24(13) when X is compact). Show that $B^0(X, Y)$ is complete iff Y is complete. Prove a similar result for $C^0(X, Y)$ in case X is compact.
(6) Show that the metric space (X, d) is complete if d is the discrete metric.
(7) Let Y be a subspace of the metric space (X, d). Show that if Y is complete in the induced metric then Y is a closed subset of X. If Y is a closed subset of (X, d), need Y be complete in the induced metric?
(8) Let (X, d), $(Y, \bar{d})$ be metric spaces and $F : X \to Y$ be an isometry. Show that (X, d) is complete iff $(Y, \bar{d})$ is complete.
(9) Let $d, \bar{d}$ be equivalent metrics on X (see Exercises 7.1.9(11) for the definition of equivalent metric). Show that (X, d) is complete iff $(X, \bar{d})$ is complete.
(10) Let (X, d) be a metric space. Construct a completion $(\hat{X}, \hat{d})$ along the lines of section "Appendix: Construction of $\mathbb{R}$ Revisited". That is, let C denote the set of all Cauchy sequences of points of X and partition C by the equivalence relation $(x_n) \sim (y_n)$ iff $\lim_{n\to\infty} d(x_n, y_n) = 0$. Let $\hat{X}$ denote the set of equivalence classes. Show that there is a natural way to define a metric $\hat{d}$ on $\hat{X}$ so that $(\hat{X}, \hat{d})$ is a completion of (X, d).
(11) Show that a nowhere dense set has no isolated points.
(12) Show that it is not possible to find a countable subset $\{a_n \mid n \in \mathbb{N}\}$ of $\mathbb{R}$ such that $\cup_{n\geq 1} a_n + \mathbf{C} \supset [0, 1]$. Here $a_n + \mathbf{C} = \{a_n + x \mid x \in \mathbf{C}\}$ is $\mathbf{C}$ translated by a_n. (See also Exercises 8.3.2(8).)

(13) Let X be a metric space. Suppose that $F_1 \supset F_2 \supset \cdots$ is a decreasing
sequence of non-empty closed subsets of X such that $\lim_{n \to \infty} D(F_n) = 0$,
where $D(F_n)$ denotes the diameter of F_n. Show that if X is complete then
$\cap_{n \geq 1} F_n$ is nonempty and consists of precisely one point. Show, by means of
examples, that if we omit either of the conditions (a) F_n closed, or (b) (X, d)
complete, then the intersection may be empty. (Hint for the first part: see the
proof of Theorem 7.15.20.)

7.16 Equicontinuity and the Arzelà–Ascoli Theorem

Let (X, d) be a compact[1] metric space and let $(C^0(X), \rho)$ denote the space of
continuous real-valued functions on X with the uniform metric

$$\rho(f, g) = \sup_{x \in X} |f(x) - g(x)|, \ f, g \in C^0(X).$$

As we showed in the previous section, $(C^0(X), \rho)$ is a complete metric space. In
this section we give a characterization of the compact subsets of $C^0(X)$. Although
we restrict to real-valued continuous maps on X, what we say generalizes easily to
the metric space of continuous maps $f : X \to Y$, where Y is a *complete* metric
space (see Exercises 7.13.24(13) for the definition and properties of the uniform
metric on $C^0(X, Y)$ and note that if Y is complete so is $C^0(X, Y)$). The methods
we use are a synthesis of many of the ideas and results we have developed for the
study and description of metric spaces. We conclude the section with an application
of our main result (the Arzelà–Ascoli theorem) to the existence theory of ordinary
differential equations.

   We start by recalling Theorem 7.13.6, the generalization of the Bolzano–
Weierstrass theorem to $\mathbb{R}^n$: a subset Z of $\mathbb{R}^n$ is compact iff Z is closed and bounded
(relative to the Euclidean norm on $\mathbb{R}^n$). Our aim in this section is to obtain an
analogous characterization of compact subsets of $C^0(X)$. First, however, we remark
that it is easy to see that a closed and bounded subset of $C^0(X)$ need not be
sequentially compact.

Example 7.16.1 Let $X = [0, 1] \subset \mathbb{R}$ (induced metric) and define $E = \overline{D}_1(0)$.
Certainly E is a closed and bounded subset of $C^0(X)$. If we define $f_n(x) = x^n$,
$n \in \mathbb{N}$, then $(f_n) \subset E$. Since the pointwise limit of f_n is discontinuous, there are no
convergent subsequences of (f_n) (in the uniform metric). ♠

Definition 7.16.2 Let E be a subset of $C^0(X)$.

(1) E is *pointwise bounded* if for every $x \in X$, there exists an $M_x \geq 0$ such that

$$|f(x)| \leq M_x, \ \text{for all} f \in E.$$

---

[1]Throughout this section, 'compact' is to be understood as sequentially compact.

(2) E is *uniformly bounded* if there exists an $R \geq 0$ such that

$$f \in \overline{D}_R(0), \quad \text{for all } f \in E.$$

Obviously if E is uniformly bounded then E is pointwise bounded. The converse is easily seen to be false.

Example 7.16.3 Take $X = [0, 1]$ and for $n \in \mathbb{N}$ define

$$f_n(x) = \begin{cases} 0, & 0 \leq x \leq \frac{1}{n+1}, \\ 2n^2(n+1)(x - \frac{1}{n+1}), & \frac{1}{n+1} \leq x \leq \frac{1}{2}(\frac{1}{n} + \frac{1}{n+1}), \\ 2n - 2n^2(n+1)(x - \frac{1}{n+1}), & \frac{1}{2}(\frac{1}{n} + \frac{1}{n+1}) \leq x \leq \frac{1}{n}, \\ 0, & \frac{1}{n} \leq x \leq 1. \end{cases}$$

The function f_n takes its maximum value n at the midpoint of $[\frac{1}{n+1}, \frac{1}{n}]$ and is zero on the complement of $[\frac{1}{n+1}, \frac{1}{n}]$. Hence $(f_n) \subset C^0(X)$ is pointwise bounded but not uniformly bounded.   ♠

Next we introduce a definition that plays a crucial role in our description of compact subsets of $C^0(X)$.

Definition 7.16.4 A subset E of $C^0(X)$ is *equicontinuous* (on X) if for every $\varepsilon > 0$, there exists a $\delta > 0$ such that if $d(x, y) < \delta$ then

$$|f(x) - f(y)| < \varepsilon, \quad \text{for all } f \in E.$$

Examples 7.16.5

(1) If E consists of a single function f, equicontinuity is automatic by the uniform continuity of f (Theorem 7.13.8(2)). Consequently, any finite subset of $C^0(X)$ is automatically equicontinuous. Viewed in this way, equicontinuity is the natural generalization of uniform continuity to an infinite family of continuous functions.

(2) An equicontinuous set E need not be pointwise bounded. For example, choose $f \in C^0(X)$ and let $E = \{f + n \mid n \in \mathbb{N}\}$.

(3) Let $\alpha > 0$. A function $f : X \to \mathbb{R}$ is *Hölder* continuous with exponent α if there exists a $K \geq 0$ such that $|f(x) - f(y)| \leq Kd(x, y)^\alpha$ for all $x, y \in X$. If $\alpha = 1$, f is *Lipschitz*. Suppose that E is a subset of $C^0(X)$ consisting of Hölder continuous functions all with the same exponent α and same bound $|f(x) - f(y)| \leq Kd(x, y)^\alpha$ (that is, $K > 0$ independent of $f \in E$). Then E is equicontinuous. Indeed, given $\varepsilon > 0$, take $\delta = (\varepsilon/K)^{\frac{1}{\alpha}}$. Since C^1 functions are Lipschitz on compact subsets of $\mathbb{R}$ (and $\mathbb{R}^n$) by the mean value theorem, we see that adding a little regularity to our functions can result in big equicontinuous sets.   ♠

Lemma 7.16.6 *Let E be a subset of $C^0(X)$.*

(1) *If E is uniformly bounded, then $\overline{E}$ is uniformly bounded.*
(2) *If E is equicontinuous, then $\overline{E}$ is equicontinuous.*

Proof Left to the exercises.   □

We now state the main theorem of this section.

Theorem 7.16.7 (Arzelà–Ascoli) *A subset E of $(C^0(X), \rho)$ is compact iff*

(1) *E is pointwise bounded,*
(2) *E is equicontinuous,*
(3) *E is closed.*

We start by proving a number of preliminary results.

Lemma 7.16.8 *Let E be an equicontinuous subset of $C^0(X)$. Then E is pointwise bounded iff E is uniformly bounded.*

Proof Suppose that E is pointwise bounded (the converse is trivial). Since E is equicontinuous, given $\varepsilon = 1$, we can choose $\delta > 0$ such that for all $x, y \in X$ with $d(x, y) < \delta$ we have

$$|f(x) - f(y)| < 1. \tag{7.6}$$

Since X is compact, we can choose a finite subset $\mathcal{P}$ of X such that for every $x \in X$, there exists a $p \in \mathcal{P}$ such that $d(x, p) < \delta$. Since E is pointwise bounded, for each $p \in \mathcal{P}$ there exists an $M_p \geq 0$ such that $|f(p)| \leq M_p$ for all $f \in E$. Set $M = \max_{p \in \mathcal{P}} M_p$. Given $x \in X$, choose $p \in \mathcal{P}$ such that $d(x, p) < \delta$. By (7.6), $|f(x)| \leq 1 + |f(p)| \leq 1 + M$. Hence E is uniformly bounded.   □

Lemma 7.16.9 *If (f_n) is a uniformly convergent sequence in $C^0(X)$, then $\{f_n \mid n \in \mathbb{N}\}$ is equicontinuous on X.*

Proof Let $\varepsilon > 0$. We must find $\delta > 0$ so that $|f_n(x) - f_n(y)| < \varepsilon$ whenever $d(x, y) < \delta$. Since (f_n) is uniformly convergent, (f_n) is a Cauchy sequence in $C^0(X)$ and so there exists an $N \in \mathbb{N}$ such that

$$|f_n(x) - f_N(x)| < \varepsilon/3, \ n \geq N, x \in X. \tag{7.7}$$

Since continuous functions on a compact metric space are uniformly continuous, there exists a $\delta > 0$ such that for $1 \leq i \leq N$ we have

$$|f_i(x) - f_i(y)| < \varepsilon/3 < \varepsilon, \ \forall x, y \in X \text{ satisfying } d(x, y) < \delta. \tag{7.8}$$

If $n > N$ and $d(x, y) < \delta$, we have by (7.7) and (7.8) with $i = N$,

$$|f_n(x) - f_n(y)| \leq |f_n(x) - f_N(x)| + |f_N(x) - f_N(y)| + |f_N(y) - f_n(y)| < \varepsilon.$$

Together with (7.8), this completes the proof that $\{f_n \mid n \in \mathbb{N}\}$ is equicontinuous on X.   □

Lemma 7.16.10 *Let Q be a dense subset of X and $(f_n) \subset C^0(X)$ be a sequence satisfying*

(1) $\{f_n \mid n \in \mathbb{N}\}$ *is equicontinuous on X.*
(2) $(f_n(q))$ *is convergent for all $q \in Q$.*

Then (f_n) is uniformly convergent.

Proof It suffices to prove that (f_n) is a Cauchy sequence in $C^0(X)$. That is, given $\varepsilon > 0$, we claim there exists an $N \in \mathbb{N}$ such that $\rho(f_m, f_n) < \varepsilon$ for all $m, n \geq N$. Since (f_n) is equicontinuous on X, we may choose $\delta > 0$ so that

$$|f_n(x) - f_n(y)| < \varepsilon/3, \text{ for all } x, y \in X \text{ such that } d(x, y) < \delta. \tag{7.9}$$

Since Q is a dense subset of the compact space X, we can choose a finite subset $\mathcal{Q}$ of Q satisfying $d(x, \mathcal{Q}) < \delta$ for all $x \in X$. Since $(f_n(q))$ is convergent for all $q \in \mathcal{Q}$, we may choose $N \in \mathbb{N}$ so that for all $m, n \geq N$ we have

$$|f_m(q) - f_n(q)| < \varepsilon/3, \text{ for all } q \in \mathcal{Q}. \tag{7.10}$$

Given $x \in X$, choose $q \in \mathcal{Q}$ so that $d(x, q) < \delta$. We have

$$|f_m(x) - f_n(x)| \leq |f_m(x) - f_m(q)| + |f_m(q) - f_n(q)| + |f_n(q) - f_n(x)|,$$
$$< \frac{\varepsilon}{3} + \frac{\varepsilon}{3} + \frac{\varepsilon}{3} = \varepsilon, \text{ if } n, m \geq N,$$

where the second inequality follows from (7.9), (7.10). $\square$

The final result we need before we prove the Arzelà–Ascoli theorem shows that we can always construct a subsequence satisfying the second condition of Lemma 7.16.10. This result uses neither continuity nor a metric.

Lemma 7.16.11 *Let (f_n) be a pointwise bounded sequence of real-valued functions defined on a countable set Q. Then there exists a subsequence (f_{n_k}) of (f_n) which is pointwise convergent on Q.*

Proof Since Q is countable, we may write $Q = \{q_1, q_2, \cdots\}$. For $k \geq 1$, we construct subsequences (f_n^k) of (f_n) satisfying

(1) (f_n^ℓ) is a subsequence of (f_n^k) if $\ell > k$.
(2) $(f^k(q_i))$ is convergent for $1 \leq i \leq k$.

The construction is inductive. We start by constructing (f_n^1). Since $(f_n(q_1)) \subset \mathbb{R}$ is bounded, there exists a subsequence (f_n^1) of (f_n) such that $(f_n^1(q_1)$ is convergent. Suppose we have constructed (f_n^j) satisfying (1,2) above for $1 \leq j < k$. In order to construct (f_n^k) we repeat the construction we gave for (f_n^1) but with (f_n^{k-1}) replacing (f_n) and (q_k) replacing q_1. Finally, we construct the required subsequence (f_{n_k}) of (f_n) by taking $f_{n_k} = f_k^k$, $k \in \mathbb{N}$. With the exception of at most $k - 1$ terms, (f_k^k) is a subsequence of (f_n^k) and so for all $i \in \mathbb{N}$, $(f_k^k(q_i))$ is convergent. $\square$

Proof of Theorem 7.16.7 Suppose that $E \subset C^0(X)$ satisfies conditions (1,2,3) of Theorem 7.16.7. We must show that every sequence $(f_n) \subset E$ has a subsequence converging to a point of E. Since X is compact, X is separable (Theorem 7.13.21) and so we may pick a countable dense subset Q of X. If $(f_n) \subset E$, it follows by Lemma 7.16.11 that there exists a subsequence (f_{n_k}) of (f_n) which is pointwise convergent on Q. Applying Lemma 7.16.10, (f_{n_k}) is uniformly convergent on X. Since E is closed, $\lim_{k \to \infty} f_{n_k} \in E$ and so E is compact. For the converse, suppose that E is a compact subset of $C^0(X)$. Necessarily, E is closed and bounded (Proposition 7.13.4). Noting Lemma 7.16.8, it suffices to prove that E is equicontinuous. Suppose the contrary. Then there exists an $\varepsilon > 0$ such that for every $n \in \mathbb{N}$, there exist $f_n \in E$, $x_n, y_n \in X$, satisfying

$$|f_n(x_n) - f_n(y_n)| \ge \varepsilon, \quad d(x_n, y_n) < 1/n.$$

Since X is compact, there exist convergent subsequences (x_{n_k}), (y_{n_k}) of (x_n), (y_n) with common limit $x^\star$. Since we assume E is compact, there exists a convergent subsequence (f_{m_k}) of (f_{n_k}). Necessarily we have $\lim_{k \to \infty} f_{m_k}(x_{m_k}) = \lim_{k \to \infty} f_{m_k}(y_{m_k})$, contradicting the estimate $|f_{m_k}(x_{m_k}) - f_{m_k}(y_{m_k})| \ge \varepsilon$. Hence E is equicontinuous. $\square$

Corollary 7.16.12 *Let* $(f_n) \subset C^0(X)$. *Then* (f_n) *has a (uniformly) convergent subsequence if* $\{f_n \mid n \in \mathbb{N}\}$ *is pointwise bounded and equicontinuous.*

Proof Suppose that $E = \{f_n \mid n \in \mathbb{N}\}$ is pointwise bounded and equicontinuous. By Lemma 7.16.6, $\overline{E}$ is pointwise bounded and equicontinuous and so $\overline{E}$ is compact and (f_n) has a convergent subsequence by Theorem 7.16.7. $\square$

7.16.1 An Application to Differential Equations

Let $f : \mathbb{R} \to \mathbb{R}$ be continuous and $x_0 \in \mathbb{R}$. If f is bounded (so $f \in B^0(\mathbb{R})$) we show that the ordinary differential equation

$$\frac{dx}{dt} = f(x)$$

has a C^1 solution $\phi : [0, 1] \to \mathbb{R}$ satisfying the initial condition $\phi(0) = x_0$. That is,

$$\phi'(t) = f(\phi(t)), \ t \in [0, 1], \text{ and } \phi(0) = x_0.$$

Remarks 7.16.13

(1) We need the boundedness condition on f, else the solution may escape to infinity in finite time (see the exercises for a simple example). Of course, we can always multiply f by a bump function to obtain a bounded function which is equal to f on some preassigned interval $[-R, R]$. Thus, our result gives solutions defined on some interval $[0, \delta]$ even if f is not bounded.

(2) The choice of the interval $[0, 1]$ is only for convenience. With minor modifica-
    tions, the proof works for any closed interval $[a, b]$ containing the origin.
(3) The solution we obtain may not be unique (see the exercises). In the next section
    we show how, if we assume more regularity on f, we obtain uniqueness of
    solutions.
(4) The existence of solutions to the non-autonomous ordinary differential equation
    $x' = f(t)$ ($f : \mathbb{R} \to \mathbb{R}$ continuous) is immediate from the fundamental theorem
    of calculus: $\phi(t) = x_0 + \int_0^t f(s) \, ds$. Solutions are unique and exist for all $t \in \mathbb{R}$.
(5) In the exercises we indicate the extension of the existence result to non-
    autonomous ordinary differential equations on $\mathbb{R}^n$, $n \geq 1$.   ✱

We give a proof of the existence of solutions to $\frac{dx}{dt} = f(x)$ based on the *Euler
method* and equicontinuity. The basic idea to define a sequence of continuous
piecewise linear approximations to a solution, prove equicontinuity of the sequence
and then apply Corollary 7.16.12 to get a sequence converging to a solution of
the differential equation. (In the next section we use the easier Picard method to
construct a sequence which converges to a solution. However, for this method to
work we need greater regularity of the function $f(x)$.)

Let $n \in \mathbb{N}$. For $0 \leq i \leq n$, let $t_i = i/n$. Define the continuous piecewise linear
map $\phi_n : [0, 1] \to \mathbb{R}$ by

$$\phi_n(0) = x_0,$$

$$\phi_n'(t) = f(\phi_n(t_i)) \text{ if } t \in (t_i, t_{i+1}),$$

$$\phi_n'(t+) = f(\phi_n(t_i)) \text{ if } t = t_i,$$

$$\phi_n'(t-) = f(\phi_n(t_i)) \text{ if } t = t_{i+1}.$$

We remark that ϕ_n is piecewise C^1 with jumps in the derivative at $t = t_i$, $0 < i < n$.
Define

$$\delta_n(t) = \begin{cases} \phi_n'(t) - f(\phi_n(t)), & t \notin \{t_1, \cdots, t_{n-1}\}, \\ 0, & t \in \{t_1, \cdots, t_n\}. \end{cases}$$

Note that $\delta_n(0) = 0$. We have

$$\phi_n(t) = x_0 + \int_0^t (f(\phi_n(s)) + \delta_n(s)) \, ds, \ t \in [0, 1]. \tag{7.11}$$

Since $f : \mathbb{R} \to \mathbb{R}$ is bounded, we may choose $M \geq 0$ such that $|f(x)| \leq M$ for
all $x \in \mathbb{R}$. Let $\| \cdot \|$ denote the uniform norm on $C^0([0, 1])$ ($\|f\| = \rho(f, 0)$). We have
the following estimates

(a) $|\phi_n(t) - \phi_n(t')| \leq M|t - t'|$, $t, t' \in [0, 1]$ (by (7.11) and $|\phi_n'(t\pm)| \leq M$ for all
    $t \in [0, 1]$).
(b) $\|\delta_n\| \leq 2M$ (by (a) and the definition of δ_n).
(c) $\|\phi_n\| \leq |x_0| + M$ (by (7.11) and $|\phi_n'(t\pm)| \leq M$).

Statement (a) implies that (ϕ_n) is equicontinuous on $[0, 1]$ (see Examples 7.16.5(3)). Statement (c) implies (ϕ_n) is uniformly bounded. Applying Corollary 7.16.12, there is a uniformly convergent subsequence (ϕ_{n_k}) of (ϕ_n). The map f is uniformly continuous on $[-M-|x_0|, M+|x_0|]$ and so $(f \circ \phi_{n_k})$ converges uniformly (exercise) to $f \circ \phi$ on $[0, 1]$. Moreover, (δ_n) converges uniformly to 0 on $[0, 1]$ using the definition of δ_n and estimate (a). Now let $k \to \infty$ in

$$\phi_{n_k}(t) = x_0 + \int_0^t \left(f(\phi_{n_k}(s)) + \delta_{n_k}(s) \right) ds$$

to obtain $\phi(t) = x_0 + \int_0^t f(\phi(s)) \, ds$, $t \in [0, 1]$. By the fundamental theorem of calculus, $\phi(t)$ is a solution of $x' = f(x)$ with initial condition $\phi(0) = x_0$.

EXERCISES 7.16.14

(1) Prove Lemma 7.16.6.

(2) Suppose that $(f_n) \subset C^0(X)$ is equicontinuous and pointwise convergent. Prove that (f_n) is uniformly convergent.

(3) Let $f : \mathbb{R} \to \mathbb{R}$ be continuous and define $f_n(t) = f(nx)$, $t \in \mathbb{R}$. Show that if (f_n) is equicontinuous on $\mathbb{R}$, then f is constant.

(4) Let $\alpha \in (0, 1)$. Show that $x \mapsto x^\alpha$ is Hölder continuous, exponent α, on $[0, a]$ for all $a > 0$. (Hint: prove that $0 \le y^\alpha - x^\alpha \le (y - x)^\alpha$ for all $y \ge x \ge 0$.) Suppose that $f : [0, a] \to \mathbb{R}$ is Hölder continuous with exponent $\alpha > 1$. Show that f is constant.

(5) Let (X, d) be a compact metric space and $(Y, \bar{d})$ be a complete metric space. Let ρ denote the uniform metric on $C^0(X, Y)$. A subset E of $C^0(X, Y)$ is *pointwise bounded* if for every $x \in X$, there exist $R = R(x) > 0$, $y = y(x) \in Y$ such that $\{f(x) \mid f \in E\} \subset D_R(y)$ and E is *uniformly bounded* if we can choose $R > 0$, $y \in Y$ such that $\{f(x) \mid f \in E\} \subset D_R(y)$ for all $x \in X$. Verify that with these definitions, Lemmas 7.16.6, 7.16.8 and 7.16.9 all extend to $C^0(X, Y)$. Hence generalize the Arzelà–Ascoli theorem and Corollary 7.16.12 to $(C^0(X, Y), \rho)$.

(6) Find the solution $\phi(t)$ to $x' = x^2$ which has initial condition $\phi(0) = x_0 > 0$. Hence show that if $x_0 > 1$, the solution escapes to $+\infty$ in time $t_c = x_0^{-1} < 1$.

(7) Show that the differential equation $x' = x^{1/3}$ does not have unique solutions $\phi(t)$ with $\phi(0) = 0$. (Hint: one solution is $\phi(t) \equiv 0$; find others.)

(8) Generalize the existence theorem for ordinary differential equations to $\mathbf{x}' = f(\mathbf{x})$, $\mathbf{x} \in \mathbb{R}^m$, where $f : \mathbb{R}^m \to \mathbb{R}^m$ is continuous and bounded. Extend to non-autonomous equations $\mathbf{x}' = f(\mathbf{x}, t)$ by defining the new variable x_{m+1} satisfying $x'_{m+1} = 1$, $x_{m+1}(0) = 0$.

7.17 The Contraction Mapping Lemma

In this section we prove one of the most interesting results about self maps of
a complete metric space: the *contraction mapping lemma*. We start with some
definitions and preliminary results and, after giving a proof of the contraction
mapping lemma, describe two applications.

Definition 7.17.1 A map $f : X \rightarrow X$ of the metric space (X, d) is called a
contraction mapping or *contraction* if there exists a k, $0 \leq k < 1$, such that

$$d(f(x), f(x')) \leq kd(x, x'), \text{ for all } x, x' \in X.$$

We call k a *contraction constant* for f. The infimum of the set of all contraction
constants for f is called *the* contraction constant of f.

Example 7.17.2 Let $f : \mathbb{R} \rightarrow \mathbb{R}$ be a C^1 map and assume that $\sup_{x \in \mathbb{R}} |f'(x)| = k <$
1. Then f is a contraction mapping with contraction constant k. Indeed, by the mean
value theorem, for all $x < y \in \mathbb{R}$, $|f(x) - f(y)| = |f'(z)||x - y|$ for some $z \in [x, y]$
and so $|f(x) - f(y)| \leq k|x - y|$. Note that k is the contraction constant of f.   ♠

Lemma 7.17.3 *A contraction mapping is uniformly continuous.*

Proof Let $f : X \rightarrow X$ be a contracting mapping with contraction constant $k > 0$. For
all $\varepsilon > 0$, we have $d(f(x), f(x')) \leq \varepsilon$ if $d(x, x') \leq \delta = \varepsilon/k$. Hence f is uniformly
continuous.   □

Remark 7.17.4 Lemma 7.17.3 holds if $d(f(x), f(x') \leq kd(x, x')$ where $\infty > k \geq 0$.
Maps which satisfy this condition are called *Lipschitz* and the smallest value of k
for which the inequality holds is called the Lipschitz constant of f. In particular, a
Lipschitz map is a contraction iff it has Lipschitz constant strictly less than 1.   ✴

Definition 7.17.5 Let $f : X \rightarrow X$. A point $x^* \in X$ is a *fixed point* of f if

$$f(x^*) = x^*.$$

Theorem 7.17.6 (Contraction Mapping Lemma) *If $f : X \rightarrow X$ is a contraction
mapping of the complete metric space (X, d), then*

(1) *the map f has a unique fixed point x^*,*
(2) *given any point $x_0 \in X$, if we define the sequence (x_{n+1}) by $x_{n+1} = f(x_n)$, $n \geq 0$,
    then $\lim_{n \to \infty} x_n = x^*$.*

Proof Suppose f has contraction constant k. We start by showing that if f has a fixed
point, then the fixed point is unique. Suppose then that x^*, y^* are fixed points of f.
Since f is a contraction, and $f(x^*) = x^*, f(y^*) = y^*$, we have

$$d(x^*, y^*) = d(f(x^*), f(y^*)) \leq kd(x^*, y^*).$$

Since $k < 1$, the only way we can satisfy $d(x^\star, y^\star) \le kd(x^\star, y^\star)$ is if $d(x^\star, y^\star) = 0$. That is, $x^\star = y^\star$.

In order to prove the existence of a fixed point, it suffices to prove (2). Fix $x_0 \in X$ and define (x_n) by $x_{n+1} = f(x_n)$, $n \ge 0$. We prove that (x_n) is a Cauchy sequence. Let $n > m$. Then

$$d(x_m, x_n) \le d(x_m, x_{m+1}) + d(x_{m+1}, x_{m+2}) + \cdots + d(x_{n-1}, x_n). \qquad (7.12)$$

Now given $r \in \mathbb{N}$, we have

$$
\begin{aligned}
d(x_r, x_{r+1}) &= d(f(x_{r-1}), f(x_r)) \\
&\le kd(x_{r-1}, x_r) = kd(f(x_{r-2}), f(x_{r-1})) \\
&\le \cdots\cdots \\
&\le k^r d(x_0, x_1).
\end{aligned}
$$

Substituting this estimate in (7.12), we get

$$
\begin{aligned}
d(x_m, x_n) &\le (k^m + k^{m+1} + \cdots + k^{n-1})d(x_0, x_1), \\
&= k^m(1 + \cdots + k^{n-m-1})d(x_0, x_1), \\
&\le k^m \left(\sum_{j=0}^{\infty} k^j \right) d(x_0, x_1), \\
&= \frac{k^m}{1-k} d(x_0, x_1).
\end{aligned}
$$

Therefore, $d(x_m, x_n) \le \frac{k^m}{1-k} d(x_0, x_1)$, $n > m$, and so, since $k < 1$, $\lim_{n,m \to \infty} d(x_m, x_n) = 0$, proving that (x_n) is a Cauchy sequence. Since (X, d) is complete, (x_n) converges. Denote the limit by $x^\star$. We have $\lim_{n \to \infty} d(x_{n+1}, x_n) = d(x^\star, x^\star) = 0$. But $d(x_{n+1}, x_n) = d(f(x_n), x_n)$ and so, since f is (sequentially) continuous (Lemma 7.17.3), we have $0 = \lim_{n \to \infty} d(f(x_n), x_n) = d(f(x^\star), x^\star)$, proving that $f(x^\star) = x^\star$. □

Remark 7.17.7 An attractive feature of Theorem 7.17.6 is that the result is constructive: it gives a simple way of finding the fixed point. Take any initial point, $x_0 \in X$, and iterate by the map f. The resulting sequence is Cauchy (even if the space is not complete) and so the iteration works well on a computer. More formally, if X is not complete, let $\hat{X}$ be the completion of X. The sequence $(x_n) \subset X$ will converge to a point $x^\star \in \hat{X}$. Even though the sequence may not converge in X, the terms will get arbitrarily close to the point $x^\star$ and so the sequence will appear to converge on a computer where one works with finite precision. This situation should be contrasted with the elementary result that every continuous map $f : [0, 1] \to [0, 1]$ has a fixed point. The proof, based on the intermediate value theorem, gives little help in finding a fixed point. Iteration generally will not work. As a simple example, take the map $f : [0, 1] \to [0, 1]$ defined by $f(x) = 4x(1 - x)$, $x \in [0, 1]$. This map has two fixed points, 0 and 3/4, which are easily found by solving $4x(1 - x) = x$. However, if we try to find the fixed points by taking a general point $x_0 \in [0, 1]$ and iterating, the resulting sequence will typically not converge.   ✸

Examples 7.17.8

(1) For $y \in \mathbb{R}$, define $f : \mathbb{R} \to \mathbb{R}$ by $f(x) = \frac{1}{2} \cos x + \frac{1}{4} \tan^{-1}(x)$. We show that f has a unique fixed point.

   We have $f'(x) = -\frac{1}{2} \sin x + \frac{1}{4(1+x^2)}$. Since $|f'(x)| \leq \frac{3}{4}$ for all $x \in \mathbb{R}$, $\sup_{x \in \mathbb{R}} |f'(x)| \leq \frac{3}{4}$. Since f is C^1, it follows by the mean value theorem that

$$|f(x) - f(y)| \leq \frac{3}{4}|x - y|, \text{ for all } x, y \in \mathbb{R}.$$

   Hence f is a contraction mapping with contraction constant at most $\frac{3}{4}$. Therefore, f has a unique fixed point $x^\star \in \mathbb{R}$.

(2) Let $F(x) = \frac{1}{2} \cos x + \frac{1}{4} \tan^{-1}(x) - x$, $x \in \mathbb{R}$. We claim that there is a unique solution to the equation $F(x) = 0$. We have $F(x) = 0$ if and only if $\frac{1}{2} \cos x + \frac{1}{4} \tan^{-1}(x) = x$. Now apply the previous result.   ♠

7.17.1 Contraction Mapping Lemma with Parameters

Before we give some applications of the contraction mapping lemma, we prove that the fixed point given by Theorem 7.17.6 depends 'continuously' on the map f.

Suppose then that (X, d) and $(\Lambda, \bar{d})$ are metric spaces and $F : X \times \Lambda \to X$. Given $\lambda \in \Lambda$, we define the map $f_\lambda : X \to X$ by $f_\lambda(x) = F(x, \lambda)$. We regard $f_\lambda : X \to X$ as a family of maps *parametrized* by $\lambda \in \Lambda$.

Theorem 7.17.9 (Contraction Mapping Lemma with Parameters) *Let (X, d) be a complete metric space and $(\Lambda, \bar{d})$ be a metric space. Suppose that $F : X \times \Lambda \to X$ and that $0 \leq k < 1$. Assume that*

(1) *For every $\lambda \in \Lambda$, $f_\lambda : X \to X$ is a contraction map with contraction constant $k_\lambda \leq k$.*
(2) *For every $x \in X$, the map $F(x, \cdot) : \Lambda \to X; \lambda \mapsto F(x, \lambda)$ is continuous.*

There exists a continuous map $x : \Lambda \to X$ such that $x(\lambda)$ is the unique fixed point of f_λ for all $\lambda \in \Lambda$.

Proof We know from Theorem 7.17.6 that, for each $\lambda \in \Lambda$, $f_\lambda : X \to X$ has a unique fixed point which we denote by $x(\lambda)$. This defines a map $x : \Lambda \to X$. It remains to prove that $x : \Lambda \to X$ is continuous. Fix $\lambda \in \Lambda$. We must show that given $\varepsilon > 0$, there exists a $\delta > 0$ such that $d(x(\lambda), x(\lambda')) < \varepsilon$, if $\bar{d}(\lambda, \lambda') < \delta$. For $\lambda' \in \Lambda$,

we have

$$d(x(\lambda), x(\lambda')) = d(f_\lambda(x(\lambda)), f_{\lambda'}(x(\lambda')))$$
$$\leq d(f_\lambda(x(\lambda)), f_{\lambda'}(x(\lambda))) + d(f_{\lambda'}(x(\lambda)), f_{\lambda'}(x(\lambda')))$$
$$\leq d(f_\lambda(x(\lambda)), f_{\lambda'}(x(\lambda))) + kd(x(\lambda), x(\lambda')).$$

Hence

$$(1 - k)d(x(\lambda), x(\lambda')) \leq d(f_\lambda(x(\lambda)), f_{\lambda'}(x(\lambda))). \tag{7.13}$$

Since, for fixed x, $F(x, \cdot) : \Lambda \to X; \lambda' \mapsto F(x, \lambda')$ is continuous, we may take $x = x(\lambda)$ and choose $\delta > 0$ such that if $\bar{d}(\lambda, \lambda') < \delta$, then $d(f_\lambda(x(\lambda)), f_{\lambda'}(x(\lambda))) < (1-k)\varepsilon$. Substituting in (7.13), we see that if $\bar{d}(\lambda, \lambda') < \delta$, then $d(x(\lambda), x(\lambda')) < \varepsilon$. $\quad\square$

7.17.2 An Application of the Contraction Mapping Lemma to Ordinary Differential Equations

We show how we can use the contraction mapping lemma to prove the local existence and uniqueness theorem for autonomous ordinary differential equations. For simplicity, we consider differential equations on the real line. Later, in Chap. 9, we indicate the straightforward generalization to ordinary differential equations on $\mathbb{R}^n$.

We start by considering the non-autonomous differential equation $x' = g(t)$, where $g : \mathbb{R} \to \mathbb{R}$ is continuous. A solution of this equation with initial condition x_0 will be a differentiable map $x : \mathbb{R} \to \mathbb{R}$ such that

$$x'(t) = g(t), \ t \in \mathbb{R}$$

and $x(0) = x_0$. An application of the fundamental theorem of calculus easily gives the explicit solution

$$x(t) = x_0 + \int_0^t g(s)\, ds, \ t \in \mathbb{R}.$$

We remark that if g is C^r, then $x : \mathbb{R} \to \mathbb{R}$ is C^{r+1} (since $x'(t) = g(t)$, x' is C^r).

If $x' = f(x)$, where $f : \mathbb{R} \to \mathbb{R}$ is C^1, then we cannot use the fundamental theorem of calculus to construct solutions with specified initial conditions unless f is never zero (and then we only obtain solutions implicitly by $\int_{x_0}^{x(t)} \frac{dx}{f(x)} = \int_0^t ds = t$). Moreover, even if we can construct solutions they may not be defined for all time.

Example 7.17.10 Consider the ordinary differential equation $x' = x^2$. The solution ϕ_{x_0} to $x' = x^2$ with initial condition x_0 is given by

$$\phi_{x_0}(t) = \frac{x_0}{1 - tx_0},$$

where $t \in (-\infty, x_0^{-1})$ if $x_0 > 0$, and $t \in (x_0^{-1}, \infty)$ if $x_0 < 0$. If $x_0 \neq 0$, then the solution "blows up" in finite time.   ♠

From now on suppose that $f : \mathbb{R} \to \mathbb{R}$ is C^1 (Lipschitz suffices). We consider the ordinary differential equation

$$x' = f(x).$$

A solution of this equation with *initial condition* x_0 is a differentiable map $x : [-\delta, \delta] \to \mathbb{R}$ such that

$$x'(t) = f(x(t)), \text{ for all } t \in [-\delta, \delta].$$
$$x(0) = x_0.$$

Remark 7.17.11 A solution curve to $x' = f(x)$ is also called an *integral curve* or a *trajectory* (of the differential equation).   ✖

Theorem 7.17.12 *Let $f : \mathbb{R} \to \mathbb{R}$ be C^r, $r \geq 1$. Let $I = [a, b]$ be a bounded closed interval and suppose $C = \sup_{x \in \mathbb{R}} |f'(x)| < \infty$. Choose $\delta > 0$ such that $\delta C < 1$. Then there exists a continuous map*

$$\phi : I \to C^0([-\delta, \delta], \mathbb{R}); \; x \mapsto \phi_x$$

satisfying

(1) *for all $x_0 \in I$, ϕ_{x_0} is a solution of $x' = f(x)$ with initial condition x_0,*
(2) *$\phi_x : [-\delta, \delta] \to \mathbb{R}$ is C^{r+1}, for all $x \in I$,*
(3) *ϕ_{x_0} is unique in the sense that if $\psi : (\alpha, \beta) \supset I \to \mathbb{R}$ is a solution of $x' = f(x)$ with $\psi(0) = x_0$, then $\psi = \phi_{x_0}$ on I.*

Proof Take the uniform metric ρ on $C^0([-\delta, \delta], \mathbb{R})$. Define $T : C^0([-\delta, \delta], \mathbb{R}) \times I \to C^0([-\delta, \delta], \mathbb{R})$ by

$$T(\phi, x)(t) = x + \int_0^t f(\phi(s)) \, ds, \; (\phi, x) \in C^0([-\delta, \delta], \mathbb{R}) \times I, \; t \in [-\delta, \delta].$$

Observe that $T(\phi, x)(0) = x$, for all $\phi \in C^0([-\delta, \delta], \mathbb{R})$. Now $T(\phi, x) : [-\delta, \delta] \to \mathbb{R}$ is C^1 (fundamental theorem of calculus) and so certainly $T(\phi, x) \in C^0([-\delta, \delta], \mathbb{R})$. We claim that T satisfies the conditions of Theorem 7.17.9 with $k = \delta C$. For fixed $\phi \in C^0([-\delta, \delta], \mathbb{R})$, the map $T(\phi, \cdot) : I \to C^0([-\delta, \delta], \mathbb{R})$, $x \mapsto x + \int_0^t f(\phi(s)) \, ds$, is obviously continuous and so condition (2) of Theorem 7.17.9 is satisfied. In order to verify the contraction property, let $\phi, \psi \in C^0([-\delta, \delta], \mathbb{R})$. For (fixed) $x \in I$, we have

$$|T(\phi, x)(t) - T(\psi, x)(t)| = \left| \int_0^t f(\phi(s)) - f(\psi(s)) \, ds \right|$$

$$\leq \left| \int_0^t |f(\phi(s)) - f(\psi(s))| \, ds \right|.$$

Now $|f(\phi(s)) - f(\psi(s))| \leq C|\phi(s) - \psi(s)|$ by the mean value theorem and so

$$|T(\phi, x)(t) - T(\psi, x)(t)| \leq \left| \int_0^t C|\phi(s) - \psi(s)| \, ds \right|$$

$$\leq C \left| \int_0^t \rho(\phi, \psi) \, ds \right|$$

$$\leq C\delta\rho(\phi, \psi), \text{ if } t \in [-\delta, \delta].$$

Since we assumed that $C\delta < 1$, T satisfies condition (1) of Theorem 7.17.9 with $k = C\delta$. Let ϕ_x denote the unique fixed point of T_x. We have

$$\phi_x(t) = x + \int_0^t f(\phi_x(s)) \, ds, \ t \in [-\delta, \delta].$$

Differentiating with respect to t, it follows from the fundamental theorem of calculus that

$$\phi_x'(t) = f(\phi_x(t)), \ t \in [-\delta, \delta].$$

We have $\phi_x(0) = x$ and so we have proved part (1) of the theorem. Since $\phi_x'(t) = f(\phi_x(t))$, we see that if ϕ_x is C^s, $s \leq r$, then ϕ_x' must be C^s (chain rule) and so ϕ_x is C^{s+1}. Hence ϕ_x is C^{r+1}. Finally, suppose that $\psi : (\alpha, \beta) \supset I \to \mathbb{R}$ is a solution of $x' = f(x)$ with $\psi(0) = x_0$. If we restrict ψ to $[-\delta, \delta]$ then the restriction lies in $C^0([-\delta, \delta], \mathbb{R})$ and so equals ϕ_x by the uniqueness part of the contraction mapping lemma.   □

Remarks 7.17.13

(1) Theorem 7.17.12 goes beyond the existence of solutions. The result shows that the solutions depend continuously on the initial condition. With more work—see Chap. 9—it can be shown that solutions depend C^r on the initial condition x.

(2) Observe that the strong condition $\sup_{x \in \mathbb{R}} |f'(x)| < \infty$ can always be satisfied by multiplying f by a tabletop function which is equal to one on a big interval containing I. In more detail, if $x' = f(x)$, where $\sup_{x \in \mathbb{R}} |f'(x)| = +\infty$, and $[a, b] \subset \mathbb{R}$, we may choose an interval $[A, B] \supset [a, b]$ with $A \ll a, B \gg b$,

and a tabletop function ψ which is equal to 1 on $[A, B]$ and is zero outside $[A - 1, B + 1]$. The differential equation $x' = \psi(x)f(x)$ will then satisfy the conditions of Theorem 7.17.12. It follows that we can choose $\delta > 0$ such that for all $x \in [a, b]$, there is a unique solution $\phi_x : [-\delta, \delta] \to \mathbb{R}$ to $x' = \psi(x)f(x)$. Choosing δ smaller if necessary, we can require that $\phi_x(t) \in [A, B]$ for all $x \in [a, b], t \in [-\delta, \delta]$. Hence $\phi_x : [-\delta, \delta] \to \mathbb{R}$ will solve $x' = f(x)$, for all $x \in [a, b]$.

(3) In Theorem 7.17.12 the right-hand side of the differential equation was independent of time. If instead we consider the non-autonomous equation $x' = f(x, t)$, where $f : \mathbb{R}^2 \to \mathbb{R}$ and $\sup_{(x,t)\in\mathbb{R}^2} |\frac{\partial f}{\partial x}(x, t)| = C < \infty$, then the proof of Theorem 7.17.12 extends immediately to give unique solutions $\phi_x : [-\delta, \delta] \to \mathbb{R}$ to $x' = f(x, t)$ with initial condition $x \in [a, b]$.   ✱

Example 7.17.14 We can use the second statement of the contraction mapping lemma to construct approximate solutions to ordinary differential equations. For example, consider the equation

$$x' = x - 1, \ x \in \mathbb{R}.$$

Given $x_0 \in \mathbb{R}$, we can construct a sequence (ϕ_n) of approximate solutions with initial condition x_0. We start by defining ϕ_0 by

$$\phi_0(t) = x_0, \ t \in \mathbb{R}.$$

This is the 'simplest' approximate solution with the required initial condition. We define ϕ_1, ϕ_2 by

$$\phi_1(t) = x_0 + \int_0^t (x - 1)(\phi_0(s)) \, ds = x_0 + \int_0^t (x_0 - 1) \, ds$$

$$= x_0 + t(x_0 - 1) = 1 + (x_0 - 1)(1 + t),$$

$$\phi_2(t) = x_0 + \int_0^t (x_0 + s(x_0 - 1) - 1) \, ds = 1 + (x_0 - 1)\left(1 + t + \frac{t^2}{2}\right).$$

The solution to $x' = x - 1$ with initial condition x_0 is $\phi(t) = 1 + (x_0 - 1)e^t$. Hence ϕ_2 gives an approximate solution agreeing with the first 3 terms in the Taylor series of $1 + (x_0 - 1)e^t$ at $t = 0$.   ♠

7.17.3   An Application of the Contraction Mapping Lemma to the Inverse Function Theorem

Our second application of the contraction mapping lemma will be to the inverse function theorem. We look at the case of maps $f : \mathbb{R} \to \mathbb{R}$ and note that there are far simpler proofs making use of the natural order on $\mathbb{R}$. The proof we give, however, has the merit that it generalizes easily to maps $f : \mathbb{R}^n \to \mathbb{R}^n$ (see Chap. 9).

Suppose that $f : \mathbb{R} \to \mathbb{R}$ is C^r, $r \geq 1$, and $f'(0) \neq 0$. The inverse function theorem states that we can choose closed intervals I, J, with $0 \in \overset{\circ}{I}, f(0) \in \overset{\circ}{J}$, such that $f : I \to J$ is 1:1 onto and $f^{-1} : J \to I$ is C^r. What we shall do here is construct I, J and verify only that the inverse map $f : J \to I$ is continuous. (We address the differentiability of f^{-1} in Chap. 9.)

We start by observing that if we replace f by $f - f(0)$, then we may assume $f(0) = 0$. Replacing f by $f/f'(0)$, we may further assume that $f'(0) = 1$.

Theorem 7.17.15 *Suppose that $f : \mathbb{R} \to \mathbb{R}$ is C^1 and satisfies*

(1) $f(0) = 0$.
(2) $f'(0) = 1$.

Then there exist a closed interval I, containing 0 as an interior point, and $s > 0$, such that f maps I 1:1 onto $[-s, s]$ and the inverse map $f^{-1} : [-s, s] \to I$ is continuous.

Proof For $y \in \mathbb{R}$, define $\Phi_y : \mathbb{R} \to \mathbb{R}$ by

$$\Phi_y(x) = x - f(x) + y.$$

Observe that x is a fixed point of Φ_y iff $f(x) = y$. This suggests finding fixed points $x = x(y)$ of Φ_y using the contracting mapping lemma with parameters. The map $y \mapsto x(y)$ will then be a continuous inverse to f (since $f(x(y)) = y$).

Set $\phi(x) = x - f(x)$. We have $\phi(0) = \phi'(0) = 0$. Since ϕ is at least C^1, there exists an $r > 0$ such that $|\phi'(x)| \leq \frac{1}{2}$ for all $x \in [-r, r]$. Hence, by the mean value theorem, we have

$$|\phi(x) - \phi(x')| \leq \frac{1}{2}|x - x'|, \text{ for all } x, x' \in [-r, r]. \tag{7.14}$$

We claim that for all $y \in [-\frac{r}{2}, \frac{r}{2}]$

(a) $\Phi_y : [-r, r] \to [-r, r]$,
(b) Φ_y is a contraction map with contraction constant $\leq \frac{1}{2}$.

In order to verify (a), observe that for $|x| \leq r$ we have

$$|\Phi_y(x)| = |\phi(x) + y| \leq |\phi(x)| + |y| \leq \frac{r}{2} + \frac{r}{2} = r,$$

where the last inequality follows from (7.14) with $x' = 0$. The proof that Φ_y is a contraction is even simpler. If $x, x' \in [-r, r]$, we have

$$|\Phi_y(x) - \Phi_y(x')| = |\phi(x) - \phi(x')| \leq \frac{1}{2}|x - x'|,$$

where the last inequality again uses (7.14).

The map $\Phi : [-r, r] \times [-\frac{r}{2}, \frac{r}{2}] \to [-r, r]$ satisfies the conditions of Theorem 7.17.9 (for fixed x, $\Phi(x, \cdot)$ is obviously continuous) and so, since $[-r, r]$ is a complete metric space, it follows by Theorem 7.17.9 that there is a continuous map $x : [-\frac{r}{2}, \frac{r}{2}] \to [-r, r]$ such that

$$\Phi_y(x(y)) = x(y), \text{ for all } y \in \left[-\frac{r}{2}, \frac{r}{2}\right].$$

In particular, $f(x(y)) = y$, for all $y \in [-\frac{r}{2}, \frac{r}{2}]$. Set $\frac{r}{2} = s$ and $x(y) = g(y)$, $y \in [-s, s]$ so that $g : [-s, s] \to [-r, r]$. Since g is continuous and $g(0) = 0$ (by uniqueness of fixed points), $g([-s, s]) = I$ is a closed interval containing 0. Since $f \circ g$ is the identity map on $[-s, s]$, $f : I \to [-s, s]$ is onto. Further, by the uniqueness of fixed points, f is $1 : 1$. Hence $g = f^{-1}$. Finally, 0 is an interior point of I since $f : I \to \mathbb{R}$ is continuous and so $f^{-1}(-s, s)$ is an open subset of $\mathbb{R}$ containing 0.   □

EXERCISES 7.17.16

(1) Find a contraction map $f : X \to X$, where $X = \mathbb{R} \smallsetminus \{0\}$, which does not have a fixed point.

(2) Let $A : \mathbb{R}^n \to \mathbb{R}^n$ be linear (A can be written as an $n \times n$ matrix) and $b \in \mathbb{R}^n$. Show that a necessary condition for the affine linear map $L(x) = Ax + b$ to be a contraction is that every eigenvalue of A is of modulus less than 1. Show that if we take the Euclidean metric on $\mathbb{R}^n$, then this condition is not sufficient.

(3) Consider the ODE $x' = x$. Fix $x \in \mathbb{R}$. Taking $\phi_0 : [-a, a] \to \mathbb{R}$ to be the constant map $\phi_0(t) = x$, compute the first three iterates of $T\phi(t) = x + \int_0^t f(\phi(s)) \, ds$, starting with $\phi = \phi_0$. Make a conjecture as to the form of $\phi_n(t)$, $n > 3$, and prove your conjecture.

(4) Consider the ODE $x' = x^2$. Taking $\phi_0 : [-a, a] \to \mathbb{R}$ to be the constant map $\phi_0(t) = x$, compute the first three iterates of $T\phi(t) = x + \int_0^t f(\phi(s)) \, ds$, starting with $\phi = \phi_0$. In particular, verify that the terms in ϕ_3 of degree 1, 2 and 3 in x match those in the binomial expansion of $x/(1 - tx)$ (the solution of $x' = x^2$ with initial condition x).

(5) Provide the justification for Remarks 7.17.13(2) and show that there exists a $\delta > 0$ such that for all $x \in [a, b]$, the solution $\phi_x : [-\delta, \delta] \to \mathbb{R}$ of $x' = \psi(x)f(x)$ satisfies $\phi_x(t) \in [A, B]$, $t \in [-\delta, \delta]$.

(6) Find an example of a map $f : \mathbb{R} \to \mathbb{R}$ such that (a) $|f(x) - f(y)| < |x - y|$ for all $x, y \in \mathbb{R}$, $x \neq y$, and (b) f has no fixed point. Why does your example not contradict the contraction mapping lemma? (Hint: Look for an example of the form $f(x) = x + \phi(x)$ where $\phi(x) > 0$ for all $x \in \mathbb{R}$ and $-1 < \phi'(x) < 0$ for all $x \in \mathbb{R}$.)

(7) Suppose that (X, d) is complete and $f : X \to X$. Show that if there exists a $p \in \mathbb{N}$ such that f^p is a contraction, then f has a unique fixed point. (f^p denotes the composition of f with itself p times.)

(8) Let $F : [0, 1] \to [0, 1]$ be differentiable and $M = \sup_{x \in [0,1]} |f'(x)| < \infty$. Define $T : C^0([0, 1]) \to C^0([0, 1])$ by $T(\phi)(x) = \int_0^x F(\phi(s)) \, ds$, $\phi \in C^0([0, 1])$. Show that if $p! > M$, then T^p is a contraction and hence T has a unique fixed point.

(9) Show that if (X, d) is compact and $f : X \to X$ satisfies $d(f(x),f(y)) < d(x,y)$, for all $x \neq y \in X$, then f has a unique fixed point. (Hint: For every $r > 0$, there exists a $k \in [0, 1)$ such that if $d(x, y) \geq r$, then $d(f(x),f(y)) \leq kd(x,y)$.)

(10) Let $\eta : \mathbb{R} \to \mathbb{R}$ be Lipschitz with Lipschitz constant k ($|\eta(x)-\eta(y)| \leq k|x-y|$ for all $x, y \in \mathbb{R}$). Let $f(x) = ax + \eta(x)$, where $a \neq 0$. Show that if $k < |a|$, then $f : \mathbb{R} \to \mathbb{R}$ is 1:1 onto and f^{-1} is Lipschitz with Lipschitz constant $\leq 1/(1-k/|a|)$.

(11) Let (X, d) be a metric space. A map $f : X \to X$ is an *expansion* if there exists a $k > 1$ such that $d(f(x),f(y)) \geq kd(x, y)$ for all $x, y \in X$. Show

   (a) If f is an expansion, then f is $1 : 1$.
   (b) If $f : X \to X$ is an expansion and f has a fixed point, then the fixed point is unique.
   (c) If X is compact and contains at least two points, then there are no expansions of X.
   (d) If X is complete and $f : X \to X$ is a continuous surjective expansion, then f has a unique fixed point.

   Find examples of (a) a continuous expansion of $\mathbb{R}$, (b) a discontinuous expansion of $\mathbb{R}$, (c) an expansion of a complete metric space which is not onto and has no fixed point.

(12) Let (X, d) be a compact metric space and suppose that $f : X \to X$ satisfies $d(f(x),f(y)) \geq d(x, y)$ for all $x, y \in X$. Prove that f is an isometry. Show by means of an example that f need not have a fixed point. (Hint: Given $u, v \in X$ and $\varepsilon > 0$, show there exists an $n \in \mathbb{N}$ such that $d(u, u_n), d(v, v_n) < \varepsilon$, where $u_n = f^n(u)$, $v_n = f^n(v)$.)

7.18 Connectedness

We conclude this chapter with a discussion of connectedness—a property implicitly used in the proof of the intermediate value theorem.

Definition 7.18.1 A metric space (X, d) is *connected* if the only open and closed subsets of X are X and the empty set. If X is not connected, X is *disconnected*.

   We have a useful characterization of disconnected spaces.

Lemma 7.18.2 *A metric space (X, d) is disconnected iff we can write $X = U \cup V$ where U and V are open, non-empty disjoint subsets of X.*

Proof Immediate since the hypotheses imply that $V = X \smallsetminus U$ is open and closed.   □

   We may extend the definition of connectedness to subsets of a metric space.

Definition 7.18.3 The non-empty subset Y of the metric space (X, d) is connected iff the only open and closed subsets of (Y, d_Y) are Y and the empty set.

Remark 7.18.4 We emphasize that connectedness and disconnectedness is only defined for *non-empty* subsets of a metric space. ✠

Lemma 7.18.5 *A subset Y of the metric space (X, d) is disconnected iff there exist open subsets U, V of X such that*

(1) *$U \cap Y$ and $V \cap Y$ are disjoint non-empty subsets of Y.*
(2) *$(U \cap Y) \cup (V \cap Y) = Y$.*

In particular, Y is a connected subset of X iff (Y, d_Y) is a connected metric space.

Proof Every open subset A of (Y, d_Y) may be written $A = U \cap Y$, where U is an open subset of X, by Proposition 7.6.1. The result follows by Lemma 7.18.2 applied to (Y, d_Y). □

Remark 7.18.6 Just as was the case for compact subsets (Proposition 7.13.10), connectedness is an intrinsic property of a subset. ✠

Examples 7.18.7

(1) A metric space consisting of a single point is connected.
(2) If (X, d) has the discrete metric, then (X, d) is disconnected if X contains more than one point.
(3) A metric space containing more than one point is *totally disconnected* if every connected subset of X consists of a single point. A space with the discrete metric is totally disconnected. So also is the Cantor set **C**. We leave the verification of this to the exercises.
(4) The metric space $(\mathbb{Q}, |\cdot|)$ is totally disconnected. Indeed, suppose that $E \subset \mathbb{Q}$ is connected. Let $x, y \in E$, $x \leq y$. Suppose that $x < y$. Regard $\mathbb{Q}$ as a subspace of $\mathbb{R}$ and choose an irrational number $z \in (x, y)$. Take $U = \mathbb{Q} \cap (-\infty, z)$, $V = \mathbb{Q} \cap (z, \infty)$. Then U, V are open disjoint non-empty subsets of $\mathbb{Q}$ such that $U \cap E, V \cap E \neq \emptyset$. ♠

As the previous examples show, it is not hard to find disconnected sets. Finding non-trivial connected sets is a little trickier. It turns out that the key to finding connected sets is to classify the connected subsets of $\mathbb{R}$. Not surprisingly a subset of $\mathbb{R}$ is connected iff it is an interval. Once we have this result, we can prove some simple results that allow us to combine connected sets and in this way find many examples of connected sets.

We start by giving a characterization of an interval.

Lemma 7.18.8 *A non-empty subset A of $\mathbb{R}$ is an interval iff for every $x, y \in A$, $x \leq y$, we have $[x, y] \subset A$.*

Proof Obviously every interval $I \subset \mathbb{R}$ satisfies the condition. Conversely, if A satisfies the condition, let $x^\star = \inf\{x \mid x \in A\}$, $y^\star = \sup\{x \mid x \in A\}$. We have $-\infty \leq x^\star \leq y^\star \leq +\infty$. Observe that if $x \in (x^\star, y^\star)$, then $x \in A$ and so $A \supset (x^\star, y^\star)$. If $x^\star = -\infty$, then $A \supset (-\infty, y^\star)$. If $y^\star < \infty$, then $A = (-\infty, y^\star]$, if $y^\star \in A$, else $A = (-\infty, y^\star)$. The other cases are handled similarly. □

Theorem 7.18.9 *A subset A of $\mathbb{R}$ is connected iff A is an interval.*

Proof Suppose $A \subset \mathbb{R}$ is an interval and let U, V be open subsets of $\mathbb{R}$ such that $U \cup V \supset A$ and $(U \cap A) \cap (V \cap A) = \emptyset$. Suppose that $U \cap A, V \cap A \neq \emptyset$. Pick $x \in U \cap A, y \in V \cap A$. Without loss of generality assume $x < y$. Since A is an interval, $[x, y] \subset A$. Let $z = \sup\{x' \in [x, y] \mid x \in U\}$. Since $[x, y] \subset A$, and $x \in U$, we must have $z > x$ as U is open. Similarly $z < y$ since $y \in V$ and V is open. If $z \in V$, then z must lie in the interior of $[x, y] \cap V$, contradicting the definition of z as the supremum of points in $[x, y]$ lying in U. On the other hand if $z \in U$, then z must lie in the interior of $[x, y] \cap U$ and so z is not the supremum of points in $[x, y]$ lying in U. Contradiction. Therefore one of the sets $U \cap A, V \cap A$ must be empty.

Conversely, suppose that A is a connected subset of $\mathbb{R}$. Let $x < y \in A$. Suppose there exists a $z \in [x, y]$ that does not lie in A. Take $U = (-\infty, z), V = (z, \infty)$. Then $U \cup V \supset A, U \cap A, V \cap A \neq \emptyset$, and $(U \cap A) \cap (V \cap A) = \emptyset$, contradicting the assumption that A is connected. Hence $[x, y] \subset A$ for all $x < y \in A$ and so A is an interval. $\qquad\qquad\square$

Remark 7.18.10 The use of the supremum in the proof of Theorem 7.18.9 is not surprising in view of Examples 7.18.7(4). $\qquad\qquad\maltese$

Theorem 7.18.11 *Let $f : X \rightarrow Y$ be a continuous map between metric spaces. If X is connected, then $f(X)$ is a connected subset of Y. More generally, if E is a connected subset of X, then $f(E)$ is a connected subset of Y.*

Proof In order to prove that $f(X)$ is connected, it suffices to show that if U, V are open subsets of Y such that $U \cup V \supset f(X)$ and $(U \cap f(X)) \cap (V \cap f(X)) = \emptyset$, then either $U \cap f(X)$ or $V \cap f(X)$ is the empty set. Since f is continuous, $f^{-1}(U)$, $f^{-1}(V)$ are open subsets of X. Since $U \cup V \supset Y, f^{-1}(U) \cup f^{-1}(V) = X$. Since $(U \cap f(X)) \cap (V \cap f(X)) = \emptyset, f^{-1}(U) \cap f^{-1}(V) = \emptyset$. Therefore, one of $f^{-1}(U)$, $f^{-1}(V)$ is empty since X is connected and so either $U \cap f(X)$ or $V \cap f(X)$ is the empty set. The result when E is a connected subset follows by replacing (X, d) by (E, d_E) and f by the restriction of f to E. $\qquad\qquad\square$

Example 7.18.12 If $\phi : (a, b) \rightarrow \mathbb{R}^n$ is continuous—so ϕ defines a continuous curve in $\mathbb{R}^n$—then $\phi(a, b)$ is a connected subset of $\mathbb{R}^n$. $\qquad\qquad\spadesuit$

Proposition 7.18.13 *The closure of a connected subset of a metric space is connected.*

Proof Let E be a connected subset of the metric space X. It suffices to show that if U, V are open subsets of X such that $U \cup V \supset \overline{E}$ and $(U \cap \overline{E}) \cap (V \cap \overline{E}) = \emptyset$, then either $U \cap \overline{E} = \emptyset$ or $V \cap \overline{E} = \emptyset$. If these conditions hold then $U \cup V \supset E$, since $\overline{E} \supset E$, and $(U \cap E) \cap (V \cap E) = \emptyset$. Therefore, since E is assumed connected, one of $U \cap E, V \cap E$ must be the empty set. Without loss of generality, suppose that $U \cap E = \emptyset$. This implies that $U \cap \overline{E} = \emptyset$ (for every $x \in U$, there exists an $r > 0$ such that $D_r(x) \cap E = \emptyset$). $\qquad\qquad\square$

Example 7.18.14 Let $E = \{(x, \sin(1/x)) \mid x > 0\} \subset \mathbb{R}^2$. Since $\{x \mid x > 0\}$ is connected and $x \mapsto ((x, \sin(1/x))$ is continuous for $x > 0$, E is a connected subset of $\mathbb{R}^2$. We have $\overline{E} = \{(0, y) \mid -1 \leq y \leq 1\} \cup E$ and this set is connected by the previous proposition. Later we give a simple example to show that the interior of a connected set need not be connected.   ♠

Theorem 7.18.15 *Let $\{E_i \mid i \in I\}$ be a family of connected subsets of X. If $E_i \cap E_j \neq \emptyset$ for all $i, j \in I$, then $\cup_{i \in I} E_i$ is a connected subset of X.*

Proof Set $E = \cup_{i \in I} E_i$. Let U, V be open subsets of X such that $U \cup V \supset E$ and $(E \cap U) \cap (E \cap V) = \emptyset$. It suffices to prove one of $E \cap U, E \cap V$ is the empty-set. Observe that

$$E \cap U = \cup_{i \in I} E_i \cap U.$$

Suppose that for some $i \in I$, $E_i \cap U \neq \emptyset$. Then $E_i \cap U = E_i$, since E_i is connected. Therefore since $E_i \cap E_j \neq \emptyset$ for all $j \in I$, we must have $E_j \cap U \neq \emptyset$ for all $j \in I$ and so $E_j \subset U$ for all $j \in I$. Since $E_j \cap V = \emptyset$ for all $j \in I$, we must have $V \cap E = \emptyset$. □

Definition 7.18.16 A metric space (X, d) is *path-connected* if for all for $x, y \in X$, there exists a continuous curve $\phi : [0, 1] \to X$ such that $\phi(0) = x$, $\phi(1) = y$.

Proposition 7.18.17 *A path-connected metric space is connected.*

Proof Let X be path-connected. Fix $x \in X$. For each $y \in X$, there exists a continuous curve $\phi_y : [0, 1] \to X$ such that $\phi_y(0) = x$, $\phi_y(1) = y$. Set $E_y = \phi_y([0, 1])$. Since ϕ_y is continuous, E_y is a connected subset of X. Since $x \in E_y \cap E_z$ for all $y, z \in X$, $\cup_{y \in X} E_y = X$ is connected by Theorem 7.18.15.   □

Examples 7.18.18

(1) For $m \geq 1$, $\mathbb{R}^m$ is path-connected and therefore connected.
(2) For all $x \in \mathbb{R}^m$, the open and closed disks $D_r(x), \overline{D}_r(x)$ are path-connected since every point $y \in D_r(x)$ (or $\overline{D}_r(x)$) can be joined to the centre x by the continuous path $\phi(t) = ty + (1 - t)x, t \in [0, 1]$.
(3) The unit sphere S^2 in $\mathbb{R}^3$ is connected. For this, observe that S^2 is the image of the continuous map $P : \mathbb{R}^2 \to \mathbb{R}^3$ defined by $S(\theta, \phi) = (\cos\theta \sin\phi, \sin\theta \sin\phi, \cos\phi)$ (spherical polar coordinate map with $r = 1$). Alternatively, one can show that $S^2 \smallsetminus \{x\}$ is the continuous image of $\mathbb{R}^2$ (x any point of S^2). Then by Proposition 7.18.13, $\overline{S^2 \smallsetminus \{x\}} = S^2$ is connected. This approach has the merit that it generalizes to prove that the n-sphere is connected for all $n \geq 1$—see the exercises.
(4) A connected set need not be path connected. For example, although the graph E of $\sin(1/x)$, $x > 0$, is path connected, $\overline{E}$ is not path connected though it is connected (see Example 7.18.14).   ♠

Example 7.18.19 The interior of a connected subset of a metric space need not be connected. Take the Euclidean metric on $\mathbb{R}^2$, and define $E = \overline{D}_1(0,0) \cup \overline{D}_1(2,0) \subset \mathbb{R}^2$. We have $\overset{\circ}{E} = D_1(0,0) \cup D_1(2,0)$, which is not connected (take $U = D_1(0,0)$, $V = D_1(2,0)$ in the definition of disconnected). ♠

EXERCISES 7.18.20

(1) Suppose that E, F are connected subsets of X. If $E \cap F \neq \emptyset$, need $E \cap F$ be connected?

(2) Show that the intermediate value theorem follows from the connectedness of an interval and Theorem 7.18.11.

(3) Suppose $f : X \to Y$ is continuous and E is a connected (respectively, path connected) subset of Y. Must $f^{-1}(E)$ be a connected (respectively, path connected) subset of X?

(4) Suppose that the metric space (X, d) is connected. Show that if $f : X \to \mathbb{R}$ is continuous then $f(X)$ is an interval. In particular, if $a, b \in f(X)$, then f takes every value between a and b. Show that if additionally X is compact, then there exist $m < M \in \mathbb{R}$ such that $f(X) = [m, M]$ (version of Theorem 2.4.10 for metric spaces).

(5) Let (X, d) be a metric space. Show that if X is countable then X is connected if and only if X consists of a single point. (Hint. Let $x_0 \in X$ and consider $f : X \to \mathbb{R}$ defined by $f(x) = d(x, x_0)$. Note the result is false for general topological spaces.)

(6) Show that a metric space (X, d) is connected iff for every proper non-empty subset E of X, $\partial E \neq \emptyset$.

(7) Prove that the middle thirds Cantor set is totally disconnected (Examples 7.18.7(3)).

(8) Suppose that E is a connected subset of the metric space (X, d). Show that E' is connected. (Hint: reduce to the case where E is closed.)

(9) Suppose that E is a connected subset of the metric space X and that E does not consist of a single point. Show that $E \cup \{z\}$ is connected iff $z \in E'$.

(10) Non-empty subsets A, B of (X, d) are *separated* if $\overline{A} \cap B, A \cap \overline{B} = \emptyset$.

    (a) Suppose that $X = A \cup B$ where A, B are separated. Show that A, B are open and closed in X and X is disconnected.

    (b) Show that X is disconnected iff we can write $X = A \cup B$, where A, B are separated.

    (c) Show that a subset E of X is disconnected iff we can write $E = A \cup B$, where A, B are separated.

    (d) Show that disjoint subsets A, B of X are separated iff no point of A is a limit point of B and no point of B is a limit point of A.

(Connectedness is often defined in terms of separated sets (for example, see [18, 27]). The definition in terms of separation is equivalent to our definition by (a). That connectedness is an intrinsic property follows from (d). Although the definition of connectedness in terms of separated sets is a little more complicated it works well when considering connectedness of subsets.)

(11) Prove that $\overline{E} = \overline{\{(x, \sin(1/x)) \mid x > 0\} \subset \mathbb{R}^2\}}$ is not path connected (see Example 7.18.14). (Hint: If $\phi : [0, 1] \to \overline{E}$ is continuous, then ϕ is uniformly continuous since $[0, 1]$ is compact.)

(12) Which of the following sets are connected? Why?

  (a) The unit circle $S^1 = \{(x, y) \mid x^2 + y^2 = 1\} \subset \mathbb{R}^2$.
  (b) The paraboloid $P = \{(x, y, z) \mid x^2 + y^2 = z\} \subset \mathbb{R}^3$.
  (c) The surface $H = \{(x, y, z) \mid x^2 - y^2 = 1\} \subset \mathbb{R}^3$.
  (d) The *cone* on the middle thirds Cantor set **C**: $\{tX + (1 - t)(0, 1) \mid t \in [0, 1], X \in \mathbf{C}\} \subset \mathbb{R}^2$, where $\mathbf{C} \subset \mathbb{R}$ is the Cantor set and we regard $\mathbb{R} \subset \mathbb{R}^2$ as the x-axis.

(13) Show that if $f : X \to Y$ is continuous and X is path connected, then $f(X)$ is path connected.

(14) Suppose that E_i is a path connected subset of the metric space $(X_i, d_i), i = 1, 2$. Show that $E_1 \times E_2$ is a path connected subset of $(X_1 \times X_2, D)$ where D is the product metric on $X_1 \times X_2$ $(D((x_1, x_2), (y_1, y_2)) = \max\{d_1(x_1, y_1), d_2(x_2, y_2))\})$.

(15) Let E be a connected subset of (X, d) and for $r > 0$, let $E(r) = \{x \in X \mid d(x, E) \le r\}$. Must $E(r)$ be connected? Would your answer change if $X = \mathbb{R}^n$?

(16) Suppose that E_n are connected subsets of a metric space such that $E_n \cap E_{n+1} \ne \emptyset, n \ge 1$. Prove that $\cup_{n \ge 1} E_n$ is connected.

(17) Let (X, d) be a metric space and let $x \in X$. Let C_x denote the union of all connected subsets of X which contain x. Show that

  (a) C_x is a closed connected subset of X.
  (b) If $A \supset C_x$ is connected then $A = C_x$.
  (c) $\{C_x \mid x \in X\}$ defines a partition of X (that is, if $x, x' \in X$, either $C_x = C_{x'}$ or $C_x \cap C_{x'} = \emptyset$).

We call the sets C_x the *connected components* of X. Show that we can also define the connected path-components of a metric space and that we obtain a partition of X into path-components. What are the path-components for the set $\overline{E}$, where E is defined in Example 7.18.14? Need path-components be closed?

(18) Let E be a non-empty closed subset of the metric space X. Show that the connected components of E are closed subsets of A. Are the connected components of an open subset A of X always open?

(19) Define the 1:1 map $F : \mathbb{R}^m \to \mathbb{R}^{m+1} = \mathbb{R}^m \times \mathbb{R}$ by

$$F(x) = \left(\frac{2x}{\|x\|^2 + 1}, \frac{\|x\|^2 - 1}{\|x\|^2 + 1} \right),$$

where $x = (x_1, \cdots, x_m) \in \mathbb{R}^m$. Verify that $F(\mathbb{R}^m) = S^m \smallsetminus \{(0, \cdots, 0, 1)\}$ and deduce that the unit sphere in $\mathbb{R}^{m+1}$ is connected. (When $m = 2$, the inverse map $F^{-1} : S^2 \smallsetminus \{0, 0, 1\}$ is *stereographic projection*: if $X \in S^2 \smallsetminus \{0, 0, 1\}$, $F^{-1}(X)$ is the unique point of intersection of the x, y-plane with the line through $\{0, 0, 1\}$ and X.)

(20) Let E be a compact subset of the metric space (X, d). Show that

   (a) E is disconnected iff and only if E can be written as the union of two disjoint (non-empty) compact subsets of E.

   (b) E is disconnected iff we can find disjoint open subsets U, V of X such that $U \cap E, V \cap E \neq \emptyset$ and $E \subset U \cup V$.

(Hint for (b): if A, B are disjoint compact subsets of the metric space X, then $\inf_{a \in A, b \in B} d(a, b) > 0$.)

(21) Suppose that $E_1 \supset E_2 \supset E_3 \supset \cdots$ is a decreasing sequence of (sequentially) compact connected sets. Show that $E = \cap_{n=1}^{\infty} E_j$ is connected. (Hint: use part (b) of the previous question.)

(22) Show, by looking for an example in $\mathbb{R}^2$, that the previous result may fail if the E_j are closed but not compact (assume $\cap_{n=1}^{\infty} E_j \neq \emptyset$.) What happens if $E_j \subset \mathbb{R}$?

(23) Suppose that $E_1 \supset E_2 \supset E_3 \supset \cdots$ is a decreasing sequence of compact path connected sets. Is $E = \cap_{n=1}^{\infty} E_j$ path connected? (Hint: Look for a sequence (E_n) which has intersection the set $\overline{E}$ of Example 7.18.14.)

(24) Let $Y = \{0, 1\}$ with the discrete metric. Show that (X, d) is connected iff every continuous function $f : X \to Y$ is constant. Use this result to show that if (X_1, d_1), (X_2, d_2) are connected metric spaces then the product $(X_1 \times X_2, d)$ is connected (where we take the product metric $d((x_1, x_2), (y_1, y_2)) = \max\{d_1(x_1, y_1), d_2(x_2, y_2))\}$ on $X_1 \times X_2$).

(25) If X has topology of open sets $\mathcal{U}$, we say X is connected iff the only open and closed subsets of X are X and $\emptyset$. More generally, if E is a non-empty subset of X, we define a topology $\mathcal{U}_E$ of open sets of E by $\mathcal{U}_E = \{U \cap E \mid U \in \mathcal{U}\}$. We say E is connected iff the only open and closed subsets of E are E and the empty set. Take the Zariski topology on $\mathbb{R}$: the open subsets of $\mathbb{R}$ are either $\mathbb{R}$, $\emptyset$ or $\mathbb{R} \smallsetminus F$ where F is finite.

   (a) Show that $\mathbb{R}$ is connected (in the Zariski topology).

   (b) Show that $\mathbb{Z} \subset \mathbb{R}$ is a connected subset of $\mathbb{R}$ (in the Zariski topology).

   (c) Classify the connected subsets of $\mathbb{R}$ (in the Zariski topology). In particular, show that every non-empty open subset of $\mathbb{R}$ is connected, as is the Cantor set, and that a finite set is connected iff it consists of just one point.

   (d) Is $\mathbb{Z}$ a path connected subset of $\mathbb{R}$? What about $[a, b]$?

(Moral: Connectedness appears to be relatively intuitive for metric spaces; for general topological spaces what connectedness detects can seem un-geometric and non-intuitive—though there is usually a mathematically significant interpretation of connectedness.)

(26) If $f : \mathbb{R} \to \mathbb{R}$ is continuous, then the graph of f, $\{(x, f(x)) \mid x \in \mathbb{R}\}$, is a closed connected subset of $\mathbb{R}^2$. Suppose that the graph of $f : \mathbb{R} \to \mathbb{R}$ is a closed subset of $\mathbb{R}^2$. Does it follow that f is continuous? What about if the graph of $f : \mathbb{R} \to \mathbb{R}$ is connected? Prove that if the graph of $f : \mathbb{R} \to \mathbb{R}$ is closed *and* connected, then f is continuous.

(27) Suppose that $f : \mathbb{R}^2 \to \mathbb{R}$ is continuous. If $f^{-1}(0)$ is connected, does it follow that $f^{-1}(0)$ is path connected? (Hint: Theorem 7.12.4. If f is a polynomial it can be shown that $f^{-1}(0)$ is connected iff $f^{-1}(0)$ is path connected.)

Chapter 8
Fractals and Iterated Function Systems

As motivation for what we do in this chapter, we start by taking another look at the middle thirds Cantor set $\mathbf{C}$ constructed in Chap. 7. Define affine linear maps $R_1, R_2 : \mathbb{R} \to \mathbb{R}$ by

$$R_1(x) = \frac{x}{3},$$

$$R_2(x) = 1 - \frac{x}{3}.$$

Following the same notation we used in our discussion of the Cantor set in Sect. 7.14, observe that $R_1(I_0) = [0, \frac{1}{3}] = I_{11}$ and $R_2(I_0) = [\frac{2}{3}, 1] = I_{12}$. Hence $R_1(I_0) \cup R_2(I_0) = I_1$. Similarly, $R_1(I_1) \cup R_2(I_1) = I_2$ and, in general, $R_1(I_n) \cup R_2(I_n) = I_{n+1}$, for all $n \geq 0$.

We can abstract this process in the following way. Let $\mathcal{H}(\mathbb{R})$ denote the set of all compact subsets of $\mathbb{R}$. Define an operator $\mathcal{R} : \mathcal{H}(\mathbb{R}) \to \mathcal{H}(\mathbb{R})$ by

$$\mathcal{R}(X) = R_1(X) \cup R_2(X), \ X \in \mathcal{H}(\mathbb{R}).$$

Observe that if $X \in \mathcal{H}(\mathbb{R})$, then $\mathcal{R}(X) \in \mathcal{H}(\mathbb{R})$ since $R_1(X)$ and $R_2(X)$ are compact subsets of $\mathbb{R}$ (R_1, R_2 are continuous) and so $R_1(X) \cup R_2(X)$ is compact (either by Exercises 7.13.24(6) or the Bolzano–Weierstrass theorem).

The Cantor set $\mathbf{C}$ is a fixed point of the operator $\mathcal{R}$. This is a consequence of the self-similarity of the Cantor set since $T(\mathbf{C} \cap [0, \frac{1}{3}]) = \mathbf{C}$, and $T(\mathbf{C} \cap [\frac{2}{3}, 1]) = \mathbf{C}$, by Corollary 7.14.10 (see Sect. 7.14 for the definition of $T : [0, 1] \to [0, 1]$). Hence $R_1(\mathbf{C}) \cup R_2(\mathbf{C}) = \mathbf{C}$ (R_1, R_2 are the *inverse* maps of T on $[0, \frac{1}{3}]$ and $[\frac{2}{3}, 1]$, respectively).

These observations suggest the natural question as to whether we can find a complete metric on $\mathcal{H}(\mathbb{R})$ with respect to which $\mathcal{R}$ is a *contraction mapping*. If we can do this, then $\lim_{n \to \infty} \mathcal{R}^n(X) = \mathbf{C}$ for every compact subset X of $\mathbb{R}$. For example, take X to be a single point and then iterate by $\mathcal{R}$ to get the Cantor set!

© Springer International Publishing AG 2017
M. Field, *Essential Real Analysis*, Springer Undergraduate Mathematics Series,
https://doi.org/10.1007/978-3-319-67546-6_8

In this chapter we develop these ideas in the setting of the space of compact subsets of $\mathbb{R}^n$. There are two issues. First we need to find a complete metric on the space $\mathcal{H}(\mathbb{R}^n)$ of all compact subsets of $\mathbb{R}^n$. Then we need to identify an interesting class of maps which lead naturally to contraction mappings on $\mathcal{H}(\mathbb{R}^n)$.

8.1   The Space $\mathcal{H}(\mathbb{R}^n)$

Let $\mathcal{H}(\mathbb{R}^n)$ denote the set of all (non-empty) compact subsets of $\mathbb{R}^n$. Since a set $X \subset \mathbb{R}^n$ is compact iff X is *closed* and *bounded*, $\mathcal{H}(\mathbb{R}^n)$ is the set of all closed and bounded subsets of $\mathbb{R}^n$. Note that $\mathcal{H}(\mathbb{R}^n)$ contains all finite subsets of $\mathbb{R}^n$. In particular, we can regard $\mathbb{R}^n$ as a subset of $\mathcal{H}(\mathbb{R}^n)$ by the map $(x_1, \cdots, x_n) \mapsto \{(x_1, \cdots, x_n)\}$.

If $X, Y \in \mathcal{H}(\mathbb{R}^n)$ then $X \cup Y \in \mathcal{H}(\mathbb{R}^n)$ and, if $X \cap Y \neq \emptyset$, we also have $X \cap Y \in \mathcal{H}(\mathbb{R}^n)$.

Our aim in this section is to define a metric h on $\mathcal{H}(\mathbb{R}^n)$ for which $(\mathcal{H}(\mathbb{R}^n), h)$ is a *complete* metric space. The metric we construct is known as the *Hausdorff metric* (named after Felix Hausdorff, 1868–1942, one of the founders of topology). The main issues are finding a natural definition for h and the verification of the metric properties. In order to define h, we start by defining a positive function $\rho(A, B)$, $A, B \in \mathcal{H}(\mathbb{R}^n)$, such that $\rho(A, B) = 0$ iff $A \subset B$. Since ρ detects inclusion one way, it is natural to define $h(A, B) = \max\{\rho(A, B), \rho(B, A)\}$. We see that $h(A, B) = 0$ iff $A \subset B$ and $B \subset A$. That is, iff $A = B$. Roughly speaking, $\rho(A, B)$ will measure the greatest distance of points of A from the set B. This will be zero if $A \subset B$. See Fig. 8.1 where we illustrate the situation $B \subsetneq A$, where $\rho(A, B) > 0$ and $\rho(B, A) = 0$.

Now for the details. Let d denote the Euclidean metric on $\mathbb{R}^n$. We recall from Sect. 7.2 that if B is a non-empty subset of $\mathbb{R}^n$ and $a \in \mathbb{R}^n$, then the distance $d(a, B)$ from a to B is defined by

$$d(a, B) = \inf_{x \in B} d(a, x).$$

Fig. 8.1 Measuring the distance between sets

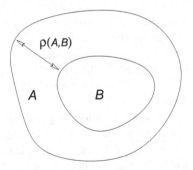

Since $\inf_{x \in B} d(a,x) = \inf\{d(a,x) \mid x \in B\}$ is bounded below by 0, $d(a,B)$ is defined and finite for every (non-empty) subset B of $\mathbb{R}^n$. Moreover, $d(a,B)$ continuous as a function of a (Proposition 7.12.3).

Now suppose $B \in \mathcal{H}(\mathbb{R}^n)$. Since $d(a,\cdot) : B \to \mathbb{R}, x \mapsto d(a,x)$, is continuous and B is compact, it follows from Theorem 7.13.9 that there exists an $x_0 \in B$ such that

$$d(a,B) = d(a,x_0).$$

In general, x_0 will not be unique. For future reference, note that

$$d(a,B) = d(a,x_0) \le d(a,x), \quad \text{for all } x \in B. \tag{8.1}$$

Lemma 8.1.1 *Let $B \in \mathcal{H}(\mathbb{R}^n)$ and $a \in \mathbb{R}^n$. Then $d(a,B) = 0$ iff $a \in B$.*

Proof Since B is closed, $a \in B$ iff $d(a,B) = 0$. □

Lemma 8.1.2 *Let $B \in \mathcal{H}(\mathbb{R}^n)$. Then $d(x,B)$ is a uniformly continuous function of $x \in \mathbb{R}^n$.*

Proof It follows from Proposition 7.2.1(4) that for all $x, \bar{x} \in \mathbb{R}^n$, we have

$$|d(x,B) - d(\bar{x},B)| \le d(\bar{x},x).$$

Hence $d(x,B)$ is uniformly continuous (given $\varepsilon > 0$, take $\delta = \varepsilon$). □

Given $A, B \in \mathcal{H}(\mathbb{R}^n)$, define

$$\rho(A,B) = \sup_{a \in A} d(a,B)$$
$$= \sup_{a \in A} \inf_{b \in B} d(a,b).$$

(For the finiteness of ρ it suffices that A is compact, B is closed.)

We can use ρ to detect inclusion of compact sets.

Lemma 8.1.3 *Let $A, B \in \mathcal{H}(\mathbb{R}^n)$. Then $\rho(A,B) = 0$ iff $A \subset B$.*

Proof Suppose $A \subset B$. Then $d(a,B) = 0$ for all $a \in A$ and so $\rho(A,B) = 0$. Conversely, suppose $\rho(A,B) = 0$. Then $\sup_{a \in A} d(a,B) = 0$. Hence $d(a,B) = 0$ for all $a \in A$. Hence $A \subset B$ (Lemma 8.1.1). □

Lemma 8.1.3 holds if A and B are closed (not necessarily compact). The next lemma makes essential use of compactness and fails if A and B are not compact.

Lemma 8.1.4 *Let $A, B \in \mathcal{H}(\mathbb{R}^n)$. Then there exist $a_0 \in A$, $b_0 \in B$ such that*

$$\rho(A,B) = d(a_0,b_0).$$

Moreover,

(a) $\rho(A,B) = d(a_0,b_0) \le d(a_0,b)$, *for all $b \in B$,*
(b) $\rho(A,B) = d(a_0,b_0) \ge d(a,b_0)$, *for all $a \in A$.*

Fig. 8.2 Lemma 8.1.4, parts
(a) and (b)

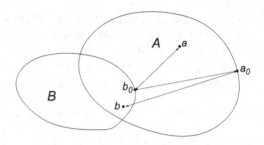

Proof Since $d(x, B)$ is continuous on A and A is compact, there exists an $a_0 \in A$ such that

$$\rho(A, B) = d(a_0, B).$$

By (8.1), there exists a $b_0 \in B$ such that

$$d(a_0, B) = d(a_0, b_0).$$

Hence $\rho(A, B) = d(a_0, b_0)$. The remaining statements are immediate from the definitions (see Fig. 8.2).   □

Lemma 8.1.5 *Let $A, X, Y \in \mathcal{H}(\mathbb{R}^n)$. If $X \subset Y$, then*

$$\rho(A, X) \geq \rho(A, Y).$$

Proof For all $a \in A$, $d(a, X) \geq d(a, Y)$ (since $X \subset Y$). Therefore, for all $a \in A$, we have

$$d(a, Y) \leq d(a, X) \leq \sup_{\bar{a} \in A} d(\bar{a}, X) = \rho(A, X).$$

That is, $\rho(A, X)$ is an upper bound for $d(a, Y)$, $a \in A$, and so

$$\rho(A, X) \geq \sup_{a \in A} d(a, Y) = \rho(A, Y).$$ □

The next result will be crucial for our main applications.

Lemma 8.1.6 *If $A, B, C, D \in \mathcal{H}(\mathbb{R}^n)$, then*

$$\rho(A \cup B, C \cup D) \leq \max\{\rho(A, C), \rho(B, D)\}.$$

Proof We claim that

$$\rho(A \cup B, C \cup D) = \max\{\rho(A, C \cup D), \rho(B, C \cup D)\}.$$

For this observe that

$$
\begin{aligned}
\rho(A \cup B, C \cup D) &= \sup_{x \in A \cup B} d(x, C \cup D) \\
&= d(x_0, C \cup D) \text{ for some } x_0 \in A \cup B \\
&= \rho(A, C \cup D) \text{ if } x_0 \in A \\
&= \rho(B, C \cup D) \text{ if } x_0 \in B \\
&= \max\{\rho(A, C \cup D), \rho(B, C \cup D)\}.
\end{aligned}
$$

Now by Lemma 8.1.5,

$$
\rho(A, C \cup D) \le \rho(A, C),
$$
$$
\rho(B, C \cup D) \le \rho(B, D).
$$

Hence $\rho(A \cup B, C \cup D) \le \max\{\rho(A, C), \rho(B, D)\}$. $\qquad\square$

Remark 8.1.7 For a more symmetric version of the lemma, see Exercises 8.1.13(2).

✠

Lemma 8.1.8 *Given $A, B, C \in \mathcal{H}(\mathbb{R}^n)$, we have*

$$
\rho(A, B) + \rho(B, C) \ge \rho(A, C).
$$

Proof By Lemma 8.1.4, there exist $a_0, a_1 \in A$, $b_0, b_1 \in B$ and $c_0, c_1 \in C$ such that

$$
\rho(A, B) = d(a_1, b_0), \quad \rho(B, C) = d(b_1, c_1), \quad \rho(A, C) = d(a_0, c_0).
$$

By Lemma 8.1.4(a), we have

$$
\rho(A, C) = d(a_0, c_0) \le d(a_0, c_1).
$$

By the triangle inequality for d, we have

$$
d(a_0, c_1) \le d(a_0, b_0) + d(b_0, c_1).
$$

By Lemma 8.1.4(b), we have

$$
\rho(A, B) = d(a_1, b_0) \ge d(a_0, b_0),
$$
$$
\rho(B, C) = d(b_1, c_1) \ge d(b_0, c_1).
$$

Hence

$$
\begin{aligned}
\rho(A, C) \le d(a_0, c_1) &\le d(a_0, b_0) + d(b_0, c_1) \\
&\le d(a_1, b_0) + d(b_1, c_1) \\
&= \rho(A, B) + \rho(B, C).
\end{aligned}
$$
$\qquad\square$

We define the *Hausdorff metric* h on $\mathcal{H}(\mathbb{R}^n)$ by

$$h(X, Y) = \max\{\rho(X, Y), \rho(Y, X)\}, \ X, Y \in \mathcal{H}(\mathbb{R}^n).$$

Theorem 8.1.9 *h defines a metric on $\mathcal{H}(\mathbb{R}^n)$. Moreover, for all $A, B, C, D \in \mathcal{H}(\mathbb{R}^n)$ we have*

$$h(A \cup B, C \cup D) \le \max\{h(A, C), h(B, D)\}.$$

Proof Obviously $h(X, Y) \ge 0$ for all $X, Y \in \mathcal{H}(\mathbb{R}^n)$.

If $h(X, Y) = 0$, then $\rho(X, Y) = \rho(Y, X) = 0$. If $\rho(X, Y) = 0$ then $X \subset Y$ (Lemma 8.1.3). Similarly, if $\rho(Y, X) = 0$, then $Y \subset X$. Hence if $h(X, Y) = 0$, then $X \subset Y \subset X$ and so $X = Y$.

It is immediate from the definition of h that $h(X, Y) = h(Y, X)$. It remains to prove the triangle inequality. For $X, Y, Z \in \mathcal{H}(\mathbb{R}^n)$, we have by Lemma 8.1.8,

$$\rho(X, Z) \le \rho(X, Y) + \rho(Y, Z),$$
$$\rho(Z, X) \le \rho(Z, Y) + \rho(Y, X)$$
$$= \rho(Y, X) + \rho(Z, Y).$$

Hence

$$\rho(X, Z), \rho(Z, X) \le \max\{\rho(X, Y), \rho(Y, X)\} + \max\{\rho(Y, Z), \rho(Z, Y)\},$$

and so

$$h(X, Z) = \max\{\rho(X, Z), \rho(Z, X)\}$$
$$\le \max\{\rho(X, Y), \rho(Y, X)\} + \max\{\rho(Y, Z), \rho(Z, Y)\}$$
$$= h(X, Y) + h(Y, Z).$$

The estimate for $h(A \cup B, C \cup D)$ follows from the corresponding result for ρ (Lemma 8.1.6). □

8.1.1 Completeness of $(\mathcal{H}(\mathbb{R}^n), h)$

We start by defining a useful family of closed compact neighbourhoods (in $\mathbb{R}^n$) of a point $X \in \mathcal{H}(\mathbb{R}^n)$. Suppose then that $X \in \mathcal{H}(\mathbb{R}^n)$ and let $r > 0$. Define $X(r) = \{x \in \mathbb{R}^n \mid d(x, X) \le r\}$. The set $X(r)$ is a closed neighbourhood of X regarded as a subset of $\mathbb{R}^n$. In particular, if $X = \{x_0\}$, $X(r) = \overline{D}_r(x_0)$ (closed r disk, centre x_0).

Lemma 8.1.10 *Let $X, Y \in \mathcal{H}(\mathbb{R}^n)$ and $r > 0$. Then $h(X, Y) \leq r$ iff $Y \subset X(r)$ and $X \subset Y(r)$.*

Proof Left to the exercises. $\qquad\qquad\square$

Remark 8.1.11 If we define $D_r(X) = \{Y \in \mathcal{H}(\mathbb{R}^n) \mid h(X, Y) \leq r\}$, then Lemma 8.1.10 implies that

$$D_r(X) = \{Y \in \mathcal{H}(\mathbb{R}^n) \mid Y \subset X(r), \ X \subset Y(r)\}. \qquad \maltese$$

We can use Lemma 8.1.10 to get a better understanding of *convergence* in $\mathcal{H}(\mathbb{R}^n)$. Suppose that $(X_n) \subset \mathcal{H}(\mathbb{R}^n)$ converges to X. This means that, given $\varepsilon > 0$, there exists an $N \in \mathbb{N}$ such that

$$h(X_n, X) < \varepsilon, \ n \geq N.$$

By Lemma 8.1.10, $X_n \subset X(\varepsilon)$, for all $n \geq N$. In particular, for $n \geq N$ large, X_n will be an ε-approximation to X: if you can only resolve detail to within ε, X_n will be indistinguishable from X for $n \geq N$.

Theorem 8.1.12 *$(\mathcal{H}(\mathbb{R}^n), h)$ is a complete metric space.*

Proof Suppose that (X_n) is a Cauchy sequence in $\mathcal{H}(\mathbb{R}^n)$, Since (X_n) is Cauchy, there exists an $N \in \mathbb{N}$ such that $h(X_n, X_m) \leq 1$, $n, m \geq N$. By Lemma 8.1.10, we have

$$X_n \subset X_N(1), \ \text{for all } n \geq N.$$

This means that we can assume all the X_n are subsets of some fixed compact subset Z of $\mathbb{R}^n$. Specifically $Z = X_1 \cup \cdots X_{N-1} \cup X_N(1)$.

We now follow the same strategy we used for Cauchy sequences in $\mathbb{R}^n$. We know the sequence (X_n) is bounded, so we look at the set of limit points. The next definition should look familiar. Define

$$\Lambda = \cap_{n \geq 1} \overline{\cup_{m \geq n} X_m}.$$

Each of the sets $\overline{\cup_{m \geq n} X_m}$ is compact, since every $X_m \subset Z$ and so for all $n \geq 1$, $\overline{\cup_{m \geq n} X_m} \subset Z$. Since Z is bounded and $\overline{\cup_{m \geq n} X_m}$ is closed, it follows by Bolzano–Weierstrass that $\overline{\cup_{m \geq n} X_m}$ is compact.

Now $\overline{\cup_{m \geq 1} X_m} \supset \overline{\cup_{m \geq 2} X_m} \supset \cdots$ is a decreasing sequence of non-empty compact subsets of $\mathbb{R}^n$ and so Λ is a non-empty compact subset of $\mathbb{R}^n$. It suffices to show that

$$\lim_{n \to \infty} X_n = \Lambda.$$

Choose $\varepsilon > 0$. Since (X_n) is Cauchy, there exists an $N_1 \in \mathbb{N}$ such that $h(X_n, X_m) \leq \varepsilon$, for all $n, m \geq N_1$. Since $X_m \subset X_n(\varepsilon)$, for all $n, m \geq N_1$ and $\Lambda = \cap_{n \geq p} \overline{\cup_{m \geq n} X_m}$, all

$p \geq 1$, we certainly have $\Lambda \subset X_n(\varepsilon)$, $n \geq N_1$. We claim that we can find $N_2 \in \mathbb{N}$ such that $X_n \subset \Lambda(\varepsilon)$, $n \geq N_2$. Assuming the claim, we then have $X_n \subset \Lambda(\varepsilon)$, $\Lambda \subset X_n(\varepsilon)$, for all $n \geq N = \max\{N_1, N_2\}$ and so, by Lemma 8.1.10, $h(X_n, \Lambda) \leq \varepsilon$, $n \geq N$, proving the convergence of (X_n) to Λ.

It remains to prove the claim. Suppose the contrary. Then for each $p \in \mathbb{N}$, there exists an $n \geq p$ such that $X_n \not\subset \Lambda(\varepsilon)$. Hence there exists an $x_n \in X_n$ such that $d(x_n, \Lambda) > \varepsilon$. Using this observation, we may construct a sequence (x_{n_k}) such that $x_{n_k} \in X_{n_k}$ and $d(x_{n_k}, \Lambda) > \varepsilon$, $k \in \mathbb{N}$. Since $(x_{n_k}) \subset Z$, it follows that (x_{n_k}) has a convergent subsequence (x_{m_k}) with limit $z \in Z$. By construction, $d(z, \Lambda) = \lim_{k \to \infty} d(x_{m_k}, \Lambda) \geq \varepsilon$. But $z \in \cap_{n \geq 1} \overline{\cup_{m \geq n} X_m} = \Lambda$ and so $d(z, \Lambda) = 0$. Contradiction. $\qquad\square$

EXERCISES 8.1.13

(1) Let $A, B \in \mathcal{H}(\mathbb{R}^n)$. Recall that

$$\rho(A, B) = \sup_{a \in A} \inf_{b \in B} d(a, b).$$

    (a) Show, by means of a (simple) example, that $\inf_{b \in B} \sup_{a \in A} d(a, b)$ does not generally equal $\rho(A, B)$.

    (b) Suppose $\inf_{b \in B} \sup_{a \in A} d(a, b) = 0$. What does this say about A and B?

(2) Show that if $A, B, C, D \in \mathcal{H}(\mathbb{R}^n)$, then

$$\rho(A \cup B, C \cup D) \leq \max\{\min\{\rho(A, C), \rho(A, D)\}, \min\{\rho(B, C), \rho(B, D)\}\}.$$

    (Hint: use the argument of the proof of Lemma 8.1.6.)

(3) Complete the proof of Lemma 8.1.10 by showing that if $A, B \in \mathcal{H}(\mathbb{R}^n)$, then $h(A, B) = \inf\{r \mid A(r) \subset B \text{ and } B(r) \subset A\}$.

(4) Prove that $(\mathcal{H}(\mathbb{R}^n), h)$ is a separable metric space. (Hint: define a countable dense subset E of $(\mathcal{H}(\mathbb{R}^n), h)$ which consists of finite sets—for example, $E = \{X \subset \mathbb{Q}^n \mid X \text{ finite}\}$.)

(5) Let (X, d) be a complete metric space and $\mathcal{H}(X)$ denote the set of compact subsets of X. Show how to define the Hausdorff metric h^X on $\mathcal{H}(X)$ and verify that $(\mathcal{H}(X), h^X)$ is complete.

(6) Let (X, d) be a metric space. Show that $(\mathcal{H}(X), h^X)$ is compact iff (X, d) is compact. (Caution: this needs the open cover definition of compactness—every open cover has a finite subcover—as there is no assumption that X is separable.)

(7) Let (X, d) be a metric space and suppose that $(x_n) \subset X$ is a Cauchy sequence. For $n \geq 1$, set $A_n = \{x_i \mid 1 \leq i \leq n\} \subset X$. Show that (A_n) is a Cauchy sequence in $(\mathcal{H}(X), h^X)$ (notation of previous example). Deduce that $(\mathcal{H}(X), h^X)$ is complete if and only if X is complete.

8.2 Iterated Function Systems

Recall from Chap. 7 that a map $f : \mathbb{R}^n \to \mathbb{R}^n$ is a contraction if there exists a $k \in [0, 1)$ such that

$$d(f(x), f(y)) \leq kd(x, y), \text{ for all } x, y \in \mathbb{R}^n,$$

and that we call the smallest value of k for which this estimate holds the contraction constant of f.

Lemma 8.2.1 *Let* $f : \mathbb{R}^n \to \mathbb{R}^n$ *be a contraction map with contraction constant* k. *If we define* $\mathcal{F} : \mathcal{H}(\mathbb{R}^n) \to \mathcal{H}(\mathbb{R}^n)$ *by* $\mathcal{F}(X) = f(X)$, *then* $\mathcal{F}$ *is a contraction mapping with contraction constant* k.

Proof We have to show $h(f(X), f(Y)) \leq kh(X, Y)$ for all $X, Y \in \mathcal{H}(\mathbb{R}^n)$. Since $h(X, Y) = \max\{\rho(X, Y), \rho(Y, X)\}$, it suffices to show $\rho(f(X), f(Y)) \leq k\rho(X, Y)$ for all $X, Y \in \mathcal{H}(\mathbb{R}^n)$. We have

$$
\begin{aligned}
\rho(f(X), f(Y)) &= \sup_{x \in X} \inf_{y \in Y} d(f(x), f(y)) \\
&\leq \sup_{x \in X} \inf_{y \in Y} kd(x, y) \\
&= k \sup_{x \in X} \inf_{y \in Y} d(x, y) \\
&= k\rho(X, Y).
\end{aligned}
$$

If we take X, Y to be the point sets $\{x\}$, $\{y\}$, we see that k is the contraction constant of $\mathcal{F}$. $\qquad\square$

Suppose that we are given continuous functions $f_1, \cdots, f_p : \mathbb{R}^n \to \mathbb{R}^n$. We define the operator[1] $\mathcal{F} : \mathcal{H}(\mathbb{R}^n) \to \mathcal{H}(\mathbb{R}^n)$ by

$$\mathcal{F}(X) = f_1(X) \cup \cdots \cup f_p(X), \ X \in \mathcal{H}(\mathbb{R}^n).$$

Note that $\mathcal{F}$ does take values in $\mathcal{H}(\mathbb{R}^n)$. Indeed, since the f_i are assumed continuous, each $f_i(X)$ is a compact subset of $\mathbb{R}^n$. Since we have a finite union of compact sets, $\mathcal{F}(X)$ is compact and so $\mathcal{F}(X) \in \mathcal{H}(\mathbb{R}^n)$ for all $X \in \mathcal{H}(\mathbb{R}^n)$ (Exercises 7.13.24(6)).

Now assume that each f_i is a contraction. If f_i has a contraction constant $k_i < 1$, then taking $k = \max_i k_i < 1$, we can assume the f_i have a common contraction constant k.

Proposition 8.2.2 (Notation and Assumptions as Above) *The operator* $\mathcal{F} :$ $\mathcal{H}(\mathbb{R}^n) \to \mathcal{H}(\mathbb{R}^n)$ *is a contraction map.*

---

[1] We prefer to use the term 'operator' rather than 'map'.

Proof Let $X, Y \in \mathcal{H}(\mathbb{R}^n)$. We have

$$h(\mathcal{F}(X), \mathcal{F}(Y)) = h(f_1(X) \cup \cdots \cup f_p(X), f_1(Y) \cup \cdots \cup f_p(Y)).$$

We have $h(A \cup B, C \cup D) \leq \max\{h(A, C), h(B, D)\}$ for all $A, B, C, D \in \mathcal{H}(\mathbb{R}^n)$. Applying this result repeatedly to the right-hand side of the expression for $h(\mathcal{F}(X), \mathcal{F}(Y))$ gives

$$h(\mathcal{F}(X), \mathcal{F}(Y)) \leq \max_{1 \leq i \leq p} h(f_i(X), f_i(Y)).$$

By Lemma 8.2.1, $h(f_i(X), f_i(Y)) \leq kh(X, Y)$, $1 \leq i \leq p$, and so we have shown $h(\mathcal{F}(X), \mathcal{F}(Y)) \leq kh(X, Y)$. $\square$

Corollary 8.2.3 (Notation and Assumptions as Above) *The operator* $\mathcal{F}$: $\mathcal{H}(\mathbb{R}^n) \to \mathcal{H}(\mathbb{R}^n)$ *has a* unique *fixed point* $X^* \in \mathcal{H}(\mathbb{R}^n)$. *Moreover,* $\mathcal{F}(X^*) = X^*$ *iff*

$$X^* = f_1(X^*) \cup \cdots \cup f_p(X^*). \tag{8.2}$$

Proof Apply the contraction mapping lemma. $\square$

Remarks 8.2.4

(1) Suppose we have a finite set of contraction maps $f_1, \cdots, f_p$ of $\mathbb{R}^n$. Start with any compact subset X of $\mathbb{R}^n$ (for example a single point). Iterate $\mathcal{F}$ and define $X_n = \mathcal{F}^n(X)$. Then X_n always converges to the same compact subset of $\mathbb{R}^n$, independent of the initial set X.

(2) Equation (8.2) shows that the fixed point X^* of the operator $\mathcal{F}$ has the property that it is the union of scaled-down copies of itself. This property is a form of self-similarity. We have already seen self-similarity in the Cantor set **C** and we shall shortly give some striking visual examples of self-similarity. We remark that sets that exhibit self-similarity at all scales are often called *fractals*.

(3) Proposition 8.2.2 applies equally well to the set $\mathcal{H}(X)$ of compact subsets of any complete metric space X. We refer to a finite set $\{f_i\}$ of contractions of X as an *iterated function system* or *IFS*. John Hutchinson showed in 1981 [15] that the operator associated to an IFS had a unique fixed point. Subsequently, iterated function systems, and their associated fractals, were popularized in Michael Barnsley's book *Fractals Everywhere* [2].   ✠

EXERCISES 8.2.5

(1) Suppose that $f_1, \cdots, f_p$ are contractions of $\mathbb{R}^n$. Let f_j have fixed point $x_j^* \in \mathbb{R}^n$, $j \in \mathbf{p}$. Show that if $X^* \in \mathcal{H}(\mathbb{R}^n)$ is the fixed point of $\mathcal{F}$ given by Corollary 8.2.3, then $x_j^* \in X^*$, for all $j \in \mathbf{p}$. Deduce that if $a_1, \cdots, a_n \in \mathbf{p}$, then for all $j \in \mathbf{p}$, $f_{a_n} \cdots f_{a_1}(x_j^*) \in X^*$.

(2) (Notation of previous question). Show that if $p = 2$, then it is possible to choose f_1, f_2 so that X^* consists of exactly two points. Let $m > 2$. Can we choose f_1, f_2 so that X^* consists of m points? What about if f_1, f_2 are affine linear contractions?

8.3 Examples of Iterated Function Systems

An *affine linear map* of $\mathbb{R}^n$ is a mapping $L : \mathbb{R}^n \to \mathbb{R}^n$ which can be written in the form

$$Lx = Ax + \mathbf{b}, \ \mathbf{x} \in \mathbb{R}^n,$$

where A is a linear mapping of $\mathbb{R}^n$ ($n \times n$ matrix) and $\mathbf{b} \in \mathbb{R}^n$.

Lemma 8.3.1

(1) *An affine linear map $Lx = ax + b$ of $\mathbb{R}$ is a contraction iff $|a| < 1$.*
(2) *The affine linear map of $(\mathbb{R}^2, d_2)$ given by*

$$L(x, y) = \begin{pmatrix} a & b \\ c & d \end{pmatrix} \begin{pmatrix} x \\ y \end{pmatrix} + \begin{pmatrix} e \\ f \end{pmatrix}$$

is a contraction iff

$$a^2 + c^2 < 1,$$
$$b^2 + d^2 < 1,$$
$$a^2 + b^2 + c^2 + d^2 < 1 + (ad - bc)^2.$$

Proof

(1) For all $x, y \in \mathbb{R}$, we have $|Lx - Ly| = |a(x - y)| = |a||x - y|$. Hence L is a contraction iff $|a| < 1$.
(2) Since $d(Lx_1, Lx_2) = \|A(x_1 - x_2)\|$, $x_1, x_2 \in \mathbb{R}^2$, L is a contraction iff A is a contraction. The linear map A is a contraction iff given $(x, y) \neq (0, 0)$ we have

$$\|A(x, y)\|^2 = (ax + by)^2 + (cx + dy)^2 < x^2 + y^2.$$

The contraction constant of A will then be $\sup\{\|A(x, y)\| \mid x^2 + y^2 = 1\}$. Now $(ax + by)^2 + (cx + dy)^2 < x^2 + y^2$ iff

$$x^2(a^2 + c^2 - 1) + 2xy(ab + cd) + y^2(b^2 + d^2 - 1) < 0.$$

This condition holds for all $(x, y) \neq (0, 0)$ iff $a^2 + c^2 < 1$, $b^2 + d^2 < 1$ and $(a^2 + c^2 - 1)(b^2 + d^2 - 1) - (ab + cd)^2 > 0$. The last condition simplifies to $a^2 + b^2 + c^2 + d^2 < 1 + (ad - bc)^2$. $\qquad\square$

Fig. 8.3 The IFS
$\{\rho_A, \rho_B, \rho_C\}$

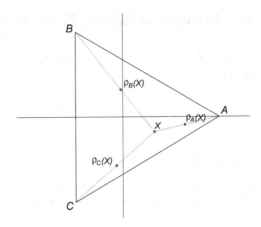

8.3.1   The Sierpiński Triangle (or Gasket)

Fix the equilateral triangle $\triangle ABC$ in the plane with vertices $A = (1, 0)$, $B = (-\frac{1}{2}, \frac{\sqrt{3}}{2})$, $C = (-\frac{1}{2}, -\frac{\sqrt{3}}{2})$, and note that the centre of this triangle is the origin of $\mathbb{R}^2$ (see Fig. 8.3).

We define affine linear contractions ρ_A, ρ_B and ρ_C of $\mathbb{R}^2$. Let $X = (x, y) \in \mathbb{R}^2$. Define

$$\rho_A(X) = \frac{X}{2} + \left(\frac{1}{2}, 0\right).$$

Observe that ρ_A is a contraction with contraction constant $k_A = \frac{1}{2}$ and unique fixed point A. Similarly, define

$$\rho_B(X) = \frac{X}{2} + \left(-\frac{1}{4}, \frac{\sqrt{3}}{4}\right),$$

which has fixed point B, and

$$\rho_C(X) = \frac{X}{2} - \left(\frac{1}{4}, \frac{\sqrt{3}}{4}\right),$$

which has fixed point C. Each of these maps moves X exactly halfway to the corresponding vertex (see Fig. 8.3) and all the maps have the same contraction constant $\frac{1}{2}$.

Let $S : \mathcal{H}(\mathbb{R}^2) \to \mathcal{H}(\mathbb{R}^2)$ be the operator defined by the IFS $\{\rho_A, \rho_B, \rho_C\}$. It follows from Corollary 8.2.3 that S has a unique fixed point. In Fig. 8.4 we show the first two iterates of the map S where as our initial point we have taken the (filled-in) triangle $\triangle ABC$.

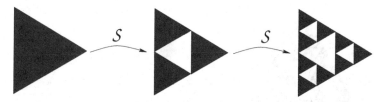

Fig. 8.4 The first two iterates of $\mathcal{S}$

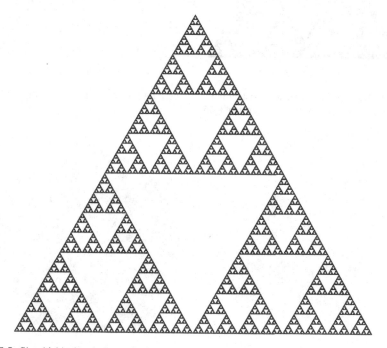

Fig. 8.5 Sierpiński triangle (or gasket)

In Fig. 8.5 we show a visualization of the fixed point. This compact subset of $\mathbb{R}^2$ is known as the Sierpiński triangle or Sierpiński gasket. Just as for the Cantor set, the Sierpiński triangle is self-similar. Each of the little triangles making up the Sierpiński triangle is a scaled down copy of the Sierpiński triangle.

8.3.2   Four Variations on the Sierpiński Triangle

In Fig. 8.6a, we show the effect of increasing the contraction constant from 0.5 to 0.55. Observe that there is now an overlap occurring if we iterate the filled-in triangle $\triangle ABC$. On the other hand if we decrease the contraction constant from 0.5 to 0.45, we get the effect shown in Fig. 8.6b. Finally, in Fig. 8.7, we show the effect

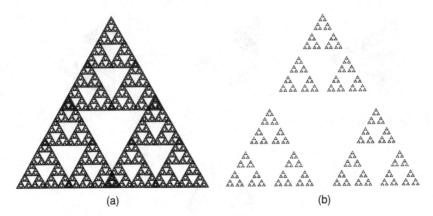

Fig. 8.6 Varying the contraction constant in the IFS used for the Sierpiński triangle. (a) Contraction constant 0.55, (b) contraction constant 0.45.

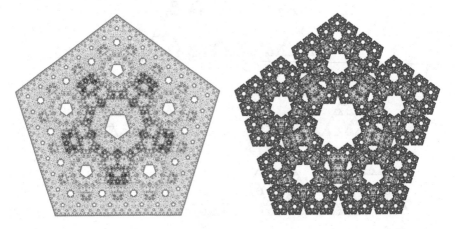

Fig. 8.7 Sierpiński pentagons: (**a**) contraction constant 0.5, (**b**) contraction constant 0.45.

of increasing the number of elements in the IFS to five. For both images shown in Fig. 8.7, we have taken five contractions, one for each vertex of a regular pentagon. In Fig. 8.7a, the contraction constants were all 0.5; in Fig. 8.7b, the contraction constants were all 0.45. We explain the "grey scale" colouring used in Fig. 8.7 in the paragraph on *random iteration* in Sect. 8.4.1.

EXERCISES 8.3.2

(1) Define

$$L_1(x) = \frac{x}{3}, \quad L_2(x) = \frac{x}{3} + \frac{2}{3}, \ x \in \mathbb{R}.$$

Show that $\{L_1, L_2\}$ is an iterated function system with fixed point the middle thirds Cantor set $\mathbf{C}$. (This IFS contracts by $1/3$ from the points $0, 1$—compare with the IFS giving the Sierpiński triangle.)

(2) Let $\{f_1, f_2\}$ be the IFS given by

$$f_1(x, y) = \begin{pmatrix} 0.4000 & -0.3733 \\ 0.0600 & 0.6000 \end{pmatrix} \begin{pmatrix} x \\ y \end{pmatrix} + \begin{pmatrix} 0.3533 \\ 0.0000 \end{pmatrix},$$

$$f_2(x, y) = \begin{pmatrix} -0.8000 & -0.1867 \\ 0.1371 & 0.8000 \end{pmatrix} \begin{pmatrix} x \\ y \end{pmatrix} + \begin{pmatrix} 1.1000 \\ 0.1000 \end{pmatrix}.$$

Verify that f_1 and f_2 are contractions. If you have access to a computer with *Matlab* or *Mathematica*, plot the resulting image you get with this IFS. (Use random iteration—see the next section.)

(3) Prove that the Sierpiński triangle is connected. What about the fractals in Figs. 8.6, 8.7? (Hint: Exercises 7.18.20(21).)

(4) Show that the Sierpiński triangle is path connected.

(5) Suppose that instead of the Euclidean metric on $\mathbb{R}^n$, we use the metric

$$d_\infty(\mathbf{x}, \mathbf{y}) = \max_{1 \le i \le n} |x_i - y_i|, \quad \mathbf{x} = (x_1, \cdots, x_n), \mathbf{y} = (y_1, \cdots, y_n).$$

Show that the affine linear map $L\mathbf{x} = A\mathbf{x} + \mathbf{b}$ is a contraction (with respect to d_∞) iff

$$\max_{1 \le i \le n} \left(\sum_{j=1}^n |a_{ij}| \right) < 1.$$

Find an example of an affine linear map of $\mathbb{R}^2$ which is a contraction with respect to d_∞ but not the Euclidean metric d_2.

(6) Suppose that $\{f_1, \cdots, f_k\}$ is a set of affine linear maps of $\mathbb{R}^n$ which are contractions of $(\mathbb{R}^n, d_\infty)$. Show that the operator $\mathcal{F} : \mathcal{H}(\mathbb{R}^n) \to \mathcal{H}(\mathbb{R}^n)$ defined by $\mathcal{F}(X) = \cup_{i=1}^p f_i(X)$ has a unique fixed point even though $\mathcal{F}$ may not be a contraction of $(\mathcal{H}(\mathbb{R}^n), h)$. (Hint: change the metric to 'h_∞'.)

(7) A necessary condition for an affine linear map $L\mathbf{x} = A\mathbf{x} + \mathbf{b}$ of $\mathbb{R}^n$ to be a contraction is that all the eigenvalues of A have modulus less than 1 (see Exercises 7.17.16(2)). Conversely, if this condition holds it can be shown (using Jordan normal form) that there exists a norm on $\mathbb{R}^n$ with respect to which L is a contraction. This suggests that if we have a finite set of affine linear maps $L_i\mathbf{x} = A_i\mathbf{x} + \mathbf{b}_i$ such that each A_i has all eigenvalues of modulus less than 1, then the corresponding IFS has a unique fixed point that can be obtained by iteration. Find an example in $\mathbb{R}^2$ with just two maps that shows this conclusion is false. (Hint: take A_1 to be the composition of rotation through $\pi/2$ with the diagonal matrix $[d_1, d_2]$ where $d_1 d_2 \in (0, 1)$ and $d_1 < 1 < d_2$ and A_2 to be

the composition of the rotation through $\pi/2$ with the diagonal matrix $[d_2, d_1]$. These ideas have implications in control theory—see [23, Chap. 1].)

(8) Show that the product $\mathbf{C}^2 = \{(c_1, c_2) \mid c_1, c_2 \in \mathbf{C}\}$ of two middle-thirds Cantor sets can be represented as the unique fixed point of the iterated function system $\mathbf{I} = \{L_{ij} \mid i, j \in \{0, 1\}\}$, where $L_{ij}(\mathbf{x}) = \frac{1}{3}(\mathbf{x} - \mathbf{v}_{ij}) + \mathbf{v}_{ij}$, $\mathbf{x} \in \mathbb{R}^2$, and $\mathbf{v}_{ij} = (i, j)$ are vertices of the unit square $[0, 1]^2$ in $\mathbb{R}^2$. Let $\mathcal{F} : \mathcal{H}(\mathbb{R}^2) \to \mathcal{H}(\mathbb{R}^2)$ denote the operator determined by $\mathbf{I}$. Let $a \in [0, 2]$ and ℓ_a denote the line $x + y = a$. Show that $\ell_a \cap \mathcal{F}^n([0, 1]^2) \neq \emptyset$ for all $a \in [0, 2]$, $n \geq 1$. Deduce that $\mathbf{C} + \mathbf{C} = [0, 2]$. (Hints: use exercise (1) and show that it is enough to prove $\ell_a \cap \mathcal{F}([0, 1]^2) \neq \emptyset$ for all $a \in [0, 2]$.)

(9) Let $r \in (0, \frac{1}{2}]$ and $\mathbf{C}_r$ denote the set defined by the iterated function system $\{L_i^r \mid i \in \{0, 1\}\}$, where $L_i^r(x) = r(x - i) + i$, $i \in \{0, 1\}$. Show that

   (a) $\mathbf{C}_r$ is a Cantor set (Definition 7.14.19) if and only if $r < \frac{1}{2}$. What is $\mathbf{C}_{\frac{1}{2}}$?
   (b) $\mathbf{C}_r + \mathbf{C}_r \subset [0, 2]$ with equality if and only if $r \in [\frac{1}{3}, \frac{1}{2}]$.
   (c) If $r < \frac{1}{3}$, then $\mathbf{C}_r + \mathbf{C}_r \subset [0, 2]$ is a Cantor set and $[0, 2] \setminus (\mathbf{C}_r + \mathbf{C}_r)$ is a disjoint union of open intervals of total length 2.

8.4 Concluding Remarks

8.4.1 Computing the Fixed Point of an IFS

Suppose we are given an IFS $\{f_i \mid i = 1, \cdots, p\}$, where each $f_i : \mathbb{R}^2 \to \mathbb{R}^2$ is an affine linear contraction with contraction constant k_i. Set $k = \max_i k_i$ and let $\mathcal{F} : \mathcal{H}(\mathbb{R}^2) \to \mathcal{H}(\mathbb{R}^2)$ be the contraction induced by the IFS. It follows from the second part of the contraction mapping lemma that in order to compute the fixed point $X^\star$ of $\mathcal{F}$, we can start with any initial $X_0 \in \mathcal{H}(\mathbb{R}^2)$ and iterate by $\mathcal{F}$. If $h(X_0, X^\star) = C$, then after n-iterations we have the estimate $h(\mathcal{F}^n(X), X^\star) \leq k^n C$. In particular, we can take $X_0 = \{x_0\}$, a point in $\mathbb{R}^2$. We have $h(x_0, X^\star) = \sup_{x \in X^\star} d(x_0, x)$, where d is the Euclidean metric on $\mathbb{R}^2$. Now $X_1 = \mathcal{F}(X_0)$ consists of (at most) p-points, $X_2 = \mathcal{F}^2(X_0)$ at most p^2 points, and so on. After n iterations, we get a compact subset X_n of $\mathbb{R}^2$ containing at most p^n points. This process works reasonably well for the Sierpiński triangle $\mathbf{S}$ where the associated IFS has $p = 3$ and $k = 1/2$. If we want to approximate the triangle to within 10^{-4} and we start with $X_0 = \{x_0\}$, where $h(x_0, \mathbf{S}) = 1$, we need to choose n so that $(1/2)^n < 10^{-4}$. The set X_n will then consist of at most 3^n points. Computing we find that it suffices to take $n = 14$ and then the number of points in X_{14} will be at most $4,782,969$. Although this is not hard to work out on a computer, note how the size of the array of numbers we need to store triples at every step. On the other hand, suppose that p is larger, say $p = 10$ and the contraction rate k is bigger, say $k = 0.9$. To get an approximation within 10^{-4}, we need to choose n so that $(9/10)^n < 10^{-4}$—that is, $9^n < 10^{4-n}$—and the number of points at the nth step will then be 10^n. Computing we find that, $n = 68$ and the number of points at step n is 10^{68}. This is now completely unrealistic to simulate on a computer. While it is possible to refine this technique for computing the limit set, we prefer to emphasize an alternative approach based on the idea of random iteration.

Random Iteration There is another way to compute the fixed point of an IFS that is computationally economical and fast. What we do is perform a *random iteration*. Fix an initial point $x_0 \in \mathbb{R}^2$. Suppose there are p functions in the IFS. Successively pick elements of the IFS with equal probability $\frac{1}{p}$. Supposing we get the random sequence $f_{i_1}, f_{i_2}, \cdots, f_{i_k}, \cdots$ of functions, we define the sequence $(x_n) \subset \mathbb{R}^2$ by $x_n = f_{i_n}(x_{n-1})$, $n \geq 1$. It may be proved that, with probability 1, the set of limit points of the sequence (x_n) is equal to the fixed point set of the IFS. In practice, this scheme often converges very rapidly. All we do is throw away the first few points as transient, and then keep iterating and plotting until the image has stabilized. We can sometimes improve the rate of convergence by choosing f_i with probability $p_i \in (0, 1)$, where $\sum_i p_i = 1$ and we do not necessarily assume $p_i = 1/p$. All the images of fractals shown thus far in this chapter were computed using random iteration. The grey scale colouring of the fractals in Fig. 8.7 gives a representation of the frequency with which points of the iteration visit regions of the fractal. For example, in Fig. 8.7a, the dark interior region is frequently visited, while the boundary of the fractal is infreqently visited. We refer the reader to the references at the end of the chapter for more information and examples.

8.4.2 The Collage Theorem

The collage theorem gives a constructive scheme for approximating a compact subset of $\mathbb{R}^2$ (more generally, $\mathbb{R}^n$) arbitrarily closely by the fixed point of an IFS consisting of affine linear contractions. More precisely, given $X \in \mathcal{H}(\mathbb{R}^2)$ and $\varepsilon > 0$, there exists an IFS $\{f_1, \cdots, f_N\}$ such that $h(X, X^\star) < \varepsilon$, where $X^\star$ is the fixed point of the IFS. Since it is computationally cheap to generate the fixed point of an IFS using random iteration, these ideas have been used in image compression. We refer to Barnsley's book [2] for details on the mathematical theory. In Fig. 8.8 we show a 'fractal fern', this was computed using the IFS $\{f_1, f_2, f_3, f_4\}$ where

$$f_1(x, y) = \begin{pmatrix} 0.7 & 0 \\ 0 & 0.7 \end{pmatrix} \begin{pmatrix} x \\ y \end{pmatrix} + \begin{pmatrix} 0.1496 \\ -0.2962 \end{pmatrix},$$

$$f_2(x, y) = \begin{pmatrix} 0.1 & 0.433 \\ -0.1732 & 0.25 \end{pmatrix} \begin{pmatrix} x \\ y \end{pmatrix} + \begin{pmatrix} 0.4478 \\ -0.0014 \end{pmatrix},$$

$$f_3(x, y) = \begin{pmatrix} 0.1 & -0.433 \\ 0.1732 & 0.25 \end{pmatrix} \begin{pmatrix} x \\ y \end{pmatrix} + \begin{pmatrix} 0.4445 \\ -0.1559 \end{pmatrix},$$

$$f_4(x, y) = \begin{pmatrix} 0 & 0 \\ 0 & 0.3 \end{pmatrix} \begin{pmatrix} x \\ y \end{pmatrix} + \begin{pmatrix} 0.4987 \\ -0.007 \end{pmatrix}.$$

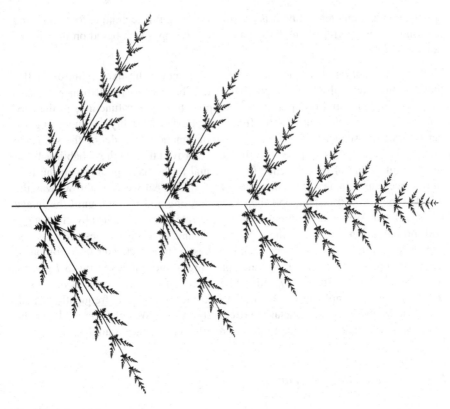

Fig. 8.8 A fractal fern

Other sources on fractals include the classic book by Benoit Mandelbrot, *The Fractal Geometry of Nature* [24], and the book by Heinz-Otto Peitgen and Peter H. Richter, *The Beauty of Fractals* [26]. These books show some of the potential for fractal-based artwork. Techniques used for making fractal landscapes and images have been used to create special effects scenes in a number of Hollywood movies, most notably in the *Star Wars* series; *Star Trek: The Wrath of Khan*; and in the *Lord of the Rings* trilogy. For a mix of fractals and symmetry, and some mathematics, we refer to *Symmetry in Chaos* [11, Chap. 7]. For an introduction to the mathematical theory of fractals, we suggest the book by Falconer, *Fractal Geometry: Mathematical Foundations and Applications* [8].

8.4.3   The Power of Abstraction and Generalization

In Chap. 2, we gave a proof of the foundational theorem from Calculus that every continuous function on a closed interval is bounded and takes all values between its upper and lower bounds.[2] The proof was tricky—it depended crucially on properties of the real numbers. In the previous chapter on metric space, we developed an abstract framework for the study of results of this type and introduced a range of new concepts, such as compactness and connectedness, which abstracted the key properties of the closed interval and real numbers needed for the proof of the foundational theorem. These concepts defined the precise *structure* needed for the proof of general results. The power of this approach can be seen in the present chapter. We have progressed from the relatively mundane study of real-valued functions on the real line to the analysis of operators defined on spaces whose points are compact subsets of $\mathbb{R}^n$. Fixed points are now compact sets rather than points on the real line. The spaces we deal with—such as spaces of compact sets or spaces of functions—may be infinite-dimensional and beyond simple visualization.

The moral is that problems in mathematics (and science) that are simple to state often require methods and concepts that are of great generality and abstraction for their solution.[3] This is the nature and power of mathematics. Finding the underlying structure—the crucial ideas—and then developing an abstract framework which includes the essential and excludes the inessential.

---

[2]This result appears in some form, often without proof, in every undergraduate or high school text on Calculus.

[3]*The paradox is now fully established that the utmost abstractions are the true weapons with which to control our thought of concrete fact.* Alfred North Whitehead, from *Science and the Modern World* [29].

Chapter 9
Differential Calculus on $\mathbb{R}^m$

In this chapter we develop the differential calculus on $\mathbb{R}^m$. The key concept is that of the *derivative*, which we view as the 'best linear approximation' to a function rather than as the limit of a quotient (as is done in the theory of differentiable maps $f : \mathbb{R} \to \mathbb{R}$). All of what we do is independent of norm and choice of coordinate system on $\mathbb{R}^m$. Linear (and multi-linear) maps between normed vector spaces play a central role in the theory. Consequently, we start by developing and reviewing the theory of continuous linear maps between finite-dimensional normed vector spaces. Proofs of some additional properties of finite-dimensional normed vector spaces, including the equivalence of all norms on a given finite-dimensional vector space, are given in an appendix at the end of the chapter. With these preliminaries out of the way, we develop in a coordinate-free way the theory of the derivative. Next, using the contraction mapping lemma, we prove the C^1 versions of the implicit and inverse function theorems, the rank theorem, and the existence and uniqueness theorem for ordinary differential equations. So as to simplify the notation, we initially assume functions are defined on $\mathbb{R}^m$, rather than on an open subset of $\mathbb{R}^m$—all definitions and results extend without difficulty to functions defined on open subsets of $\mathbb{R}^m$. In the remainder of the chapter, we develop the theory of higher derivatives and prove C^r versions of the chain rule, the inverse and implicit function theorems, and Taylor's theorem. All of this will require some preliminaries on multi-linear maps and polynomial maps between vector spaces. We conclude with the C^r version of the existence theorem for ordinary differential equations—including the C^r dependence on initial conditions.

9.1 Normed Vector Spaces

Suppose that V is a finite-dimensional real vector space. If the dimension of V is m, we set $\dim(V) = m$. The choice of a basis $\{\mathbf{v}_1, \cdots, \mathbf{v}_m\}$ for V uniquely determines a linear isomorphism $A : V \to \mathbb{R}^m$ by $A\mathbf{v}_i = \mathbf{e}_i$, $1 \leq i \leq m$, where $\{\mathbf{e}_1, \cdots, \mathbf{e}_m\}$

© Springer International Publishing AG 2017

M. Field, *Essential Real Analysis*, Springer Undergraduate Mathematics Series,
https://doi.org/10.1007/978-3-319-67546-6_9

denotes the standard basis of $\mathbb{R}^m$ consisting of unit vectors along each coordinate axis.[1] In terms of coordinates, if $\mathbf{x} = \sum_{i=1}^m x_i \mathbf{v}_i \in V$, then $A\mathbf{x}$ has coordinates $(x_1, \ldots, x_m) \in \mathbb{R}^m$.

We recall the definition of a norm on V.

Definition 9.1.1 Let V be a vector space. A *norm* on V is a map $\|\cdot\| : V \to \mathbb{R}$ satisfying

(1) $\|\mathbf{v}\| \geq 0$ for all $\mathbf{v} \in V$.
(2) $\|\mathbf{v}\| = 0$ iff $\mathbf{v} = 0$.
(3) $\|a\mathbf{v}\| = |a| \|\mathbf{v}\|$ for all $a \in \mathbb{R}$ and $\mathbf{v} \in V$.
(4) $\|\mathbf{v} + \mathbf{w}\| \leq \|\mathbf{v}\| + \|\mathbf{w}\|$ for all $\mathbf{v}, \mathbf{w} \in V$ (triangle inequality).

We call $(V, \|\cdot\|)$ a *normed vector space*.

If $(V, \|\cdot\|)$ is a normed vector space, we define the associated metric d on V by

$$d(\mathbf{v}, \mathbf{w}) = \|\mathbf{v} - \mathbf{w}\|, \quad \mathbf{v}, \mathbf{w} \in V.$$

It is conceivable that different norms on V could define metrics which have different topologies. While this certainly can and does happen if V is infinite-dimensional (see Exercises 7.1.9(11)), it turns out that all norms define the same topology on a finite-dimensional vector space. Before we state the precise result, we need a definition.

Definition 9.1.2 Two norms $\|\cdot\|_1$ and $\|\cdot\|_2$ on a vector space V are *equivalent* if there exists a $C \geq 1$ such that

$$C^{-1}\|\mathbf{v}\|_1 \leq \|\mathbf{v}\|_2 \leq C\|\mathbf{v}\|_1, \quad \text{for all } \mathbf{v} \in V.$$

Remarks 9.1.3

(1) Observe that if the condition of the definition holds, then $C^{-1}\|\mathbf{v}\|_2 \leq \|\mathbf{v}\|_1 \leq C\|\mathbf{v}\|_2$ and so the definition is symmetrical in the two norms. It is also clear that if we can find $c', C' > 0$ such that $c'\|\mathbf{v}\|_1 \leq \|\mathbf{v}\|_2 \leq C'\|\mathbf{v}\|_1$ for all $\mathbf{v} \in V$, then the conditions of the definition are satisfied with $C = \max\{C', 1/c'\}$.
(2) Equivalent norms on V define equivalent metrics on V. Consequently, equivalent norms define the same topology of open subsets of V (Exercises 7.4.27(6)) and so have the same continuous functions.   ✳

We give the proof of the next theorem in the appendix at the end of the chapter.

Theorem 9.1.4 *Any two norms on a finite-dimensional vector space V are equivalent. In particular,*

(1) *all norms define the same topology on V,*
(2) *$(V, \|\cdot\|)$ is complete with respect to any norm on V.*

---

[1] We write $A\mathbf{x}$ rather than $A(\mathbf{x})$ when A is a linear map.

Lemma 9.1.5 *Let* $(V, \| \cdot \|_V)$ *be an m-dimensional normed vector space and* $A :$ $V \to \mathbb{R}^m$ *be a linear isomorphism. If we define* $\|\mathbf{x}\| = \|A^{-1}\mathbf{x}\|_V$, $\mathbf{x} \in \mathbb{R}^m$, *then* $\| \cdot \|$ *is a norm on* $\mathbb{R}^m$ *and the topology of open sets on* $\mathbb{R}^m$ *defined by* $\| \cdot \|$ *is the same as that defined by the Euclidean norm on* $\mathbb{R}^m$. *Moreover,* $A : V \to \mathbb{R}^m$ *is a norm-preserving linear homeomorphism:*

$$\|\mathbf{x}\|_V = \|A\mathbf{x}\|, \text{ for all } \mathbf{x} \in V.$$

Proof We leave the verification that $\| \cdot \|$ defines a norm on $\mathbb{R}^m$ as an exercise for the reader. The statement about the topology of open sets on $\mathbb{R}^m$ follows from Theorem 9.1.4. $\qquad\square$

Example 9.1.6 Let V have basis $\{\mathbf{v}_1, \cdots, \mathbf{v}_m\}$ and define the linear isomorphism $A : V \to \mathbb{R}^m$ by $A\mathbf{v}_i = \mathbf{e}_i$, $1 \le i \le m$. Every norm $\| \cdot \|$ on $\mathbb{R}^m$ uniquely determines a norm $\| \cdot \|^\star$ on V by

$$\|\mathbf{v}\|^\star = \|A\mathbf{v}\|, \quad \mathbf{v} \in V.$$

Obviously $A : V \to \mathbb{R}^m$ is norm-preserving. If $\| \cdot \|$ is the Euclidean norm on $\mathbb{R}^m$, then $\|\mathbf{x}\|^\star = (A\mathbf{x}, A\mathbf{x})^{1/2}$, where $(\cdot, \cdot)$ is the Euclidean inner product on $\mathbb{R}^m$. Consequently, $\| \cdot \|^\star$ is defined by the inner product $(\mathbf{x}, \mathbf{y})^\star = (A\mathbf{x}, A\mathbf{y})$ on V. $\qquad\spadesuit$

Given an m-dimensional normed vector space $(V, \|\cdot\|_V)$, we can always fix a basis of V and identify V with $\mathbb{R}^m$ (as in Example 9.1.6). Moreover, Theorem 9.1.4 implies that the Euclidean norm $\| \cdot \|_2$ on $\mathbb{R}^m$ is equivalent to the norm induced from $\|\cdot\|_V$ on $\mathbb{R}^m$. The metric topology on $\mathbb{R}^m$ will be the same whether we use the Euclidean norm $\| \cdot \|_2$ or the induced norm (Theorem 9.1.4(1)). Consequently, as far as continuity properties are concerned, there is no loss of generality in working with $(\mathbb{R}^m, \|\cdot\|_2)$— but note that this statement does depend on the non-trivial Theorem 9.1.4. From a formal point of view it is easier to work at the abstract level of maps $f : V \to W$ between general finite-dimensional normed vector spaces. However, when it comes to examples, especially computations, we usually have to choose a coordinate system—now we are looking at maps $f : \mathbb{R}^m \to \mathbb{R}^n$. We compromise by looking at maps $f : \mathbb{R}^m \to \mathbb{R}^n$ between spaces with the Euclidean norm but present arguments that generalize to the abstract setting $f : V \to W$ by simply changing $\mathbb{R}^m$ to V and $\mathbb{R}^n$ to W. There is precisely one point in the development of the theory where we have to choose a coordinate system and implicitly make use of Theorem 9.1.4. Later, when we come to higher derivatives, we will work at the abstract level of maps $f : V \to W$. We do this to avoid burying the ideas in the complex notation that results from using coordinates.

Summary of Conventions Let $\mathbb{R}^m$ denote m-dimensional Euclidean space. Denote vectors in $\mathbb{R}^m$ (or any normed space) using boldface: $\mathbf{x}, \mathbf{y} \in \mathbb{R}^m$. Let $(x_1, \cdots, x_m)$ denote the coordinates of $\mathbf{x} \in \mathbb{R}^m$ (relative to the standard basis $\{\mathbf{e}_1, \cdots, \mathbf{e}_m\}$ of $\mathbb{R}^m$). Denote the Euclidean norm on $\mathbb{R}^m$ by $\| \cdot \|$ and recall that $\|\mathbf{x}\|^2 = (\mathbf{x}, \mathbf{x})$ where $(\cdot, \cdot)$ denotes the inner or 'dot' product on $\mathbb{R}^m$ $((\mathbf{x}, \mathbf{y}) = \mathbf{x} \cdot \mathbf{y})$. Denote the unit sphere of $\mathbb{R}^m$ by S^{m-1}. That is, $S^{m-1} = \{\mathbf{x} \in \mathbb{R}^m \mid \|\mathbf{x}\| = 1\}$. If

$A : \mathbb{R}^m \to \mathbb{R}^n$ is linear and $\mathbf{x} \in \mathbb{R}^m$, we usually write $A\mathbf{x}$, rather than $A(\mathbf{x})$, for the value of A at $\mathbf{x}$. Using a matrix representation for A, it is not hard to verify that every linear map $A : \mathbb{R}^m \to \mathbb{R}^n$ is continuous (relative to the topology defined by the Euclidean norm—we give a formal proof shortly).

EXERCISES 9.1.7

(1) Let $(V, \| \cdot \|)$ be a normed vector space and let d denote the associated metric on V. Verify that

  (a) $d(\mathbf{x} + \mathbf{z}, \mathbf{y} + \mathbf{z}) = d(\mathbf{x}, \mathbf{y})$, for all $\mathbf{x}, \mathbf{y}, \mathbf{z} \in V$ ('translation invariance' of d).
  (b) $d(k\mathbf{x}, k\mathbf{y}) = |k| d(\mathbf{x}, \mathbf{y})$ for all $\mathbf{x}, \mathbf{y} \in V$, $k \in \mathbb{R}$ ('scalar invariance' of d).

(2) For $p \geq 1$, define the *p-norm* $\| \cdot \|_p$ on $\mathbb{R}^n$ by

$$\| (x_1, \cdots, x_n) \|_p = \left(\sum_{i=1}^{n} x_i^p \right)^{1/p}.$$

It is easy to verify that $\| \cdot \|_p$ satisfies (1–3) of Definition 9.1.1. The triangle inequality is *Minkowski's inequality*:

$$\left(\sum_{i=1}^{n} |x_i + y_i|^p \right)^{1/p} \leq \left(\sum_{i=1}^{n} |x_i|^p \right)^{1/p} + \left(\sum_{i=1}^{n} |y_i|^p \right)^{1/p}.$$

This is easy to prove if $p = 1, 2$ (in case $p = 2$ we have the Euclidean norm). For the remainder of this exercise we indicate the steps needed to prove the general case.

  (a) Let $f(x, y) = \alpha x + \beta y - x^\alpha y^\beta$, where $x, y \geq 0$, $\alpha, \beta \in (0, 1)$ and $\alpha + \beta = 1$. By finding the minimum value of f for a fixed value of y, show that $\alpha x + \beta y \geq x^\alpha y^\beta$ for all $x, y \geq 0$.
  (b) Let $(a_n), (b_n)$ be real sequences consisting of n terms. Let $p, q > 1$ satisfy $1/p + 1/q = 1$. Set $A_m = a_m / (\sum_{i=1}^{n} |a_i|^p)^{1/p}$, $B_m = b_m / (\sum_{i=1}^{n} |b_i|^q)^{1/q}$, $1 \leq i \leq m$. Using (a), show that

$$|A_m B_m| \leq |A_m|^p/p + |B_m|^q/q,$$

  and hence, by summing over m, that

$$\sum_{i=1}^{n} |A_i B_i| \leq 1 \leq \left(\sum_{i=1}^{n} |A_i|^p \right)^{1/p} \left(\sum_{i=1}^{n} |B_i|^q \right)^{1/q}.$$

Deduce *Hölder's inequality*:

$$\sum_{i=1}^{n} |a_i b_i| \leq \left(\sum_{i=1}^{n} |a_i|^p \right)^{1/p} \left(\sum_{i=1}^{n} |b_i|^q \right)^{1/q}.$$

(c) Under the assumptions of (b) show that

$$\sum_{i=1}^{n} |a_i + b_i|^p \leq \sum_{i=1}^{n} |a_i + b_i|^{p-1} |a_i| + \sum_{i=1}^{n} |a_i + b_i|^{p-1} |b_i|,$$

and apply Hölder's inequality to deduce Minkowski's inequality.

(3) Show that the product norm $\|(x_1, \cdots, x_n)\|_\infty = \max_i |x_i|$ may be regarded as $\lim_{p \to \infty} \|x\|_p$.
(4) What goes wrong if we try to define $\| \cdot \|_p$ for $p < 1$?
(5) Define $\| \cdot \|_P : \mathbb{R}^{n+1} \to \mathbb{R}$ by

$$\|(x_1, \cdots, x_{n+1})\|_P = \max\{x_1, \cdots, x_{n+1}\} - \min\{x_1, \cdots, x_{n+1}\}.$$

(a) Show that $\| \cdot \|_P$ defines a norm on the hyperplane $x_1 + \cdots + x_{n+1} = 0$.
(b) Let $n = 2$. Show that the unit 'circle' defined by $\|x\|_P = 1$ on the hyperplane $x_1 + x_2 + x_3 = 0$ is a regular hexagon and find the vertices of the hexagon.

9.2 Linear Maps

In this section we cover some elementary results on linear maps that we need for the development of the differential calculus on $\mathbb{R}^m$. As far as possible we do this in a 'coordinate-free' way.

Let $A : \mathbb{R}^m \to \mathbb{R}^n$ be linear. Although we can represent A as a matrix, conceptually it is easiest to regard A as a map $A : \mathbb{R}^m \to \mathbb{R}^n$ which is linear. That is,

$$A(\mathbf{x} + \lambda \mathbf{y}) = A\mathbf{x} + \lambda A\mathbf{y}, \text{ for all } \mathbf{x}, \mathbf{y} \in \mathbb{R}^m, \lambda \in \mathbb{R}.$$

We start by showing that every linear map $A : \mathbb{R}^m \to \mathbb{R}^n$ is continuous. One way of doing this is by using the matrix representation of A and writing $A\mathbf{x}$ in coordinates. However, we give a proof that suggests the real issue is the finite-dimensionality of the vector spaces $\mathbb{R}^m$, $\mathbb{R}^n$. Indeed, linear maps defined on an infinite-dimensional normed vector space need not be continuous (see the exercises for an example).

Lemma 9.2.1 *Let $A : \mathbb{R}^m \to \mathbb{R}^n$ be linear. If A is continuous at $\mathbf{x} = \mathbf{0}$, then A is continuous on $\mathbb{R}^m$.*

Proof Let $\mathbf{x}_0 \in \mathbb{R}^m$ and $\varepsilon > 0$. Since A is continuous at $\mathbf{x} = \mathbf{0}$, there exists a $\delta > 0$ such that $\|A\mathbf{z} - A\mathbf{0}\| = \|A\mathbf{z}\| < \varepsilon$, for all $\mathbf{z}$ such that $\|\mathbf{z}\| < \delta$. Observe that $\|A\mathbf{x}_0 - A\mathbf{x}\| = \|A(\mathbf{x}_0 - \mathbf{x})\|$ (linearity) and so, taking $\mathbf{z} = \mathbf{x}_0 - \mathbf{x}$, we have $\|A\mathbf{x}_0 - A\mathbf{x}\| < \varepsilon$, if $\|\mathbf{x}_0 - \mathbf{x}\| < \delta$, proving continuity of A at $\mathbf{x}_0$. $\qquad\square$

Remark 9.2.2 A consequence of the proof of Lemma 9.2.1 is that if A is continuous then A is uniformly continuous. ✠

Lemma 9.2.3 *Let $A : \mathbb{R}^m \to \mathbb{R}^n$ be linear. Then A is continuous at $\mathbf{0}$ if A is bounded on the unit sphere S^{m-1} of $\mathbb{R}^m$. That is, if there exists a $C \geq 0$ such that $\|A\mathbf{u}\| \leq C$ for all $\mathbf{u} \in S^{m-1}$.*

Proof We are given that $\|A\mathbf{u}\| \leq C$, for all $\mathbf{u} \in S^{m-1}$. If $\mathbf{x} \in \mathbb{R}^m$ is non-zero, then $\mathbf{x}/\|\mathbf{x}\| \in S^{m-1}$ and so $\|A(\mathbf{x}/\|\mathbf{x}\|)\| \leq C$. By linearity, $A(\mathbf{x}/\|\mathbf{x}\|) = \frac{1}{\|\mathbf{x}\|}A\mathbf{x}$ and so $\|\frac{1}{\|\mathbf{x}\|}A\mathbf{x}\| = \frac{1}{\|\mathbf{x}\|}\|A\mathbf{x}\|$. Hence

$$\|A\mathbf{x}\| \leq C\|\mathbf{x}\|, \text{ for all } \mathbf{x} \in \mathbb{R}^m.$$

Let $\varepsilon > 0$ and take $\delta = \varepsilon/\max\{C, 1\}$. Our estimate on $\|A\mathbf{x}\|$ implies that $\|A\mathbf{x}\| < \varepsilon$ whenever $\|\mathbf{x}\| < \delta$ and so A is continuous at $\mathbf{x} = 0$. $\qquad\square$

Lemma 9.2.4 *If $A : \mathbb{R}^m \to \mathbb{R}^n$ is linear, then A is bounded on S^{m-1}.*

Proof Every point $\mathbf{u} \in S^{m-1}$ may be written uniquely as $\mathbf{u} = \sum_{j=1}^{m} u_j \mathbf{e}_j$, where $\{\mathbf{e}_1, \cdots, \mathbf{e}_m\}$ is the standard basis of $\mathbb{R}^m$ and $\sum_{j=1}^{m} u_j^2 = 1$. By linearity of A, we have

$$A\mathbf{u} = \sum_{j=1}^{m} u_j A\mathbf{e}_j,$$

and so, by the triangle inequality

$$\|A\mathbf{u}\| \leq \sum_{j=1}^{m} |u_j| \|A\mathbf{e}_j\|$$

$$\leq M \sum_{j=1}^{m} |u_j| \leq mM,$$

where $M = \max\{\|A\mathbf{e}_j\|\}$ and we have used $|u_j| \leq \|\mathbf{u}\| = 1$, $1 \leq j \leq m$. Hence $\|A\mathbf{u}\| \leq mM$, for all $\mathbf{u} \in S^{m-1}$. $\qquad\square$

Remark 9.2.5 Notice that if A is continuous, then A is bounded on S^{m-1} since S^{m-1} is a compact subset of $\mathbb{R}^m$. ✠

Proposition 9.2.6 *Every linear map $A : \mathbb{R}^m \to \mathbb{R}^n$ is continuous.*

Proof Immediate from Lemmas 9.2.1, 9.2.3 and 9.2.4. $\qquad\square$

Remark 9.2.7 Note the point in the proof of Lemma 9.2.4 where we use the finite-dimensionality of $\mathbb{R}^m$ and the Euclidean norm. Proposition 9.2.6 holds for linear maps $A : V \to W$ between normed vector spaces *provided that V is finite-dimensional.* For this we need Theorem 9.1.4. ✱

9.2.1 Normed Vector Spaces of Linear Maps

If $A : \mathbb{R}^m \to \mathbb{R}^n$ is a linear map, define

$$\|A\| = \sup_{\|\mathbf{u}\|=1} \|A\mathbf{u}\| = \sup_{\mathbf{u}\in S^{m-1}} \|A\mathbf{u}\|.$$

Since A is continuous and S^{m-1} is closed and bounded (therefore compact), we have $\|A\| < \infty$ (alternatively, use Lemma 9.2.4). We refer to $\|A\|$ as the *norm* or *operator norm* of A.

Examples 9.2.8

(1) Let $I : \mathbb{R}^m \to \mathbb{R}^m$ denote the *identity map* of $\mathbb{R}^m$. Then $\|I\| = 1$.
(2) If $A : \mathbb{R}^2 \to \mathbb{R}^2$ is the linear map with matrix $[A]$ given by

$$[A] = \begin{pmatrix} \alpha & -\beta \\ \beta & \alpha \end{pmatrix},$$

then $\|A\| = \sqrt{\alpha^2 + \beta^2}$. The hardest way of seeing this is by using Lagrange multipliers to find the maximum value of $\|A\mathbf{u}\|$ on the unit circle in $\mathbb{R}^2$. A much easier way is to identify $\mathbb{R}^2$ with $\mathbb{C}$ ($(x, y) \sim x + \imath y$) and observe that $A(x, y)$ corresponds to complex multiplication by $\alpha + \imath \beta$. That is, $Az = (\alpha + \imath \beta)z$ and so $\|Az\| = |(\alpha + \imath \beta)z|$ (modulus on the right-hand side). The claimed result follows since $|(\alpha + \imath \beta)z| = |\alpha + \imath \beta||z| = \sqrt{\alpha^2 + \beta^2} |z|$. ♠

Let $L(\mathbb{R}^m, \mathbb{R}^n)$ denote the (vector) space of all linear maps from $\mathbb{R}^m$ to $\mathbb{R}^n$ and let $\mathbf{0}$ (or $\mathbf{0}_{m,n}$) denote the zero linear map.

Theorem 9.2.9 *Let* $A, B \in L(\mathbb{R}^m, \mathbb{R}^n)$. *We have*

(1) $\|A\| \geq 0$ *and* $\|A\| = 0$ *iff* $A = \mathbf{0}$.
(2) $\|aA\| = |a|\|A\|$ *for all* $a \in \mathbb{R}$.
(3) $\|A + B\| \leq \|A\| + \|B\|$ *for all* $A, B \in L(\mathbb{R}^m, \mathbb{R}^n)$.

In particular, $(L(\mathbb{R}^m, \mathbb{R}^n), \|\cdot\|)$ *has the structure of a normed vector space.*

Proof

(1) Obviously $\|A\| \geq 0$ for all $A \in L(\mathbb{R}^m, \mathbb{R}^n)$. If $\|A\| = 0$, then $A\mathbf{u} = 0$ for all unit vectors $\mathbf{u} \in \mathbb{R}^m$. Since every non-zero vector in $\mathbb{R}^m$ is a scalar multiple of a unit vector, it follows by the linearity of A that $A\mathbf{x} = \mathbf{0}$ for all $\mathbf{x} \in \mathbb{R}^m$ and so $A = \mathbf{0}$.

(2) We have $\sup_{\|\mathbf{u}\|=1} \|aA\mathbf{u}\| = \sup_{\|\mathbf{u}\|=1} |a|\|A\mathbf{u}\| = |a|\|A\|$.

(3) Suppose $A, B \in L(\mathbb{R}^m, \mathbb{R}^n)$. We have

$$
\begin{aligned}
\|A + B\| &= \sup_{\|\mathbf{u}\|=1} \|A\mathbf{u} + B\mathbf{u}\| \\
&\leq \sup_{\|\mathbf{u}\|=1} (\|A\mathbf{u}\| + \|B\mathbf{u}\|) \\
&\leq \sup_{\|\mathbf{u}\|=1} \|A\mathbf{u}\| + \sup_{\|\mathbf{u}\|=1} \|B\mathbf{u}\| \\
&= \|A\| + \|B\|,
\end{aligned}
$$

proving the triangle inequality. $\square$

Remark 9.2.10 The operator norm defines the Euclidean norm on $L(\mathbb{R}^m, \mathbb{R}^n) \cong \mathbb{R}^{mn}$ iff $n = 1$ or $m = 1$—see the exercises at the end of the section for the isomorphism between $L(\mathbb{R}^m, \mathbb{R}^n)$ and $\mathbb{R}^{mn}$.   ✽

Proposition 9.2.11 (Additional Properties of $\|\cdot\|$)

(a) $\|A\mathbf{x}\| \leq \|A\|\|\mathbf{x}\|$, *for all $A \in L(\mathbb{R}^m, \mathbb{R}^n)$ and $\mathbf{x} \in \mathbb{R}^m$.*

(b) *If we define $d(A, B) = \|A - B\|$, $A, B \in L(\mathbb{R}^m, \mathbb{R}^n)$, then d defines a (complete) metric on $L(\mathbb{R}^m, \mathbb{R}^n)$.*

(c) *If $L : \mathbb{R}^m \to \mathbb{R}^n$, $M : \mathbb{R}^n \to \mathbb{R}^p$, then $\|ML\| \leq \|M\|\|L\|$.*

(d) *If $A \in L(\mathbb{R}^m, \mathbb{R}^n)$ is invertible (so A^{-1} exists and $m = n$), then $\|A^{-1}\| \geq 1/\|A\|$.*

Proof All the statements are quite elementary. We prove (a,c) and leave (b,d) to the exercises.

   (a) If $\mathbf{x} = \mathbf{0}$, then certainly $0 = \|A\mathbf{x}\| \leq \|A\|\|\mathbf{x}\| = 0$. So suppose $\mathbf{x} \neq \mathbf{0}$ and set $\mathbf{u} = \frac{\mathbf{x}}{\|\mathbf{x}\|}$. By definition of $\|A\|$, we have $\|A\mathbf{u}\| \leq \|A\|$ ($\mathbf{u}$ is a unit vector). Since $\mathbf{u} = \frac{\mathbf{x}}{\|\mathbf{x}\|}$, it follows by linearity of A that $A\mathbf{u} = A(\frac{\mathbf{x}}{\|\mathbf{x}\|}) = \frac{1}{\|\mathbf{x}\|}A\mathbf{x}$ and so $\frac{1}{\|\mathbf{x}\|}A\mathbf{x} \leq \|A\|$. Multiplying through by $\|\mathbf{x}\|$ gives the result.

   (c) Suppose $L : \mathbb{R}^m \to \mathbb{R}^n$, $M : \mathbb{R}^n \to \mathbb{R}^p$. Let $\mathbf{u} \in S^{m-1}$. We have

$$
\|ML\mathbf{u}\| = \|M(L\mathbf{u})\| \leq \|M\|\|L\mathbf{u}\| \leq \|M\|\|L\|,
$$

where the first inequality follows by (a) and the second inequality either by (a) or the definition of $\|L\|$. Since this estimate holds for all $\mathbf{u} \in S^{m-1}$, $\|ML\| = \sup_{u \in S^{m-1}} \|ML\mathbf{u}\| \leq \|M\|\|L\|$. $\square$

Remark 9.2.12 An important consequence of Proposition 9.2.11(a) is that for all $\mathbf{x}, \mathbf{y} \in \mathbb{R}^m$ we have the estimate

$$
\|A\mathbf{x} - A\mathbf{y}\| \leq \|A\|\|\mathbf{x} - \mathbf{y}\|.
$$

This estimate plays an absolutely crucial role in our analysis of linear maps. The mean value theorem for differentiable maps $f : \mathbb{R}^m \to \mathbb{R}^n$ is of the same form and it is this that often enables us to attack problems about non-linear maps using techniques of linear analysis. ✠

EXERCISES 9.2.13

(1) The space $L(\mathbb{R}^m, \mathbb{R}^n)$ is isomorphic to $\mathbb{R}^{mn}$ (map the matrix $[a_{ij}]$ of $A \in L(\mathbb{R}^m, \mathbb{R}^n)$ to $(a_{11}, a_{12}, \ldots, a_{1n}, \ldots, a_{mn}) \in \mathbb{R}^{mn}$. Show that the operator norm $\| \cdot \|$ induced on $\mathbb{R}^{mn}$ is equal to the Euclidean norm on $\mathbb{R}^{mn}$ iff $n = 1$ or $m = 1$ and that the norms are always equivalent (without recourse to Theorem 9.1.4).

(2) Prove statements (b,d) of Theorem 9.2.11

(3) Suppose $A \in L(\mathbb{R}^m, \mathbb{R}^n)$. Let $A^t \in L(\mathbb{R}^n, \mathbb{R}^m)$ denote the transpose of A (if the matrix of A is $[a_{ij}]$ then the matrix of $[A^t]$ is $[a_{ij}^t] = [a_{ji}]$). Define $|A|^2 = \text{trace}(AA^t)$ (the trace of a matrix is the sum of the diagonal elements—note that AA^t is an $n \times n$ matrix).

(a) Show that $\text{trace}(AA^t) = \text{trace}(A^tA)$.
(b) $|A| \geq 0$ for all $A \in L(\mathbb{R}^m, \mathbb{R}^n)$ and $|A| = 0$ iff $A = 0$.
(c) $| \cdot |$ defines a norm on $L(\mathbb{R}^m, \mathbb{R}^n)$ (you will need to 'verify' the triangle inequality.)
(d) Show that there is an inner product $\langle \cdot, \cdot \rangle$ on $L(\mathbb{R}^n, \mathbb{R}^m)$ which defines $| \cdot |$.

(This gives a natural norm on linear maps that depends only on the inner product structures on $\mathbb{R}^m, \mathbb{R}^n$—inner products are needed to define the transpose in a coordinate-free way.)

(4) Let **m** denote the space of all infinite sequences $\mathbf{x} = (x_i)_{i=1}^{\infty}$ of real numbers such that all but finitely many of the x_i are equal to zero. Thus if $\mathbf{x} \in \mathbf{m}$, there exists an $N \in \mathbb{N}$ such that $x_i = 0$ for all $i \geq N$.

(a) Verify that **m** has the structure of a vector space if we define vector space addition and scalar multiplication coordinate-wise.
(b) Show that if we define $\|\mathbf{x}\| = \max_i |x_i|$, then $\| \cdot \|$ defines a norm on **m**.
(c) Define $f : \mathbf{m} \to \mathbb{R}$ by $f(\mathbf{x}) = \sum_{n=1}^{\infty} nx_n$. Verify that f is linear but not continuous with respect to the topology defined by $\| \cdot \|$. (Hint: It is enough to show f is not bounded on the closed unit ball of **m**. Why?)

(It is easy to see that $(\mathbf{m}, \| \cdot \|)$ is not complete. Examples of discontinuous linear maps can be defined on infinite-dimensional complete normed vector spaces, such as $C^0([0, 1])$ with the uniform metric, but they are harder to construct.)

(5) Suppose $(V, \| \cdot \|)$ and $(W, \| \cdot \|)$ are normed vector spaces and set $S(V) = \{\mathbf{u} \in V \mid \|\mathbf{u}\| = 1\}$. Show that if we everywhere replace $(\mathbb{R}^m, \| \cdot \|_2)$ by $(V, \| \cdot \|)$ and $(\mathbb{R}^n, \| \cdot \|_2)$ by $(W, \| \cdot \|)$, then:

(a) Lemmas 9.2.1, 9.2.3 remain true and $A : V \to W$ is continuous iff A is bounded on $S(V)$. (No assumption on the finite-dimensionality of V or W.)
(b) If we let $L(V, W)$ denote the space of *continuous* linear maps from V to W, then Theorem 9.2.1 and Proposition 9.2.11 are true. (Of course, Theorem 9.1.4 implies that every linear map $A : V \to W$ is continuous if $\dim(V) < \infty$.)

9.3   The Derivative

For functions $f : \mathbb{R} \to \mathbb{R}$, differentiability at $x_0 \in \mathbb{R}$ is most easily described in terms of the existence of a unique tangent line to the graph of f at x_0. The tangent line is constructed using limiting chords to the graph. The derivative, if it exists, is then defined to be the slope of the tangent line and is a real number. This approach does not generalize naturally to functions $f : \mathbb{R}^m \to \mathbb{R}^n$, $m > 1$. The difficulty lies with defining the analog of the tangent line. What we require is a unique tangent plane to the graph of f at x_0 but there is no obvious analogy of the limiting chords construction used for functions of one variable. Of course, one can define partial derivatives but these depend on choosing a coordinate system on $\mathbb{R}^m$. Whatever the derivative of a function *is*, it should surely not depend on the choice of a coordinate system—just as we do not need a coordinate system to define a linear map. Our goal then is to give a natural coordinate-free definition of differentiability and the derivative. The way forward is to realize that the tangent line is the graph of an affine linear function and the tangent plane to the graph, if it exists, will be the graph of an affine linear map. Instead of thinking of the derivative as a scalar (or vector), we regard the derivative as a function—more precisely the linear part of the affine linear map that determines the graph of the tangent plane. To make all this precise requires ideas of approximation.

Roughly speaking, a function $f : \mathbb{R}^m \to \mathbb{R}^n$ is differentiable at $\mathbf{x}_0 \in \mathbb{R}^m$ if we have a good affine linear approximation to f near $\mathbf{x}_0$. We need to make precise the meaning of 'good approximation' and 'affine linear map'. We start with the easier definition (see also Chap. 8). The map $G : \mathbb{R}^m \to \mathbb{R}^n$ is an *affine linear map* if we can write

$$G(\mathbf{x}) = A\mathbf{x} + \mathbf{b},$$

where $A : \mathbb{R}^m \to \mathbb{R}^n$ is a linear map and $\mathbf{b} \in \mathbb{R}^n$ is a constant vector. If $m = n = 1$, then an affine linear map may be written in the familiar form $y = mx + c$, where $m, c \in \mathbb{R}$.

The definition of what is meant by a good approximation to f near $\mathbf{x}_0$ is trickier. Suppose that $G(\mathbf{x}) = A\mathbf{x} + \mathbf{b}$ is an affine linear map. If G is to be a good approximation to f near $\mathbf{x}_0$, we certainly want $G(\mathbf{x}_0) = f(\mathbf{x}_0)$. Hence $A\mathbf{x}_0 + \mathbf{b} = f(\mathbf{x}_0)$. However, this condition says nothing about how the values $A\mathbf{x} + \mathbf{b}$ compare with $f(\mathbf{x})$ at points $\mathbf{x}$ near $\mathbf{x}_0$. For this we need an estimate on $\|f(\mathbf{x}) - (A\mathbf{x} + \mathbf{b})\|$, when $\mathbf{x}$ is close to $\mathbf{x}_0$. As a first attempt we might ask that $\lim_{\mathbf{x} \to \mathbf{x}_0} \|f(\mathbf{x}) - (A\mathbf{x} + \mathbf{b})\| = 0$. However, since we have $f(\mathbf{x}_0) = A\mathbf{x}_0 + \mathbf{b}$ (if $G(\mathbf{x}_0) = f(\mathbf{x}_0)$), this condition is equivalent to the continuity of f at $\mathbf{x}_0$. A stronger condition is needed for differentiability. What we shall require is that $\|f(\mathbf{x}) - (A\mathbf{x} + \mathbf{b})\|$ goes to zero faster than $\|\mathbf{x} - \mathbf{x}_0\|$ as $\|\mathbf{x} - \mathbf{x}_0\| \to 0$. That is,

$$\lim_{\mathbf{x} \to \mathbf{x}_0} \frac{\|f(\mathbf{x}) - (A\mathbf{x} + \mathbf{b})\|}{\|\mathbf{x} - \mathbf{x}_0\|} = 0. \qquad (9.1)$$

As we shall see, this condition implies that $A\mathbf{x} + \mathbf{b}$ is the best possible affine linear approximation to f at $\mathbf{x}_0$. If we write $\mathbf{x} = \mathbf{x}_0 + \mathbf{h}$, then $A(\mathbf{x}_0 + \mathbf{h}) + \mathbf{b} = f(\mathbf{x}_0) + A\mathbf{h}$ and (9.1) is equivalent to

$$\lim_{\mathbf{h}\to 0} \frac{\|f(\mathbf{x}_0 + \mathbf{h}) - f(\mathbf{x}_0) - A\mathbf{h}\|}{\|\mathbf{h}\|} = 0. \tag{9.2}$$

Equation (9.2) is reminiscent of the definition of the derivative of a map $f : \mathbb{R} \to \mathbb{R}$. Since we cannot divide by vectors, we take norms of vectors instead. There is another way of looking at (9.2) that avoids division and explicit mention of limits. If we define the *remainder* or *error* term $r(\mathbf{h})$ by $r(\mathbf{h}) = f(\mathbf{x}_0 + \mathbf{h}) - (f(\mathbf{x}_0) + A\mathbf{h})$, then (9.2) can be rewritten as

$$f(\mathbf{x}_0 + \mathbf{h}) - (f(\mathbf{x}_0) + A\mathbf{h}) = r(\mathbf{h}),$$

where $\lim_{\mathbf{h}\to 0} \frac{\|r(\mathbf{h})\|}{\|\mathbf{h}\|} = 0$. As we shall frequently encounter this condition on $r(\mathbf{h})$, we introduce the economical 'small o' notation: write $r(\mathbf{h}) = o(\mathbf{h})$ if $r(0) = 0$ and $\lim_{\mathbf{h}\to 0} \frac{\|r(\mathbf{h})\|}{\|\mathbf{h}\|} = 0$. Equivalently, $r(\mathbf{h}) = o(\mathbf{h})$ if for every $\varepsilon > 0$, there exists a $\delta > 0$ such that $\|r(\mathbf{h})\| < \varepsilon\|\mathbf{h}\|$ whenever $\|\mathbf{h}\| < \delta$.

Example 9.3.1 Let $f : \mathbb{R} \to \mathbb{R}$ be differentiable at x_0. The affine linear map $g(h) = f(x_0) + ah$ is the tangent line to the graph of f at x_0 iff $f(x_0 + h) - f(x_0) - ah = o(h)$. Referring to Fig. 9.1, the error $r(h) = o(h)$ goes to zero faster than $|h|$ for the tangent line. For the line L, the error goes to zero like $|h|$. If $r(h) = o(h)$, then $g(h) = f(x_0) + f'(x_0)h$ ($a = f'(x_0)$ in the figure). ♠

We can now give a formal definition of what it means for a function to be differentiable at a point.

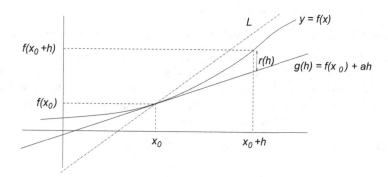

Fig. 9.1 Remainder term $r(h)$ for $f : \mathbb{R} \to \mathbb{R}$

Definition 9.3.2 The map $f : \mathbb{R}^m \to \mathbb{R}^n$ is *differentiable* at $\mathbf{x}_0 \in \mathbb{R}^m$ if there exists a linear map $A : \mathbb{R}^m \to \mathbb{R}^n$ such that

$$f(\mathbf{x}_0 + \mathbf{h}) = f(\mathbf{x}_0) + A\mathbf{h} + r(\mathbf{h}),$$

where $r(\mathbf{h}) = o(\mathbf{h})$.

Remarks 9.3.3

(1) We shall shortly show that if we can find a linear map A satisfying Definition 9.3.2 then A is unique. Naturally we call A the *derivative* of f at $\mathbf{x}_0$. We denote the derivative of f at $\mathbf{x}_0$ either by $Df_{\mathbf{x}_0}$ or $Df(\mathbf{x}_0)$ (we usually use the first notation). Thus, differentiability at $\mathbf{x}_0$ means that there exists a linear map $Df_{\mathbf{x}_0} : \mathbb{R}^m \to \mathbb{R}^n$ such that

$$f(\mathbf{x}_0 + \mathbf{h}) = f(\mathbf{x}_0) + Df_{\mathbf{x}_0}(\mathbf{h}) + r(\mathbf{h}),$$

where $r(\mathbf{h}) = o(\mathbf{h})$. Alternatively, if we write $\mathbf{x} = \mathbf{x}_0 + \mathbf{h}$,

$$f(\mathbf{x}) = f(\mathbf{x}_0) + Df_{\mathbf{x}_0}(\mathbf{x} - \mathbf{x}_0) + r(\mathbf{x} - \mathbf{x}_0),$$

where $r(\mathbf{x} - \mathbf{x}_0) = o(\mathbf{x} - \mathbf{x}_0)$. We emphasize that the definition implies that f has a good affine linear approximation at $\mathbf{x}_0$. That is, the error $r(\mathbf{x} - \mathbf{x}_0)$ we get by replacing $f(\mathbf{x})$ near $\mathbf{x}_0$ by the (affine) linear map $f(\mathbf{x}_0) + Df_{\mathbf{x}_0}(\mathbf{x} - \mathbf{x}_0)$ goes to zero faster than $\|\mathbf{x} - \mathbf{x}_0\|$:

$$f(\mathbf{x}) - (f(\mathbf{x}_0) + Df_{\mathbf{x}_0}(\mathbf{x} - \mathbf{x}_0)) = o(\mathbf{x} - \mathbf{x}_0).$$

(2) We develop properties of the small o notation, and introduce the big O notation, in the exercises at the end of the section. ✠

Lemma 9.3.4 *If $f : \mathbb{R}^m \to \mathbb{R}^n$ is differentiable at $\mathbf{x}_0$, then f is continuous at $\mathbf{x}_0$.*

Proof If f is differentiable at $\mathbf{x}_0$, then there exists a linear map $A : \mathbb{R}^m \to \mathbb{R}^n$ such that $f(\mathbf{x}) = f(\mathbf{x}_0) + A(\mathbf{x} - \mathbf{x}_0) + r(\mathbf{x} - \mathbf{x}_0)$, where $r(\mathbf{x} - \mathbf{x}_0) = o(\mathbf{x} - \mathbf{x}_0)$. Since A is linear, A is continuous and so $\lim_{\mathbf{x}\to\mathbf{x}_0} A(\mathbf{x} - \mathbf{x}_0) = A(\mathbf{0}) = \mathbf{0}$. Since $r(\mathbf{x} - \mathbf{x}_0) = o(\mathbf{x} - \mathbf{x}_0)$, we also have $\lim_{\mathbf{x}\to\mathbf{x}_0} r(\mathbf{x} - \mathbf{x}_0) = r(\mathbf{0}) = 0$. Therefore $\lim_{\mathbf{x}\to\mathbf{x}_0} f(\mathbf{x}) = f(\mathbf{x}_0)$ and f is continuous at $\mathbf{x}_0$. □

Definition 9.3.5 If $f : \mathbb{R} \to \mathbb{R}^n$ is differentiable at $x_0 \in \mathbb{R}$, we define $f'(x_0) \in \mathbb{R}^n$ by

$$f'(x_0) = Df_{x_0}(1).$$

We end with an example that shows the connection between Definition 9.3.5 and the limit definition for functions of one variable. We continue to assume the derivative is unique—a result that is well-known and easy in the one variable case.

Example 9.3.6 Suppose $f : \mathbb{R} \to \mathbb{R}^n$ is differentiable at $x_0 \in \mathbb{R}$ in the sense of Definition 9.3.2. Following Definition 9.3.5, set $f'(x_0) = Df_{x_0}(1) \in \mathbb{R}^n$. We claim that

$$\lim_{h \to 0} \frac{f(x_0 + h) - f(x_0)}{h} = f'(x_0).$$

To see this, observe that

$$f(x_0 + h) = f(x_0) + Df_{x_0}(h) + r(h)$$
$$= f(x_0) + hf'(x_0) + r(h).$$

Hence $\lim_{h \to 0} \left(\frac{f(x_0+h)-f(x_0)}{h} - f'(x_0) \right) = \lim_{h \to 0} \frac{r(h)}{h} = 0.$ ♠

EXERCISES 9.3.7

(1) Let $\mathbf{x} \in \mathbb{R}^m$. Define $g : \mathbb{R} \to \mathbb{R}^m$ by $g(t) = t\mathbf{x}$. What is $g'(t) = Dg_t(1)$?

(2) Working from Definition 9.3.2, show that the Euclidean norm $\| \cdot \|_2 : \mathbb{R}^n \to \mathbb{R}$ is never differentiable at $\mathbf{x} = \mathbf{0}$. Using Theorem 9.1.4, deduce that every norm $\| \cdot \|$ on $\mathbb{R}^n$ is not differentiable at $\mathbf{x} = \mathbf{0}$.

(3) Suppose that $r : \mathbb{R}^m \to \mathbb{R}^n$ and that $r(\mathbf{x}) = o(\mathbf{x})$. Show that r is differentiable at $\mathbf{0}$ and that $Dr_0 = \mathbf{0}$.

(4) Suppose that $f, g : \mathbb{R}^m \to \mathbb{R}^n$ and that $f(\mathbf{x}), g(\mathbf{x}) = o(\mathbf{x})$. Verify that $(f \pm g)(\mathbf{x}) = o(\mathbf{x})$. If $n = 1$, show that $f(\mathbf{x})g(\mathbf{x}) = o(\mathbf{x})$.

(5) Let $f : \mathbb{R}^m \to \mathbb{R}^n$. We write $f(\mathbf{x}) = O(\mathbf{x})$ and say f is $O(\mathbf{x})$ ('big zero $\mathbf{x}$') if there exist $r > 0$, $C > 0$ such that $\| f(\mathbf{x}) \| \leq C \|\mathbf{x}\|$ for all $\mathbf{x} \in \overline{D}_r(\mathbf{0})$. Verify that

   (a) If f is $O(\mathbf{x})$ then $f(\mathbf{0}) = \mathbf{0}$ and f is continuous at $\mathbf{x} = \mathbf{0}$.

   (b) If f is differentiable at $\mathbf{x} = 0$ and $f(\mathbf{0}) = \mathbf{0}$, then f is $O(\mathbf{x})$. Find an example to show that if $f(\mathbf{0}) = \mathbf{0}$ and f is $O(\mathbf{x})$, then f may not be differentiable at $\mathbf{x} = \mathbf{0}$.

   (c) Let $f, g : \mathbb{R}^m \to \mathbb{R}^n$. Suppose that f, g are $O(\mathbf{x})$. What can be said about $f \pm g$? Suppose $n = 1$. What can be said about fg? Deduce that if f, g are $O(\mathbf{x})$, then fg is differentiable at $\mathbf{x} = \mathbf{0}$ and find $D(fg)_0$.

(6) Suppose we follow the assumptions of Exercises 9.2.13(5)—in particular, $L(V, W)$ consists of continuous linear maps. Show that all results and definitions of the section continue to apply.

9.4 Properties of the Derivative

Lemma 9.4.1 *If f is differentiable at x_0, the derivative is unique.*

Proof Suppose the linear maps $A, B : \mathbb{R}^m \to \mathbb{R}^n$ both satisfy the defining equation for the derivative

$$f(x_0 + h) = f(x_0) + Ah + r_1(h)$$
$$= f(x_0) + Bh + r_2(h).$$

Subtract the second equation from the first to get

$$(A - B)(h) = r_2(h) - r_1(h) = o(h),$$

since $r_1(h), r_2(h) = o(h)$. Hence

$$\lim_{h \to 0} \frac{\|(A - B)(h)\|}{\|h\|} = 0.$$

But $\frac{\|(A-B)(h)\|}{\|h\|} = \|(A - B)(u)\|$ where $u = \frac{h}{\|h\|}$. Since every unit vector $u \in \mathbb{R}^m$ can be written in the form $u = \frac{h}{\|h\|}$ for arbitrarily small vectors h, it follows that $\|A - B\| = 0$. Hence, by Theorem 9.2.9(1), $A = B$. $\qquad\qquad \square$

Examples 9.4.2

(1) If $f : \mathbb{R}^m \to \mathbb{R}^n$ is linear, then f is differentiable at all points $x \in \mathbb{R}^m$ and $Df_x = f$. (This corresponds to the 1-variable result that if $f(x) = ax$, then f' is constant, equal to a.)

(2) Define $\phi : \mathbb{R}^n \to \mathbb{R}$ by $\phi(x) = \|x\|^2 = (x, x)$. We claim that ϕ is differentiable on $\mathbb{R}^n$ and $D\phi_x(h) = 2(x, h)$ for all $x, h \in \mathbb{R}^n$. For $x, h \in \mathbb{R}^n$ we have

$$\phi(x+h) = (x+h, x+h) = (x, x) + 2(x, h) + (h, h) = \phi(x) + 2(x, h) + \|h\|^2.$$

Since $\lim_{h \to 0} \frac{\|h\|^2}{\|h\|} = 0$, ϕ is differentiable and $D\phi_x(h) = 2(x, h)$. We remark that $D\phi_x(h) = 2(x, h) = 0$ iff $(x, h) = 0$. That is, $D\phi_x(h) = 0$ iff $h \perp x$. If $\|h\| = 1$, then for $x \neq 0$, $D\phi_x(h)$ takes its maximal value when $h = x/\|x\|$ and its minimal value when $h = -x/\|x\|$.

(3) Let $A_1, \cdots, A_p \in L(\mathbb{R}^m, \mathbb{R})$. Define $F : \mathbb{R}^m \to \mathbb{R}$ by $F(x) = A_1(x) \cdots A_p(x)$, $x \in \mathbb{R}^m$. We claim that F is differentiable at $x = 0$ with $DF_0 = 0$, if $p > 1$, and $DF_0 = A_1$ if $p = 1$. The statement for $p = 1$ is the first example above so suppose $p > 1$. We have

$$F(0 + h) = F(0) + A_1(h) \cdots A_p(h) = A_1(h) \cdots A_p(h).$$

Hence $\|f(\mathbf{h})\| \leq \|A_1\| \cdots \|A_n\| \|\mathbf{h}\|^p$. Since $p > 1$, $\lim_{\mathbf{h}\to 0} \frac{\|F(\mathbf{h})\|}{\|\mathbf{h}\|} = 0$ and so F is differentiable at $\mathbf{x} = \mathbf{0}$ with $DF_0 = \mathbf{0}$. Note that every coordinate functional $\mathbf{x} \mapsto x_i$ is linear and so this example implies that the monomials $F(\mathbf{x}) = x_1^{a_1} \cdots x_m^{a_m}$ are differentiable at $\mathbf{x} = \mathbf{0}$ and $DF_0 = \mathbf{0}$ if $a_1 + \cdots + a_m > 1$.
♠

Definition 9.4.3 Let $f : \mathbb{R}^m \to \mathbb{R}^n$.

(1) f is *differentiable* if f is differentiable at all points of $\mathbb{R}^m$.
(2) f is *continuously differentiable*, or C^1, if f is differentiable and the derivative map $Df : \mathbb{R}^m \to L(\mathbb{R}^m, \mathbb{R}^n)$ is continuous.

Remarks 9.4.4

(1) The continuity in (2) is relative to the metric on $L(\mathbb{R}^m, \mathbb{R}^n)$ given in Proposition 9.2.11(b)—indeed the metric associated to any norm on $L(\mathbb{R}^m, \mathbb{R}^n)$ (Theorem 9.1.4).
(2) In ε, δ terms, to say that f is C^1 means that given $\mathbf{x}_0 \in \mathbb{R}^m$, $\varepsilon > 0$, there exists a $\delta > 0$ such that $\|Df_{\mathbf{x}} - Df_{\mathbf{x}_0}\| < \varepsilon$, $\|\mathbf{x} - \mathbf{x}_0\| < \delta$.
✱

Examples 9.4.5

(1) If $f : \mathbb{R}^m \to \mathbb{R}^n$ is linear then f is C^1. Indeed, $Df_{\mathbf{x}} = f$ for all $\mathbf{x} \in \mathbb{R}^m$ and so $Df : \mathbb{R}^m \to L(\mathbb{R}^m, \mathbb{R}^n)$ is constant and obviously continuous.
(2) Define $\phi : \mathbb{R}^n \to \mathbb{R}$ by $\phi(\mathbf{x}) = \|\mathbf{x}\|^2 = (\mathbf{x}, \mathbf{x})$. As we showed in Examples 9.4.2(2), ϕ is differentiable on $\mathbb{R}^n$ and $D\phi_{\mathbf{x}}(\mathbf{h}) = 2(\mathbf{x}, \mathbf{h})$. In order to prove that ϕ is C^1, we need to estimate $\|D\phi_{\mathbf{x}} - D\phi_{\mathbf{x}'}\|$. We have

$$\|D\phi_{\mathbf{x}} - D\phi_{\mathbf{x}'}\| = \sup_{\|\mathbf{u}\|=1} |D\phi_{\mathbf{x}}(\mathbf{u}) - D\phi_{\mathbf{x}'}(\mathbf{u})| = \sup_{\|\mathbf{u}\|=1} |2(\mathbf{x}, \mathbf{u}) - 2(\mathbf{x}', \mathbf{u})|.$$

Now $|2(\mathbf{x}, \mathbf{u}) - 2(\mathbf{x}', \mathbf{u})| = 2|(\mathbf{x} - \mathbf{x}', \mathbf{u})| \leq 2\|\mathbf{x} - \mathbf{x}'\| \|\mathbf{u}\|$, by the Cauchy–Schwarz inequality. Therefore, $\|D\phi_{\mathbf{x}} - D\phi_{\mathbf{x}'}\| \leq 2\|\mathbf{x} - \mathbf{x}'\|$ and so $D\phi$ is continuous at $\mathbf{x}$ (given $\varepsilon > 0$, take $\delta = \varepsilon/2$).
♠

9.4.1 Directional Derivative

Let $f : \mathbb{R}^m \to \mathbb{R}^n$ be differentiable and suppose $\mathbf{u} \in S^{m-1}$ (a unit vector). We define the *directional derivative* of f at $\mathbf{x}_0$ in direction $\mathbf{u}$ to be the vector $D_{\mathbf{u}} f_{\mathbf{x}_0} \in \mathbb{R}^n$ defined by

$$D_{\mathbf{u}} f_{\mathbf{x}_0} = Df_{\mathbf{x}_0}(\mathbf{u}).$$

We also denote the directional derivative at $\mathbf{x}_0$ by $\frac{\partial f}{\partial \mathbf{u}}(\mathbf{x}_0)$. In particular, we set $D_{\mathbf{e}_j} f_{\mathbf{x}_0} = \frac{\partial f}{\partial x_j}(\mathbf{x}_0)$, where $\{\mathbf{e}_1, \cdots, \mathbf{e}_m\}$ denotes the standard basis of $\mathbb{R}^m$.

Lemma 9.4.6 *If f is differentiable at $\mathbf{x}_0 \in \mathbb{R}^m$ and $\mathbf{u} \in S^{m-1}$, then*

$$D_{\mathbf{u}} f_{\mathbf{x}_0} = \frac{d}{dt} f(\mathbf{x}_0 + t\mathbf{u})|_{t=0}.$$

Proof By definition of the derivative of f at $\mathbf{x}_0$ we have

$$f(\mathbf{x}_0 + \mathbf{h}) = f(\mathbf{x}_0) + Df_{\mathbf{x}_0}(\mathbf{h}) + r(\mathbf{h}).$$

Now set $\mathbf{h} = t\mathbf{u}$ to obtain

$$f(\mathbf{x}_0 + t\mathbf{u}) = f(\mathbf{x}_0) + t Df_{\mathbf{x}_0}(\mathbf{u}) + r(t\mathbf{u}).$$

Dividing by $t \neq 0$ gives

$$\frac{f(\mathbf{x}_0 + t\mathbf{u}) - f(\mathbf{x}_0)}{t} = Df_{\mathbf{x}_0}(\mathbf{u}) + \frac{r(t\mathbf{u})}{t}.$$

Since $r(\mathbf{h}) = o(\mathbf{h})$, we have $r(t\mathbf{u}) = o(t)$. Letting $t \to 0$, we see that $f(\mathbf{x}_0 + t\mathbf{u})$ is differentiable as a function of t at $t = 0$ with derivative $Df_{\mathbf{x}_0}(\mathbf{u}) \overset{\text{def}}{=} D_{\mathbf{u}} f_{\mathbf{x}_0}$.   □

Example 9.4.7 The function $f(\mathbf{x}_0 + t\mathbf{u})$ may be differentiable as a function of t for all $\mathbf{u} \in S^{m-1}$ without f being differentiable at $\mathbf{x}_0$. As a simple example, take $f(x_1, x_2) = x_1^2 x_2 / (x_1^2 + x_2^2)$, $(x_1, x_2) \neq (0,0)$, and $f(0,0) = 0$. Set $(0,0) = \mathbf{0}$. It is easy to check that f is continuous, all directional derivatives exist at $\mathbf{0}$, and $\frac{\partial f}{\partial x_i}(\mathbf{0}) = 0$, $i = 1, 2$. Hence, if f is differentiable at $\mathbf{0}$, then the derivative must be the zero linear map $\mathbf{0} : \mathbb{R}^2 \to \mathbb{R}$. But this is absurd since $D_{\mathbf{u}} f(\mathbf{0}) \neq 0$ if $\mathbf{u} \notin \{\pm \mathbf{e}_1, \pm \mathbf{e}_2\}$ (alternatively, if $Df_0 = \mathbf{0}$, then $f(\mathbf{h}) = r(\mathbf{h})$ and it is easy to verify that $f(\mathbf{h}) \neq o(\mathbf{h})$).   ♠

9.4.2 Partial Derivatives

If $f = (f_1, \cdots, f_n) : \mathbb{R}^m \to \mathbb{R}^n$ is differentiable and we take $u = \mathbf{e}_j \in \mathbb{R}^m$, $1 \leq j \leq m$, then

$$D_{\mathbf{e}_j} f(\mathbf{x}_0) = \frac{\partial f}{\partial x_j}(\mathbf{x}_0) = \left(\frac{\partial f_1}{\partial x_j}, \cdots, \frac{\partial f_n}{\partial x_j} \right)(x_0).$$

The $m \times n$ matrix $[\frac{\partial f_i}{\partial x_j}]$ of partial derivatives of f at $\mathbf{x}_0$ is then equal to the matrix of the derivative $Df_{\mathbf{x}_0}$. We refer to $[\frac{\partial f_i}{\partial x_j}]$ as the *Jacobian matrix* of f at $\mathbf{x}_0$ ($Df_{\mathbf{x}_0}$ is sometimes called the *Jacobian* of f at $\mathbf{x}_0$).

Observe that in order to compute the partial derivatives we need to choose coordinate systems on $\mathbb{R}^m$, $\mathbb{R}^n$. Thus, if $f : V \to W$ is differentiable at $\mathbf{x}_0$, the derivative $Df_{\mathbf{x}_0} \in L(V, W)$ (see Exercises 9.3.7(6)). In order to define the partial derivatives of f at $\mathbf{x}_0$, we need to choose bases for V and W and thereby identify V with $\mathbb{R}^m$ and W with $\mathbb{R}^n$. The matrix of $Df_{\mathbf{x}_0}$ relative to the coordinate systems on V and W will then be the matrix $[\frac{\partial f_i}{\partial x_j}]$ of partial derivatives of f at $\mathbf{x}_0$.

Provided that $f : \mathbb{R}^m \to \mathbb{R}^n$ is differentiable at $\mathbf{x}_0$, we can always compute the partial derivatives of f. As we see later, the converse is more subtle and requires some continuity of the partial derivatives of f. As Example 9.4.7 shows, we cannot deduce differentiability just from the existence of partial derivatives.

9.4.3   The Chain Rule

The chain rule is one of the most useful results about derivatives. Simply put, the chain rule asserts that the best affine linear approximation to a composite $g \circ f$ of differentiable functions is the composite of the best affine linear approximations of g and f. Viewed in this way, the proof that we give is quite natural: we verify that the composite of the derivatives does give the best approximation to the composite of the maps.

Theorem 9.4.8 (Chain Rule) *Let* $f : \mathbb{R}^m \to \mathbb{R}^n$ *and* $g : \mathbb{R}^n \to \mathbb{R}^p$. *Suppose* $\mathbf{x} \in \mathbb{R}^m$ *and set* $\mathbf{y} = f(\mathbf{x}) \in \mathbb{R}^n$. *If* f *is differentiable at* $\mathbf{x}$ *and* g *is differentiable at* $\mathbf{y}$, *then* $g \circ f$ *is differentiable at* $\mathbf{x}$ *and*

$$D(g \circ f)_{\mathbf{x}} = Dg_{\mathbf{y}} \circ Df_{\mathbf{x}}.$$

Proof It is enough to show that $Dg_{\mathbf{y}} \circ Df_{\mathbf{x}} \in L(\mathbb{R}^m, \mathbb{R}^p)$ satisfies the defining condition for the differentiability of $g \circ f$ at $\mathbf{x}$. That is,

$$g \circ f(\mathbf{x} + \mathbf{h}) = g \circ f(\mathbf{x}) + Dg_{\mathbf{y}} \circ Df_{\mathbf{x}}(\mathbf{h}) + R(\mathbf{h}),$$

where $R(\mathbf{h}) = o(\mathbf{h})$. We start by writing down the differentiability assumptions we are given on f and g.

$$f(\mathbf{x} + \mathbf{h}) = f(\mathbf{x}) + Df_{\mathbf{x}}(\mathbf{h}) + r(\mathbf{h}),$$
$$g(\mathbf{y} + \mathbf{k}) = g(\mathbf{y}) + Dg_{\mathbf{y}}(\mathbf{k}) + s(\mathbf{k}),$$

where $r(\mathbf{h}) = o(\mathbf{h})$ and $s(\mathbf{k}) = o(\mathbf{k})$. Taking $\mathbf{k} = Df_{\mathbf{x}}(\mathbf{h}) + r(\mathbf{h})$ and substituting in the right-hand side of the formula for $g(\mathbf{y} + \mathbf{k}) = g(f(\mathbf{x}) + Df_{\mathbf{x}}(\mathbf{h}) + r(\mathbf{h})) = g \circ f(\mathbf{x} + \mathbf{h})$ gives

$$g(f(\mathbf{x} + \mathbf{h})) = gf(\mathbf{x}) + Dg_{\mathbf{y}}(Df_{\mathbf{x}}(\mathbf{h}) + r(\mathbf{h})) + s(Df_{\mathbf{x}}(\mathbf{h}) + r(\mathbf{h}))$$
$$= g \circ f(\mathbf{x}) + Dg_{\mathbf{y}} \circ Df_{\mathbf{x}}(\mathbf{h})$$
$$+ Dg_{\mathbf{y}}(r(\mathbf{h})) + s(Df_{\mathbf{x}}(\mathbf{h}) + r(\mathbf{h})).$$

Therefore, $R(\mathbf{h}) = Dg_\mathbf{y}(r(\mathbf{h})) + s(Df_\mathbf{x}(\mathbf{h}) + r(\mathbf{h}))$. To complete the proof, we show that $Dg_\mathbf{y}(r(\mathbf{h})) = o(\mathbf{h})$ and $s(Df_\mathbf{x}(\mathbf{h}) + r(\mathbf{h})) = o(\mathbf{h})$.

(1) $Dg_\mathbf{y}(r(\mathbf{h})) = o(\mathbf{h})$. For $\mathbf{h} \neq \mathbf{0}$, we have $\frac{\|Dg_\mathbf{y}(r(\mathbf{h}))\|}{\|\mathbf{h}\|} \leq \frac{\|Dg_\mathbf{y}\|\|r(\mathbf{h})\|}{\|\mathbf{h}\|}$. Since $r(\mathbf{h}) = o(\mathbf{h})$, $\lim_{\mathbf{h} \to 0} \frac{\|Dg_\mathbf{y}\|\|r(\mathbf{h})\|}{\|\mathbf{h}\|} = 0$ and so $\|Dg_\mathbf{y}(r(\mathbf{h}))\| = o(\mathbf{h})$.

(2) $s(Df_\mathbf{x}(\mathbf{h}) + r(\mathbf{h})) = o(\mathbf{h})$. Since $r(\mathbf{h}) = o(\mathbf{h})$, we can choose $\delta_1 > 0$ such that $\|r(\mathbf{h})\| \leq \|\mathbf{h}\|$, if $\|\mathbf{h}\| \leq \delta_1$. Hence

$$\|Df_\mathbf{x}(\mathbf{h}) + r(\mathbf{h})\| \leq \|Df_\mathbf{x}(\mathbf{h})\| + \|r(\mathbf{h})\| \leq (\|Df_\mathbf{x}\| + 1)\|\mathbf{h}\|, \quad \|\mathbf{h}\| \leq \delta_1.$$

Since $s(\mathbf{k}) = o(\mathbf{k})$, given $\varepsilon > 0$, we can choose $\delta_2 > 0$ such that

$$\|s(\mathbf{k})\| \leq \frac{\varepsilon}{(\|Df_\mathbf{x}\| + 1)}\|\mathbf{k}\|, \quad \|\mathbf{k}\| \leq \delta_2.$$

Set $\delta = \min\{\delta_1, \frac{\delta_2}{(\|Df_\mathbf{x}\|+1)}\}$. Then if $\|\mathbf{h}\| \leq \delta$, we have

$$\|Df_\mathbf{x}(\mathbf{h}) + r(\mathbf{h})\| \leq (\|Df_\mathbf{x}\| + 1)\|\mathbf{h}\| \leq \delta_2,$$

and so

$$\|s(Df_\mathbf{x}(\mathbf{h}) + r(\mathbf{h}))\| \leq \frac{\varepsilon}{(\|Df_\mathbf{x}\| + 1)}(\|Df_\mathbf{x}\| + 1)\|\mathbf{h}\| = \varepsilon\|\mathbf{h}\|.$$

Hence $s(Df_\mathbf{x}(\mathbf{h}) + r(\mathbf{h})) = o(\mathbf{h})$. $\qquad\qquad\qquad\qquad\qquad\qquad\qquad \square$

Remarks 9.4.9

(1) Notice how well this proof using approximation avoids the difficulties encountered using the $\lim_{h \to 0} \frac{g(f(x+h)) - g(f(x))}{h}$ definition from the 1-variable theory.

(2) If we assume f and g have respective domains the open subsets U of $\mathbb{R}^m$ and V of $\mathbb{R}^n$, then $g \circ f$ is defined on the open set $U \cap f^{-1}(V) \subset \mathbb{R}^m$ and Theorem 9.4.8 applies with the proviso that $\mathbf{x} \in U \cap f^{-1}(V)$.   ✸

Examples 9.4.10

(1) Let $V : \mathbb{R}^m \to \mathbb{R}$ and $\phi = (\phi_1, \cdots, \phi_m) : \mathbb{R} \to \mathbb{R}^m$ be differentiable. We claim that $V \circ \phi : \mathbb{R} \to \mathbb{R}$ is differentiable and $(V \circ \phi)'(t) = DV_{\phi(t)}(\phi'(t))$ (see Definition 9.3.5 for the notation ϕ', $(V \circ \phi)'$). In terms of partial derivatives,

$$\frac{d}{dt}V(\phi(t)) = \sum_{i=1}^{m} \frac{\partial V}{\partial x_i}(\phi(t))\phi_i'(t).$$

To verify the claim, apply the chain rule to get $D(V \circ \phi)_t = DV_{\phi(t)} \circ D\phi_t$. Either side of the equation defines a linear map from $\mathbb{R}$ to $\mathbb{R}$. Evaluate at 1 to get $(V \circ \phi)'(t) = DV_{\phi(t)}(\phi'(t))$. Now $\phi'(t) = \sum_{i=1}^{m} \phi_i'(t)\mathbf{e}_i$ and so we have

$$DV_{\phi(t)}(\phi'(t)) = DV_{\phi(t)}\left(\sum_{i=1}^{m} \phi_i'(t)\mathbf{e}_i\right)$$

$$= \sum_{i=1}^{m} \phi_i'(t)DV_{\phi(t)}(\mathbf{e}_i)$$

$$= \sum_{i=1}^{m} \phi_i'(t)\frac{\partial V}{\partial x_i}(\phi(t)),$$

where the last line follows by definition of the partial derivative.

(2) Let $U \subset \mathbb{R}^m$ and $V \subset \mathbb{R}^n$ be open sets and suppose that $f : U \to V$ is 1:1 onto and both f and $f^{-1} : V \to U$ are differentiable. Then

  (a) For all $\mathbf{x} \in U$, $Df_{\mathbf{x}} : \mathbb{R}^m \to \mathbb{R}^n$ is a linear isomorphism and $(Df_{\mathbf{x}})^{-1} = Df^{-1}_{f(\mathbf{x})}$.
  (b) $m = n$.

(a) We have $f^{-1} \circ f = I_U$, where I_U is the identity map of U. It follows by the chain rule that if $\mathbf{x} \in U$, then

$$Df^{-1}_{f(\mathbf{x})} \circ Df_{\mathbf{x}} = I \in L(\mathbb{R}^m, \mathbb{R}^m).$$

Hence the linear map $Df_{\mathbf{x}}$ is invertible with inverse $Df^{-1}_{f(\mathbf{x})}$.

(b) If we have a linear isomorphism $A : \mathbb{R}^m \to \mathbb{R}^n$ then (by linear algebra—look at bases), $m = n$.   ♠

Remark 9.4.11 It is natural to ask if $m = n$ when $f : U \to V$ is 1:1 onto but f and f^{-1} are only continuous, that is, f is a homeomorphism. The answer is yes but the proof is tricky and depends on results from topology, specifically the "invariance of domain theorem".   ✠

9.4.4 The Mean Value Theorem

We recall that the mean value theorem for maps $F : [a, b] \to \mathbb{R}$, continuous on $[a, b]$ and differentiable on (a, b), states that there exists a $c \in (a, b)$ such that

$$F(b) - F(a) = F'(c)(b - a). \tag{9.3}$$

Before we state the version of the mean value theorem appropriate for vector spaces, we need some notation. Given $\mathbf{x} \neq \mathbf{y} \in \mathbb{R}^m$, let $[\mathbf{x}, \mathbf{y}] \subset \mathbb{R}^m$ be the line segment joining $\mathbf{x}$ and $\mathbf{y}$. That is,

$$[\mathbf{x}, \mathbf{y}] = \{(1 - t)\mathbf{x} + t\mathbf{y} \mid t \in [0, 1]\}.$$

Theorem 9.4.12 (The Mean Value Theorem) *Let* $U \subset \mathbb{R}^m$ *be open and* $f : U \to \mathbb{R}^n$ *be differentiable. Given* $\mathbf{x}, \mathbf{y} \in U$ *such that* $[\mathbf{x}, \mathbf{y}] \subset U$, *we have*

$$\|f(\mathbf{x}) - f(\mathbf{y})\| \leq \sup_{\mathbf{z} \in (\mathbf{x}, \mathbf{y})} \|Df_{\mathbf{z}}\| \|\mathbf{x} - \mathbf{y}\|.$$

(We allow $\sup_{\mathbf{z} \in (\mathbf{x}, \mathbf{y})} \|Df_{\mathbf{z}}\| = +\infty$.)

Proof We prove the result in two steps. If $n = 1$, we deduce the theorem from the 1-variable version of the mean value theorem. If $n > 1$, we reduce to the $n = 1$ case by projecting $\mathbb{R}^n$ along the line defined by the vector $f(\mathbf{y}) - f(\mathbf{x})$. Now for the details.

Suppose $n = 1$. Define $F : [0, 1] \to \mathbb{R}$ by

$$F(t) = f((1 - t)\mathbf{x} + t\mathbf{y}), \ t \in [0, 1].$$

Observe that $F(0) = f(\mathbf{x})$, $F(1) = f(\mathbf{y})$. Apply the 1-dimensional version of the mean value theorem (9.3) to get

$$f(\mathbf{y}) - f(\mathbf{x}) = F(1) - F(0) = F'(c),$$

for some point $c \in (0, 1)$. Since $F(t) = f((1 - t)\mathbf{x} + t\mathbf{y})$ it follows by the chain rule that

$$F'(t) = Df_{(1-t)\mathbf{x}+t\mathbf{y}}(-\mathbf{x} + \mathbf{y}),$$

and so

$$f(\mathbf{y}) - f(\mathbf{x}) = Df_{\tilde{\mathbf{z}}}(\mathbf{y} - \mathbf{x}),$$

where $\tilde{\mathbf{z}} = (1 - c)\mathbf{x} + c\mathbf{y}$. Hence

$$\|f(\mathbf{y}) - f(\mathbf{x})\| = \|Df_{\tilde{\mathbf{z}}}(\mathbf{y} - \mathbf{x})\| \leq \|Df_{\tilde{\mathbf{z}}}\| \|\mathbf{y} - \mathbf{x}\| \leq \sup_{\mathbf{z} \in (\mathbf{x}, \mathbf{y})} \|Df_{\mathbf{z}}\| \|\mathbf{x} - \mathbf{y}\|.$$

Next suppose $n > 1$. Since the result is obvious if $f(\mathbf{x}) = f(\mathbf{y})$, we may suppose that $f(\mathbf{x}) \neq f(\mathbf{y})$. Set $\mathbf{u} = \frac{f(\mathbf{y}) - f(\mathbf{x})}{\|f(\mathbf{y}) - f(\mathbf{x})\|} \in S^{n-1}$. Define the linear map $\psi : \mathbb{R}^n \to \mathbb{R}$ by $\psi(\mathbf{w}) = (\mathbf{u}, \mathbf{w})$, $\mathbf{w} \in \mathbb{R}^n$. The map ψ gives the component of the orthogonal projection of $\mathbb{R}^n$ along the line $\{t\mathbf{u} \mid t \in \mathbb{R}\}$. Since $|\psi(\mathbf{w})| = |(\mathbf{u}, \mathbf{w})| \leq \|\mathbf{u}\| \|\mathbf{w}\| = \|\mathbf{w}\|$, we have (take $\mathbf{w} = \mathbf{u}$)

$$\|\psi\| = 1. \tag{9.4}$$

Define $G : [0, 1] \to \mathbb{R}$ by

$$G(t) = \psi(F(t)) = (\mathbf{u}, F(t)), \ t \in [0, 1].$$

Observe that

$$G(1) - G(0) = \left(\frac{f(\mathbf{y}) - f(\mathbf{x})}{\|f(\mathbf{y}) - f(\mathbf{x})\|}, F(1) - F(0) \right)$$

$$= \left(\frac{f(\mathbf{y}) - f(\mathbf{x})}{\|f(\mathbf{y}) - f(\mathbf{x})\|}, f(\mathbf{y}) - f(\mathbf{x}) \right) = \|f(\mathbf{y}) - f(\mathbf{x})\|.$$

Applying the 1-dimensional version of the mean value theorem (9.3) to G we get

$$\|f(\mathbf{y}) - f(\mathbf{x})\| = G(1) - G(0) = G'(c),$$

for some $c \in (0, 1)$. It remain to compute $G'(c)$. We apply the chain rule to $G = \psi \circ F$. Since ψ is linear, $D\psi = \psi$ and so

$$G'(t) = \psi(F'(t)) = \psi(Df_{(1-t)\mathbf{x}+t\mathbf{y}}(-\mathbf{x} + \mathbf{y})).$$

Setting $t = c$, $\tilde{\mathbf{z}} = (1 - c)\mathbf{x} + c\mathbf{y}$, gives

$$\|f(\mathbf{y}) - f(\mathbf{x})\| = G(1) - G(0) = G'(c) = \psi(Df_{\tilde{\mathbf{z}}}(\mathbf{y} - \mathbf{x})).$$

Since $G'(c) = \|f(\mathbf{y}) - f(\mathbf{x})\|$, $G'(c) > 0$ and so

$$\|f(\mathbf{y}) - f(\mathbf{x})\| = |G'(c)| = |\psi(Df_{\tilde{\mathbf{z}}}(\mathbf{y} - \mathbf{x}))|$$

$$\leq \|\psi\| \|Df_{\tilde{\mathbf{z}}}\| \|\mathbf{y} - \mathbf{x}\| = \|Df_{\tilde{\mathbf{z}}}\| \|\mathbf{y} - \mathbf{x}\|,$$

where we have used (9.4). □

Remarks 9.4.13

(1) The multivariable form of the mean value theorem is written as an inequality. For maps into $\mathbb{R}^n$, $n > 1$, it is generally not possible to write $f(\mathbf{y}) - f(\mathbf{x}) = Df_{\mathbf{z}}(\mathbf{y} - \mathbf{x})$ for some $\mathbf{z} \in [\mathbf{x}, \mathbf{y}]$. It is not hard to construct examples—see the exercises at the end of the section.

(2) The mean value theorem is easily the most important foundational result in the differential calculus. It estimates $\|f(\mathbf{y}) - f(\mathbf{x})\|$ as though f were a *linear* map— if f is linear, then $\|f(\mathbf{y}) - f(\mathbf{x})\| \leq \|f\| \|\mathbf{y} - \mathbf{x}\|$. Having inequality, rather than equality, is no loss: it is rare (outside of contrived problems) that explicit use is made of the value c in (9.3). ✱

Corollary 9.4.14 *Let $U \subset \mathbb{R}^m$ be open and connected. Suppose that $f : U \to \mathbb{R}^n$ is differentiable and $Df_{\mathbf{x}} = 0$ for all $\mathbf{x} \in U$. Then f is constant.*

Proof Given $\mathbf{x} \in U$, choose $r > 0$ such that $D_r(\mathbf{x}) \subset U$. For every $\mathbf{y} \in D_r(\mathbf{x})$, $[\mathbf{x}, \mathbf{y}] \subset D_r(\mathbf{x}) \subset U$. Hence if $\mathbf{y} \in D_r(\mathbf{x})$, we can apply the mean value theorem to get

$$\|f(\mathbf{y}) - f(\mathbf{x})\| \leq \sup_{\mathbf{z} \in [\mathbf{x}, \mathbf{y}]} \|Df_{\mathbf{z}}\| \|\mathbf{y} - \mathbf{x}\| = 0,$$

since $Df = 0$. Hence $f(\mathbf{y}) = f(\mathbf{x})$, for all $\mathbf{y} \in D_r(\mathbf{x})$. Fix $\mathbf{x}_0 \in U$ and define

$$W = \{\mathbf{y} \in U \mid f(\mathbf{y}) = f(\mathbf{x}_0)\}.$$

Since f is continuous (because f is differentiable), W is a closed subset of U. On the other hand, it follows from the argument above that W is open. Since $\mathbf{x}_0 \in W$, $W \neq \emptyset$ and so, by the connectivity of U, $W = U$. $\square$

EXERCISES 9.4.15

(1) Suppose that $f, g : \mathbb{R}^m \to \mathbb{R}^n$ are both differentiable at x. Show that $f \pm g : \mathbb{R}^m \to \mathbb{R}^n$ is differentiable at x with derivative $Df_x \pm Dg_x$.

(2) Let $Q : \mathbb{R}^m \to \mathbb{R}^n$ be differentiable and suppose that $Q(t\mathbf{x}) = t^d Q(\mathbf{x})$ for all $t \in \mathbb{R}$, $\mathbf{x} \in \mathbb{R}^m$ (d is assumed to be a strictly positive integer).

   (a) Show that $Q(\mathbf{0}) = 0$.
   (b) $DQ_{\mathbf{x}}(\mathbf{x}) = dQ(\mathbf{x})$, for all $\mathbf{x} \in \mathbb{R}^n$ (Euler's theorem). (Hint: Apply the chain rule to $Q \circ g : \mathbb{R} \to \mathbb{R}^n$ where $g(t) = t\mathbf{x}$.)

(3) Let $f : \mathbb{R}^m \to \mathbb{R}^n$ be C^1 and $f(\mathbf{0}) = \mathbf{0}$. Show that $f(\mathbf{x}) = \int_0^1 Df_{t\mathbf{x}}(\mathbf{x}) \, dt$ and deduce that we can write the components f_i of f in the form

$$f_i(\mathbf{x}) = \sum_{j=1}^m x_j g_{ji}(\mathbf{x}), \ 1 \leq i \leq n,$$

where $g_{ji}(\mathbf{x}) = \int_0^1 \frac{\partial f_i}{\partial x_j}(t\mathbf{x}) \, dt$. (Hint: $f(\mathbf{x}) = \int_0^1 \frac{d}{dt} f(t\mathbf{x}) \, dt$. The integral of an $\mathbb{R}^n$-valued function is defined component-wise: $\int (f_1, \cdots, f_n) = (\int f_1, \cdots, \int f_n)$.)

(4) Recall that the derivative of the norm squared function $\phi(x) = \|x\|^2$ is given by $D\phi_x(h) = 2(x, h)$. Using this, together with the chain rule, show that $\psi(x) = \|x\|^{\frac{5}{2}}$ is C^1 on $\mathbb{R}^m$ and find $D\psi_x \in L(\mathbb{R}^m, \mathbb{R})$. (You may assume that $g : \mathbb{R}_+ \to \mathbb{R}$ defined by $g(t) = t^a$ is C^1 with derivative at^{a-1}, provided that $a \geq 1$. Start by writing ψ as a composition.)

(5) Let $f, g : \mathbb{R}^m \to \mathbb{R}$ be differentiable functions. Find, from first principles (using the definition of differentiable), the derivative $D(fg)_{\mathbf{x}} \in L(\mathbb{R}^m, \mathbb{R})$, $\mathbf{x} \in \mathbb{R}^m$, in terms of $Df_{\mathbf{x}}, Dg_{\mathbf{x}}$ and the values of f and g at $\mathbf{x}$. (Note that $fg = f \times g$.) Deduce a general formula for the derivative at $\mathbf{x}$ of a product of p differentiable functions. Hence show that if $f(\mathbf{x}) = A_1(\mathbf{x}) \cdots A_p(\mathbf{x})$, where each $A_i : \mathbb{R}^n \to \mathbb{R}$ is linear, then f is C^1. Deduce that every monomial $M : \mathbb{R}^n \to \mathbb{R}$, $M(\mathbf{x}) =$

$x_1^{a_1} \cdots x_n^{a_n}, a_1, \cdots, a_n \in \mathbb{Z}_+$, is C^1 and hence that every real-valued polynomial on $\mathbb{R}^n$ is C^1.

(6) Let $f, g : \mathbb{R}^m \to \mathbb{R}$ be C^1. Verify the Leibniz law:

$$D(fg)_\mathbf{x} = g(\mathbf{x})Df_\mathbf{x} + f(\mathbf{x})Dg_\mathbf{x}, \quad \mathbf{x} \in \mathbb{R}^m.$$

($fg : \mathbb{R}^m \to \mathbb{R}$ is defined by $f \times g(\mathbf{x}) = f(\mathbf{x})g(\mathbf{x}), \mathbf{x} \in \mathbb{R}^m$.)

(7) Define $f : \mathbb{R} \times \mathbb{R} = \mathbb{R}^2 \to \mathbb{R}$ by

$$f(x_1, x_2) = \begin{cases} \frac{x_1 x_2^3}{x_1^2 + x_2^4}, & (x_1, x_2) \neq (0, 0), \\ 0, & (x_1, x_2) = (0, 0). \end{cases}$$

Show that

(a) f is continuous on $\mathbb{R}^2$.
(b) The directional derivatives $D_\mathbf{u} f_\mathbf{x}$ exist on $\mathbb{R}^2$ for all unit vectors $\mathbf{u}$. Compute them at $(x, y) = (0, 0)$.
(c) f is not differentiable at $(0, 0)$.

(Hints: for (a) you may find the arithmetic-geometric mean inequality $((A + B \geq 2\sqrt{AB}, A, B \geq 0)$ helpful. For (c), look at what happens on a curve (h^2, h), $0 < h \leq 1$, and note that if the derivative exists it may be found using (b). Now use the chain rule.)

(8) Find an example of a C^1 map $f : \mathbb{R} \to \mathbb{R}^2$ such that

$$f(1) - f(0) \neq f'(t), \quad \text{for all } t \in [0, 1].$$

(Failure of the mean value theorem as an *equality* for maps into $\mathbb{R}^n$, $n > 1$. Note that if the result were true then it would apply to the components and so for $j = 1, 2, f_j(1) - f_j(0) = f_j'(t)$ at the *same* point $t \in (0, 1)$. So it is enough to find two functions $f_1, f_2 : \mathbb{R} \to \mathbb{R}$ where the equality occurs at *different* points.)

(9) Let $U \subset \mathbb{R}^m$ be open and suppose that $f : U \to \mathbb{R}$. Given $\mathbf{x}_0 \in U$, we say $f(\mathbf{x}_0)$ is a *local maximum value* of f if there exists an $r > 0$ such that

$$f(\mathbf{x}) \leq f(\mathbf{x}_0), \quad \text{whenever } \|\mathbf{x} - \mathbf{x}_0\| < r.$$

Show that if f is differentiable on U then a necessary condition for $f(\mathbf{x}_0)$ to be a local maximum value of f is $Df_{\mathbf{x}_0} = 0$ (as a linear map from $\mathbb{R}^m$ to $\mathbb{R}$). You should work from the definition of differentiable map (and there should be no mention of partial derivatives).

(10) Let U be an open subset of $\mathbb{R}^m$ and let $f : U \to \mathbb{R}^n$. Suppose that there is a continuous map $u : U \to L(\mathbb{R}^m, \mathbb{R}^n)$ such that for all $\mathbf{y} \in \mathbb{R}^m$ and all $\mathbf{x} \in U$,

we have

$$\lim_{t \to 0} \frac{f(\mathbf{x} + t\mathbf{y}) - f(\mathbf{x})}{t} = u(\mathbf{x})(\mathbf{y}).$$

By applying the mean value theorem to the map $g(t) = f(\mathbf{x} + t\mathbf{y}) - u(\mathbf{x})(t\mathbf{y})$, prove that f is C^1 on U and $Df_{\mathbf{x}} = u(\mathbf{x})$ for all $\mathbf{x} \in U$. Rephrase this result in terms of directional derivatives of f.

9.5   Maps to and from Products

We start with the easier case of a map to a product. Suppose then that $f = (f_1, f_2)$: $\mathbb{R}^m \to \mathbb{R}^{n_1} \times \mathbb{R}^{n_2}$. If f_1, f_2 are continuous so is f and conversely. Before giving the result for differentiability of maps $f : \mathbb{R}^m \to \mathbb{R}^{n_1} \times \mathbb{R}^{n_2}$, note the natural linear isomorphism

$$L(\mathbb{R}^m, \mathbb{R}^{n_1} \times \mathbb{R}^{n_2}) \approx L(\mathbb{R}^m, \mathbb{R}^{n_1}) \times L(\mathbb{R}^m, \mathbb{R}^{n_2})$$

defined by taking the components A_1, A_2 of $A : \mathbb{R}^m \to \mathbb{R}^{n_1} \times \mathbb{R}^{n_2}$.

Proposition 9.5.1 *Let $U \subset \mathbb{R}^m$ be open and suppose $f = (f_1, f_2) : U \subset \mathbb{R}^m \to \mathbb{R}^{n_1} \times \mathbb{R}^{n_2}$.*

(1) *If $\mathbf{x} \in U$, then f is differentiable at $\mathbf{x}$ iff both f_1 and f_2 are differentiable at $\mathbf{x}$, and the derivatives of f and f_1, f_2 at $\mathbf{x}$ are related by*

$$Df_{\mathbf{x}}(\mathbf{h}) = (Df_{1,\mathbf{x}}, Df_{2,\mathbf{x}})(\mathbf{h}) = (Df_{1,\mathbf{x}}(\mathbf{h}), Df_{2,\mathbf{x}}(\mathbf{h})), \ \mathbf{h} \in \mathbb{R}^m.$$

(2) *f is differentiable (respectively, C^1) on U iff both f_1, f_2 are differentiable (respectively, C^1) on U.*

Proof We prove (1). If f_1, f_2 are differentiable at $\mathbf{x}$ and $\mathbf{h} \in \mathbb{R}^m$, we have

$$f_1(\mathbf{x} + \mathbf{h}) = f_1(\mathbf{x}) + Df_{1,\mathbf{x}}(\mathbf{h}) + r_1(\mathbf{h}),$$
$$f_2(\mathbf{x} + \mathbf{h}) = f_2(\mathbf{x}) + Df_{2,\mathbf{x}}(\mathbf{h}) + r_2(\mathbf{h}),$$

where $r_1(\mathbf{h}), r_2(\mathbf{h}) = o(\mathbf{h})$. Hence,

$$f(\mathbf{x} + \mathbf{h}) = f(\mathbf{x}) + (Df_{1,\mathbf{x}}, Df_{2,\mathbf{x}})(\mathbf{h}) + (r_1(\mathbf{h}), r_2(\mathbf{h})), \ \mathbf{h} \in \mathbb{R}^m.$$

Taking the (Euclidean norm) $\|(\mathbf{v}_1, \mathbf{v}_2)\| = \|\mathbf{v}_1\| + \|\mathbf{v}_2\|$ on the product $\mathbb{R}^{n_1} \times \mathbb{R}^{n_2} \approx \mathbb{R}^{n_1 + n_2}$, we have $\|(r_1(\mathbf{h}), r_2(\mathbf{h}))\| = \|r_1(\mathbf{h})\| + \|r_2(\mathbf{h})\|$ and so $(r_1(\mathbf{h}), r_2(\mathbf{h})) = o(\mathbf{h})$. Hence f is differentiable at $\mathbf{x}$ and $Df_{\mathbf{x}} = (Df_{1,\mathbf{x}}, Df_{2,\mathbf{x}})$. For the converse, reverse the argument. $\square$

Corollary 9.5.2 *Let $U \subset \mathbb{R}^m$ be open and $f = (f_1, \cdots, f_n) : U \subset \mathbb{R}^m \to \mathbb{R}^n$ be differentiable at $\mathbf{x}_0 \in U$. Then*

$$Df_{\mathbf{x}} = (Df_{1,\mathbf{x}}, \cdots Df_{n,\mathbf{x}}) \in \times^n L(\mathbb{R}^m, \mathbb{R}).$$

Proof A straightforward induction using Proposition 9.5.1.   □

Next we look at maps from a product. Suppose that $f : \mathbb{R}^p \times \mathbb{R}^q \to \mathbb{R}^n$, $p, q \in \mathbb{N}$. It does not follow that separate continuity implies continuity. That is, if both $\mathbf{x} \mapsto f(\mathbf{x}, \mathbf{y})$ ($\mathbf{y}$ fixed) and $\mathbf{y} \mapsto f(\mathbf{x}, \mathbf{y})$ ($\mathbf{x}$ fixed) are continuous, f need not be continuous (for example, define $f(0, 0) = 0$ and $f(x, y) = xy/(x^2 + y^2)$, $(x, y) \neq (0, 0)$). This suggests that inferring results about the differentiability of $f(\mathbf{x}, \mathbf{y})$ from the separate differentiability of f in $\mathbf{x}$ and $\mathbf{y}$ may not be so straightforward.

In order to relate derivatives in $\mathbf{x}$ and $\mathbf{y}$ with the derivative at $(\mathbf{x}, \mathbf{y})$, we make use of the natural isomorphism

$$L(\mathbb{R}^p, \mathbb{R}^n) \times L(\mathbb{R}^q, \mathbb{R}^n) \approx L(\mathbb{R}^p \times \mathbb{R}^q, \mathbb{R}^n); \ (A_1, A_2) \mapsto A, \tag{9.5}$$

where $A(\mathbf{u}, \mathbf{v}) = A_1(\mathbf{u}) + A_2(\mathbf{v})$, $(\mathbf{u}, \mathbf{v}) \in \mathbb{R}^p \times \mathbb{R}^q$.

Let $\mathbf{X}_0 = (\mathbf{x}_0, \mathbf{y}_0) \in \mathbb{R}^p \times \mathbb{R}^q$. Suppose that the map $\mathbf{x} \mapsto f(\mathbf{x}, \mathbf{y}_0)$ is differentiable at $\mathbf{x} = \mathbf{x}_0$. We denote the derivative at $\mathbf{x}_0$ by $D_1 f_{(\mathbf{x}_0, \mathbf{y}_0)} = D_1 f_{\mathbf{X}_0}$. Note that this notation emphasizes that the derivative in $\mathbf{x}$-variables depends on $\mathbf{y}$. We similarly let $D_2 f_{\mathbf{X}_0}$ denote the derivative of $\mathbf{y} \mapsto f(\mathbf{x}_0, \mathbf{y})$ at $\mathbf{y}_0$.

Since $D_1 f_{\mathbf{X}_0} \in L(\mathbb{R}^p, \mathbb{R}^n)$ and $D_2 f_{\mathbf{X}_0} \in L(\mathbb{R}^q, \mathbb{R}^n)$, the linear map $A \in L(\mathbb{R}^p \times \mathbb{R}^q, \mathbb{R}^n)$ determined by the natural isomorphism (9.5) is given by

$$A(\mathbf{u}, \mathbf{v}) = D_1 f_{\mathbf{X}_0}(\mathbf{u}) + D_2 f_{\mathbf{X}_0}(\mathbf{v}), \ (\mathbf{u}, \mathbf{v}) \in \mathbb{R}^p \times \mathbb{R}^q.$$

We show that if f is differentiable at $\mathbf{X}_0$, then $Df_{\mathbf{X}_0} = A$.

Proposition 9.5.3 *If $f : \mathbb{R}^p \times \mathbb{R}^q \to \mathbb{R}^n$ is differentiable at $\mathbf{X}_0$, then f is differentiable with respect to $\mathbf{x}$ and $\mathbf{y}$ at $\mathbf{X}_0$ and*

$$D_1 f_{\mathbf{X}_0}(\mathbf{h}) = Df_{\mathbf{X}_0}(\mathbf{h}, \mathbf{0}), \ \mathbf{h} \in \mathbb{R}^p,$$

$$D_2 f_{\mathbf{X}_0}(\mathbf{k}) = Df_{\mathbf{X}_0}(\mathbf{0}, \mathbf{k}), \ \mathbf{k} \in \mathbb{R}^q,$$

$$Df_{\mathbf{X}_0}(\mathbf{h}, \mathbf{k}) = D_1 f_{\mathbf{X}_0}(\mathbf{h}) + D_2 f_{\mathbf{X}_0}(\mathbf{k}), \ (\mathbf{h}, \mathbf{k}) \in \mathbb{R}^p \times \mathbb{R}^q.$$

Proof Since f is differentiable at $\mathbf{X}_0$,

$$f(\mathbf{x}_0 + \mathbf{h}, \mathbf{y}_0 + \mathbf{k}) = f(\mathbf{x}_0, \mathbf{y}_0) + Df_{\mathbf{X}_0}(\mathbf{h}, \mathbf{k}) + R(\mathbf{h}, \mathbf{k}), \tag{9.6}$$

where $R(\mathbf{h}, \mathbf{k}) = o(\mathbf{h}, \mathbf{k})$. Taking $\mathbf{k} = \mathbf{0}$, we have $\lim_{\mathbf{h} \to 0} \frac{\|R(\mathbf{h}, 0)\|}{\|(\mathbf{h}, 0)\|} = 0$ and so $r(\mathbf{h}) = R(\mathbf{h}, 0) = o(\mathbf{h})$. Similarly, $s(\mathbf{k}) = R(0, \mathbf{k}) = o(\mathbf{k})$. Taking $\mathbf{k} = \mathbf{0}$ in (9.6) gives

$$f(\mathbf{x}_0 + \mathbf{h}, \mathbf{y}_0) = f(\mathbf{x}_0, \mathbf{y}_0) + Df_{\mathbf{X}_0}(\mathbf{h}, \mathbf{0}) + r(\mathbf{h}),$$

and so $\mathbf{x} \mapsto f(\mathbf{x}, \mathbf{y}_0)$ is differentiable at $\mathbf{x}_0$ with derivative $D_1 f_{\mathbf{X}_0}$ given by $D_1 f_{\mathbf{X}_0}(\mathbf{h}) = Df_{(\mathbf{X}_0)}(\mathbf{h}, \mathbf{0})$. The result for $D_2 f_{\mathbf{X}_0}$ is proved similarly. The final statement is immediate. □

Now we look at the converse of Proposition 9.5.3.

Theorem 9.5.4 *Let $U \subset \mathbb{R}^{m_1} \times \mathbb{R}^{m_2}$ be open and suppose $f : U \subset \mathbb{R}^{m_1} \times \mathbb{R}^{m_2} \to \mathbb{R}^n$. Then f is C^1 on U iff f is separately continuously differentiable (that is, iff the maps $D_j f : U \to L(\mathbb{R}^{m_j}, \mathbb{R}^n)$ exist and are continuous, $j = 1, 2$).*

Proof Let $(\mathbf{h}, \mathbf{k}) \in \mathbb{R}^{m_1} \times \mathbb{R}^{m_2}$, $\mathbf{X} = (\mathbf{x}, \mathbf{y}) \in U$. We need to show that

$$f(\mathbf{x} + \mathbf{h}, \mathbf{y} + \mathbf{k}) - f(\mathbf{x}, \mathbf{y}) = D_1 f_{\mathbf{X}}(\mathbf{h}) + D_2 f_{\mathbf{X}}(\mathbf{k}) + R(\mathbf{h}, \mathbf{k}),$$

where $R(\mathbf{h}, \mathbf{k}) = o(\mathbf{h}, \mathbf{k})$. We have

$$\begin{aligned} f(\mathbf{x} + \mathbf{h}, \mathbf{y} + \mathbf{k}) - f(\mathbf{x}, \mathbf{y}) &= (f(\mathbf{x} + \mathbf{h}, \mathbf{y} + \mathbf{k}) - f(\mathbf{x} + \mathbf{h}, \mathbf{y})) \\ &\quad + (f(\mathbf{x} + \mathbf{h}, \mathbf{y}) - f(\mathbf{x}, \mathbf{y})). \end{aligned}$$

We start by considering the second term on the right-hand side. Since f is assumed differentiable in the first variable at $\mathbf{X}$ we have

$$f(\mathbf{x} + \mathbf{h}, \mathbf{y}) - f(\mathbf{x}, \mathbf{y}) = D_1 f_{\mathbf{X}}(\mathbf{h}) + r(\mathbf{h}),$$

where $r(\mathbf{h}) = o(\mathbf{h})$. For all $\mathbf{k} \in \mathbb{R}^q$ we have $\|(\mathbf{h}, \mathbf{k})\| \geq \|(\mathbf{h}, 0)\| = \|\mathbf{h}\|$ (Euclidean norms). Hence

$$\lim_{(\mathbf{h}, \mathbf{k}) \to 0} \frac{\|r(\mathbf{h})\|}{\|(\mathbf{h}, \mathbf{k})\|} = 0, \tag{9.7}$$

and so $r(\mathbf{h}) = o(\mathbf{h}, \mathbf{k})$. Now we turn to the less straightforward analysis of the first term. For fixed $\mathbf{h}$, define

$$g(\mathbf{k}) = f(\mathbf{x} + \mathbf{h}, \mathbf{y} + \mathbf{k}) - f(\mathbf{x} + \mathbf{h}, \mathbf{y}) - D_2 f_{\mathbf{X}}(\mathbf{k}).$$

We need to show that $g(\mathbf{k}) = o(\mathbf{h}, \mathbf{k})$. Since f is assumed differentiable with respect to the $\mathbf{y}$-variable on U, $g(\mathbf{k})$ is differentiable for $(\mathbf{h}, \mathbf{k}) \in D_r(0, 0)$, where $\mathbf{X} + D_r(0, 0) \subset U$. The derivative of g is given by

$$Dg_{\mathbf{k}} = D_2 f_{\mathbf{X} + (\mathbf{h}, \mathbf{k})} - D_2 f_{\mathbf{X}}.$$

(Recall the derivative of a linear map is constant, equal to the linear map.) Now $D_2 f$ is continuous at $\mathbf{X}$ and so given $\varepsilon > 0$, there exists a $\delta > 0$ such that

$$\|Dg_\mathbf{k}\| = \|D_2 f_{\mathbf{X}+(\mathbf{h},\mathbf{k})} - D_2 f_\mathbf{X}\| \le \varepsilon, \quad \|(\mathbf{h},\mathbf{k})\| \le \delta. \tag{9.8}$$

Next we apply the mean value theorem to $g(\mathbf{k})$ to obtain

$$\|g(\mathbf{k}) - g(\mathbf{0})\| \le \sup_{0<t<1} \|Dg_{t\mathbf{k}}\|\|\mathbf{k}\|.$$

Since $g(\mathbf{0}) = \mathbf{0}$, this gives us the estimate

$$\|g(\mathbf{k})\| \le \sup_{0<t<1} \|Dg_{t\mathbf{k}}\|\|\mathbf{k}\|.$$

By (9.8) we therefore have

$$\|g(\mathbf{k})\| \le \varepsilon\|\mathbf{k}\|, \quad \|(\mathbf{h},\mathbf{k})\| \le \delta.$$

That is, given $\varepsilon > 0$, we have shown there exists a $\delta > 0$ such that $\|g(\mathbf{k})\| \le \varepsilon\|\mathbf{k}\| \le \varepsilon\|(\mathbf{h},\mathbf{k})\|$, whenever $\|(\mathbf{h},\mathbf{k})\| \le \delta$. As a consequence, we have $\lim_{(\mathbf{h},\mathbf{k})\to 0} \frac{\|g(\mathbf{k})\|}{\|(\mathbf{h},\mathbf{k})\|} = 0$. Summarizing, we have shown that

$$f(\mathbf{x}+\mathbf{h},\mathbf{y}+\mathbf{k}) - f(\mathbf{x},\mathbf{y}) - D_1 f_\mathbf{X}(\mathbf{h}) - D_2 f_\mathbf{X}(\mathbf{k}) = r(\mathbf{h}) + g(\mathbf{k}),$$

where $r(\mathbf{h}) + g(\mathbf{k}) = o(\mathbf{h},\mathbf{k})$. Hence f is differentiable at $\mathbf{X}$ with $Df_\mathbf{X}(\mathbf{h},\mathbf{k}) = D_1 f_\mathbf{X}(\mathbf{h}) + D_2 f_\mathbf{X}(\mathbf{k})$. □

Remarks 9.5.5

(1) The result is false without some continuity assumption on the 'partial' derivatives $D_j f$. Inspection of the proof given shows that f will be differentiable at $\mathbf{X}$ provided that (a) $D_1 f_\mathbf{X}$ exists, (b) $D_2 f$ exists on an open disk centred at $\mathbf{X}$, and (c) $D_2 f$ is continuous at $\mathbf{X}$.
(2) The proof of the result makes essential use of the mean value theorem. ✶

Corollary 9.5.6 *Let $U \subset \mathbb{R}^m$ be open and $f = (f_1, \cdots, f_n) : U \to \mathbb{R}^n$. Then f is C^1 on U iff all partial derivatives $\partial f_i/\partial x_j : U \to \mathbb{R}$ exist and are continuous on U.*

Proof If f is C^1, then the partial derivatives exist and are continuous on U (Sect. 9.4.2). We prove the converse by induction on m. The result is immediate from Theorem 9.5.4 if $m = 2$ (with $p = q = 1$). Suppose the result is proved for $m - 1$, $m > 2$. Take $p = m - 1$, $q = 1$ in Theorem 9.5.4. By the inductive assumption, $D_1 f : U \to L(\mathbb{R}^p, \mathbb{R}^n)$ exists and is continuous and, by assumption, $D_2 f : U \to L(\mathbb{R}^1, \mathbb{R}^n)$ exists and is continuous. By Theorem 9.5.4, f is C^1 on U. □

9.6   Inverse and Implicit Function Theorems

We start with the definition of the general linear group and then discuss the process of inverting a linear map.

9.6.1   The General Linear Group

If we let $GL(\mathbb{R}, n)$ denote the set of invertible linear maps $A : \mathbb{R}^n \to \mathbb{R}^n$, then

$$GL(\mathbb{R}, n) = \{A \in L(\mathbb{R}^n, \mathbb{R}^n) \mid \det(A) \neq 0\},$$

where $\det(A)$ denotes the *determinant* of A. The set $GL(\mathbb{R}, n)$ has the structure of a *group* since for all $A, B \in GL(\mathbb{R}, n)$ the composition $AB^{-1} \in GL(\mathbb{R}, n)$. The group identity is $I = I_n$—the identity map of $\mathbb{R}^n$. We refer to $GL(\mathbb{R}, n)$ as the *general linear group* (of degree n). We give $GL(\mathbb{R}, n)$ the metric determined by the operator norm on $L(\mathbb{R}^n, \mathbb{R}^n)$. We can define coordinates on $GL(\mathbb{R}, n)$ by identifying $L(\mathbb{R}^n, \mathbb{R}^n)$ with $\mathbb{R}^{n^2}$ ($[a_{ij}] \leftrightarrow (a_{ij})$). The topology on $GL(\mathbb{R}, n)$ determined by the Euclidean metric on $\mathbb{R}^{n^2}$ is identical to that given by the operator norm (Exercises 9.2.13(1)).

Lemma 9.6.1 $GL(\mathbb{R}, n)$ *is an open and dense subset of* $L(\mathbb{R}^n, \mathbb{R}^n)$.

Proof The determinant map $\det : L(\mathbb{R}^n, \mathbb{R}^n) \to \mathbb{R}$ is continuous—it is a polynomial map in matrix coordinates. Hence $\det^{-1}(0) \subset L(\mathbb{R}^n, \mathbb{R}^n)$ is closed and so $GL(\mathbb{R}, n) = L(\mathbb{R}^n, \mathbb{R}^n) \smallsetminus \det^{-1}(0)$ is open. For density, recall that $\det(A) = 0$ iff A has a zero eigenvalue. If $A \notin GL(\mathbb{R}, n)$, then $A + \varepsilon I \in GL(\mathbb{R}, n)$ for all sufficiently small $\varepsilon > 0$ since the eigenvalues of $A + \varepsilon I$ are ε-translates of the eigenvalues of A and there are at most n distinct eigenvalues of A. $\square$

Lemma 9.6.2 *If we define* $\beta : GL(\mathbb{R}, n) \to GL(\mathbb{R}, n)$ *by* $\beta(A) = A^{-1}$, *then* β *is* C^1.

Proof If we denote the *ij*-component of $[\beta(A)]$ by $\beta(A)_{ij}$, then $\beta(A)_{ij} = A_{ij}/\det(A)$, where A_{ij} is $(-1)^{i+j}$ times the determinant of the $(n-1) \times (n-1)$ matrix defined by removing the *i*th row and *j*th column from A (the (i, j)-*cofactor* of A). This function is clearly C^∞ in the matrix entries a_{ij} and so is certainly C^1 by Corollary 9.5.6. $\square$

9.6.2   Diffeomorphisms and the Inverse Function Theorem

Definition 9.6.3 Let $U \subset \mathbb{R}^m$, $V \subset \mathbb{R}^n$ be non-empty open sets. A map $f : U \to V$ is a C^1 *diffeomorphism* (of U onto V) if

(1) f is 1:1 onto.
(2) Both f and f^{-1} are C^1.

Remark 9.6.4 As we showed earlier, it follows from the chain rule that if $f : U \to V$ is a C^1 diffeomorphism then $Df_{f(x)}^{-1} = (Df_x)^{-1}$ for all $x \in U$. In particular, Df_x is a linear isomorphism and $m = n$. ✠

It is difficult to give *sufficient* conditions on the derivative of a C^1 map $f : U \to V$ for it to be a diffeomorphism unless $n = 1$ (then it is enough that f' is of constant sign and U is connected). However, there is a very useful result that shows when f is a *local* diffeomorphism. This result—the inverse function theorem—states that if Df_x is invertible, then f will restrict to a diffeomorphism on a sufficiently small open neighbourhood of x. In this section we only prove C^1 results. However, all our results extend easily to C^r-maps and we indicate proofs in Sect. 9.11 after we have defined higher derivatives.

Theorem 9.6.5 (The Inverse Function Theorem) *Let W be an open subset of $\mathbb{R}^m$ and $f : W \to \mathbb{R}^m$ be C^1. Suppose that Df_{x_0} is invertible at $x_0 \in W$. Then we can find open neighbourhoods $U \subset W$ of x_0, and V of $f(x_0)$ such that*

(1) f maps U 1:1 onto V. In particular, $V = f(U)$ is open.
(2) $f : U \to V$ is a C^1 diffeomorphism.

Proof We start by proving a special case of the theorem. Assume that $f : \mathbb{R}^m \to \mathbb{R}^m$ and

$$f(0) = 0, \quad Df_0 = I.$$

We show that for all $y \in \mathbb{R}^m$ sufficiently close to the origin of $\mathbb{R}^m$ we can find $x = x(y) \in \mathbb{R}^m$ such that $f(x) = y$. The point x will be our candidate for $f^{-1}(y)$. That is, if we set $f^{-1}(y) = x$, then $f(f^{-1}(y)) = y$. We construct the point x using the contraction mapping lemma with parameters (Theorem 7.17.9). A bonus of this approach is that it gives us an iterative scheme for constructing $f^{-1}(y)$.

Given $y \in \mathbb{R}^m$, define $\Psi_y : \mathbb{R}^m \to \mathbb{R}^m$ by

$$\Psi_y(x) = x - f(x) + y.$$

Observe that $\Psi_y(x) = x$ iff $x - f(x) + y = x$ iff $f(x) = y$. That is, every fixed point x of Ψ_y gives a solution of $f(x) = y$ (and conversely).

If $f = I$ (so $f(x) = x$ for all x), then $\Psi_y(x) = y$ and so the (unique) fixed point is $x = y$. In this trivial case the inverse is the identity map. We are assuming that $f(0) = 0$ and $Df_0 = I$ and so although $x - f(x)$ will not generally be zero, it should be small, at least for small $\|x\|$. Our first steps will be to quantify the size of the term $x - f(x)$. To this end, set $\phi(x) = x - f(x)$. We claim there exists an $r > 0$ such that

(a) $\|\phi(x)\| \le \frac{1}{2}\|x\|$, $\|x\| \le r$.
(b) $\|\phi(x_1) - \phi(x_2)\| \le \frac{1}{2}\|x_1 - x_2\|$, $\|x_1\|, \|x_2\| \le r$.

Since $\phi(0) = 0$, (b) $\Longrightarrow$ (a) and so it suffices to prove (b). Since $D\phi_0 = I - Df_0 = 0$ and f, therefore ϕ, is C^1, we can choose $r > 0$ such that $\|D\phi_x\| \le \frac{1}{2}$, all $\|x\| \le r$. Let

$\overline{D}_r$ denote the closed disk, centre $\mathbf{0}$, radius r in $\mathbb{R}^m$. If $\mathbf{x}_1, \mathbf{x}_2 \in \overline{D}_r$, then $[\mathbf{x}_1, \mathbf{x}_2] \subset \overline{D}_r$. Hence we may apply the mean value theorem to get

$$\|\phi(\mathbf{x}_1) - \phi(\mathbf{x}_2)\| \leq \sup_{\mathbf{z} \in [\mathbf{x}_1, \mathbf{x}_2]} \|D\phi_{\mathbf{z}}\| \|\mathbf{x}_1 - \mathbf{x}_2\| \leq \frac{1}{2}\|\mathbf{x}_1 - \mathbf{x}_2\|, \quad \mathbf{x}_1, \mathbf{x}_2 \in \overline{D}_r,$$

proving (b).

We now show that for all $\mathbf{y} \in \overline{D}_{r/2}$,

(1) $\Psi_{\mathbf{y}} : \overline{D}_r \to \overline{D}_r$,
(2) $\Psi_{\mathbf{y}}$ is a contraction mapping, contraction constant $\frac{1}{2}$.

Suppose $\|\mathbf{y}\| \leq r/2$ and $\|\mathbf{x}\| \leq r$. We have

$$\begin{aligned}
\|\Psi_{\mathbf{y}}(\mathbf{x})\| &= \|\phi(\mathbf{x}) + \mathbf{y}\| \\
&\leq \|\phi(\mathbf{x})\| + \|\mathbf{y}\| \\
&\leq \|\phi(\mathbf{x})\| + \frac{r}{2} \text{ (since } \|\mathbf{y}\| \leq r/2) \\
&\leq \frac{r}{2} + \frac{r}{2} = r \text{ (by estimate (a) on } \phi).
\end{aligned}$$

This proves (1). Turning to (2), suppose $\|\mathbf{x}_1\|, \|\mathbf{x}_2\| \leq r$ and $\|\mathbf{y}\| \leq r/2$. We have

$$\begin{aligned}
\|\Psi_{\mathbf{y}}(\mathbf{x}_1) - \Psi_{\mathbf{y}}(\mathbf{x}_2)\| &= \|\phi(\mathbf{x}_1) - \phi(\mathbf{x}_2)\| \text{ (the } \mathbf{y}'s \text{ cancel)} \\
&= \leq \frac{1}{2}\|\mathbf{x}_1 - \mathbf{x}_2\| \text{ (by estimate (b) on } \phi).
\end{aligned}$$

Hence $\Psi_{\mathbf{y}} : \overline{D}_r \to \overline{D}_r$ is a contraction map with contraction constant $k = \frac{1}{2}$.

Apply Theorem 7.17.9 with $X = \overline{D}_r$, $\Lambda = \overline{D}_{r/2}$, $F(\mathbf{x}, \mathbf{y}) = \Psi_{\mathbf{y}}(\mathbf{x})$ and $k = \frac{1}{2}$ to deduce that there is a continuous map $\tilde{\tilde{x}} : \overline{D}_{r/2} \to \overline{D}_r$ such that $\mathbf{x} = \tilde{\tilde{x}}(\mathbf{y})$ is the unique solution in $\overline{D}_r$ to $f(\mathbf{x}) = \mathbf{y}$, $\mathbf{y} \in \overline{D}_{r/2}$. Since $\tilde{\tilde{x}}(\mathbf{0}) = \mathbf{0}$ and $\tilde{\tilde{x}}$ is continuous, we may choose $0 < s \leq r/2$ such that $\tilde{\tilde{x}}(D_s) \subset D_r$. Set $V = D_s$ and $U = f^{-1}(V) \cap D_r$. Since f is continuous, U is an open subset of $\mathbb{R}^m$. We claim that $\tilde{\tilde{x}} : V \to U$ and is 1:1 onto. Since $\tilde{\tilde{x}}(\mathbf{y}) = f^{-1}(\mathbf{y}) \cap D_r$, for all $\mathbf{y} \in V$, we have $\tilde{\tilde{x}}(V) = U$ proving that $\tilde{\tilde{x}}$ maps V onto U. Moreover, if $\mathbf{y}, \mathbf{y}' \in \overline{D}_{r/2}$, then $\tilde{\tilde{x}}(\mathbf{y}) = \tilde{\tilde{x}}(\mathbf{y}') = \mathbf{x} \in \overline{D}_r$ iff $f(\mathbf{x}) = \mathbf{y} = \mathbf{y}'$. Hence $\tilde{\tilde{x}}$ is 1:1. It follows that $\tilde{\tilde{x}} : V \to U$ is the inverse of $f : U \to V$.

Set $f^{-1} = \tilde{\tilde{x}} : V \to U$. Since $\tilde{\tilde{x}}$ is continuous, it only remains to prove that f^{-1} is C^1.

Let $\mathbf{y} \in V$ and set $\mathbf{x} = f^{-1}(\mathbf{y})$. We know that if f^{-1} is differentiable at $\mathbf{y}$, then $Df_{\mathbf{y}}^{-1} = (Df_{\mathbf{x}})^{-1}$. If we define $s(\mathbf{k})$ by

$$f^{-1}(\mathbf{y} + \mathbf{k}) = f^{-1}(\mathbf{y}) + (Df_{\mathbf{x}})^{-1}(\mathbf{k}) + s(\mathbf{k}),$$

we have to show that $s(\mathbf{k}) = o(\mathbf{k})$. Set $f^{-1}(\mathbf{y} + \mathbf{k}) = \mathbf{x} + \mathbf{h}$. We have

$$f(\mathbf{x} + \mathbf{h}) = f(\mathbf{x}) + Df_{\mathbf{x}}(\mathbf{h}) + r(\mathbf{h}),$$

where $r(\mathbf{h}) = o(\mathbf{h})$. Since $f(\mathbf{x} + \mathbf{h}) = f(\mathbf{x}) + \mathbf{k}$, it follows that

$$\mathbf{k} = Df_{\mathbf{x}}(\mathbf{h}) + r(\mathbf{h}),$$

and so $(Df_{\mathbf{x}})^{-1}(\mathbf{k}) = \mathbf{h} + (Df_{\mathbf{x}})^{-1}(r(\mathbf{h}))$. Since $r(\mathbf{h}) = o(\mathbf{h})$ and $\|(Df_{\mathbf{x}})^{-1}\| > 0$, we can choose $\delta > 0$ such that

$$\|k\| \geq \frac{1}{2}\|(Df_{\mathbf{x}})^{-1}\|^{-1}\|\mathbf{h}\|, \text{ if } \|\mathbf{h}\| \leq \delta. \tag{9.9}$$

Estimating $\|s(\mathbf{k})\|$ we have

$$\begin{aligned}
\|s(\mathbf{k})\| &= \|f^{-1}(\mathbf{y} + \mathbf{k}) - f^{-1}(\mathbf{y}) - (Df_{\mathbf{x}})^{-1}(\mathbf{k})\| \\
&= \|(Df_{\mathbf{x}})^{-1}(Df_{\mathbf{x}}(f^{-1}(\mathbf{y} + \mathbf{k}) - f^{-1}(\mathbf{y})) - \mathbf{k})\| \\
&\leq \|(Df_{\mathbf{x}})^{-1}\|\|Df_{\mathbf{x}}(f^{-1}(\mathbf{y} + \mathbf{k}) - f^{-1}(\mathbf{y})) - \mathbf{k}\| \\
&= \|(Df_{\mathbf{x}})^{-1}\|\|r(\mathbf{h})\|.
\end{aligned}$$

It follows from (9.9) that if $\|\mathbf{h}\| \leq \delta$, then

$$\frac{\|s(\mathbf{k})\|}{\|\mathbf{k}\|} \leq \|(Df_{\mathbf{x}})^{-1}\|\frac{\|r(\mathbf{h})\|}{\frac{1}{2}\|(Df_{\mathbf{x}})^{-1}\|^{-1}\|\mathbf{h}\|} = 2\|(Df_{\mathbf{x}})^{-1}\|^2\frac{\|r(\mathbf{h})\|}{\|\mathbf{h}\|}.$$

Since $\mathbf{k}$ depends continuously on $\mathbf{h}$ ($\mathbf{k} = f(\mathbf{x}+\mathbf{h}) - f(\mathbf{x})$) and $r(\mathbf{h}) = o(\mathbf{h})$, it follows that $s(\mathbf{k}) = o(\mathbf{k})$, proving that f^{-1} is differentiable at $\mathbf{y}$ with derivative $(Df_{\mathbf{x}})^{-1}$.

The derivative map $Df^{-1} : V \to L(\mathbb{R}^m, \mathbb{R}^m)$ is the composite

$$V \xrightarrow{f^{-1}} U \xrightarrow{Df} L(\mathbb{R}^m, \mathbb{R}^m) \xrightarrow{\beta} L(\mathbb{R}^m, \mathbb{R}^m),$$

where $\beta(A) = A^{-1}$. Since these maps are continuous, Df^{-1} is continuous and so f^{-1} is C^1.

Finally, we need to address the case of general maps f. Suppose $Df_{\mathbf{x}_0}$ is a linear isomorphism. Define $\tilde{f}(\mathbf{x}) = (Df_{\mathbf{x}_0})^{-1}(f(\mathbf{x} + \mathbf{x}_0) - f(\mathbf{x}_0))$. We have $\tilde{f}(\mathbf{0}) = \mathbf{0}$ and $D\tilde{f}_0 = I$ and so the previous analysis applies to $\tilde{f}$ to give open neighbourhoods $\tilde{U}, \tilde{V}$ of the origin such that $\tilde{f} : \tilde{U} \to \tilde{V}$ is a C^1 diffeomorphism. If we set $U = \tilde{U} + \mathbf{x}_0$, $V = Df_{\mathbf{x}_0}(\tilde{V}) + f(\mathbf{x}_0)$, then $f(U) = V$ and $f : U \to V$ is a C^1 diffeomorphism. □

9.6.3   The Implicit Function Theorem

We prove next an extension of the inverse function theorem: the *implicit function theorem*. The theorem gives sufficient conditions for the local solvability of an equation. The result follows easily from the inverse function theorem. We give a C^1 version. The C^r version is an immediate consequence of the C^r version of the inverse function theorem which we prove in the section on higher derivatives.

Theorem 9.6.6 (The Implicit Function Theorem) *Let $f : \mathbb{R}^m \times \mathbb{R}^n \to \mathbb{R}^n$ be C^1. Suppose that*

(1) $f(0, 0) = 0$.
(2) $D_2 f_0 \in L(\mathbb{R}^n, \mathbb{R}^n)$ *is a linear isomorphism.*

Then there exist open neighbourhoods U of $0 \in \mathbb{R}^m$, V of $0 \in \mathbb{R}^n$, W of $(0, 0) \in \mathbb{R}^m \times \mathbb{R}^n$, and a C^1 diffeomorphism $H : U \times V \to W$ such that

(a) *H preserves the first coordinate: $H(\mathbf{x}, \mathbf{y}) = (\mathbf{x}, h(\mathbf{x}, \mathbf{y}))$, all $(\mathbf{x}, \mathbf{y}) \in U \times V$.*
(b) *$f(\mathbf{x}, h(\mathbf{x}, \mathbf{y})) = \mathbf{y}$ for all $(\mathbf{x}, \mathbf{y}) \in U \times V$. In particular, if $\mathbf{y} = 0$ and we set $h(\mathbf{x}, 0) = u(\mathbf{x})$ then*

$$f(\mathbf{x}, u(\mathbf{x})) = 0, \text{ for all } \mathbf{x} \in U.$$

The derivative of u is given by

$$Du_{\mathbf{x}} = -(D_2 f_{(\mathbf{x}, u(\mathbf{x}))})^{-1} D_1 f_{(\mathbf{x}, u(\mathbf{x}))}, \ \mathbf{x} \in U.$$

(c) *If $(\mathbf{x}, \mathbf{z}) \in W$, $\mathbf{y} \in W$, then $f(\mathbf{x}, \mathbf{z}) = \mathbf{y}$ iff $\mathbf{z} = h(\mathbf{x}, \mathbf{y})$.*

In Fig. 9.2, we indicate how the map $H^{-1} : W \to U \times V$ 'straightens out' each solution set $f^{-1}(\mathbf{z}) \cap W$ onto the open subset $U \times \{\mathbf{z}\}$ of $\mathbb{R}^m \times \{\mathbf{z}\}$. Condition (a) of the theorem implies that H shears W parallel to the $\mathbb{R}^n$ direction.

Statement (b) of the theorem shows that we can solve $f(\mathbf{x}, \mathbf{y}) = 0$ near the origin obtaining $\mathbf{y}$ as a C^1 function of $\mathbf{x}$. That is, with $\mathbf{y} = u(\mathbf{x})$, $f(\mathbf{x}, u(\mathbf{x})) = 0$, $\mathbf{x} \in U$. However, the theorem goes far beyond this. Statement (b) implies that for all $\mathbf{y} \in V$, we can solve $f(\mathbf{x}, \mathbf{y}) = \mathbf{y}$ by the C^1 function $u_{\mathbf{y}} : U \to \mathbb{R}^n$ defined by $u_{\mathbf{y}}(\mathbf{x}) = h(\mathbf{x}, \mathbf{y})$. Moreover, by (c), we find *all* the solutions to $f(\mathbf{x}, \mathbf{y}) = \mathbf{y}$ in W.

Note that the result is obvious if $f : \mathbb{R}^m \times \mathbb{R}^n \to \mathbb{R}^n$ is the projection p on the second factor (that is, $f(\mathbf{x}, \mathbf{y}) = \mathbf{y}$). Indeed, we may then take $h(\mathbf{x}, \mathbf{y}) = \mathbf{y}$, $U = \mathbb{R}^m$, $V = \mathbb{R}^n$ and $W = \mathbb{R}^m \times \mathbb{R}^n$. Viewed in this way, statement (b) of the theorem says that we can make a local change of coordinates on $\mathbb{R}^m \times \mathbb{R}^n$ so that in the new coordinates, f is (locally, near the origin) the projection on the second factor. That is, $f \circ H : U \times V \to V$ is the projection on $\mathbb{R}^n$: $f \circ H(\mathbf{x}, \mathbf{y}) = \mathbf{y}$. As always with local results in the differential calculus, we show that subject to certain conditions, maps locally look linear. See Fig. 9.2.

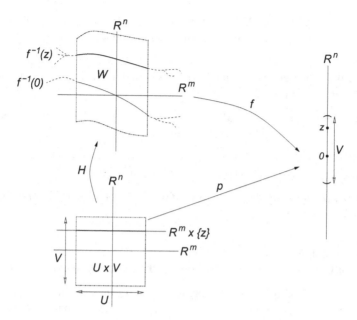

Fig. 9.2 Implicit function theorem

Proof of Theorem 9.6.6 Define $G(\mathbf{x}, \mathbf{y}) = (\mathbf{x}, f(\mathbf{x}, \mathbf{y}))$. Observe that $G(\mathbf{0}, \mathbf{0}) = (\mathbf{0}, \mathbf{0})$ and

$$DG_{(0,0)} = \begin{pmatrix} I_m & \mathbf{0} \\ D_1 f_{(0,0)} & D_2 f_{(0,0)} \end{pmatrix} \in L(\mathbb{R}^m \times \mathbb{R}^n, \mathbb{R}^m \times \mathbb{R}^n),$$

where I_m is the identity map of $\mathbb{R}^m$. Since $D_2 f_{(0,0)}$ is a linear isomorphism of $\mathbb{R}^n$, $DG_{(0,0)}$ is a linear isomorphism. Hence we may apply the inverse function theorem to find open neighbourhoods U, V, W of the origins in $\mathbb{R}^m$, $\mathbb{R}^n$ and $\mathbb{R}^m \times \mathbb{R}^n$ respectively such that $G : W \to U \times V$ is a C^1 diffeomorphism. Set $H = G^{-1}$ and let $p : \mathbb{R}^m \times \mathbb{R}^n \to \mathbb{R}^n$ denote the projection on $\mathbb{R}^n$. Observe that H preserves the first coordinate and so we may write $H(\mathbf{x}, \mathbf{y}) = (\mathbf{x}, h(\mathbf{x}, \mathbf{y}))$ for some C^1 function $h : U \times V \to \mathbb{R}^n$ (write $H(\mathbf{x}, \mathbf{y}) = (\bar{\mathbf{x}}, \bar{\mathbf{y}})$ and apply G to both sides). We have

$$p \circ G = f,$$

as maps of W into $\mathbb{R}^n$. Now compose on the right by $H = G^{-1}$ to obtain

$$p = f \circ H.$$

Here $p = f \circ H$ maps $U \times V$ onto $V \subset \mathbb{R}^n$. The equation $p = f \circ H$ is equivalent to statement (b) of the theorem. Statement (c) follows because we know all the solutions to $p(\mathbf{x}, \mathbf{y}) = \mathbf{y}$ on $U \times V$—$p(\mathbf{x}, \mathbf{y}) = \mathbf{y}$ iff $f(\mathbf{x}, h(\mathbf{x}, \mathbf{y})) = \mathbf{y}$. The formula for the derivative of u follows by differentiating the identity $f(\mathbf{x}, u(\mathbf{x})) = 0$, $\mathbf{x} \in U$.

$\square$

Remark 9.6.7 Theorem 9.6.6 extends to maps with domain a proper open subset of $\mathbb{R}^m \times \mathbb{R}^n$—the proof is unchanged.   ✸

Corollary 9.6.8 *Let* $f : U \subset \mathbb{R}^p \to \mathbb{R}^q$ *be* C^1 *and* $q \leq p$. *Suppose that* $\mathbf{x}_0 \in U$ *and*

(1) $f(\mathbf{x}_0) = \mathbf{0}$,
(2) $Df_{\mathbf{x}_0} \in L(\mathbb{R}^p, \mathbb{R}^q)$ *is onto* ($Df_{\mathbf{x}_0}$ *has maximal rank equal to* q).

Then there exist open neighbourhoods U *of* $\mathbf{0} \in \mathbb{R}^{p-q}$, W *of* $\mathbf{x}_0 \in \mathbb{R}^p$, *and a* C^1 *map* $u : U \to W$ *such that*

(a) $u(\mathbf{0}) = \mathbf{x}_0$.
(b) $f(u(\mathbf{x})) = \mathbf{0}$, *for all* $\mathbf{x} \in U$.
(c) *The only solutions of the equation* $f(\mathbf{x}) = \mathbf{0}$ *in* W *are those given by (a,b).*

Proof Replacing $f(\mathbf{x})$ by $f(\mathbf{x}+\mathbf{x}_0)$, there is no loss of generality in assuming $\mathbf{x}_0 = \mathbf{0}$. If we set $K = \ker(Df_0)$, then $K \cong \mathbb{R}^{p-q}$. Let F be a vector space complement to K in $\mathbb{R}^p$. Then $F \cong \mathbb{R}^q$ and $Df_0 : F \to \mathbb{R}^q$ is an isomorphism. Choose a basis $\{e_i\}$ for $\mathbb{R}^p$ so that $\{e_1, \cdots, e_{p-q}\}$ spans K and $\{e_{p-q+1}, \cdots, e_p\}$ spans F. With these conventions, $\mathbb{R}^p \cong \mathbb{R}^{p-q} \times \mathbb{R}^q$ and f satisfies the hypotheses of Theorem 9.6.6. The result is now immediate from Theorem 9.6.6(b). $\square$

Examples 9.6.9

(1) Let $f : \mathbb{R}^2 \to \mathbb{R}$ be defined by

$$f(x, y) = x^2 \cos y + \sin y.$$

We have $f(0, 0) = 0$ and $\frac{\partial f}{\partial y}(0, 0) = 1$. Hence Theorem 9.6.6 applies and there exists an open interval I containing $0 \in \mathbb{R}$, an open neighbourhood W of $(0, 0) \in \mathbb{R}^2$ and a C^1 map $u : J \to \mathbb{R}$ such that $u(0) = 0$ and

$$f(x, u(x)) = x^2 \cos u(x) + \sin u(x) = 0, \ x \in I.$$

These are the only solutions to $f(x, y) = 0$ in W. We have $u'(0) = 0$ (Theorem 9.6.6(c)).

(2) Consider the equation

$$F(x, y, z) = z^3 + (x^4 + y^4)z + 1 = 0.$$

We claim we can find a unique solution $z = f(x, y)$ to this equation which is defined on all of $\mathbb{R}^2$ and such that f is C^1. First note that for fixed x, y, the equation has at least one real root since the sign of $z^3 + (x^4 + y^4)z + 1$ is that of z for $|z|$ sufficiently large. We have $\frac{\partial F}{\partial z}(x, y, z) = 3z^2 + (x^4 + y^4) = 0$ iff $x = y = z = 0$. Since $F(0, 0, 0) \neq 0$ and elsewhere $\frac{\partial F}{\partial z}(x, y, z) > 0$, it follows that there is exactly one real root of the equation $F(x, y, z) = 0$ for each $(x, y) \in \mathbb{R}^2$. Let $f : \mathbb{R}^2 \to \mathbb{R}$ be the function giving this root. By Theorem 9.6.6, f is C^1. Using Theorem 9.6.6(c) to compute the partial derivatives of f at $(x, y) \in \mathbb{R}^2$, we find that $\frac{\partial f}{\partial x}(0, 0) = \frac{\partial f}{\partial y}(0, 0) = 0$ and

$$\frac{\partial f}{\partial x}(x, y) = -\frac{4x^3 z}{3z^2 + x^4}, \quad (x, y) \neq (0, 0),$$

$$\frac{\partial f}{\partial y}(x, y) = -\frac{4y^3 z}{3z^2 + y^4}, \quad (x, y) \neq (0, 0). \qquad \spadesuit$$

9.6.4   A Dual Version of the Implicit Function Theorem

The implicit function theorem gives conditions that allow us to show that under a local change of coordinates a map is a projection and therefore locally onto. We now look at the case of maps that are locally injective and give conditions that imply a map is locally an inclusion (after a change of coordinates on the range).

Theorem 9.6.10 *Let $U \subset \mathbb{R}^m$ be an open neighbourhood of $\mathbf{0} \in \mathbb{R}^m$ and $f : U \subset \mathbb{R}^m \to \mathbb{R}^m \times \mathbb{R}^n$ be a C^1 map satisfying*

(1) $f(\mathbf{0}) = (\mathbf{0}, \mathbf{0})$.
(2) *Df_0 is a linear isomorphism of $\mathbb{R}^m$ onto $\mathbb{R}^m \times \{\mathbf{0}\} \subset \mathbb{R}^m \times \mathbb{R}^n$.*

There exist open neighbourhoods $V_1 \subset U$ of $\mathbf{0} \in \mathbb{R}^m$, V_2 of $\mathbf{0} \in \mathbb{R}^n$, W of $(\mathbf{0}, \mathbf{0}) \in \mathbb{R}^m \times \mathbb{R}^n$ and a C^1 diffeomorphism $g : W \to V_1 \times V_2$ such that the composite $gf : V_1 \to \mathbb{R}^m \times \mathbb{R}^n$ is the restriction to V_1 of the inclusion i_1 of $\mathbb{R}^m$ onto the subspace $\mathbb{R}^m \times \{\mathbf{0}\}$ of $\mathbb{R}^m \times \mathbb{R}^n$. That is,

$$gf(\mathbf{x}) = i_1(\mathbf{x}) = (\mathbf{x}, \mathbf{0}), \quad \mathbf{x} \in V_1.$$

Remark 9.6.11 The final statement of Theorem 9.6.10 implies that $f : V_1 \to \mathbb{R}^m \times \mathbb{R}^n$ is injective.   ✠
We illustrate Theorem 9.6.10 in Fig. 9.3. The map g 'straightens out' the image $f(V_1)$ of $f : V_1 \to \mathbb{R}^m \times \mathbb{R}^n$.

Proof of Theorem 9.6.10 Define $\phi : U \times \mathbb{R}^n \to \mathbb{R}^m \times \mathbb{R}^n$ by

$$\phi(\mathbf{x}, \mathbf{y}) = f(\mathbf{x}) + (\mathbf{0}, \mathbf{y}), \quad \mathbf{x} \in U, \ \mathbf{y} \in \mathbb{R}^n.$$

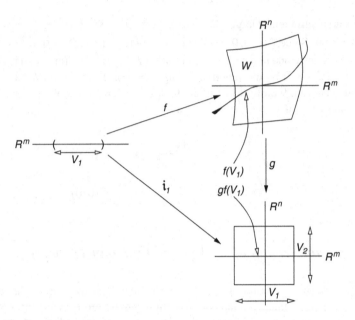

Fig. 9.3 Dual implicit function theorem

Observe that $\phi = fp_1 + p_2$, where $p_1 : \mathbb{R}^m \times \mathbb{R}^n \to \mathbb{R}^m$ is the projection on $\mathbb{R}^m$ and $p_2 : \mathbb{R}^m \times \mathbb{R}^n \to \mathbb{R}^n$ is the projection on $\mathbb{R}^n$. It follows from the chain rule that ϕ is C^1 and

$$D\phi_0 = Df_0\, p_1 + p_2 = \begin{pmatrix} Df_0 & 0 \\ 0 & I_n \end{pmatrix} \in L(\mathbb{R}^m \times \mathbb{R}^n, \mathbb{R}^m \times \mathbb{R}^n).$$

Since Df_0 is a linear isomorphism of $\mathbb{R}^m$ onto $\mathbb{R}^m \times \{0\}$, $D\phi_0$ is a linear isomorphism of $\mathbb{R}^m \times \mathbb{R}^n$. Hence, by the inverse function theorem, we may find an open neighbourhood $V_1 \times V_2$ of $(0,0) \in \mathbb{R}^m \times \mathbb{R}^n$ such that $\phi(V_1 \times V_2) = W$ is an open neighbourhood of $(0,0) \in \mathbb{R}^m \times \mathbb{R}^n$ and $\phi : V_1 \times V_2 \to \phi(V_1 \times V_2)$ is a C^1 diffeomorphism. Set $\phi^{-1} = g : W \to V_1 \times V_2$. We claim that g satisfies the conditions of the theorem. Indeed, on V_1, we have $\phi\, i_1 = f$. Composing on the left by $\phi^{-1} = g$, we obtain $gf = i_1|V_1$. $\qquad\square$

Remark 9.6.12 If f preserves the first coordinate (that is, $f(\mathbf{x}) = (\mathbf{x}, f_2(\mathbf{x}))$), then so does ϕ and therefore g. ✸

Example 9.6.13 Let $f : \mathbb{R}^2 \to \mathbb{R}^3$ be defined by

$$f(x, y) = (ye^x \cos y, x + xy^2, x\sin(xy^2)).$$

We claim we can choose a neighbourhood V of $(0,0) \in \mathbb{R}^2$ such that $f : V \to \mathbb{R}^3$ is injective and the only solution to $f(x, y) = (0, 0, 0)$ in V is $(x, y) = (0, 0)$. Clearly

$(0, 0)$ is a solution of $f(x, y) = (0, 0, 0)$. Computing we find that

$$Df_{(0,0)} = \begin{pmatrix} 0 & 1 \\ 1 & 0 \\ 0 & 0 \end{pmatrix}.$$

Hence $Df_{(0,0)}$ defines a linear isomorphism of $\mathbb{R}^2$ onto $\mathbb{R}^2 \times \{0\} \subset \mathbb{R}^2 \times \mathbb{R}$. The claim now follows from Theorem 9.6.10.   ♠

9.6.5   The Rank Theorem

The final result of the section includes the inverse and both implicit function theorems as special cases.

Throughout we will be especially careful with notation so as to improve the clarity of the exposition. In particular, we often use $\mathbf{0}_p$ (rather than $\mathbf{0}$) to denote the origin of $\mathbb{R}^p$ and $\mathbf{0}_{p,q}$ (rather than $\mathbf{0}$) to denote the zero map from $\mathbb{R}^p$ to $\mathbb{R}^q$.

Before we state the rank theorem, we give an outline of the general idea. Suppose then that $f : \mathbb{R}^m \to \mathbb{R}^n$ is C^1 and $f(\mathbf{0}) = \mathbf{0}$. Let W be an open neighbourhood of $\mathbf{0} \in \mathbb{R}^m$ and suppose that for all $\mathbf{X} \in W$, $Df_{\mathbf{X}}$ has constant rank $q \geq 1$. That is, $\dim(Df_{\mathbf{X}}(\mathbb{R}^m)) = q$ for all $\mathbf{X} \in W$. The inverse function theorem applies if $m = n = q$, the implicit function if $q = n > m$, and the dual implicit function if $q = m < n$. In these cases, $q \in \{m, n\}$ and we need only assume $\mathrm{rank}(Df_{\mathbf{0}}) = q$ since it can be shown that $\mathrm{rank}(Df_{\mathbf{X}}) = q$ for all $\mathbf{X}$ in a neighbourhood of $\mathbf{0}_m$. If $q \notin \{m, n\}$, then $\mathrm{rank}(Df_{\mathbf{X}}) \geq q$ on a neighbourhood of $\mathbf{0}$ but equality does not usually hold.

If $Df_{\mathbf{X}}$ has constant rank q, $\mathbf{X} \in W$, then the rank theorem gives the local structure of both image and level sets. Referring to Fig. 9.4, we can find open neighbourhoods $W_1 \subset W$ of $\mathbf{0}_m$, W_2 of $\mathbf{0}_n$, such that $f(W_1)$ is the diffeomorphic image of an open subset of $\mathbb{R}^q$ and, setting $f = f|W_1$, the level sets $\{f^{-1}(\mathbf{z}) \mid \mathbf{z} \in f(W_1)\}$ form a C^1-family of sets which is the diffeomorphic image of a trivial q-dimensional family

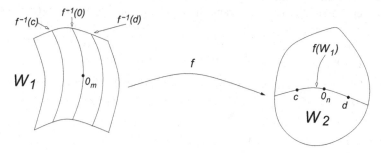

Fig. 9.4 Geometry of the rank theorem

$\{\{\mathbf{x}\} \times V_2 \mid \mathbf{x} \in V_1\}$, where V_1 is an open neighbourhood of $\mathbf{0}_q$, V_2 of $\mathbf{0}_{m-q}$. In particular, each level set is $m - q$-dimensional.

Now for the formal statement of the rank theorem.

Theorem 9.6.14 (The Rank Theorem) *Let W be an open neighbourhood of $\mathbf{0}_m \in \mathbb{R}^m$ and $f : W \to \mathbb{R}^n$ be C^1 with $f(\mathbf{0}_m) = \mathbf{0}_n$. Suppose that $Df_{\mathbf{x}}$ is of constant rank q on W. Then there exist open neighbourhoods $W_1 \subset W$ of $\mathbf{0}_m$, W_2 of $\mathbf{0}_n$, open neighbourhoods $V_1 \subset \mathbb{R}^q$ of $\mathbf{0}_q$, $V_2 \subset \mathbb{R}^{m-q}$ of $\mathbf{0}_{m-q}$, $U_2 \subset \mathbb{R}^{n-q}$ of $\mathbf{0}_{n-q}$, and C^1 diffeomorphisms $h : V_1 \times V_2 \to W_1$, $g : W_2 \to V_1 \times U_2$ such that*

$$gfh(\mathbf{x}, \mathbf{y}) = (\mathbf{x}, \mathbf{0}_{n-q}), \quad \mathbf{x} \in V_1, \mathbf{y} \in V_2.$$

In particular, $f^{-1}(\mathbf{0})) \cap W_1$ is mapped by h^{-1} homeomorphically onto the open set $\{\mathbf{0}_q\} \times V_2 \subset \{\mathbf{0}_q\} \times \mathbb{R}^{m-q}$.

Although the proof of the rank theorem is not that difficult, it does require careful preparation.

Notational Conventions and Assumptions Fix bases of $\mathbb{R}^m$ and $\mathbb{R}^n$ so that $\ker(Df_0) = \{\mathbf{0}_q\} \times \mathbb{R}^{m-q}$ and $Df_0(\mathbb{R}^q \times \{\mathbf{0}_{m-q}\}) = \mathbb{R}^q \times \{\mathbf{0}_q\}$. Denote the subspace $\mathbb{R}^q \times \{\mathbf{0}_{m-q}\}$ of $\mathbb{R}^q \times \mathbb{R}^{m-q}$ by $\mathbb{R}^q$ (so that $Df_0(\mathbb{R}^q) = \mathbb{R}^q \times \{\mathbf{0}_{n-q}\} \subset \mathbb{R}^n$). Let $p_1 : \mathbb{R}^q \times \mathbb{R}^{m-q} \to \mathbb{R}^q$ and $p_1' : \mathbb{R}^q \times \mathbb{R}^{n-q} \to \mathbb{R}^q$ denote the projections on the first factors and p_2, p_2' denote the corresponding projections on the second factors. Let $i_1 : \mathbb{R}^q \to \mathbb{R}^q \times \mathbb{R}^{m-q}$ denote the inclusion on the first factor and similarly define i_1', i_2 and i_2'. For reasons of clarity, we adopt for the remainder of the section the convention that composition of *linear* maps is indicated by $\circ$. Thus the expression $(Df)g \circ Dg$ arising from the chain rule will be a linear-map-valued function whose value at the point $\mathbf{x}$ in the domain of f is the linear map $Df_{g(\mathbf{x})} \circ Dg_{\mathbf{x}}$.

We illustrate the geometric content of the rank theorem in Fig. 9.5. The map g straightens out $f(W_1)$ onto the open subset $V_1 \times \{\mathbf{0}_{n-q}\}$ of $\mathbb{R}^q \times \{\mathbf{0}_{n-q}\}$ and h^{-1} straightens out the family of inverse images $\{f^{-1}(\mathbf{z}) \cap W_1 \mid \mathbf{z} \in f(W_1)\}$ onto the family $\{\{p_1' g(\mathbf{z})\} \times V_2 \mid \mathbf{z} \in f(W_1)\}$. The map $u : V_1 \to W_2$ given by $u(\mathbf{x}) = hf(\mathbf{x}, \mathbf{0}_{m-q})$ is injective and maps onto $f(W_1)$. In essence, the rank theorem asserts that we can make local differentiable (non-linear) changes of coordinate at $\mathbf{0}_m$ and $\mathbf{0}_n$ so that in the new coordinates $f = i_1' \circ p_1$.

Our first result provides a key step in the proof of the rank theorem and shows that, under the conditions of the rank theorem, we can represent $Df_{\mathbf{x}}(\mathbb{R}^m)$ as the graph of a linear map $\sigma_{\mathbf{x}} \in L(\mathbb{R}^q, \mathbb{R}^{n-q})$, for $\mathbf{X}$ in some neighbourhood $\widetilde{W}$ of $\mathbf{0}$. That is, $Df_{\mathbf{x}}(\mathbb{R}^m) = \text{graph}(\sigma_{\mathbf{x}})(\mathbb{R}^q)$, where $\text{graph}(\sigma_{\mathbf{x}})(\mathbf{u}) = (\mathbf{u}, \sigma_{\mathbf{x}}(\mathbf{u}))$, $\mathbf{u} \in \mathbb{R}^m$. Indeed, more is shown: if $f = (f_1, f_2)$, then $Df_{2,\mathbf{x}} = \sigma_{\mathbf{x}} \circ Df_{1,\mathbf{x}}$, $\mathbf{X} \in \widetilde{W}$.

Lemma 9.6.15 (Notation and Assumptions as Above) *We may choose an open neighbourhood $\widetilde{W} \subset W$ of $\mathbf{0} \in \mathbb{R}^m$ and a continuous map $\gamma : \widetilde{W} \to L(\mathbb{R}^q, \mathbb{R}^q \times \mathbb{R}^{n-q})$, $\mathbf{X} \mapsto \gamma_{\mathbf{X}}$, such that for all $\mathbf{X} \in \widetilde{W}$*

(a) $\gamma_{\mathbf{X}} = (I_q, \sigma_{\mathbf{X}})$, *where $\sigma : \widetilde{W} \to L(\mathbb{R}^q, \mathbb{R}^{n-q})$ and I_q is the identity map of $\mathbb{R}^q$.*
(b) $\gamma_{\mathbf{X}} \circ p_1' \circ Df_{\mathbf{X}} = Df_{\mathbf{X}}$.

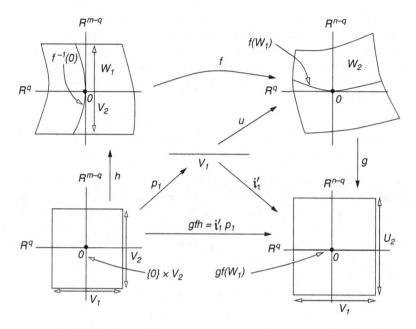

Fig. 9.5 The rank theorem: mappings and geometry

Proof Define $\mu : W \to L(\mathbb{R}^q, \mathbb{R}^q)$ by $\mu_{\mathbf{X}} = p'_1 \circ Df_{\mathbf{X}} \circ i_1$. Since $Df_0(\mathbb{R}^q) = \mathbb{R}^q \times \{0\}$, $\mu_0 \in \mathrm{GL}(\mathbb{R}, q)$. Now $\mathrm{GL}(\mathbb{R}, q)$ is an open subset of $L(\mathbb{R}^q, \mathbb{R}^q)$ and μ is continuous, so we can choose an open neighbourhood $\widetilde{W} \subset W$ of $\mathbf{0}_m$ such that $\mu_{\mathbf{X}} \in \mathrm{GL}(\mathbb{R}, q)$ for all $\mathbf{X} \in \widetilde{W}$. It follows that $Df_{\mathbf{X}}(\mathbb{R}^q)$ is a q-dimensional subspace of $\mathbb{R}^q \times \mathbb{R}^{n-q}$ for all $\mathbf{X} \in \widetilde{W}$. We define $\sigma_{\mathbf{X}}$ by requiring that

$$Df_{\mathbf{X}} \circ i_1 = (I_q, \sigma_{\mathbf{X}}) \circ p'_1 \circ Df_{\mathbf{X}} \circ i_1 = (I_q, \sigma_{\mathbf{X}}) \circ \mu_{\mathbf{X}}, \ \mathbf{X} \in \widetilde{W}.$$

Since $p'_1 \circ Df_{\mathbf{X}} \circ i_1 = \mu_{\mathbf{X}}$, this condition is satisfied iff $\sigma_{\mathbf{X}} = p'_2 \circ Df_{\mathbf{X}} \circ i_1 \circ \mu_{\mathbf{X}}^{-1}$. Now f is C^1 and μ is continuous, and so $\sigma : \widetilde{W} \to L(\mathbb{R}^q, \mathbb{R}^{n-q})$ is continuous, proving (a). For (b), observe that since $Df_{\mathbf{X}}$ has rank q for all $\mathbf{X} \in \widetilde{W}$ and $Df_{\mathbf{X}} \circ i_1(\mathbb{R}^q)$ is q-dimensional, we have $Df_{\mathbf{X}}(\mathbb{R}^m) = Df_{\mathbf{X}} \circ i_1(\mathbb{R}^q)$ for all $\mathbf{X} \in \widetilde{W}$. Given $\mathbf{X} \in \widetilde{W}$, $\mathbf{e} \in \mathbb{R}^m$, we may write $\mathbf{e}$ uniquely as $\mathbf{e}_q + \mathbf{k}$, where $\mathbf{e}_q \in \mathbb{R}^q$ and $\mathbf{k} \in \ker(Df_{\mathbf{X}})$. Since $\gamma_{\mathbf{X}} \circ p'_1 \circ Df_{\mathbf{X}} \circ i_1 = Df_{\mathbf{X}} \circ i_1$, it follows that $\gamma_{\mathbf{X}} \circ p'_1 \circ Df_{\mathbf{X}} = Df_{\mathbf{X}}$. □

Proof of Theorem 9.6.14 The proof of the rank theorem is a combination of the proofs of the implicit and dual implicit function theorems. Define the C^1 map $\phi : W \to \mathbb{R}^q \times \mathbb{R}^{m-q}$ by

$$\phi(\mathbf{x}) = p'_1 f(\mathbf{X}) + p_2(\mathbf{X}), \ \mathbf{X} \in W.$$

As we shall see, for a small enough neighbourhood $W' \subset \widetilde{W}$ of $\mathbf{0}_m$, ϕ will map each level set $L_\mathbf{z} = f^{-1}(\mathbf{z}) \cap W'$ of $f|W'$ diffeomorphically onto an open neighbourhood of the origin in $\{\mathbf{x}\} \times \mathbb{R}^{m-q}$, where $\mathbf{x} = \mathbf{x}(\mathbf{z}) \in \mathbb{R}^q$ is the unique intersection of $L_\mathbf{z}$ with $\mathbb{R}^q \times \{\mathbf{0}_{m-q}\}$. That is, ϕ locally straightens out the level sets of f. Note this is obvious if $f(\mathbf{x}, \mathbf{y}) = (\mathbf{x}, \mathbf{0}) \in \mathbb{R}^q \times \mathbb{R}^{n-q}$, $(\mathbf{x}, \mathbf{y}) \in \mathbb{R}^q \times \mathbb{R}^{m-q}$, as $\phi = I_m$.

Computing, $D\phi_0$ we find

$$D\phi_0 = \begin{pmatrix} p_1' \circ Df_0 \circ i_1 & p_1' \circ Df_0 \circ i_2 \\ \mathbf{0}_{m,m-q} & I_{m-q} \end{pmatrix} \in L(\mathbb{R}^q \times \mathbb{R}^{m-q}, \mathbb{R}^q \times \mathbb{R}^{m-q}).$$

Hence $D\phi_0 \in GL(\mathbb{R}, m)$, since $p_1' \circ Df_0 \circ i_1 \in GL(\mathbb{R}, q)$. Applying the inverse function theorem, there exist connected open neighbourhoods $\widetilde{V}_1 \times V_2$ of $(\mathbf{0}_q, \mathbf{0}_{m-q})$ and $W' \subset \widetilde{W}$ of $\mathbf{0}_m$ such that $\phi : W' \to \widetilde{V}_1 \times V_2$ is a C^1 diffeomorphism. Set $h = \phi^{-1}$ and define $F = fh : \widetilde{V}_1 \times V_2 \subset \mathbb{R}^q \times \mathbb{R}^{m-q} \to \mathbb{R}^n$. We claim that F is independent of the V_2 variable. We show this by proving $D_2 F = 0$ on $\widetilde{V}_1 \times V_2$. We have $f = F\phi$. Differentiating this identity we obtain

$$Df = (DF)\phi \circ D\phi$$

$$= (D_1 F)\phi \circ p_1' \circ Df + (D_2 F)\phi \circ p_2,$$

since $D\phi = p_1' \circ Df + p_2$. Hence

$$(D_2 F)\phi \circ p_2 = Df - (D_1 F)\phi \circ p_1' \circ Df$$

$$= (\gamma - (D_1 F)\phi) \circ p_1' \circ Df,$$

where the last line follows by Lemma 9.6.15(b). It follows that

$$((\gamma_\mathbf{X} - D_1 F_{\phi(\mathbf{X})}) \circ p_1' \circ Df_\mathbf{X})(\mathbf{e}) = 0, \quad \mathbf{e} \in \mathbb{R}^q, \quad \mathbf{X} \in W'.$$

But $p_1' \circ Df_\mathbf{X}(\mathbb{R}^m) = \mathbb{R}^q$ and so

$$\sigma - (D_1 F)\phi \equiv 0.$$

Hence $(D_2 F)\phi \circ p_2 = 0$ and $D_2 F = 0$ on $\phi(W') = \widetilde{V}_1 \times V_2$.

Define the C^1 map $u = (u^1, u^2) : \widetilde{V}_1 \to \mathbb{R}^q \times \mathbb{R}^{n-q}$ by

$$F(\mathbf{x}, \mathbf{y}) = (u^1(\mathbf{x}), u^2(\mathbf{x})), \quad (\mathbf{x}, \mathbf{y}) \in \widetilde{V}_1 \times V_2.$$

We have $u(\widetilde{V}_1) = f(W')$.

Now $Du_0^1 \in GL(\mathbb{R}, q)$, since $F = fh = u^1$ on $\widetilde{V}_1 \times \{0\}$, and $Du_0^2 = \mathbf{0}_{q,n-q}$. Hence we may apply the dual implicit function theorem (Theorem 9.6.10) to u :

$\tilde{V}_1 \to \mathbb{R}^q \times \mathbb{R}^{n-q}$ to obtain open neighbourhoods $V_1 \subset \tilde{V}_1$ of $\mathbf{0}_q$, $W_2 \subset \mathbb{R}^n$ of $\mathbf{0}_n$, $U_2 \subset \mathbb{R}^{n-q}$ of $\mathbf{0}_{n-q}$, and a C^1 diffeomorphism $g : W_2 \to V_1 \times U_2$ such that

$$gu(\mathbf{x}) = (\mathbf{x}, \mathbf{0}), \ \mathbf{x} \in V_1.$$

Setting $V_2 = \tilde{V}_2$, $W_1 = h(V_1 \times V_2)$, completes the proof of the rank theorem. □

EXERCISES 9.6.16

(1) Show that a map f which is C^1 and a homeomorphism (f is bijective and f^{-1} is continuous) need not have a differentiable inverse.

(2) Find an example of a C^1 map $f : \mathbb{R}^2 \to \mathbb{R}^2$ such that $Df_{\mathbf{x}} \in GL(\mathbb{R}, 2)$ for all $\mathbf{x} \in \mathbb{R}^2$ and f is *not* 1:1.

(3) Let $f : \mathbb{R}^{n+1} \to \mathbb{R}$ be C^1 and suppose that at the point $\mathbf{a} = (a_1, \cdots, a_n, b) \in \mathbb{R}^{n+1}$, $\frac{\partial f}{\partial x_{n+1}}(\mathbf{a}) \neq 0$, Show that there exist an open neighbourhood V of $(a_1, \cdots, a_n)$, an open neighbourhood W of $\mathbf{a}$ and a C^1 map $u : V \to \mathbb{R}$ such that

$$f(x_1, \cdots, x_n, u(x_1, \cdots, x_n)) = 0, \text{ for all } (x_1, \cdots, x_n) \in V,$$

and these are the only solutions to $f(x_1, \cdots, x_n, y) = 0$ in W. Show that the partial derivatives of u are given by

$$\frac{\partial u}{\partial x_i}(\mathbf{x}) = -\frac{\partial f}{\partial x_i}(\mathbf{x}, u(\mathbf{x})) / \frac{\partial f}{\partial x_{n+1}}(\mathbf{x}, u(\mathbf{x})), \ \mathbf{x} \in V, \ i \in \mathbf{n}.$$

(4) Consider the simultaneous equations in the unknown functions f and g

$$f(x, y)^3 + xg(x, y)^2 + y = 0,$$
$$g(x, y)^3 + yg(x, y) + f(x, y)^2 - x = 0.$$

Show that the solution set of these equations is given by the zero set of the function $F : \mathbb{R}^4 \to \mathbb{R}^2$ defined by

$$F(x, y, u, v) = (u^3 + xv^2 + y, v^3 + yv + u^2 - x).$$

Verify that there is an open neighbourhood V of $(1, 1) \in \mathbb{R}^2$ and C^1 functions $f, g : V \to \mathbb{R}$ such that $f(1, 1) = -1$, $g(1, 1) = 0$ and

$$F(x, y, f(x, y), g(x, y)) = 0, \ (x, y) \in V,$$

and that these are the only solutions to $F = 0$ on some neighbourhood of $(1, 1, -1, 0)$.

(5) Show that there exist C^1 functions f and g defined on some neighbourhood of $(0,0) \in \mathbb{R}^2$ that satisfy the equations

$$(8 + x^2)f(x,y) - (y+1)g(x,y)^3 + y^2 = 0,$$
$$g(x,y)^2 - (y+1)f(x,y)g(x,y) - 2 = 0,$$

subject to the condition that $f(0,0) = 1$ and $g(0,0) = 2$.

(6) Let $n \leq m$. Show that the subset of $L(\mathbb{R}^m, \mathbb{R}^n)$ consisting of surjective linear maps is an open dense subset of $L(\mathbb{R}^m, \mathbb{R}^n)$. State and prove an analogous result in case $n \geq m$.

(7) Suppose that $\phi : \mathbb{R} \rightarrow \mathbb{R}^n$ ($n > 1$) is C^1 and suppose that (a) $\phi'(t) \neq \mathbf{0}$ for all $t \in \mathbb{R}$, and (b) ϕ is 1:1. Show, by means of examples, that ϕ need not map $\mathbb{R}$ homeomorphically onto $\phi(\mathbb{R})$ (induced topology on $\phi(\mathbb{R})$). (Hint: find examples satisfying (a,b) such that $\phi(\mathbb{R})$ (1) is, and (2) is not, a closed subset of $\mathbb{R}^n$.)

(8) A map $f : \mathbb{R}^m \rightarrow \mathbb{R}^n$ is *proper* if $f^{-1}(K)$ is compact whenever K is a compact subset of $\mathbb{R}^n$.

   (a) If f is proper, show that $f(\mathbb{R}^m)$ must be an unbounded subset of $\mathbb{R}^n$.

   (b) If f is proper and continuous, show that f is a *closed* map: f maps closed sets to closed sets. (Hint: sequential arguments make this easy.)

   (c) Suppose that the map ϕ of the previous question satisfies (a,b) and is also proper. Verify that ϕ maps $\mathbb{R}$ homeomorphically onto $\phi(\mathbb{R})$ (induced topology).

   (d) Generalize (c) to maps $f : \mathbb{R}^m \rightarrow \mathbb{R}^n$.

(9) Let $n > m$. Suppose that $f : \mathbb{R}^m \rightarrow \mathbb{R}^n$ is (a) C^1, (b) 1:1, (c) $Df_{\mathbf{x}} : \mathbb{R}^m \rightarrow \mathbb{R}^n$ is of rank m (injective) for all $\mathbf{x} \in \mathbb{R}^m$, and (d) f is proper. Show that given any $\mathbf{y} \in f(\mathbb{R}^m) \subset \mathbb{R}^n$, there exists an open neighbourhood U of $\mathbf{y} \in \mathbb{R}^n$ and a C^1 diffeomorphism ψ of U onto a product open neighbourhood $V \times W$ of $\mathbf{0} \in \mathbb{R}^m \times \mathbb{R}^{n-m}$ such that $\psi(f(\mathbb{R}^m) \cap U) = V \times \{0\}$. (Hint: dual implicit function theorem and take care with 'multiple' intersections of $f(\mathbb{R}^m)$ with U.) Show by means of an example that the result may fail if f is not proper. (If conditions (a–d) hold, we say that f is a C^1 *embedding* and $\phi(\mathbb{R}^m)$ has the structure of a (C^1) submanifold of $\mathbb{R}^n$.)

(10) Show, by means of examples, that the constant rank condition in the statement of the rank theorem cannot be weakened.

(11) Show that a C^1 map $f : \mathbb{R}^m \rightarrow \mathbb{R}^n$ can only be injective if $m \leq n$. (Hint: find a non-empty open subset of $\mathbb{R}^m$ on which the rank of $Df_{\mathbf{x}}$ is maximal and use the rank theorem.)

(12) Let U be an open subset of $\mathbb{R}^p$ and suppose $f : U \rightarrow L(\mathbb{R}^m, \mathbb{R}^n)$ is C^1 and that $f(\mathbf{x})$ is of rank q for all $\mathbf{x} \in U$. Given $\mathbf{x}_0 \in U$, show that we can choose an open neighbourhood U_0 of $\mathbf{x}_0$ and C^1 maps $\alpha : U_0 \rightarrow GL(\mathbb{R}, m)$, $\beta : U_0 \rightarrow GL(\mathbb{R}, n)$ such that

$$\beta \circ f \circ \alpha : U_0 \rightarrow L(\mathbb{R}^m, \mathbb{R}^n)$$

is the constant map π_q defined by $\pi_q(x_1, \cdots, x_m) = (x_1, \cdots, x_q, 0, \cdots, 0)$. (Hints: Using the rank theorem, study the map

$$F : U \times (L(\mathbb{R}^m, \mathbb{R}^m) \times L(\mathbb{R}^m, \mathbb{R}^m)) \to U \times L(\mathbb{R}^m, \mathbb{R}^n)$$

defined by $F(\mathbf{x}, A, B) = (\mathbf{x}, B \circ f(\mathbf{x}) \circ A)$. Alternatively, consider the map $\tilde{f} : U \times \mathbb{R}^m \to U \times \mathbb{R}^n$ defined by $\tilde{f}(\mathbf{x}, \mathbf{e}) = (\mathbf{x}, f(\mathbf{x})(\mathbf{e}))$.)

9.7 Local Existence and Uniqueness Theorem for Ordinary Differential Equations

Let $f : \mathbb{R}^m \to \mathbb{R}^m$ be a C^1 vector field on $\mathbb{R}^m$ (we could just as well assume f is defined on an open subset of $\mathbb{R}^m$ but to keep notation simple, we assume the domain is all of $\mathbb{R}^m$). We consider the ordinary differential equation (or 'ODE')

$$\mathbf{x}' = f(\mathbf{x}). \tag{9.10}$$

In coordinates, (9.10) corresponds to the system

$$x_i' = f_i(x_1, \cdots, x_m), \ 1 \leq i \leq m,$$

of ODEs. We recall that a solution of (9.10) with *initial condition* $\mathbf{x}_0$ consists of a C^1 map $\phi : I \to \mathbb{R}^m$, where I is an open interval in $\mathbb{R}$ containing the origin, such that

$$\phi'(t) = f(\phi(t)), \ t \in I, \ \text{and} \ \phi(0) = \mathbf{x}_0.$$

Remarks 9.7.1

(1) Since $\phi' = f \circ \phi$ and both f and ϕ are C^1, ϕ' must be C^1 by the chain rule and so ϕ is actually C^2. This is characteristic: solutions of ODEs are one order of differentiability more regular than the vector field defining the ODE.

(2) We assume the equation is autonomous—f does not depend on t. However, if f does depend on t, $\mathbf{x}' = f(\mathbf{x}, t)$, we can make the equation autonomous by introducing a new variable τ and considering the system $\mathbf{x}' = f(\mathbf{x}, \tau)$, $\tau' = 1$. Alternatively, the proof of the existence and uniqueness theorem continues to work under the assumption that f depends on t—changes required are minimal. For future reference, we give an existence result for a class of linear non-autonomous ODEs at the end of the section (the vector field f will be independent of $\mathbf{x}$). ✠

We are going to prove a theorem that gives the existence and uniqueness of local solutions to (9.10) and the continuous dependence of solutions on the initial conditions. Later, in Sect. 9.15, we strengthen this result and show that solutions are C^1 in time *and* space.

Before stating the main result, it is helpful to introduce some new terminology. Suppose that there exists an open neighbourhood U of $\mathbf{x}_0 \in \mathbb{R}^m$, an open interval $I = (-\delta, \delta) \subset \mathbb{R}$ and a C^0 map

$$\phi : U \times I \to \mathbb{R}^m$$

such that if $\mathbf{x} \in U$ and we set $\phi_{\mathbf{x}}(t) = \phi(\mathbf{x}, t), t \in I$, then $\phi_{\mathbf{x}}$ is a solution to $\mathbf{x}' = f(\mathbf{x})$ with initial condition $\mathbf{x}$. The requirement that $\phi_{\mathbf{x}}$ equals $\mathbf{x}$ at $t = 0$ implies that $\phi(\mathbf{x}, 0) = \mathbf{x}$ for all $\mathbf{x} \in U$. If we can find $\phi : U \times I \to \mathbb{R}^m$ satisfying these conditions we say that ϕ is a C^0 *local flow* for $\mathbf{x}' = f(\mathbf{x})$ (on a neighbourhood of $\mathbf{x}_0$).

Theorem 9.7.2 (Existence of Local Flows) *Suppose that f is a C^1 vector field on* $\mathbb{R}^m$. *Then $\mathbf{x}' = f(\mathbf{x})$ has a C^0 local flow on a neighbourhood of every point in $\mathbb{R}^m$. Moreover, the solutions to $\mathbf{x}' = f(\mathbf{x})$ are unique. That is, if $\phi_{\mathbf{x}} : I \to \mathbb{R}^m$ is the solution with initial condition $\mathbf{x}$ given by the local flow and $\delta : J \to \mathbb{R}^m$ is any other solution with initial condition $\mathbf{x}$ (so $0 \in J$), then $\phi_{\mathbf{x}} = \delta$ on $I \cap J$.*

Proof We prove the existence of a local flow on a neighbourhood of the origin of $\mathbb{R}^m$. For $s > 0$, let $\overline{D}_s$ denote the closed s-disk with centre $\mathbf{0}$ in $\mathbb{R}^m$. Given $r, a > 0$, let $C^0([-a, a], \overline{D}_{2r})$ denote the set of all continuous maps $f : [-a, a] \to \overline{D}_{2r} \subset \mathbb{R}^m$. Recall that if we define the uniform metric ρ on $C^0([-a, a], \mathbb{R}^m)$ by $\rho(f, g) = \sup_{t \in [-a,a]} \|f(t) - g(t)\|$, then $(C^0([-a, a], \mathbb{R}^m), \rho)$ is a complete metric space (Theorem 7.15.9) and $C^0([-a, a], \overline{D}_{2r})$ is complete subspace of $(C^0([-a, a], \mathbb{R}^m), \rho)$ (Exercises 7.15.23(5)).

For $\mathbf{x} \in \overline{D}_r$, $\psi \in C^0([-a, a], \overline{D}_{2r})$, define $T_{\mathbf{x}}(\psi) \in C^0([-a, a], \mathbb{R}^m)$ by

$$T_{\mathbf{x}}(\psi)(t) = \mathbf{x} + \int_0^t f(\psi(s)) \, ds, \ t \in [-a, a].$$

Observe that $T_{\mathbf{x}}(\psi)(0) = \mathbf{x}$ and $T_{\mathbf{x}}(\psi)$ is differentiable with

$$T_{\mathbf{x}}(\psi)'(t) = f(\psi(t)), \ t \in [-a, a].$$

In particular, if $T_{\mathbf{x}}(\psi) = \psi$, ψ will be a solution of $\mathbf{x}' = f(\mathbf{x})$ with initial condition $\mathbf{x}$.

Let $M_1 = \sup_{\mathbf{x} \in \overline{D}_{2r}} \|f(\mathbf{x})\| < \infty$ (since f is continuous and $\overline{D}_{2r}$ is compact). We claim that if $a \le r/M_1$, then $T_{\mathbf{x}}(\psi) \in C^0([-a, a], \overline{D}_{2r})$. This follows since if $\mathbf{x} \in \overline{D}_r$, $t \in [-a, a]$, we have

$$\|T_{\mathbf{x}}(\psi)(t)\| = \left\| \mathbf{x} + \int_0^t f(\psi(s)) \, ds \right\|$$

$$\le \|\mathbf{x}\| + \left\| \int_0^t f(\psi(s)) \, ds \right\|$$

$$\leq \|\mathbf{x}\| + \left| \int_0^t \|f(\psi(s))\| \, ds \right|$$

$$\leq \|\mathbf{x}\| + aM_1 \leq 2r.$$

Let $M_2 = \sup_{\mathbf{x} \in \overline{D}_{2r}} \|Df_\mathbf{x}\| < \infty$ (since f is C^1 and $\overline{D}_{2r}$ is compact) and set $a = \min\{\frac{r}{M_1}, \frac{1}{2M_2}\}$. We claim that $T : C^0([-a, a], \overline{D}_{2r}) \times \overline{D}_r \to C^0([-a, a], \overline{D}_{2r})$, $(\mathbf{x}, \psi) \mapsto T_\mathbf{x}(\psi)$, is a family of contraction mappings satisfying the hypotheses of the contraction mapping lemma with parameters.

First of all note that T is well defined as a map to $C^0([-a, a], \overline{D}_{2r})$ since $a \leq r/M_1$.

Let $\mathbf{x} \in \overline{D}_r$ and $\psi, \eta \in C^0([-a, a], \mathbb{R}^m)$. We have

$$d(T_\mathbf{x}(\psi), T_\mathbf{x}(\eta)) = \sup_{t \in [-a,a]} \left\| \mathbf{x} + \int_0^t f(\psi(s)) \, ds - \left(\mathbf{x} + \int_0^t f(\eta(s)) \, ds \right) \right\|$$

$$= \sup_{t \in [-a,a]} \left\| \int_0^t [f(\psi(s)) - f(\eta(s))] \, ds \right\|$$

$$\leq \sup_{t \in [-a,a]} \left| \int_0^t \|f(\psi(s)) - f(\eta(s))\| \, ds \right|.$$

It follows by the mean value theorem that $\|f(\psi(s)) - f(\eta(s))\| \leq M_2 \|\psi(s) - \eta(s)\|$, $s \in [-a, a]$, and so

$$d(T_\mathbf{x}(\psi), T_\mathbf{x}(\eta)) \leq \sup_{t \in [-a,a]} \left| \int_0^t M\|\psi(s) - \eta(s)\| \, ds \right|$$

$$\leq aM_2 \sup_{s \in [-a,a]} \|\psi(s) - \eta(s)\|$$

$$= aM_2 d(\psi, \eta) \leq \frac{1}{2} d(\psi, \eta),$$

where the last line follows by the definition of a. This estimate holds for all $\mathbf{x} \in \overline{D}_r$. Since $\mathbf{x} \mapsto T_\mathbf{x}(\psi)$ is obviously continuous on $\overline{D}_r$ for fixed ψ, the conditions of the contraction mapping lemma with parameters hold and we obtain a continuous map

$$\phi : U \times (-a, a) \to \mathbb{R}^m,$$

such that for all $\mathbf{x} \in U$, $\phi_\mathbf{x} : (-a, a) \to \mathbb{R}^m$ is a solution to $\mathbf{x}' = f(\mathbf{x})$ with initial condition $\mathbf{x}$.

Observe that the proof continues to work for any closed interval $K \subset [-a, a]$ containing the origin. So if $\delta : J \to \mathbb{R}^m$ is another solution with initial condition $\mathbf{x}$, take $K = J \cap (-a, a)$ and use uniqueness of fixed points to get $\delta = \phi_\mathbf{x}$ on the overlap.   □

Remark 9.7.3 Theorem 9.7.2 continues to hold if the vector field f is defined on a proper open subset U of $\mathbb{R}^m$ with the minor change that we additionally require $\overline{D}_{2r} \subset U$.   ✠

Lemma 9.7.4 (Uniqueness of Solutions) *Let f be C^1 and I_1, I_2 be open intervals containing $0 \in \mathbb{R}$. If $\phi_i : I_i \to \mathbb{R}^m$ are solutions of $\mathbf{x}' = f(\mathbf{x})$ with the same initial condition, then*

$$\phi_1(t) = \phi_2(t) \text{ for all } t \in I_1 \cap I_2.$$

Proof Set $I_1 \cap I_2 = (a, b)$ and define $X = \{t \in (a, b) \mid \phi_1(t) = \phi_2(t)\}$. Since $0 \in I$, $X \neq \emptyset$. As ϕ_1, ϕ_2 are continuous, X is a closed subset of (a, b). Suppose $s \in X$ and set $\phi_1(s) = \phi_2(s) = \mathbf{y}$. Differentiating with respect to t, we see that $\psi_i(t) = \phi_i(s + t)$, $t \in (a - s, b - s)$, are both solutions to $\mathbf{x}' = f(\mathbf{x})$ with initial condition $\mathbf{y}$. By the uniqueness part of Theorem 9.7.2, $\psi_1 = \psi_2$ on some interval $(-\delta, \delta)$ containing $t = 0$ and so $\phi_1(t) = \phi_2(t)$ for $t \in (s - \delta, s + \delta) \cap (a, b)$. Therefore X is open. Since (a, b) is connected and $X \neq \emptyset$, $X = (a, b)$.   □

If we assume f depends on t and is independent of $\mathbf{x}$, results are much easier to prove. We use the next result later in Sect. 9.15 when we strengthen Theorem 9.7.2 and show ϕ is C^1 (in $(\mathbf{x}, t)$).

Proposition 9.7.5 *Let $M > 0$ and define*

$$\Lambda = \{G \in C^0(\mathbb{R}, L(\mathbb{R}^m, \mathbb{R}^m)) \mid |G| = \sup_{t \in \mathbb{R}} \|G(t)\| \leq M\}.$$

Given $G \in \Lambda$, consider the ordinary differential equation

$$\mathbf{A}' = G(t) \circ \mathbf{A}(t) \tag{9.11}$$

defined on $L(\mathbb{R}^m, \mathbb{R}^m)$. With $a = 1/(2M)$, there exists a C^1 map $\phi : L(\mathbb{R}^m, \mathbb{R}^m) \times [-a, a] \to L(\mathbb{R}^m, \mathbb{R}^m)$ such that $\mathbf{A}(t) = \phi(\mathbf{A}_0, t)$ is the unique solution to (9.11) with initial condition $\mathbf{A}_0$. Furthermore, if $\overline{G} \in \Lambda$, and we denote the corresponding family of solutions to (9.11) by $\overline{\mathbf{A}}(t)$, then for solutions with the same initial condition we have the estimate

$$\|\mathbf{A} - \overline{\mathbf{A}}\| \leq 2a|G - \overline{G}|\|\mathbf{A}\|,$$

where $\|\mathbf{A}\| = \sup_{t \in [-a, a]} \|\mathbf{A}(t)\|$.

Proof Let $X = C^0([-a, a], L(\mathbb{R}^m, \mathbb{R}^m))$ (uniform metric) and define $T : X \times L(\mathbb{R}^m, \mathbb{R}^m) \to X$ by

$$T(\psi, \mathbf{A}_0)(t) = \mathbf{A}_0 + \int_0^t G(s) \circ \psi(s) \, ds, \quad \psi \in X, \ t \in [-a, a].$$

The proof of existence follows that of Theorem 9.7.2. The estimate in $|G - \overline{G}|$ uses the integral form of the solution and $a = 1/(2M)$. Note that the estimate is a quantitative version of continuous dependence on parameters—in this case, the parameter is G. We leave the straightforward details to the exercises. $\square$

EXERCISES 9.7.6

(1) Complete the proof of Proposition 9.7.5.
(2) Let $A : \mathbb{R} \to L(\mathbb{R}^m, \mathbb{R}^m)$ be continuous and consider the linear system $\mathbf{x}' = A(t)\mathbf{x}$. State and prove an existence and uniqueness theorem for this system.

9.8 Higher Derivatives as Approximations

We motivated the idea of derivative in terms of affine linear approximation. It is natural to try to extend this idea to define higher-order derivatives. For example, if $f : U \subset \mathbb{R}^m \to \mathbb{R}^n$ is C^1, then we might define f to be twice differentiable at $\mathbf{x}_0 \in U$ if there exists a homogeneous quadratic polynomial $Q : \mathbb{R}^m \to \mathbb{R}^n$ such that if we define the remainder term $r(\mathbf{h})$ by

$$f(\mathbf{x}_0 + \mathbf{h}) = f(\mathbf{x}_0) + Df_{\mathbf{x}_0}(\mathbf{h}) + Q(\mathbf{h}) + r(\mathbf{h}),$$

then $r(\mathbf{h}) = o(\|\mathbf{h}\|^2)$. That is, $\|r(\mathbf{h})\| \to 0$ as $\mathbf{h} \to 0$ faster than $\|\mathbf{h}\|^2$. Ignoring for now the definition of a homogeneous quadratic polynomial, what we are saying is that f is twice differentiable at $\mathbf{x}_0$ if f has a 'good' quadratic approximation near $\mathbf{x}_0$. Exactly this type of condition holds if $f : \mathbb{R} \to \mathbb{R}$ is twice differentiable at x_0: $f(x_0 + h) = f(x)_0 + f'(x_0)h + f''(x_0)h^2/2 + r(h)$, where $r(h) = o(h^2)$. This follows from Taylor's theorem (Theorem 2.7.10). Observe that $f(x_0) + f'(x_0)h + f''(x_0)h^2/2$ is the 'best possible' quadratic approximation to f near x_0. There is, however, another way of approaching the theory of higher derivatives. If $f : U \subset \mathbb{R}^m \to \mathbb{R}^n$ is C^1, then it is natural to say that f is twice differentiable at $\mathbf{x}_0$ if $Df : U \to L(\mathbb{R}^m, \mathbb{R}^n)$ is differentiable at $\mathbf{x}_0$. Viewed this way, the second derivative of f at $\mathbf{x}_0$ will be a linear map $D^2 f_{\mathbf{x}_0} : \mathbb{R}^m \to L(\mathbb{R}^m, \mathbb{R}^n)$. That is, $D^2 f_{\mathbf{x}_0} \in L(\mathbb{R}^m, L(\mathbb{R}^m, \mathbb{R}^n))$. Assuming f is twice continuously differentiable on U, the third derivative of f at $\mathbf{x}_0$ would be defined as a linear map $D^3 f_{\mathbf{x}_0} \in L(\mathbb{R}^m, L(\mathbb{R}^m, L(\mathbb{R}^m, \mathbb{R}^n)))$ and so on. This, of course, looks complicated but the situation is saved because there is a natural way of going from elements of $L(\mathbb{R}^m, \cdots L(\mathbb{R}^m, \mathbb{R}^n) \cdots)$ to polynomial maps $\mathbb{R}^m \to \mathbb{R}^n$ and this will give us the connection between higher derivatives and approximation. The reality is that the details for higher-order derivatives for vector valued maps defined on a vector space may appear to be complicated but most of the difficulties lie with keeping the notation under control. For example, to specify the matrix of the derivative linear map from $\mathbb{R}^m$ to $\mathbb{R}^n$ we need nm partial derivatives. For the second derivative we need a total of $m^2 n$ partial derivatives; for the pth derivative, $m^p n$ partial derivatives. Writing all of this out in coordinates is both daunting and unhelpful. One of our goals is to develop a good 'language' so that the results mirror those of the one-variable theory in a transparent way. Turning to the details, we shall start our work on higher derivatives with an extended discussion of polynomial and multi-linear maps between vector spaces. We then define *symmetric* multi-linear maps and show that there is a natural bijective correspondence between polynomials and symmetric multi-linear maps. With these preliminaries out of the way, we show how higher derivatives define symmetric multi-linear maps and so determine polynomial maps. We also show how the inverse and implicit functions generalize easily to C^r-maps, $r > 1$. We conclude our work on differentiation with a number of more advanced results about higher-order derivatives of products (Leibniz law) and compositions (Faà di Bruno's formula).

9.9 Multi-Linear Maps and Polynomials

9.9.1 Preliminaries on Normed Vector Spaces

It is useful to work with general normed vector spaces $(V, \| \cdot \|_V)$ rather than restricting to $(\mathbb{R}^m, \| \cdot \|_2)$. In part this is because when we consider spaces of linear and multi-linear maps, the operator norm will not usually be the Euclidean norm. Another consideration is that working in greater generality simplifies the notation and helps to reveal the relationships between spaces of linear and multi-linear maps. We usually drop the subscript V from the norm symbol and denote the norm on V by $\| \cdot \|$ (if we need to emphasize the space V, we write $\| \cdot \|_V$).

We assume all vector spaces are finite-dimensional (and so isomorphic to $\mathbb{R}^n$ for some n). Moreover, all norms on a finite-dimensional vector space are equivalent and define the same topology as the Euclidean norm (Theorem 9.1.4).

Let $L(V, W)$ denote the vector space of linear maps from V to W. Since we assume V, W are finite-dimensional normed vector spaces, $L(V, W)$ consists of continuous linear maps (see Sect. 9.2 and note this only needs the finite-dimensionality of V). As shown in Sect. 9.2.1, continuous linear maps are bounded on the unit disk, centre the origin, and we define the *operator norm* on $L(V, W)$ by

$$\|A\| = \sup_{\|\mathbf{v}\|=1} \|A\mathbf{v}\|, \quad A \in L(V, W).$$

All the results proved in Sect. 9.2 extend immediate to general finite-dimensional normed vector spaces. In particular, $\|A\mathbf{x}\| \le \|A\| \|\mathbf{x}\|$, $\mathbf{x} \in V, A \in L(V, W)$ (see also Exercises 9.2.13).

9.9.2 Multi-Linear Maps

Let $(V, \|\cdot\|)$ be a normed vector space and $p \in \mathbb{N}$. We define the *p-fold* product $(V^p, \|\cdot\|)$ to be the normed vector space which is the product of p copies of V with *product norm* defined by

$$\|(\mathbf{v}_1, \cdots, \mathbf{v}_p)\| = \max\{\|\mathbf{v}_i\| \mid 1 \leq i \leq p\}, \ (\mathbf{v}_1, \cdots, \mathbf{v}_p) \in V^p.$$

Remarks 9.9.1

(1) In what follows we make no use of the vector space structure on V^p which is defined by coordinate-wise addition and scalar multiplication in the usual way. Note that the *topology* on V^p is uniquely defined, independently of the choice of norm on V (Theorem 9.1.4).
(2) An alternative notation for V^p is $\times^p V$.
(3) If the dimension of V is m, then the dimension of V^p is pm. ✖

Definition 9.9.2 Let V, W be normed vector spaces and $p \in \mathbb{N}$. A map $T : V^p \to W$ is called *p-linear* or *multi-linear* if T is linear in each variable separately. That is, for every $i \in \{1, \ldots, p\}$, and all $\mathbf{u}, \mathbf{v} \in V, \lambda \in \mathbb{R}$, and $\mathbf{x}_j \in V, j \neq i$, we have

$$T(\mathbf{x}_1, \cdots, \mathbf{x}_{i-1}, \lambda\mathbf{u} + \mathbf{v}, \cdots, \mathbf{x}_p) = \lambda T(\mathbf{x}_1, \cdots, \mathbf{x}_{i-1}, \mathbf{u}, \cdots, \mathbf{x}_p)$$
$$+ T(\mathbf{x}_1, \cdots, \mathbf{x}_{i-1}, \mathbf{v}, \cdots, \mathbf{x}_p).$$

For $p \geq 0$, let $L^p(V; W)$ denote the space of all p-linear maps $T : V^p \to W$ (define $L^0(V; W) = W$). Clearly, $L^p(V; W)$ inherits the structure of a vector space from W: $(\lambda T + S)(\mathbf{X}) = \lambda T(\mathbf{X}) + S(\mathbf{X}), T, S \in L^p(V; W), \lambda \in \mathbb{R}, \mathbf{X} \in V^p$.

Examples 9.9.3

(1) Since $L^1(V; W) = L(V, W)$, every 1-linear map is linear. If $T \in L^2(V; W)$, T is called *bilinear*. For example, an inner product $\langle \cdot, \cdot \rangle$ on V defines a bilinear map $\langle \cdot, \cdot \rangle : V^2 \to \mathbb{R}$.
(2) If $p > 1$, a p-linear map $T : V^p \to W$ is *not* linear with respect to the vector space structure on V^p (note Remarks 9.9.1(1)).
(3) Suppose $T : V^2 \to W$ is bilinear. Let $\{\mathbf{e}_i\}_{i=1}^m$ and $\{\mathbf{f}_\ell\}_{\ell=1}^n$ be bases for V and W, respectively. Denote the associated coordinates on V by $(x_1, \cdots, x_m)$. Relative to these bases, we may write T in coordinate form as

$$T((x_1, \cdots, x_m), (y_1, \cdots, y_m)) = \left(\sum_{i,j=1}^m a_{ij}^1 x_i y_j, \cdots, \sum_{i,j=1}^m a_{ij}^n x_i y_j \right),$$

where the coefficients a_{ij}^ℓ are uniquely determined by T according to $T(\mathbf{e}_i, \mathbf{e}_j) = \sum_{\ell=1}^n a_{ij}^\ell \mathbf{f}_\ell, 1 \leq i, j \leq m, 1 \leq \ell \leq n$. It follows that $\dim(L^2(V; W)) =$

$\dim(V)^2 \times \dim(W)$. Similar expressions and results hold if $p > 2$. In particular, $\dim(L^p(V; W)) = \dim(V)^p \dim(W)$.   ♠

Lemma 9.9.4 *A multi-linear map between finite-dimensional normed vector spaces is continuous.*

Proof Let $T \in L^p(V; W)$. Choosing bases for V and W, we may write T in coordinate form as we did in Examples 9.9.3(3) for the case $p = 2$. Since each component T_ℓ of $T = (T_1, \cdots, T_n)$ may be written as a finite sum of continuous monomials $a^\ell_{i_1 \cdots i_p} x^1_{i_1} \cdots x^p_{i_p}$, where $1 \leq i_1, \cdots, i_p \leq m$ and $(x^j_1, \cdots, x^j_m)$ are the coordinates of a point in the jth factor of V^p, T_ℓ is continuous, $1 \leq \ell \leq n$. Hence T is continuous.   □

Remark 9.9.5 At the cost of some extra work, it is possible to avoid the coordinate computations of the previous lemma. Specifically, given $p \in \mathbb{N}$, it can be shown that there exists a (unique up to isomorphism) finite-dimensional vector space $\otimes^p V$ and natural continuous p-linear map $j : V^p \to \otimes^p V$ such that every p-linear map $T : V^p \to W$ can be uniquely factored through $\otimes^p V$ as the composite $\hat{T} j$, where $\hat{T} \in L(\otimes^p V, W)$. The space $\otimes^p V$ is the *p-fold tensor product* of V and is of dimension $\dim(V)^p$. Since $L(\otimes^p V, W)$ consists of continuous linear maps and $j : V^p \to \otimes^p V$ is continuous, every p-linear map $T : V^p \to W$ is continuous.   ✸

Theorem 9.9.6 *Let $(V, \| \cdot \|), (W, \| \cdot \|)$ be normed vector spaces and $p \in \mathbb{N}$. Then $L^p(V; W)$ has the structure of a normed vector space with norm defined by*

$$\|T\| = \sup_{\|(\mathbf{x}_1, \cdots, \mathbf{x}_p)\| = 1} \|T(\mathbf{x}_1, \cdots, \mathbf{x}_p)\|.$$

Moreover, for all $T \in L^p(V; W)$, we have

$$\|T(\mathbf{x}_1, \cdots, \mathbf{x}_p)\| \leq \|T\| \|\mathbf{x}_1\| \cdots \|\mathbf{x}_p\|, \quad (\mathbf{x}_1, \cdots, \mathbf{x}_p) \in V^p. \tag{9.12}$$

Proof Since V^p is a finite-dimensional vector space, the topology on V^p is uniquely defined independently of the choice of norm on V (see Remarks 9.9.1(1)). Hence $S(V^p) = \{\mathbf{u} \in V^p \mid \|\mathbf{u}\| = 1\}$ is a compact subset of V^p and so $\|T\| < \infty$ for all $T \in L^p(V; W)$. Standard arguments show that $\| \cdot \|$ defines a norm on $L^p(V; W)$. Finally, suppose that $(\mathbf{x}_1, \cdots, \mathbf{x}_p) \in V^p$. If any one of the vectors $\mathbf{x}_i = 0$, then $T(\mathbf{x}_1, \cdots, \mathbf{x}_p) = 0$ (by p-linearity) and so (9.12) holds trivially. So suppose $\mathbf{x}_i \neq 0$, $1 \leq i \leq p$. Set $\mathbf{X} = (\frac{\mathbf{x}_1}{\|\mathbf{x}_1\|}, \cdots, \frac{\mathbf{x}_p}{\|\mathbf{x}_p\|})$. By definition of the product norm we have $\|\mathbf{X}\| = 1$ and so

$$\left\| T\left(\frac{\mathbf{x}_1}{\|\mathbf{x}_1\|}, \cdots, \frac{\mathbf{x}_p}{\|\mathbf{x}_p\|} \right) \right\| \leq \|T\|.$$

Now

$$T\left(\frac{\mathbf{x}_1}{\|\mathbf{x}_1\|}, \cdots, \frac{\mathbf{x}_p}{\|\mathbf{x}_p\|}\right) = \frac{1}{\|\mathbf{x}_1\| \cdots \|\mathbf{x}_p\|} T(\mathbf{x}_1, \cdots, \mathbf{x}_p).$$

by the p-linearity of T. Taking the norm of both sides and multiplying by $\|\mathbf{x}_1\| \cdots \|\mathbf{x}_p\|$ gives (9.12). □

Example 9.9.7 Let $(\cdot, \cdot) : \mathbb{R}^n \times \mathbb{R}^n \to \mathbb{R}$ denote the Euclidean inner product on $\mathbb{R}^n$. We have $\|(\cdot, \cdot)\| = 1$ (by the Cauchy–Schwarz inequality). ♠

Lemma 9.9.8 *Let V, W be normed vector spaces. There is a natural norm-preserving linear isomorphism $L(V, L(V, W)) \approx L^2(V; W)$.*

Proof Given $T \in L(V, L(V, W))$, define $\hat{T} \in L^2(V; W)$ by

$$\hat{T}(\mathbf{x}_1, \mathbf{x}_2) = T(\mathbf{x}_1)(\mathbf{x}_2), \quad \mathbf{x}_1, \mathbf{x}_2 \in V.$$

Since T and $T(\mathbf{x}_1)$ are linear, $\hat{T}$ is bilinear. We have

$$
\begin{aligned}
\|\hat{T}\| &= \sup_{\|\mathbf{x}_1\|, \|\mathbf{x}_2\|=1} \|\hat{T}(\mathbf{x}_1, \mathbf{x}_2)\| \\
&= \sup_{\|\mathbf{x}_1\|, \|\mathbf{x}_2\|=1} \|T(\mathbf{x}_1)(\mathbf{x}_2)\| \\
&= \sup_{\|\mathbf{x}_1\|=1} \|T(\mathbf{x}_1)\| \\
&= \|T\|,
\end{aligned}
$$

where the last two lines follow from the definition of the operator norm. The map $T \mapsto \hat{T}$ is obviously linear: $\widehat{\lambda T + S} = \lambda \hat{T} + \hat{S}$ for all $T, S \in L(V, L(V, W))$, $\lambda \in \mathbb{R}$. Since $\|\hat{T}\| = 0$ iff $\|T\| = 0$, we see that $T \mapsto \hat{T}$ must be 1:1. Since $L(V, L(V, W))$ and $L^2(V; W)$ have the same dimension, $T \mapsto \hat{T}$ must be a linear isomorphism between $L(V, L(V, W))$ and $L^2(V; W)$. Alternatively, we may construct the inverse map: given $S \in L^2(V; W)$, define $S' \in L(V, L(V, W))$ by $S'(\mathbf{x}_1)(\mathbf{x}_2) = S(\mathbf{x}_1, \mathbf{x}_2)$, $\mathbf{x}_1, \mathbf{x}_2 \in V$. We leave it to the reader to verify that this formula defines a linear inverse to the map $T \mapsto \hat{T}$. □

Remark 9.9.9 We use the word 'natural' in Lemma 9.9.8 in the sense that the isomorphism we construct does not depend on choosing bases for either V or W. Indeed, the construction works just as well if V, W are infinite-dimensional normed vector spaces and we consider spaces of *continuous* linear and multi-linear maps. ✱

Theorem 9.9.10 *Let V, W be normed vector spaces. For $p \geq 1$, there are natural norm-preserving isomorphisms*

$$L(V, L^{p-1}(V; W)) \approx L^p(V; W),$$

$$L(V, L(V, \cdots, L(V, W) \cdots)) \approx L^p(V; W),$$

where there are p copies of V on the left-hand side of the second isomorphism.

Proof Let $T \in L(V, L^{p-1}(V; W))$ and $\mathbf{x}_1, \mathbf{x}_2, \cdots, \mathbf{x}_p \in V$. We define $\hat{T} \in L^p(V; W)$ by

$$\hat{T}(\mathbf{x}_1, \mathbf{x}_2, \cdots, \mathbf{x}_p) = T(\mathbf{x}_1)(\mathbf{x}_2, \cdots, \mathbf{x}_p).$$

Exactly as in the proof of Lemma 9.9.8, we verify that $T \mapsto \hat{T}$ is a norm-preserving linear isomorphism. The proof of the second statement is similar and may either be proved directly or by induction.   □

EXERCISES 9.9.11

(1) Let $T : V \times V = V^2 \to V$ be defined as vector space addition: $T(\mathbf{x}, \mathbf{y}) = \mathbf{x} + \mathbf{y}$. Verify that T is linear (with respect to the linear structure defined on V^2, Remarks 9.9.1(1)) and find $\|T\|$.

(2) The product norm on $\mathbb{R}^p = \times^p \mathbb{R}$ is usually denoted by $\| \cdot \|_\infty$. What is $\|(x_1, \cdots, x_p)\|$? (See Exercises 7.1.9(3) for the definition of the corresponding d_∞ metric when $p = 2$.)

(3) Take the norm $\| \cdot \|_\infty$ on $\mathbb{R}^m, \mathbb{R}^n$. Let $A = [a_{ij}] \in L(\mathbb{R}^m, \mathbb{R}^n)$. Verify that

$$\|A\mathbf{x}\|_\infty \leq \left(\max_i \left(\sum_{j=1}^m |a_{ij}| \right) \right) \|\mathbf{x}\|_\infty, \quad \mathbf{x} \in \mathbb{R}^m,$$

and that $\|A\| = \max_i (\sum_{j=1}^m |a_{ij}|)$.

(4) Let $(V_1, \| \cdot \|_1), \cdots, (V_p, \| \cdot \|_p)$ be (finite-dimensional) normed vector spaces. Define the product norm on $V_1 \times \cdots \times V_p = \times_{i=1}^p V_i$ by $\|(\mathbf{v}_1, \cdots, \mathbf{v}_p)\| = \max_i \|\mathbf{v}_i\|$. Verify that $(\times_{i=1}^p V_i, \| \cdot \|)$ has the structure of a normed vector space.

(5) Continuing with the assumptions of the preceding exercise, define the space $L^p(V_1, \cdots, V_p; W)$ of p-linear multi-linear maps from $\times_{i=1}^p V_i$ to W. Show that $L^p(V_1, \cdots, V_p; W)$ has the structure of a normed vector space such that given $T \in L^p(V_1, \cdots, V_p; W)$ we have

$$\|T(\mathbf{v}_1, \cdots, \mathbf{v}_p)\| \leq \|T\| \|\mathbf{v}_1\| \cdots \|\mathbf{v}_p\|, \quad (\mathbf{v}_1, \cdots, \mathbf{v}_p) \in \times_{i=1}^p V_i.$$

(6) Show that scalar multiplication on V defines a bilinear map $S \in L^2(V, \mathbb{R}; V)$. What is $\|S\|$?

(7) Let U, V, W be finite-dimensional normed vector spaces. Show that

(a) The map $\mathbf{C} : L(V, W) \times L(U, V) \to L(U, W)$ defined by $C(A, B) = A \circ B$ (composition) is bilinear.

(b) $\|\mathbf{C}\| = 1$. (It is easy, by Proposition 9.2.11(c), to show that $\|\mathbf{C}\| \leq 1$.)

(8) Let V_1, V_2, W be finite-dimensional normed vector spaces. Verify we have a natural norm-preserving linear isomorphism $L(V_1, L(V_2, W)) \approx L(V_1, V_2; W)$. Extend to the case of p normed vector spaces $V_1, \cdots, V_p$.

(9) Let $V_1, \cdots, V_p, W_1, \cdots, W_q$ be finite-dimensional normed vector spaces. Show there is a natural linear isomorphism

$$L(\times_{i=1}^p V_i, \times_{j=1}^q W_j) \approx \times_{i,j} L(V_i, W_j).$$

Verify that if $p = 1$, the isomorphism is norm-preserving. What happens if $p > 1$?

9.9.3 Symmetric Multi-Linear Maps and Polynomials

For $n \in \mathbb{N}$, let S_n denote the group of all permutations of $\{1, \cdots, n\}$. We refer to S_n as the *symmetric group* on n symbols and recall that the order of the group S_n is $n!$

Definition 9.9.12 Let V, W be normed vector spaces. A p-linear map $T \in L^p(V; W)$ is *symmetric* if for all $(\mathbf{x}_1, \cdots, \mathbf{x}_p) \in V^p$ we have

$$T(\mathbf{x}_{\sigma(1)}, \cdots, \mathbf{x}_{\sigma(p)}) = T(\mathbf{x}_1, \cdots, \mathbf{x}_p), \text{ for all } \sigma \in S_p.$$

We denote the set of all symmetric p-linear maps from V to W by $L_s^p(V; W)$; as for p-linear maps, we take $L_s^0(V; W) = W$.

Examples 9.9.13

(1) Let $\langle \cdot, \cdot \rangle$ be an inner product on the vector space V. Then $\langle \cdot, \cdot \rangle \in L_s^2(V; \mathbb{R})$. This is a consequence of the symmetry property of an inner product: $\langle \mathbf{u}, \mathbf{v} \rangle = \langle \mathbf{v}, \mathbf{u} \rangle$, for all $\mathbf{u}, \mathbf{v} \in V$.

(2) If V is 1-dimensional, then $L_s^p(V; W) = L^p(V; W) \cong W$, for all $p \in \mathbb{N}$. For this observe that if $\{\mathbf{v}\}$ is a basis of V and we set $T(\mathbf{v}, \cdots, \mathbf{v}) = \mathbf{w}$, then $T(x_1\mathbf{v}, \cdots, x_p\mathbf{v}) = x_1 \cdots x_p\mathbf{w}$. This expression is obviously symmetric in $x_1, \cdots, x_p$. ♠

Proposition 9.9.14 *If V, W are finite-dimensional normed vector spaces and $p \in \mathbb{Z}_+$, then $L_s^p(V; W)$ is a vector subspace of $L^p(V; W)$. In particular, $L_s^p(V; W)$ inherits the structure of a normed vector space from the norm on $L^p(V; W)$.*

Proof Left to the reader. □

Our definition of a homogeneous polynomial on a normed vector space is given in terms of symmetric multi-linear maps.

Definition 9.9.15 Let V, W be finite-dimensional normed vector spaces and $d \in \mathbb{Z}_+$. A map $p : V \to W$ is a *homogeneous polynomial of degree d* if there exists a $T \in L_s^d(V; W)$ such that

$$p(\mathbf{x}) = T(\mathbf{x}, \mathbf{x}, \cdots, \mathbf{x}), \; \mathbf{x} \in V.$$

We denote the set of all homogeneous polynomial maps of degree d from V to W by $P^{(d)}(V, W)$. Note that $P^{(0)}(V, W) = W$.

Remarks 9.9.16

(1) Since multi-linear maps are continuous, polynomials are continuous (vector spaces are assumed finite-dimensional).
(2) Rather than writing $P(\mathbf{x}) = T(\mathbf{x}, \mathbf{x}, \cdots, \mathbf{x})$, it is often more convenient and suggestive to write $P(\mathbf{x}) = T(\mathbf{x}^d)$, where it is understood that $\mathbf{x}^d$ is shorthand for $(\mathbf{x}, \cdots, \mathbf{x}) \in V^d$ and does *not* refer to the product of $\mathbf{x}$ with itself d-times (this is not defined on a general vector space). We may generalize this notation in the obvious way and define $T(\mathbf{x}^a, \mathbf{y}^b)$ for $a + b = d$, $\mathbf{x}, \mathbf{y} \in V$. Since T is symmetric, there is no ambiguity with this notation.
(3) The definition of a homogeneous polynomial p allows the possibility of there being more than one choice of symmetric multi-linear map T defining p. In due course, we show that the choice is unique and that there is a linear isomorphism between $P^{(d)}(V, W)$ and $L_s^d(V; W)$.   ✠

Examples 9.9.17

(1) Let $p \in P^{(d)}(V, W)$. We have $p(\lambda\mathbf{x}) = \lambda^d p(\mathbf{x})$ for all $\mathbf{x} \in V$, $\lambda \in \mathbb{R}$. This homogeneity condition does not (quite) imply p is a polynomial. For example, if $V = \mathbb{R}^2$, $W = \mathbb{R}$, and we define $f(x, y) = \frac{x^2 y}{x^2 + y^2}$, $(x, y) \neq (0, 0)$ and $f(0, 0) = (0, 0)$, then f is continuous and homogeneous of degree 1 but f is not a polynomial. On the other hand, if f homogeneous of degree d and d times differentiable at $\mathbf{x} = \mathbf{0}$, then f is a homogeneous polynomial of degree d. This is not hard to show using the vector-valued version of Taylor's theorem which we prove later.
(2) We can associate a polynomial to every $T \in L^d(V; W)$ (no symmetry assumed). To do this, define

$$p(\mathbf{x}) = T(\mathbf{x}^n), \; \mathbf{x} \in V.$$

Observe that p is a polynomial in the sense of Definition 9.9.15 since if we define $T_s \in L_s^d(V; W)$ by

$$T_s(\mathbf{x}_1, \cdots, \mathbf{x}_d) = \frac{1}{d!} \sum_{\sigma \in S_d} T(\mathbf{x}_{\sigma(1)}, \cdots, \mathbf{x}_{\sigma(d)}), \quad (9.13)$$

then T_s is symmetric and $T(\mathbf{x}^d) = T_s(\mathbf{x}^d)$ for all $\mathbf{x} \in V$. We refer to T_s as the *symmetrization* of T. Note that the symmetrization map $L^d(V; W) \to L^d_s(V; W)$, $T \mapsto T_s$, is linear and onto.

(3) If $p \in P^{(d)}(V, W)$ and $p(\mathbf{x}) = T(\mathbf{x}^d)$, where $T \in L^d_s(V; W)$, then p is C^1 with derivative given by

$$Dp_\mathbf{x}(\mathbf{e}) = dT(\mathbf{x}^{d-1}, \mathbf{e}), \quad \mathbf{x}, \mathbf{e} \in V.$$

This follows since $p(\mathbf{x} + \mathbf{h}) = T((\mathbf{x} + \mathbf{h})^d) = T(\mathbf{x}^d) + dT(\mathbf{x}^{d-1}, \mathbf{h}) + R(\mathbf{x}, \mathbf{h})$, where we have used the symmetry of T and $R(\mathbf{x}, \mathbf{h})$ is a sum of terms of the form $T(\mathbf{x}^r, \mathbf{h}^s)$, $r + s = d$, $s \geq 2$. We have $\|R(\mathbf{x}, \mathbf{h})\| \leq C\|h\|^2$, by Theorem 9.9.6, and so $R(\mathbf{x}, \mathbf{h}) = o(\|h\|)$. ♠

For $d \in \mathbb{Z}_+$, $P^{(d)}(V, W)$ has the structure of a vector space. Given $p \in P^{(d)}(V, W)$, define

$$\|p\| = \sup_{\|\mathbf{x}\|=1} \|p(\mathbf{x})\|.$$

Proposition 9.9.18 (Notation as Above)

(1) *For $d \in \mathbb{Z}_+$, $(p^{(d)}(V, W), \|\cdot\|)$ has the structure of a normed vector space.*
(2) *Given $p \in P^{(d)}(V, W)$, we have*

$$\|p(\mathbf{x})\| \leq \|p\|\|\mathbf{x}\|^d, \quad \mathbf{x} \in V.$$

Proof Exactly the same method used for the proof of the corresponding result for multi-linear maps (Theorem 9.9.6). We leave the details to the reader. □

Remark 9.9.19 If $p(\mathbf{x}) = T(\mathbf{x}^n)$, $T \in L^d_s(V; W)$, it is not true that the polynomial norm $\|p\|$ equals the multi-linear norm $\|T\|$. The relation between the norms is given in the exercises at the end of the section. ✸

Definition 9.9.20 A map $p : V \to W$ is a *polynomial of degree d* if there exist homogenous polynomials $p_j \in P^{(j)}(V, W)$, $0 \leq j \leq d$, such that

$$p(\mathbf{x}) = \sum_{j=0}^{d} p_j(\mathbf{x}), \quad \mathbf{x} \in V.$$

Let $P^d(V, W)$ denote the vector space of all polynomial maps of degree d from V to W.

9.9.4 Multi-Index Notation and Coordinate form for Polynomials

Before we give the coordinate description of polynomial maps, we need to review multi-index notation. Let $m \in \mathbb{N}$ and $\alpha \in \mathbb{Z}_+^m$. If $\alpha = (\alpha_1, \cdots, \alpha_m)$, set $|\alpha| = \sum_{i=1}^m \alpha_i$ and $\alpha! = \alpha_1! \cdots \alpha_m!$ Given $\mathbf{x} = (x_1, \cdots, x_m) \in \mathbb{R}^m$, define

$$x^\alpha = x_1^{\alpha_1} \cdots x_m^{\alpha_m}.$$

Thus x^α is a *monomial* of degree $|\alpha|$ and x^α defines a real-valued map on $\mathbb{R}^m$. It is useful to extend this notation. Suppose W is a vector space. If $\mathbf{w} \in W$, we call the map $x \mapsto x^\alpha \mathbf{w}$ a *monomial of degree d from $\mathbb{R}^m$ to W.*

More generally, suppose $\mathbf{x} \in V$ and $d \in \mathbb{N}$. We previously defined $\mathbf{x}^d = (\mathbf{x}, \cdots, \mathbf{x}) \in V^d$. If α is a multi-index, and $\mathbf{x}_1, \cdots, \mathbf{x}_m \in V$, we define $\mathbf{x}^\alpha \in V^{|\alpha|}$ to be $(\mathbf{x}_1^{\alpha_1}, \cdots, \mathbf{x}_m^{\alpha_m}) \in V^{\alpha_1} \times \cdots \times V^{\alpha_m}$.

Lemma 9.9.21 *Let $T \in L_s^d(V; W)$. For all $\mathbf{x}_1, \cdots, \mathbf{x}_m \in V$, and $\lambda_1, \cdots, \lambda_m \in \mathbb{R}$, we have*

$$T((\lambda_1 \mathbf{x}_1 + \cdots + \lambda_m \mathbf{x}_m)^d) = \sum_{\alpha:|\alpha|=d} \frac{|\alpha|!}{\alpha!} \lambda^\alpha T(\mathbf{x}^\alpha). \qquad (9.14)$$

Proof A straightforward computation that uses the symmetry and d-linearity of T together with the elementary theory of permutations and combinations. We leave the details to the exercises. $\qquad \square$

Corollary 9.9.22 *Fix a basis $\mathcal{V} = \{\mathbf{v}_i\}_{i=1}^m$ of V. If $p \in P^{(d)}(V; W)$ is defined by $p(\mathbf{x}) = T(\mathbf{x}^d)$, where $T \in L_s^d(V; W)$, then*

$$p(\mathbf{x}) = \sum_{\alpha:|\alpha|=d} \frac{|\alpha|!}{\alpha!} a_\alpha x^\alpha,$$

where $(x_1, \cdots, x_m)$ are the coordinates of $\mathbf{x} \in V$ relative to $\mathcal{V}$ and the coefficients $a_\alpha \in W$ are given by $a_\alpha = T(\mathbf{v}^\alpha)$. Conversely, any sum of this type defines a homogeneous polynomial $p : V \to W$.

Proof The first part of the corollary follows from Lemma 9.9.21. For the converse, it is enough (by linearity) to prove that any monomial $x^\alpha \mathbf{w}$, where $\mathbf{w} \in W$, is a polynomial. Observe that $x^\alpha \mathbf{w}$ may be defined by the asymmetric d-linear map

$$S(\mathbf{x}_1, \cdots, \mathbf{x}_d) = \left(\prod_{i=1}^{\alpha_1} x_{1i} \cdots \prod_{i=1}^{\alpha_m} x_{mi} \right) \mathbf{w},$$

where the coordinates of $\mathbf{x}_i$ are $(x_{1i}, \cdots, x_{mi})$. Now define T to be the symmetrization S_s of S (see (9.13)). Obviously $T(\mathbf{x}^d) = x^\alpha \mathbf{w}$. $\qquad \square$

9.9.5 The Polarization Lemma

In this section, which is not needed in the remainder of the chapter, we construct an explicit linear isomorphism between $P^{(d)}(V; W)$ and $L_s^d(V; W)$. We start with a simple example.

Example 9.9.23 Suppose that $T \in L_s^2(V; W)$ and define $p(\mathbf{x}) = T(\mathbf{x}^2)$, $\mathbf{x} \in V$. We may recover T, knowing p, using the identity

$$T(\mathbf{x}, \mathbf{y}) = \frac{1}{8}(p(\mathbf{x} + \mathbf{y}) - p(\mathbf{x} - \mathbf{y}) - p(-\mathbf{x} + \mathbf{y}) + p(-\mathbf{x} - \mathbf{y})).$$

Indeed, since $p(\mathbf{x}) = T(\mathbf{x}, \mathbf{x})$ we have, by the bilinearity of T, $p(\mathbf{x} \pm \mathbf{y}) = T(\mathbf{x}, \mathbf{x}) + T(\mathbf{y}, \mathbf{y}) \pm 2T(\mathbf{x}, \mathbf{y})$, $p(-(\mathbf{x} \pm \mathbf{y})) = p(\mathbf{x} \pm \mathbf{y})$. Of course, this is not the only expression we can use. For example, $(p(\mathbf{x} + \mathbf{y}) - p(\mathbf{x}) - p(\mathbf{y}))/2$ also defines $T(\mathbf{x}, \mathbf{y})$. However, we can and will be able to generalize the first formula to apply to homogeneous polynomials of degree $d > 2$. ♠

Before we state the polarization lemma, we introduce some new notation. Given $d \in \mathbb{N}$, let $\mathbb{S}(d)$ denote the set of all $\epsilon \in \{-1, +1\}^d$. Thus, if $\epsilon \in \mathbb{S}(d)$, we have $\epsilon = (\epsilon_1, \cdots, \epsilon_d)$, where $\epsilon_i = \pm 1$, $1 \leq i \leq d$.

Lemma 9.9.24 (Polarization Lemma) *Let $T \in L_s^d(V; W)$. For all $(\mathbf{x}_1, \cdots, \mathbf{x}_d) \in V^d$, we have*

$$T(\mathbf{x}_1, \cdots, \mathbf{x}_d) = \frac{1}{2^d d!}\left(\sum_{\epsilon \in \mathbb{S}(d)} \epsilon_1 \cdots \epsilon_d T((\epsilon_1 \mathbf{x}_1 + \cdots + \epsilon_d \mathbf{x}_d)^d)\right).$$

Proof We may assume $d > 1$. Expanding $\sum_{\epsilon \in \mathbb{S}(d)} \epsilon_1 \cdots \epsilon_d T((\epsilon_1 \mathbf{x}_1 + \cdots + \epsilon_d \mathbf{x}_d)^d)$ by Lemma 9.9.21, we have to consider the sum over $\epsilon \in \mathbb{S}(d)$ of terms of the form

$$\sum_{\alpha:|\alpha|=d} \frac{d!}{\alpha!}\epsilon_1^{\alpha_1+1} \cdots \epsilon_d^{\alpha_d+1} T(\mathbf{x}_1^{\alpha_1}, \cdots, \mathbf{x}_d^{\alpha_d}).$$

Suppose $\alpha \neq (1, \cdots, 1)$. Then it straightforward to check that at least two of the indices $\alpha_i + 1$ must be odd. Without loss of generality, suppose $\alpha_1 + 1, \alpha_2 + 1$ are odd. Sum over ϵ_1, ϵ_2, keeping the remaining ϵ_i fixed. We obtain a contribution from ϵ_1, ϵ_2 of $\sum_{\epsilon_1=\pm 1, \epsilon_2=\pm 1} \epsilon_1^{\alpha_1+1}\epsilon_2^{\alpha_2+1}$. This sum is zero since $\alpha_1 + 1, \alpha_2 + 1$ are odd. Consequently, if we fix α and sum over $\epsilon \in \mathbb{S}(d)$ we obtain zero. On the other hand, if $\alpha = (1, \cdots, 1)$, then $\epsilon_1^{\alpha_1+1} \cdots \epsilon_d^{\alpha_d+1} = 1$. Summing over $\epsilon \in \mathbb{S}(d)$, we see that the coefficient of $T(\mathbf{x}_1, \cdots, \mathbf{x}_d)$ is $d!2^q$, since $\frac{d!}{\alpha!} = d!$ and $|\mathbb{S}(d)| = 2^d$. □

Proposition 9.9.25 *The map $\chi : L_s^d(V; W) \to P^{(d)}(V, W)$ defined by $\chi(T)(\mathbf{x}) = T(\mathbf{x}^n)$ is a vector space isomorphism with inverse the map $U : P^{(d)}(V, W) \to$*

$L_s^d(V; W)$ *defined by*

$$U(p)(\mathbf{x}_1, \cdots, \mathbf{x}_d) = \frac{1}{2^d d!} \left(\sum_{\epsilon \in \mathbb{S}(d)} \epsilon_1 \cdots \epsilon_d p(\epsilon_1 \mathbf{x}_1 + \cdots + \epsilon_d \mathbf{x}_d) \right),$$

where $p \in P^{(d)}(V, W)$ and $\mathbf{x}_1, \cdots, \mathbf{x}_d \in V$.

Proof The map $\chi : L_s^d(V; W) \to P^{(d)}(V, W)$ is linear and surjective by definition. If $\chi(T) = 0$ then, by the polarization lemma, $T = 0$. Hence χ is injective. We have $U = \chi^{-1}$ by the polarization lemma.   □

EXERCISES 9.9.26

(1) Let $p \in P^d(\mathbb{R}^m, \mathbb{R})$. Show that there exist unique $a_\alpha \in \mathbb{R}$ such that

$$p(\mathbf{x}) = \sum_{|\alpha| \leq d} \frac{|\alpha|!}{\alpha!} a_\alpha x^\alpha, \ \mathbf{x} \in \mathbb{R}^m.$$

(2) Given $p \in P^{(a)}(V, \mathbb{R})$ and $q \in P^{(b)}(V, \mathbb{R})$, define $pq : V \to \mathbb{R}$ by $(pq)(\mathbf{x}) = p(\mathbf{x})q(\mathbf{x})$. Show that $pq \in P^{(a+b)}(V, \mathbb{R})$.

(3) Find symmetric multi-linear maps which define the following homogeneous polynomials on $\mathbb{R}^m$:

   (a) $p(x) = x_1^2 + \cdots + x_m^2$.
   (b) $q(x) = x_i x_j$, where $i \neq j$.

(4) Let $\{e_i\}_{i=1}^m$ denote the standard basis of $\mathbb{R}^m$. Given $T \in L_s^2(\mathbb{R}^m; \mathbb{R})$, define the $m \times m$ matrix $[a_{ij}]$ by $a_{ij} = T(e_i, e_j)$. Show that the matrix $[a_{ij}]$ is symmetric (that is, equal to its transpose).

(5) Let $(V, \| \cdot \|)$ be an m-dimensional normed vector space. Let $V^* = L(V, \mathbb{R})$ denote the dual space of V. If $\mathcal{B} = \{e\}_{i=1}^m$ is a basis of V, let $\mathcal{B}^* = \{e_j^*\}_{j=1}^m$ denote the dual basis of V^* ($e_j^*(e_i) = 0$, $i \neq j$, $e_i^*(e_i) = 1$). If $T \in L^2(V; \mathbb{R})$, let $\hat{T} \in L(V, V^*)$ be the linear map given by Lemma 9.9.8. Show that T is symmetric iff the matrix of $\hat{T}$ (relative to the bases $\mathcal{B}, \mathcal{B}^*$) is symmetric.

(6) For $T \in L_s^d(V; W)$, define the 'polynomial norm' of T by

$$\|T\|' = \sup_{\|\mathbf{x}\|=1} \|T(\mathbf{x}^d)\|.$$

Prove that $\| \cdot \|'$ is related to the norm we defined on $L^d(V; W)$ by

$$\|T\|' \leq \|T\| \leq \frac{n^n}{n!} \|T\|', \text{ for all } T \in L_s^d(V; W).$$

Deduce that $\| \cdot \|'$ defines a norm on $L_s^d(V; W)$. Does $\| \cdot \|'$ define a norm on $L^d(V; W)$, if $d > 1$? (Hint for the first part: use the polarization lemma.)

9.10 Higher-Order Derivatives

In this section, we return to our study of maps $f : U \subset \mathbb{R}^m \to \mathbb{R}^n$ defined on open subsets of $\mathbb{R}^m$. However, everything we do works perfectly well for general finite-dimensional normed vector spaces. In particular, it is a consequence of the result on equivalence of norms (Theorem 9.1.4) that we may choose any norm on $\mathbb{R}^n, \mathbb{R}^m$.

Definition 9.10.1 Let U be an open subset of $\mathbb{R}^m$ and suppose $f : U \to \mathbb{R}^n$ is C^1. We say f is *twice differentiable* at the point $\mathbf{x}_0 \in U$ if the map

$$Df : U \to L(\mathbb{R}^m, \mathbb{R}^n)$$

is differentiable at $\mathbf{x}_0$. We set $D(Df)_{\mathbf{x}_0} = D^2 f_{\mathbf{x}_0}$ and call $D^2 f_{\mathbf{x}_0}$ the *second derivative* of f at $\mathbf{x}_0$.

If f is twice differentiable at $\mathbf{x}_0$, then $D^2 f_{\mathbf{x}_0} \in L(\mathbb{R}^m, L(\mathbb{R}^m, \mathbb{R}^n)) \approx L^2(\mathbb{R}^m; \mathbb{R}^n)$, by Lemma 9.9.8. In the sequel, we almost always regard the second derivative $D^2 f_{\mathbf{x}_0}$ as defining a bilinear map in $L^2(\mathbb{R}^m; \mathbb{R}^n)$.

The differentiability of Df at $\mathbf{x}_0$ implies that we have the equation in $L(\mathbb{R}^m, \mathbb{R}^n)$

$$Df_{\mathbf{x}_0 + h} = Df_{\mathbf{x}_0} + D^2 f_{\mathbf{x}_0}(\mathbf{h}) + o(\mathbf{h}),$$

where $o(\mathbf{h}), D^2 f_{\mathbf{x}_0}(\mathbf{h}) \in L(\mathbb{R}^m, \mathbb{R}^n)$. If we evaluate the equation at $\mathbf{k} \in \mathbb{R}^m$, we obtain

$$Df_{\mathbf{x}_0 + h}(\mathbf{k}) = Df_{\mathbf{x}_0}(\mathbf{k}) + D^2 f_{\mathbf{x}_0}(\mathbf{h})(\mathbf{k}) + o(\mathbf{h})(\mathbf{k})$$
$$= Df_{\mathbf{x}_0}(\mathbf{k}) + D^2 f_{\mathbf{x}_0}(\mathbf{h}, \mathbf{k}) + o(\mathbf{h})(\mathbf{k}),$$

where we have used the natural isomorphism $L(\mathbb{R}^m, L(\mathbb{R}^m, \mathbb{R}^n)) \approx L^2(\mathbb{R}^m; \mathbb{R}^n)$. Since $o(\mathbf{h}) \in L(\mathbb{R}^m, \mathbb{R}^n)$, we have $\|o(\mathbf{h})(\mathbf{k})\| \leq \|o(\mathbf{h})\| \|\mathbf{k}\|$ and so $o(\mathbf{h})(\mathbf{k}) = o(\mathbf{h}, \mathbf{k})$ (note that $\|(\mathbf{h}, \mathbf{k})\| = \max\{\|\mathbf{h}\|, \|\mathbf{k}\|\}$ and so $o(\mathbf{h})(\mathbf{k})/\|(\mathbf{h}, \mathbf{k})\| \to 0$ as $\|(\mathbf{h}, \mathbf{k})\| \to 0$). As a result, we have the equation in $\mathbb{R}^n$

$$Df_{\mathbf{x}_0 + h}(\mathbf{k}) = Df_{\mathbf{x}_0}(\mathbf{k}) + D^2 f_{\mathbf{x}_0}(\mathbf{h}, \mathbf{k}) + o(\mathbf{h}, \mathbf{k}).$$

Definition 9.10.2 (Notation as Above) A map $f : U \subset \mathbb{R}^m \to \mathbb{R}^n$ is twice differentiable on U if f is twice differentiable at every point of U. The map f is C^2 or twice continuously differentiable (on U) if, in addition, $D^2 f : U \to L^2(\mathbb{R}^m; \mathbb{R}^n)$ is continuous.

9.10.1 Second-Order Partial Derivatives

Before we give the relationship between the second derivative and second-order partial derivatives, we prove a useful result that allows us to interchange differentiation with evaluation at a fixed vector.

Lemma 9.10.3 (Evaluation Lemma) *Let* $d, m, p, q \in \mathbb{N}$. *Suppose that* $f : U \subset \mathbb{R}^m \to L^d(\mathbb{R}^p; \mathbb{R}^q)$ *is differentiable at* $\mathbf{x}_0 \in U$. *If we fix* $\mathbf{e}_1, \cdots, \mathbf{e}_d \in \mathbb{R}^p$, *then the map* $f \cdot (\mathbf{e}_1, \cdots, \mathbf{e}_d) : U \to \mathbb{R}^q$ *defined by* $f \cdot (\mathbf{e}_1, \cdots, \mathbf{e}_d)(\mathbf{x}) = f(\mathbf{x})(\mathbf{e}_1, \cdots, \mathbf{e}_d)$ *is differentiable at* $\mathbf{x}_0$ *with derivative given by*

$$D(f \cdot (\mathbf{e}_1, \cdots, \mathbf{e}_d))_{\mathbf{x}_0} = Df_{\mathbf{x}_0} \cdot (\mathbf{e}_1, \cdots, \mathbf{e}_d).$$

If f *is* C^1 *on* U *so is* $f \cdot (\mathbf{e}_1, \cdots, \mathbf{e}_d)$.

Proof Since f is differentiable at $\mathbf{x}_0$, we have

$$f(\mathbf{x}_0 + \mathbf{h}) = f(\mathbf{x}_0) + Df_{\mathbf{x}_0}(\mathbf{h}) + o(\mathbf{h}), \quad \mathbf{x}_0 + \mathbf{h} \in U.$$

This is an equation in $L^d(\mathbb{R}^p; \mathbb{R}^q)$. Evaluating at $\mathbf{E} = (\mathbf{e}_1, \cdots, \mathbf{e}_d)$, we obtain

$$\begin{aligned}(f \cdot \mathbf{E})(\mathbf{x}_0 + \mathbf{h}) &= (f \cdot \mathbf{E})(\mathbf{x}_0) + Df_{\mathbf{x}_0}(\mathbf{h})(\mathbf{E}) + o(\mathbf{h})(\mathbf{E}) \\ &= (f \cdot \mathbf{E})(\mathbf{x}_0) + (Df_{\mathbf{x}_0} \cdot \mathbf{E})(\mathbf{h}) + o(\mathbf{h})(\mathbf{E}).\end{aligned}$$

Since $\|o(\mathbf{h})(\mathbf{E})\| = \|o(\mathbf{h})(\mathbf{e}_1, \cdots, \mathbf{e}_d)\| \leq \|o(\mathbf{h})\| \|\mathbf{e}_1\| \cdots \|\mathbf{e}_p\|$, by our results on multi-linear maps, we have $\|o(\mathbf{h})(\mathbf{e}_1, \cdots, \mathbf{e}_d)\| = o(\mathbf{h})$.   □

Lemma 9.10.4 *Suppose that* $f : U \subset \mathbb{R}^m \to \mathbb{R}^n$ *is twice differentiable at* $\mathbf{x}_0 \in U$. *Then all second-order partial derivatives of* f *exist at* $\mathbf{x}_0$ *and we have*

$$\frac{\partial^2 f}{\partial x_j \partial x_k}(\mathbf{x}_0) = D^2 f_{\mathbf{x}_0}(\mathbf{e}_j, \mathbf{e}_k), \ 1 \leq j, k \leq m,$$

where $\{\mathbf{e}_j\}$ *denotes the standard basis of* $\mathbb{R}^m$. *If* f *is* C^2 *on* U, *then all the second partial derivatives exist and are continuous on* U.

Proof Applying Lemma 9.10.3 to $Df : U \to L(\mathbb{R}^m, \mathbb{R}^n)$, we see that $Df \cdot \mathbf{e}_k$ is differentiable at $\mathbf{x}_0$ with derivative given by

$$D(Df \cdot \mathbf{e}_k)_{\mathbf{x}_0} = D^2 f_{\mathbf{x}_0} \cdot \mathbf{e}_k.$$

Evaluating at $\mathbf{e}_j$, we get $D(Df \cdot \mathbf{e}_k)_{\mathbf{x}_0}(\mathbf{e}_j) = D^2 f_{\mathbf{x}_0}(\mathbf{e}_j, \mathbf{e}_k)$. Since $Df \cdot \mathbf{e}_k = \frac{\partial f}{\partial x_k}$, and $D(Df \cdot \mathbf{e}_k)_{\mathbf{x}_0}(\mathbf{e}_j) = \frac{\partial}{\partial x_j}(\frac{\partial f}{\partial x_k}) = \frac{\partial^2 f}{\partial x_j \partial x_k}$, it follows that $\frac{\partial^2 f}{\partial x_j \partial x_k}(\mathbf{x}_0) = D^2 f_{\mathbf{x}_0}(\mathbf{e}_j, \mathbf{e}_k)$.   □

We now come to the main theorem of this section: the symmetry of the second derivative.

Theorem 9.10.5 *If* $f : U \subset \mathbb{R}^m \to \mathbb{R}^n$ *is twice differentiable at* $\mathbf{x}_0 \in U$, *then*

$$D^2 f_{\mathbf{x}_0} \in L_s^2(\mathbb{R}^m; \mathbb{R}^n).$$

Proof We have to show that $D^2 f_{\mathbf{x}_0}(\mathbf{h}, \mathbf{k}) = D^2 f_{\mathbf{x}_0}(\mathbf{k}, \mathbf{h})$ for all $\mathbf{h}, \mathbf{k} \in \mathbb{R}^m$. Fix $d > 0$ so that $\overline{D}_{2d}(\mathbf{x}_0) \subset U$ and assume in what follows that $\mathbf{h}, \mathbf{k} \in \overline{D}_d(\mathbf{0})$. Taking the product norm on $\mathbb{R}^m \times \mathbb{R}^m$, we have $\|(\mathbf{h}, \mathbf{k})\| = \max\{\|\mathbf{h}\|, \|\mathbf{k}\|\}$. Define the map $S : \overline{D}_r(\mathbf{0}) \times \overline{D}_r(\mathbf{0}) \to \mathbb{R}^n$ by

$$S(\mathbf{h}, \mathbf{k}) = f(\mathbf{x}_0 + \mathbf{h} + \mathbf{k}) - f(\mathbf{x}_0 + \mathbf{h}) - f(\mathbf{x}_0 + \mathbf{k}) + f(\mathbf{x}_0).$$

Clearly S is symmetric: $S(\mathbf{h}, \mathbf{k}) = S(\mathbf{k}, \mathbf{h})$ for all $\mathbf{h}, \mathbf{k} \in \overline{D}_d(\mathbf{0})$. We prove that if

$$\tau(\mathbf{h}, \mathbf{k}) = S(\mathbf{h}, \mathbf{k}) - D^2 f_{\mathbf{x}_0}(\mathbf{h}, \mathbf{k}),$$

then $\tau(\mathbf{h}, \mathbf{k}) = o(\|\mathbf{h}, \mathbf{k}\|^2)$. Specifically, we show that if $\varepsilon > 0$, then there exists a $d_1 \in (0, d]$ such that $\|\tau(\mathbf{h}, \mathbf{k})\| \leq 4\varepsilon \|(\mathbf{h}, \mathbf{k})\|^2$, for all $\|(\mathbf{h}, \mathbf{k})\| \leq d_1$. It will then follow easily from the symmetry of S and the bilinearity of $D^2 f_{\mathbf{x}_0}$ that $D^2 f_{\mathbf{x}_0} \in L_s^2(\mathbb{R}^m; \mathbb{R}^n)$.

Define $g : [0, 1] \to \mathbb{R}^n$ by

$$g(t) = f(\mathbf{x}_0 + \mathbf{h} + t\mathbf{k}) - f(\mathbf{x}_0 + t\mathbf{k}) - tD^2 f_{\mathbf{x}_0}(\mathbf{h}, \mathbf{k}), \ t \in [0, 1].$$

Observe that

$$g(1) - g(0) = S(\mathbf{h}, \mathbf{k}) - D^2 f_{\mathbf{x}_0}(\mathbf{h}, \mathbf{k}).$$

Since g is continuous on $[0, 1]$ and differentiable on $(0, 1)$, it follows by the mean value theorem (Theorem 9.4.12) that

$$\|g(1) - g(0)\| \leq \sup_{t \in (0,1)} \|g'(t)\|.$$

Computing $g'(t)$, we find

$$\begin{aligned}
g'(t) &= Df_{\mathbf{x}_0 + t\mathbf{k} + \mathbf{h}}(\mathbf{k}) - Df_{\mathbf{x}_0 + t\mathbf{k}}(\mathbf{k}) - D^2 f_{\mathbf{x}_0}(\mathbf{h}, \mathbf{k}) \\
&= (Df_{\mathbf{x}_0 + t\mathbf{k} + \mathbf{h}}(\mathbf{k}) - Df_{\mathbf{x}_0}(\mathbf{k})) \\
&\quad - (Df_{\mathbf{x}_0 + t\mathbf{k}}(\mathbf{k}) - Df_{\mathbf{x}_0}(\mathbf{k})) - D^2 f_{\mathbf{x}_0}(\mathbf{h}, \mathbf{k}).
\end{aligned}$$

Since Df is differentiable at $\mathbf{x}_0$, given $\varepsilon > 0$, there exists a $d_1 \in (0, d]$ such that if $\|\mathbf{u}\| \leq 2d_1$,

$$Df_{\mathbf{x}_0 + \mathbf{u}} = Df_{\mathbf{x}_0} + D^2 f_{\mathbf{x}_0}(\mathbf{u}) + r(\mathbf{u}),$$

where $\|r(\mathbf{u})\| \leq \varepsilon \|\mathbf{u}\|$. Evaluating this equation in $L(\mathbb{R}^m, \mathbb{R}^n)$ at $\mathbf{k} \in \mathbb{R}^m$, we have

$$Df_{\mathbf{x}_0 + \mathbf{u}}(\mathbf{k}) = Df_{\mathbf{x}_0}(\mathbf{k}) + D^2 f_{\mathbf{x}_0}(\mathbf{u}, \mathbf{k}) + r(\mathbf{u}, \mathbf{k}),$$

where $\|r(\mathbf{u}, \mathbf{k})\| \leq \varepsilon \|\mathbf{u}\| \|\mathbf{k}\|$, for all $\|(\mathbf{u}, \mathbf{k})\| \leq d_1$. Substituting in our expression for $g'(t)$ gives

$$g'(t) = D^2 f_{\mathbf{x}_0}(t\mathbf{k} + \mathbf{h}, \mathbf{k}) - D^2 f_{\mathbf{x}_0}(t\mathbf{k}, \mathbf{k}) - D^2 f_{\mathbf{x}_0}(\mathbf{h}, \mathbf{k})$$
$$+ r(t\mathbf{k} + \mathbf{h}, \mathbf{k}) - r(t\mathbf{k}, \mathbf{k})$$
$$= r(t\mathbf{k} + \mathbf{h}, \mathbf{k}) - r(t\mathbf{k}, \mathbf{k}),$$

where the first three terms cancel using the bilinearity of $D^2 f_{\mathbf{x}_0}$. Estimating $\|g'(t)\|$ we see that

$$\|g'(t)\| \leq \|r(t\mathbf{k} + \mathbf{h}, \mathbf{k})\| + \|r(t\mathbf{k}, \mathbf{k})\|$$
$$\leq \varepsilon \|\mathbf{k}\| (\|t\mathbf{k} + \mathbf{h}\| + \|t\mathbf{k}\|), \text{ if } \|(\mathbf{h}, \mathbf{k})\| \leq d_1$$
$$\leq 2\varepsilon \|\mathbf{k}\| (\|\mathbf{h}\| + \|\mathbf{k}\|)$$
$$\leq 4\varepsilon \|(\mathbf{h}, \mathbf{k})\|^2.$$

Since $\|S(\mathbf{h}, \mathbf{k}) - D^2 f_{\mathbf{x}_0}(\mathbf{h}, \mathbf{k})\| \leq \sup_{t \in (0,1)} \|g'(t)\|$, we have the estimate

$$\|S(\mathbf{h}, \mathbf{k}) - D^2 f_{\mathbf{x}_0}(\mathbf{h}, \mathbf{k})\| \leq 4\varepsilon \|(\mathbf{h}, \mathbf{k})\|^2, \quad \|(\mathbf{h}, \mathbf{k})\| \leq d_1.$$

Since $S(\mathbf{h}, \mathbf{k}) = S(\mathbf{k}, \mathbf{h})$, an application of the triangle inequality yields

$$\|D^2 f_{\mathbf{x}_0}(\mathbf{h}, \mathbf{k}) - D^2 f_{\mathbf{x}_0}(\mathbf{k}, \mathbf{h})\| \leq 8\varepsilon \|(\mathbf{h}, \mathbf{k})\|^2, \quad \|(\mathbf{h}, \mathbf{k})\| \leq d_1.$$

The bilinearity of $D^2 f_{\mathbf{x}_0}$ implies that this estimate holds for all $(\mathbf{h}, \mathbf{k}) \in \mathbb{R}^m \times \mathbb{R}^m$. Since $\varepsilon > 0$ was arbitrary, we have $D^2 f_{\mathbf{x}_0}(\mathbf{h}, \mathbf{k}) = D^2 f_{\mathbf{x}_0}(\mathbf{k}, \mathbf{h})$ for all $(\mathbf{h}, \mathbf{k}) \in \mathbb{R}^m \times \mathbb{R}^m$.   □

Remark 9.10.6 The assumptions of Theorem 9.10.5 are both natural and minimal: the map f is twice differentiable at $\mathbf{x}_0$. No assumptions about continuity or the existence of the second derivative on a neighbourhood of $\mathbf{x}_0$ are required. Note again the central role of the mean value theorem in the proof.   ✱

Corollary 9.10.7 *If $f = (f_1, \cdots, f_n) : U \subset \mathbb{R}^m \to \mathbb{R}^n$ is twice differentiable at $\mathbf{x}_0$, then for all $1 \leq i, j \leq m$, $1 \leq \ell \leq n$ we have*

$$\frac{\partial^2 f_\ell}{\partial x_i \partial x_j}(\mathbf{x}_0) = \frac{\partial^2 f_\ell}{\partial x_j \partial x_i}(\mathbf{x}_0).$$

(Symmetry of second-order partial derivatives.)

Proof Theorem 9.10.5, Lemma 9.10.4 and Corollary 9.5.2.   □

9.10.2 Higher-Order Derivatives: General Case

Let $1 \leq p \leq \infty$ and suppose $U \subset \mathbb{R}^m$ is open and $f : U \subset \mathbb{R}^m \to \mathbb{R}^n$. Proceeding inductively, suppose that f is $(p-1)$ times continuously differentiable on U with associated $(p-1)th$ derivative map $D^{p-1}f : U \to L^{p-1}(\mathbb{R}^m; \mathbb{R}^n)$. The map f is *p-times differentiable* at $\mathbf{x}_0 \in U$, if $D^{p-1}f : U \to L^{p-1}(\mathbb{R}^m; \mathbb{R}^n)$ is differentiable at $\mathbf{x}_0$. As usual, we regard $D(D^{p-1}f)_{\mathbf{x}_0} \in L(\mathbb{R}^m, L^{p-1}(\mathbb{R}^m; \mathbb{R}^n))$ as defining an element of $L^p(\mathbb{R}^m; \mathbb{R}^n)$ via the natural isomorphism $L(\mathbb{R}^m, L^{p-1}(\mathbb{R}^m; \mathbb{R}^n)) \approx L^p(\mathbb{R}^m; \mathbb{R}^n)$ and set $D(D^{p-1}f)_{\mathbf{x}_0} = D^p f_{\mathbf{x}_0}$. If f is p-times differentiable at every point of U, then f is p times differentiable on U and if $D^p f : U \to L^p(\mathbb{R}^m; \mathbb{R}^n)$ is continuous we say f is *p times continuously differentiable*, or C^p, on U. If f is C^p for all $p \in \mathbb{N}$, we say f is *infinitely differentiable* or C^∞.

Examples 9.10.8

(1) If $A \in L(\mathbb{R}^m, \mathbb{R}^n)$, then A is C^∞ and $D^p A = 0, p > 1$.
(2) If $T \in L^d_s(\mathbb{R}^m; \mathbb{R}^n)$ and we define $p \in P^{(d)}(\mathbb{R}^m, \mathbb{R}^n)$ by $p(\mathbf{x}) = T(\mathbf{x}^d)$, then p is C^∞ and

$$D^r p_{\mathbf{x}}(\mathbf{e}_1, \cdots, \mathbf{e}_r) = \begin{cases} \frac{d!}{(d-r)!} T(\mathbf{e}_1, \cdots, \mathbf{e}_r, \mathbf{x}^{d-r}), & r \leq d, \\ 0, & r > d. \end{cases}$$

The proof is a straightforward induction on p (the case $p = 1$ is Examples 9.9.17(3)).
(3) Let $\phi : \mathbb{R}^m \times \mathbb{R}^n \to \mathbb{R}^p$ be a bilinear map. Then ϕ is C^∞ and for $(\mathbf{x}, \mathbf{y}) \in \mathbb{R}^m \times \mathbb{R}^n$, $(\mathbf{e}_1, \mathbf{f}_1), (\mathbf{e}_2, \mathbf{f}_2) \in \mathbb{R}^m \times \mathbb{R}^n$ we have

$$D_1 \phi_{(\mathbf{x}, \mathbf{y})}(\mathbf{e}_1) = \phi(\mathbf{e}_1, \mathbf{y}),$$

$$D_2 \phi_{(\mathbf{x}, \mathbf{y})}(\mathbf{f}_1) = \phi(\mathbf{x}, \mathbf{f}_1),$$

$$D^2 \phi_{(\mathbf{x}, \mathbf{y})}((\mathbf{e}_1, \mathbf{f}_1), (\mathbf{e}_2, \mathbf{f}_2)) = \phi(\mathbf{e}_1, \mathbf{f}_2) + \phi(\mathbf{e}_2, \mathbf{f}_1),$$

$$D^r \phi_{(\mathbf{x}, \mathbf{y})} = 0, \quad r > 2.$$

The result is easily proved directly or by using Proposition 9.5.3. ♠

Lemma 9.10.9 *Let* $f : U \subset \mathbb{R}^m \to \mathbb{R}^n$ *be p-times differentiable at* $\mathbf{x}_0 \in U$ *and* $\mathbf{e}_2, \cdots, \mathbf{e}_p \in \mathbb{R}^m$. *If*

$$D^{p-1}f \cdot (\mathbf{e}_2, \cdots, \mathbf{e}_p) : U \to \mathbb{R}^n$$

is the map defined by

$$(D^{p-1}f \cdot (\mathbf{e}_2, \cdots, \mathbf{e}_p))(\mathbf{x}) = D^{p-1}f_{\mathbf{x}}(\mathbf{e}_2, \cdots, \mathbf{e}_p), \quad \mathbf{x} \in U,$$

then $D^{p-1}f \cdot (\mathbf{e}_2, \cdots, \mathbf{e}_p)$ is differentiable at $\mathbf{x}_0$ with derivative given by

$$D(D^{p-1}f \cdot (\mathbf{e}_2, \cdots, \mathbf{e}_p))_{\mathbf{x}_0}(\mathbf{e}_1) = D^p f_{\mathbf{x}_0}(\mathbf{e}_1, \cdots, \mathbf{e}_p), \quad \mathbf{e}_1 \in \mathbb{R}^m.$$

Proof Apply the evaluation lemma (Lemma 9.10.3) to $D^{p-1}f$. □

Theorem 9.10.10 (Symmetry of Higher-Order Derivatives) *If $f : U \subset \mathbb{R}^m \to \mathbb{R}^n$ is p-times differentiable at $\mathbf{x}_0$, then $D^p f_{\mathbf{x}_0} \in L_s^p(\mathbb{R}^m; \mathbb{R}^n)$.*

Proof We prove the result by induction on p. The result is true when $p = 2$—Theorem 9.10.5. Suppose the result is proved for derivatives of order less than or equal to $p-1$. Let $\mathbf{e}_1, \cdots, \mathbf{e}_p \in \mathbb{R}^m$. By Lemma 9.10.9, the map $D^{p-1}f \cdot (\mathbf{e}_2, \cdots, \mathbf{e}_p) : U \to \mathbb{R}^n$ is differentiable at $\mathbf{x}_0$ with derivative given by

$$D(D^{p-1}f \cdot (\mathbf{e}_2, \cdots, \mathbf{e}_p))_{\mathbf{x}_0}(\mathbf{e}_1) = D^p f_{\mathbf{x}_0}(\mathbf{e}_1, \cdots, \mathbf{e}_p).$$

By the inductive hypothesis, $D_{\mathbf{x}_0}^p$ is symmetric in the last $p - 1$ variables. By Lemma 9.10.9 again, the map $D^{p-2}f \cdot (\mathbf{e}_3, \cdots, \mathbf{e}_p) : U \to \mathbb{R}^n$ is twice differentiable at $\mathbf{x}_0$ and

$$D^2(D^{p-2}f \cdot (\mathbf{e}_3, \cdots, \mathbf{e}_p))_{\mathbf{x}_0}(\mathbf{e}_1, \mathbf{e}_2) = D^p f_{\mathbf{x}_0}(\mathbf{e}_1, \cdots, \mathbf{e}_p).$$

By Theorem 9.10.5, $D^p f_{\mathbf{x}_0}$ is symmetric in the first two variables. Combining this with the symmetry in the last $p - 1$ variables, it follows that $D^p f_{\mathbf{x}_0}$ is symmetric. □

Our final result in this section shows the relationship between higher-order derivatives and higher-order partial derivatives.

Theorem 9.10.11 *Let $f : U \subset \mathbb{R}^m \to \mathbb{R}^n$ be p-times differentiable at $\mathbf{x}_0$ and $\{\mathbf{e}_i\}_{i=1}^m$ denote the standard basis of $\mathbb{R}^m$. Given $\alpha \in \mathbb{N}^m$, we have*

(1) $\partial^\alpha f(\mathbf{x}_0) \stackrel{\text{def}}{=} \frac{\partial^{|\alpha|} f}{\partial x_1^{\alpha_1} \cdots \partial x_m^{\alpha_m}}(\mathbf{x}_0) = D^{|\alpha|} f_{\mathbf{x}_0}(\mathbf{e}^\alpha).$

(2) *The higher-order partial derivatives of f are independent of the order of differentiation.*

(3) *If all the partial derivatives $\partial^\alpha f$ exist and are continuous on U, $|\alpha| \le p$, then f is C^p.*

Proof Part (1) follows by definition of partial derivative, and (2,3) follow by induction on p, Corollary 9.5.6 and Theorem 9.10.10. □

9.11 Extension of Results from C^1 to C^r-Maps

In this section we extend some of the main results proved in Sects. 9.4.3 and 9.6 to C^r-maps. We conclude with statements and proofs of a version of Leibniz' law for the rth derivative of a product and a vector-valued version of Faà di Bruno's formula for the rth derivative of a composite of vector-valued functions.

We start with the C^r version of the chain rule.

Theorem 9.11.1 *Let $r \geq 1$ and suppose that $f : U \subset \mathbb{R}^m \to \mathbb{R}^n$, $g : V \subset \mathbb{R}^n \to \mathbb{R}^p$ are C^r. Then $gf : U \cap f^{-1}(V) \to \mathbb{R}^p$ is C^r.*

Proof We indicate two proofs.

Method 1: Induction on r. The case $r = 1$ is Theorem 9.4.8. Let $\phi : L(\mathbb{R}^n, \mathbb{R}^p) \times L(\mathbb{R}^m, \mathbb{R}^n) \to L(\mathbb{R}^m, \mathbb{R}^p)$ be the bilinear map defined by $\phi(A, B) = A \circ B$. Since ϕ is bilinear, ϕ is C^∞ (Examples 9.10.8(3)). Suppose the result has been proved for $r - 1$ (where $r \geq 2$). The map $(Dg)f : U \cap f^{-1}(V) \to L(\mathbb{R}^n, \mathbb{R}^p)$ is a composition of C^{r-1} maps and so, by the inductive hypothesis, $(Dg)f$ is C^{r-1}. We have $D(gf) = \phi((Dg)f, Df)$. Since ϕ is C^∞ and $(Dg)f$ and Df are C^{r-1}, it follows by the inductive hypothesis that $D(gf)$ is C^{r-1}. Hence gf is C^r.

Method 2: Coordinates and partial derivatives. An inductive argument shows that an sth order partial derivative of the composite gf will be a polynomial of degree s in the partial derivatives of f of order at most s with coefficients depending linearly on partial derivatives of g of order less than or equal s. If f and g are C^r it follows that all partial derivatives of gf of order less than or equal to r exist and are continuous. Now apply Theorem 9.10.11(3).   □

9.11.1   The Inverse and Implicit Function Theorems

Let $U \subset \mathbb{R}^m$, $V \subset \mathbb{R}^m$ be open non-empty sets. Suppose that $1 \leq r \leq \infty$. A map $f : U \to V$ is a C^r *diffeomorphism* (of U onto V) if

(1) f is 1:1 onto.
(2) Both f and f^{-1} are C^r.

As we showed earlier, if $f : U \to V$ is a C^r diffeomorphism then $Df^{-1}_{f(\mathbf{x})} = (Df_{\mathbf{x}})^{-1}$ for all $\mathbf{x} \in U$.

We start with a simple extension of Lemma 9.6.2.

Lemma 9.11.2 *The map $\beta : \mathrm{GL}(\mathbb{R}, n) \to \mathrm{GL}(\mathbb{R}, n)$, $\beta(A) = A^{-1}$, is C^∞.*

Proof The proof of Lemma 9.6.2 already shows that partial derivatives of all orders of β with respect to the components a_{ij} of A exist and are continuous. The result follows by Theorem 9.10.11.   □

Lemma 9.11.3 *Let $U \subset \mathbb{R}^m$, $V \subset \mathbb{R}^m$ be open non-empty sets. Suppose that $f : U \to V$ is a C^r-map, $r \geq 1$, which is a C^1 diffeomorphism. Then f is a C^r diffeomorphism.*

Proof The proof is by induction on r. Suppose the result is true for $r - 1$. We may write $D(f^{-1})$ as the composite $\beta(Df)f^{-1}$ (that is, $(Df^{-1})_{\mathbf{y}} = (Df_{f^{-1}(\mathbf{y})})^{-1}$ for all $\mathbf{y} \in V$). Since β is C^∞, Df is C^{r-1}, and f^{-1} is C^{r-1} (inductive hypothesis), Theorem 9.11.1 implies that $D(f^{-1})$ is C^{r-1}. Hence, f^{-1} is C^r.   □

Theorem 9.11.4 (The Inverse Function Theorem for C^r-Maps) *Let W be an open subset of $\mathbb{R}^m$ and $f : W \to \mathbb{R}^m$ be C^r, where $1 \le r \le \infty$. If $Df_{\mathbf{x}_0}$ is invertible at $\mathbf{x}_0 \in W$, then we can find open neighbourhoods $U \subset W$ of $\mathbf{x}_0$, and V of $f(\mathbf{x}_0)$ such that*

(1) f maps U 1:1 onto V. In particular, $V = f(U)$ is open.
(2) $f : U \to V$ is a C^r diffeomorphism.

Proof Immediate from the C^1 version of the inverse function theorem (Theorem 9.6.5) and Lemma 9.11.3.   □

Once we have the C^r version of the inverse function theorem, the proofs of the implicit function theorem, dual implicit function and rank theorem all extend immediately to give C^r versions of these results.

9.12 Taylor's Theorem

In this section we prove versions of Taylor's theorem for vector-valued maps.

Theorem 9.12.1 (Taylor's Theorem, Version 1) *Let $U \subset \mathbb{R}^m$ be open and suppose $f : U \to \mathbb{R}^n$ is p-times differentiable at $\mathbf{x} \in U$. If we define the remainder term $r(\mathbf{h})$ by*

$$r(\mathbf{h}) = f(\mathbf{x} + \mathbf{h}) - \sum_{r=0}^{p} \frac{1}{r!} D^r f_{\mathbf{x}}(\mathbf{h}^r),$$

then $r(\mathbf{h}) = o(\|\mathbf{h}\|^p)$. That is, the function $T^r f_{\mathbf{x}}(\mathbf{h}) = \sum_{r=0}^{p} \frac{1}{r!} D^r f_{\mathbf{x}}(\mathbf{h}^r)$ gives an approximation to f at $\mathbf{x}$ of order p.

Proof The proof is by induction on p. The result is true for $p = 1$ by the definition of derivative. Suppose the result is true for $p - 1, p \ge 2$. Fix $d > 0$ so that $D_d(\mathbf{x}) \subset U$. Define $S : D_d(\mathbf{0}) \to \mathbb{R}^n$ by

$$S(\mathbf{k}) = f(\mathbf{x} + \mathbf{k}) - f(\mathbf{x}).$$

Since S is p-times differentiable at $\mathbf{0}$ we have

$$D^r S_0 = D^r f_{\mathbf{x}}, \ 1 \le r \le p.$$

Substituting in the defining equation for $r(\mathbf{h})$, we have

$$S(\mathbf{k}) = \sum_{j=1}^{p} \frac{1}{j!} D^j S_0(\mathbf{k}^j) + r(\mathbf{k}).$$

Differentiating with respect to $\mathbf{k}$ and setting $DS_{\mathbf{h}} = g(\mathbf{h})$ gives

$$g(\mathbf{h}) = \sum_{j=0}^{p-1} \frac{1}{j!} D^j g_0(\mathbf{h}^j) + Dr_{\mathbf{h}}.$$

Hence, by the inductive hypothesis, given $\varepsilon > 0$, there exists a $\delta > 0$ such that if $\|h\| < \delta$, then $\|Dr_{\mathbf{h}}\| \leq \varepsilon \|\mathbf{h}\|^{p-1}$. By the mean value theorem we have

$$\|r(\mathbf{h})\| = \|r(\mathbf{h}) - r(\mathbf{0})\| \leq \|\mathbf{h}\| \sup_{0<t<1} \|Dr_{t\mathbf{h}}\|.$$

Since $\|Dr_{t\mathbf{h}}\| \leq \varepsilon \|t\mathbf{h}\|^{p-1} \leq \varepsilon \|\mathbf{h}\|^{p-1}$, we have

$$\|r(\mathbf{h})\| \leq \varepsilon \|\mathbf{h}\|^p, \quad \|\mathbf{h}\| \leq \delta.$$

Since $\varepsilon > 0$ was arbitrary, the result follows. $\qquad\square$

Corollary 9.12.2 *Let U be an open subset of $\mathbb{R}^m$ and suppose that $f : U \to \mathbb{R}$ is p times differentiable at $\mathbf{x}$. We have*

$$f(\mathbf{x} + \mathbf{h}) = f(\mathbf{x}) + \sum_{r=1}^{p} \left(\sum_{|\alpha|=r} \frac{|\alpha|!}{\alpha!} \partial^\alpha f(\mathbf{x}) \mathbf{h}^\alpha \right) + r(\mathbf{h}),$$

where $\mathbf{x} + \mathbf{h} \in U$ and $r(\mathbf{h}) = o(\|\mathbf{h}\|^p)$.

Remark 9.12.3 In classical partial derivative notation, the expression for $f(\mathbf{x} + \mathbf{h})$ given by Corollary 9.12.2 is

$$f(\mathbf{x}) + \sum_{r=1}^{p} \left(\sum_{|\alpha|=r} \frac{|\alpha|!}{\alpha!} \frac{\partial^r f}{\partial x_1^{\alpha_1} \cdots \partial x_m^{\alpha_m}} (\mathbf{x}) h_1^{\alpha_1} \cdots h_m^{\alpha_m} \right) + r(\mathbf{h}).$$

✠

Example 9.12.4 We find the quadratic approximation at $\mathbf{x} = \mathbf{0}$ given by Taylor's theorem to $f(x_1, x_2, x_3) = e^{x_1} \cos x_2 \sin x_3$. Listing the partial derivatives of order at most 2 at $\mathbf{x} = \mathbf{0}$ we have

$$f(\mathbf{0}) = 0,$$

$$\frac{\partial f}{\partial x_1}(\mathbf{0}) = \frac{\partial f}{\partial x_2}(\mathbf{0}) = 0, \quad \frac{\partial f}{\partial x_3}(\mathbf{0}) = 1,$$

$$\frac{\partial^2 f}{\partial x_i \partial x_j}(\mathbf{0}) = 0, \text{ if } (i,j) \neq (1,3), (3,1), \quad \frac{\partial^2 f}{\partial x_1 \partial x_3}(\mathbf{0}) = 1.$$

Applying Corollary 9.12.2, we see that

$$Q(x_1, x_2, x_3) = x_3 + x_1 x_3$$

is the quadratic approximation at $\mathbf{x} = \mathbf{0}$ given by Taylor's theorem.   ♠

Theorem 9.12.5 (Scalar-Valued Taylor's Theorem) *Let $U \subset \mathbb{R}^m$ be open and suppose $f : U \to \mathbb{R}$ is p-times differentiable on the line segment $[\mathbf{x}, \mathbf{x} + \mathbf{h}] \subset U$. Then there exists a $\theta \in (0, 1)$ such that*

$$f(\mathbf{x} + \mathbf{h}) = \sum_{r=0}^{p-1} \frac{1}{r!} D^r f_{\mathbf{x}}(\mathbf{h}^r) + \frac{1}{p!} D^p f_{\mathbf{x}+\theta\mathbf{h}}(\mathbf{h}^p).$$

Proof Define $\phi : [0, 1] \to U \subset \mathbb{R}^m$ by $\phi(t) = \mathbf{x} + t\mathbf{h}$, $t \in [0, 1]$. If we set $F = f\phi : [0, 1] \to \mathbb{R}$, then F is p-times differentiable on $(0, 1)$ and

$$F^{(r)}(t) = D^r f_{\mathbf{x}+t\mathbf{h}}(\mathbf{h}^r), \ r = 1, \cdots, p.$$

Applying the classical Taylor's theorem (Theorem 2.7.11(a)) to F gives the result.   □

We conclude this section with an explicit, and very useful, form for the remainder term in Taylor's theorem. In this case we assume f is C^p on a neighbourhood of $\mathbf{x}$.

Theorem 9.12.6 (Taylor's Theorem, Integral Remainder) *Let $U \subset \mathbb{R}^m$ be open and $f : U \to \mathbb{R}^n$ be C^p. If $[\mathbf{x}, \mathbf{x} + \mathbf{h}] \subset U$, then*

$$f(\mathbf{x} + \mathbf{h}) = \sum_{r=0}^{p-1} \frac{1}{r!} D^r f_{\mathbf{x}}(\mathbf{h}^r) + \int_0^1 \frac{(1-s)^{p-1}}{(p-1)!} D^p f_{\mathbf{x}+s\mathbf{h}}(\mathbf{h}^p) \, ds.$$

Proof The integral $\int_0^1 \frac{(1-s)^{p-1}}{(p-1)!} D^p f_{\mathbf{x}+s\mathbf{h}}(\mathbf{h}^p) \, ds$ is defined to be the integral of the components of $\frac{(1-s)^{p-1}}{(p-1)!} D^p f_{\mathbf{x}+s\mathbf{h}}(\mathbf{h}^p) \in \mathbb{R}^n$. The result is proved by a simple induction on p and uses integration by parts. We leave the details to the exercises.   □

EXERCISES 9.12.7

(1) Find the cubic approximation at $\mathbf{x} = \mathbf{0}$ given by Taylor's theorem to $f(x_1, x_2) = (\sin(x + y^2), ye^x, e^{x^2+y})$.
(2) What is the Taylor series at $\mathbf{x} = \mathbf{0}$ of $f(x, y) = \Phi(x)\Phi(y)$, where Φ is the C^∞-function defined in Proposition 5.2.3.
(3) Show that if we assume the conditions of Theorem 9.12.6 and define the remainder

$$r(\mathbf{h}) = f(\mathbf{x} + \mathbf{h}) - \left(\sum_{r=0}^{p-1} \frac{1}{r!} D^r f_{\mathbf{x}}(\mathbf{h}^r) \right),$$

then $r(\mathbf{h}) = o(\|\mathbf{h}\|^\alpha)$ for all $\alpha \in (0, p)$.
(4) Complete the proof of Theorem 9.12.6.

9.13 The Leibniz Rule and Faà di Bruno's Formula

Recall that if $f, g : \mathbb{R} \to \mathbb{R}$ are C^r and we denote the product of f and g by $f \cdot g$, then the Leibniz rule states that $f \cdot g$ is C^r with derivative given by

$$(f \cdot g)^{(r)}(x) = \sum_{j=0}^{r} \binom{r}{j} f^{(j)}(x) g^{(r-j)}(x).$$

Suppose now that $f, g : U \subset \mathbb{R}^m \to \mathbb{R}$ are C^r, $r \geq 1$. Using either Exercises 9.4.15(6) or a simple argument based on the chain rule $(f \cdot g = \phi(f(\mathbf{x}), g(\mathbf{x}))$, where $\phi(x, y) = xy)$, we find that $f \cdot g$ is differentiable with derivative

$$D(f \cdot g)_\mathbf{x} = g(\mathbf{x}) Df_\mathbf{x} + f(\mathbf{x}) Dg_\mathbf{x}, \ \mathbf{x} \in U.$$

When we come to the second derivative of $f \cdot g$, the terms involving first derivatives of f and g $(Df_\mathbf{x} \cdot Dg_\mathbf{x}$ and $Dg_\mathbf{x} \cdot Df_\mathbf{x})$ will not be symmetric. Similar problems arise for all higher derivatives. The way we handle this is to symmetrize the terms involving derivatives of both f and g.

For $p, q \geq 0$, we define the symmetrization operator $\star$: $L_s^p(\mathbb{R}^m; \mathbb{R}) \times L_s^q(\mathbb{R}^m; \mathbb{R}) \to L_s^{p+q}(\mathbb{R}^m; \mathbb{R})$ by

$$(A \star B)(\mathbf{e}_1, \cdots, \mathbf{e}_{p+q})$$

$$= \frac{1}{(p+q)!} \sum_1 A(\mathbf{e}_{\sigma(1)}, \cdots, \mathbf{e}_{\sigma(p)}) B(\mathbf{e}_{\sigma(p+1)}, \cdots, \mathbf{e}_{\sigma(p+q)})$$

$$= \frac{p!q!}{(p+q)!} \sum_2 A(\mathbf{e}_{\sigma(1)}, \cdots, \mathbf{e}_{\sigma(p)}) B(\mathbf{e}_{\sigma(p+1)}, \cdots, \mathbf{e}_{\sigma(p+q)}),$$

where $\sum_1$ is the sum over all permutations $\sigma \in S_{p+q}$ and $\sum_2$ is the sum over all permutations $\sigma \in S_{p+q}$ such that $\sigma(1) < \sigma(2) < \cdots < \sigma(p)$ and $\sigma(p+1) < \cdots < \sigma(p+q)$.

Remarks 9.13.1

(1) $A \star B = B \star A$, for all $A \in L_s^p(\mathbb{R}^m; \mathbb{R})$, $B \in L_s^q(\mathbb{R}^m; \mathbb{R})$.
(2) If $q = 0$, then $B \in \mathbb{R}$ and $A \star B = BA \in L_s^p(\mathbb{R}^m; \mathbb{R})$. ✠

We may now give the general version of the Leibniz rule.

Theorem 9.13.2 (Leibniz Rule) *Let $f, g : U \subset \mathbb{R}^m \to \mathbb{R}$ be C^r, $r \geq 1$. Then $f \cdot g$ is C^r with derivative given by*

$$D^r(f \cdot g)_\mathbf{x} = \sum_{j=0}^{r} \binom{r}{j} D^j f_\mathbf{x} \star D^{(r-j)} g_\mathbf{x}, \ \mathbf{x} \in U.$$

Proof The proof is a simple induction on r using Lemma 9.10.9. We leave the details
(and generalizations) to the exercises.   □

9.13.1 Faà di Bruno's Formula

In this section we give a formula for the rth derivative of a composite of vector-
valued maps. In the case of real-valued maps, there is the following result attributed
to Faà di Bruno [7] (see the historical notes at the end of the section).

Theorem 9.13.3 (Faà di Bruno's Formula) *Let $f, g : \mathbb{R} \to \mathbb{R}$ be C^r, $r \geq 1$. We
have*

$$(gf)^{(r)}(x) = \sum \frac{r!}{q_1! \cdots q_r!} g^{(q)}(f(x)) \prod_{j=1}^{r} \left(\frac{f^{(j)}(x)}{j!} \right)^{q_j},$$

where the sum is over all $(q_1, \cdots, q_r) \in \mathbb{Z}_+^r$ satisfying $\sum_{j=1}^{r} jq_j = r$ and $q_1 + \cdots + q_r = q$.

We shall give three versions of Faà di Bruno's formula for vector-valued maps.
Before stating these we need to define some new symmetrization operators. Fix $r \in
\mathbb{N}$ and suppose we are given integers $q_1, \cdots, q_r \geq 0$ such that $r = q_1 + 2q_2 + \cdots + rq_r$.
Set $q = q_1 + \cdots + q_r$ and note that $1 \leq q \leq r$. Set $M_0 = 0$ and for $1 \leq j \leq r$, define
$M_j = q_1 + \cdots + jq_j$ and

$$M_{ij} = M_{j-1} + ij, \ i = 0, \cdots, q_j - 1.$$

Note that $M_{i1} = i, i = 0, \cdots, q_1 - 1$.

Recall that S_r is the symmetric group on r symbols. We define two subsets of S_r.
First, let $S_r^\star(\mathbf{q})$ be the subset of S_r consisting of permutations σ such that

$$\sigma(M_{ij} + 1) < \sigma(M_{ij} + 2) < \cdots < \sigma(M_{ij} + j), \tag{9.15}$$

for $j = 1, \cdots, r - 1$ and $i = 0, \cdots, q_j - 1$.

Second, let $S_r(\mathbf{q})$ be the subset of S_r consisting of permutations σ such that

$$\sigma(M_{j-1} + 1) < \sigma(M_{j-1} + 2) < \cdots < \sigma(M_j), \tag{9.16}$$

for $j = 1, \cdots, r - 1$. Note that $S_r(\mathbf{q}) \subset S_r^\star(\mathbf{q})$. It is elementary to show that the
cardinalities of $S_r^\star(\mathbf{q})$ and $S_r(\mathbf{q})$ are given by

$$|S_r^\star(\mathbf{q})| = \frac{r!}{\prod_{j=1}^{r} (j!)^{q_j}}, \quad |S_r(\mathbf{q})| = \frac{r!}{\prod_{j=1}^{r} q_j! \prod_{j=1}^{r} (j!)^{q_j}}. \tag{9.17}$$

Suppose given multi-linear maps $A_j \in L_s^j(\mathbb{R}^m; \mathbb{R}^n)$, $1 \leq j \leq r$. For $k \in \mathbb{N}$, let A_j^k denote the k-tuple $(A_j, \cdots, A_j) \in \times^k L_s^j(\mathbb{R}^m, \mathbb{R}^n) = L_s^j(\mathbb{R}^m, \mathbb{R}^n)^k$.

Set $A = A(\mathbf{q}) = (A_1^{q_1}, \cdots, A_r^{q_r}) \in \prod_{j=1}^r L_s^j(\mathbb{R}^m; \mathbb{R}^n)^{q_j}$ (if $q_j = 0$, we omit the corresponding term from the product). We view A as the r-linear mapping from $\mathbb{R}^m$ to $\times^q \mathbb{R}^n$ defined by

$$A(\mathbf{E}_1, \cdots, \mathbf{E}_r) = (A_1^{q_1}(\mathbf{E}_1), A_2^{q_2}(\mathbf{E}_2), \cdots, A_r^{q_r}(\mathbf{E}_r)),$$

where $\mathbf{E}_j = (\mathbf{e}_{M_{j-1}+1}, \cdots, \mathbf{e}_{M_j}) \in times^{jq_j}\mathbb{R}^m$, $1 \leq j \leq r$.

If $\sigma \in S_r$, define $\sigma : \times^r \mathbb{R}^m \to \times^r \mathbb{R}^m$ by

$$\sigma(\mathbf{e}_1, \cdots, \mathbf{e}_r) = (\mathbf{e}_{\sigma(1)}, \cdots, \mathbf{e}_{\sigma(r)}), \quad (\mathbf{e}_1, \cdots, \mathbf{e}_r) \in \times^r \mathbb{R}^m.$$

Given A as above, we define σA to be the r-linear mapping from $\mathbb{R}^m$ to $\times^q \mathbb{R}^n$ defined by

$$(\sigma A)(\mathbf{e}_1, \cdots, \mathbf{e}_r) = A(\sigma(\mathbf{e}_1, \cdots, \mathbf{e}_r)), \quad (\mathbf{e}_1, \cdots, \mathbf{e}_r) \in \times^r \mathbb{R}^m.$$

In terms of the jq_j-tuples $\mathbf{E}_j = (\mathbf{e}_{M_{j-1}+1}, \cdots, \mathbf{e}_{M_j})$, if we define

$$\sigma \mathbf{E}_j = (\mathbf{e}_{\sigma(M_{j-1}+1)}, \cdots, \mathbf{e}_{\sigma(M_j)}),$$

then $(\sigma A)(\mathbf{E}_1, \cdots, \mathbf{E}_r) = A(\sigma \mathbf{E}_1, \cdots, \sigma \mathbf{E}_r)$.

Now suppose that $B \in L_s^q(\mathbb{R}^n; \mathbb{R}^p)$. We will combine A and B in three different ways to define r-linear symmetric maps from $\mathbb{R}^m$ to $\mathbb{R}^p$. Define the r-linear mappings $B \star A, B \ast A, B \circledast A$ by

$$B \star A(\mathbf{e}_1, \cdots, \mathbf{e}_r) = \frac{1}{r!} \sum_{\sigma \in S_r} B(\sigma A(\mathbf{e}_1, \cdots, \mathbf{e}_r)),$$

$$B \diamond A(\mathbf{e}_1, \cdots, \mathbf{e}_r) = \frac{1}{q_1! \cdots q_r!} \sum_{\sigma \in S_r^\star(\mathbf{q})} B(\sigma A(\mathbf{e}_1, \cdots, \mathbf{e}_r)),$$

$$B \circledast A(\mathbf{e}_1, \cdots, \mathbf{e}_r) = \sum_{\sigma \in S_r(\mathbf{q})} B(\sigma A(\mathbf{e}_1, \cdots, \mathbf{e}_r)).$$

(For the definitions of $S_r^\star(\mathbf{q})$, $S_r(\mathbf{q})$, see (9.15), (9.16).)

Lemma 9.13.4 (Assumptions and Notation as Above)

(1) $B \star A, B \diamond A, B \circledast A \in L_s^r(\mathbb{R}^m; \mathbb{R}^p)$.

(2) $B \circledast A = B \diamond A = \left(\dfrac{r!}{\prod_{j=1}^r q_j! \prod_{j=1}^r (j!)^{q_j}} \right) B \star A$.

Proof Clearly $B \star A$ is symmetric without requiring any symmetry conditions on B or A. Indeed, if $\eta \in S_r$, $(\mathbf{e}_1, \cdots, \mathbf{e}_r) \in \mathbb{R}^m$, we have

$$(B \star A)(\eta(\mathbf{e}_1, \cdots, \mathbf{e}_r)) = \frac{1}{r!} \sum_{\sigma \in S_r} B(A(\sigma\eta(\mathbf{e}_1, \cdots, \mathbf{e}_r)))$$

$$= \frac{1}{r!} \sum_{\tau \in S_r} B(A(\tau(\mathbf{e}_1, \cdots, \mathbf{e}_r)))$$

$$= (B \star A)(\mathbf{e}_1, \cdots, \mathbf{e}_r),$$

since every permutation $\tau \in S_r$ can be written uniquely as $\sigma\eta$ (take $\sigma = \tau\eta^{-1}$). Hence $B \star A \in L_s^r(\mathbb{R}^m; \mathbb{R}^p)$. For $B \diamond A$ to be symmetric, it suffices that each A_j is a symmetric j-linear map. For $B \circledast A$ to be symmetric we also need B to be a symmetric q-linear map. Finally, the relation between $B \star A$, $B \diamond A$ and $B \circledast A$ follows easily from (9.17).   □

Remark 9.13.5 Note that all the combinatorial coefficients in Faà di Bruno's formula occur in Lemma 9.13.4.   ✚

Example 9.13.6 Suppose that $f, g : \mathbb{R} \to \mathbb{R}$ are C^r. Suppose given $q_1, \cdots, q_r \geq 0$, satisfying $q_1 + \cdots q_r = q$, $q_1 + 2q_2 + \cdots + rq_r = r$. We have

$$(D^q g)f \circledast ((D^1 f)^{q_1}, \cdots, (D^r f)^{q_r})$$

$$= \frac{r!}{\prod_{j=1}^r q_j! \prod_{j=1}^r (j!)^{q_j}} (D^q g)f((D^1 f)^{q_1}, \cdots, (D^r f)^{q_r}).$$

Evaluating the expression on the right-hand side at x and 1^r gives

$$\frac{r!}{\prod_{j=1}^r q_j! \prod_{j=1}^r (j!)^{q_j}} g^{(q)}(f(x)) \prod_{j=1}^r f^{(j)}(x)^{q_j}$$

$$= \frac{r!}{q_1! \cdots q_r!} g^{(q)}(f(x)) \prod_{j=1}^r \left(\frac{f^{(j)}(x)}{j!} \right)^{q_j},$$

which is the general term in Faà di Bruno's formula.   ♠

Theorem 9.13.7 *Let* $f : \mathbb{R}^m \to \mathbb{R}^n$, $g : \mathbb{R}^n \to \mathbb{R}^p$ *be* C^r-*maps. Then*

$$D^r(gf) = \sum (D^q g)f \diamond ((D^1 f)^{q_1}, \cdots, (D^r f)^{q_r})$$

$$= \sum (D^q g)f \circledast ((D^1 f)^{q_1}, \cdots, (D^r f)^{q_r})$$

$$= \sum \frac{r!}{\prod_{j=1}^r q_j!} (D^q g)f \star \left(\left(\frac{D^1 f}{1!} \right)^{q_1}, \cdots, \left(\frac{D^r f}{r!} \right)^{q_r} \right),$$

where in each case the sum is over all $q_1, \cdots, q_r \geq 0$ such that $q_1 + 2q_2 + \cdots + rq_r = r$ and $q_1 + \cdots + q_r = q$.

Remarks 9.13.8

(1) It follows from Example 9.13.6 that the third form for $D^r(gf)$ gives Faà di Bruno's formula (Theorem 9.13.3).
(2) The second form of Theorem 9.13.7 gives the most economical and natural expression for the terms in $D^r(gf)$. For example, forms 1 and 2 give for $q = 1$ the term $(Dg)fD^rf$ while form 3 involves an unnecessary symmetrization of D^rf. Turning to the final term $q = r$ in the sum, the second form gives the term $(D^rg)f(Df)^r$ while the first and third form give $\frac{1}{r!}\sum_{\sigma \in S_r}(D^rg)f((Df(e_{\sigma(1)}), \cdots, Df(e_{\sigma(r)})))$. ✠

Proof of Theorem 9.13.7 (sketch). It follows from Lemma 9.13.4 that the three expressions for $D^r(gf)$ are equal.

It suffices to verify the third form. Although proof by induction might appear to be an attractive approach, it is quickly seen that induction does not work well. Instead, we start by observing that we can write

$$D^r(gf) = \sum a_{q_1 q_2 \cdots q_r}(D^q g)f \star \left(\left(\frac{D^1 f}{1!}\right)^{q_1}, \cdots, \left(\frac{D^r f}{r!}\right)^{q_r} \right),$$

where $q_1 + \cdots + q_r = q$, $\sum_{j=1}^{r} jq_j = r$ and the coefficients $a_{q_1 q_2 \cdots q_r}$ are rational and independent of f, g and m, n, p. The problem is to find the coefficients $a_{q_1 q_2 \cdots q_n}$. This we can do by judicious choice of g and f. Noting Example 9.13.6, it suffices to restrict to the case $n = m = p = 1$. We consider the function $H = \exp(\tau f)$, where $f : \mathbb{R} \to \mathbb{R}$ is C^∞, $\tau \in \mathbb{R}$ and $g(y) = \exp(\tau y)$ (so $H = gf$). The Taylor series of H at x is

$$T_x H(h) = \sum_{r=0}^{\infty} (\exp(\tau f))^{(r)}(x)\frac{h^r}{r!}.$$

The Taylor series of f at x is $T_x f(h) = \sum_{j=0}^{\infty} f^{(j)}(x)\frac{h^j}{j!}$. Substituting $T_x f(h)$ for $f(x + h)$ in $\exp(\tau f(x + h))$ and using the exponent law for exp gives the formal identity

$$\sum_{r=0}^{\infty} (\exp(\tau f))^{(r)}(x)\frac{h^r}{r!} = \prod_{j=0}^{\infty} \exp\left(\tau f^{(j)}(x)\frac{h^j}{j!} \right). \tag{9.18}$$

It follows from Taylor's theorem (Theorem 2.7.10) that coefficients of like powers in (9.18) are equal. In other words, we can work within the framework of *formal power series* and ignore issues of convergence. The right-hand side of (9.18) is equal to $\exp(\tau f(x)) \prod_{j=1}^{\infty} \exp(\tau f^{(j)}(x)\frac{h^j}{j!})$. Expanding the terms $\exp(\tau f^{(j)}(x)\frac{h^j}{j!}), j > 0$, and collecting like powers of h, we find that the right-hand side of (9.18) is equal to

$$\sum_{r=0}^{\infty} h^r \left[\sum_{q=0}^{r} \tau^q \exp(\tau f(x)) \sum \frac{1}{q_1! \cdots q_r!} \prod_{j=1}^{r} \left(\frac{f^{(j)}(x)}{j!} \right)^{q_j} \right],$$

where the innermost sum is over $q_1, \cdots, q_r \geq 0$, satisfying $q_1 + \cdots + q_r = q$ and $q_1 + \cdots + r q_r = r$. Comparing the coefficients of h^r in (9.18) and noting that the term in $(\exp(\tau f))^{(r)}(x)$ associated to the qth derivative of $\exp$ corresponds to the term $\tau^q \exp(\tau f(x))$ gives

$$a_{q_1 q_2 \cdots q_r} = \frac{r!}{q_1! \cdots q_r!},$$

completing the proof of Faà di Bruno's formula and Theorem 9.13.7.   □

Historical Comments Although Francesco Faà di Bruno may have been the first to publish the formula for the higher derivative of a composite of real-valued functions, he surely was not the first to discover the formula. We refer the reader to the article by Johnson [17] for more historical and mathematical details about the formula as well as the proof and relationships with Bell polynomials. The method we use is elementary and based on an "anonymous" proof published several years before that of Faà di Bruno (who only gave the result, not the proof, in his original papers). The first reference I am aware of for a formula for the pth derivative of a composite of vector-valued functions appears in Abraham and Robbin's 1967 research text [1]. The formula they give is recursive and is not quite clear as stated since the terms appear not to be symmetric (there is a similar issue with their version of Leibniz's theorem). Versions of their result appears in [9] and in [10, page 293]. There have been many publications in recent years proving various versions of Faà di Bruno's formula for vector-valued maps. We refer to Krantz's text on real analysis [19] for references. From our perspective, we find it remarkable that even though Faà di Bruno did not deal with vector-valued functions or symmetric multi-linear maps, all of the difficulties are already present in his formula.

EXERCISES 9.13.9

(1) Provide the details of the proof of Theorem 9.13.2. (Hint: Prove by induction on r, start by considering the formula for $D^{r-1}(f \cdot g)_x(e_2, \cdots, e_r)$ and use Lemma 9.10.9 and standard combinatorial identities.)

(2) Let $\alpha, \beta \in \mathbb{Z}_+^m$. Write $\beta \leq \alpha$, if $\alpha_i \leq \beta_i$, $1 \leq i \leq m$. If $\beta \leq \alpha$, define $\alpha - \beta = (\alpha_1 - \beta_1, \cdots, \alpha_m - \beta_m)$. Show that if $f, g : U \subset \mathbb{R}^m \to \mathbb{R}$ are r-times

differentiable at $\mathbf{x} \in U$ and $|\alpha| = r$ then

$$\partial^\alpha (f \cdot g)(\mathbf{x}) = \sum_{\beta \leq \alpha} \begin{pmatrix} \alpha \\ \alpha - \beta \end{pmatrix} \partial^\beta f(\mathbf{x}) \partial^{\alpha - \beta} g(\mathbf{x}),$$

where $\begin{pmatrix} \alpha \\ \alpha - \beta \end{pmatrix} = |\alpha|! / [\alpha_1! \cdots \alpha_m! (\alpha_1 - \beta_1)! \cdots (\alpha_m - \beta_m)!]$.

(3) Suppose $f : \mathbb{R}^m \to \mathbb{R}^{n_1}$, $g : \mathbb{R}^m \to \mathbb{R}^{n_2}$ and we are given a bilinear map $\phi : \mathbb{R}^{n_1} \times \mathbb{R}^{n_2} \to \mathbb{R}^n$. Define $f \cdot g : \mathbb{R}^m \to \mathbb{R}^n$ by $f \cdot g(\mathbf{x}) = \phi(f(\mathbf{x}), g(\mathbf{x}))$. State and prove a version of the Leibniz law for $f \cdot g$.

(4) Verify (9.17).

(5) Let $f, g : \mathbb{R}^m \to \mathbb{R}^m$ be smooth. Compute the first four derivatives of gf and compare with the formulas given by Theorem 9.13.7. How many terms involving $D^4 f_\mathbf{x}$ and $(Df_\mathbf{x})^2$ are there in the expression $(D^3 g) f \circledast ((Df_\mathbf{x})^2, D^4 f_\mathbf{x})$ occurring in the formula for $D^6 (gf)_\mathbf{x}(\mathbf{e}_1, \cdots, \mathbf{e}_6)$?

(6) Suppose that f, g are real-valued analytic functions of one variable. Using Faà di Bruno's formula for functions of one variable, show that the composition $f \circ g$ is analytic. (Hint: use (9.17). This provides a proof of Proposition 5.4.6 that does not use complex analysis. See also Krantz and Parks [20, §1.3].)

9.14 Smooth Functions and Uniform Approximation

In this section we give examples of smooth (C^∞) non-polynomial functions on $\mathbb{R}^m$, $m > 1$, generalizing the 'bump' and 'tabletop' functions of Chap. 5. Using these functions, we prove a variant of the Weierstrass approximation theorem that allows us to uniformly approximate C^r-functions, and their first r-derivatives, by smooth functions. This result is used later in the proof of the existence of C^r local flows for ordinary differential equations defined by a C^r vector field.

Definition 9.14.1 Let $f : \mathbb{R}^m \to \mathbb{R}^n$. The (closed) *support* of f, denoted supp(f), is defined by

$$\text{supp}(f) = \overline{\{\mathbf{x} \in \mathbb{R}^m \mid f(\mathbf{x}) \neq 0\}}.$$

The map f is of *compact support* if supp(f) is compact.

Example 9.14.2 If $p : \mathbb{R}^m \to \mathbb{R}^n$ is a homogeneous polynomial, then supp(p) is compact iff $p \equiv 0$. Indeed, since p is homogeneous, if $p(\mathbf{x}) \neq \mathbf{0}$, then $p(\lambda \mathbf{x}) = \lambda^d p(\mathbf{x}) \neq \mathbf{0}$ for all $\lambda \in \mathbb{R}$, $\lambda \neq 0$. Hence supp$(p) \supset \mathbb{R}\mathbf{x}$ and so supp(p) cannot be compact. The same result holds without assuming the homogeneity of p (see the exercises). ♠

For $0 \leq p \leq \infty$, let $C_c^p(\mathbb{R}^m, \mathbb{R}^n)$ be the set of all C^p maps $f : \mathbb{R}^m \to \mathbb{R}^n$ with compact support. Since supp$(f + g) \subset \{\mathbf{x} + \mathbf{y} \mid \mathbf{x} \in \text{supp}(f), \mathbf{y} \in \text{supp}(g)\}$, and $\operatorname{supp}(\lambda f) = \lambda \operatorname{supp}(f)$, it follows that $C_c^p(\mathbb{R}^m, \mathbb{R}^n)$ is a vector subspace of $C^p(\mathbb{R}^m, \mathbb{R}^n)$.

In order to construct C^∞-functions with compact support we make use of the theory developed in Sect. 5.2. Recall that the C^∞ map $\Phi : \mathbb{R} \to \mathbb{R}$ is defined by

$$\Phi(x) = \begin{cases} \exp(-1/x), & x > 0, \\ 0, & x \leq 0, \end{cases}$$

and that Φ is used to construct C^∞-functions on $\mathbb{R}$ with compact support. In particular, if $-\infty < a < b < +\infty$, we define the 'bump' function

$$\Psi_{a,b}(x) = \Phi(b - x)\Phi(x - a), \; x \in \mathbb{R},$$

satisfying $\Psi_{a,b} \geq 0$ and $\operatorname{supp}(\Psi_{a,b}) = [a, b]$. If $0 < r < s < \infty$, we define the 'tabletop' function

$$\Theta_{r,s}(x) = \frac{\Phi(x^2 - r^2)}{\Phi(s^2 - x^2) + \Phi(x^2 - r^2)}.$$

We have $\operatorname{supp}(\Theta_{r,s}) = [-s, s]$ and $\Theta_{r,s}(x) = 1$, for all $x \in [-r, r]$.

It is straightforward to define bump and tabletop functions on $\mathbb{R}^m$. For example, if $x \in \mathbb{R}^m$, define the tabletop function

$$\Theta_{r,s}(\mathbf{x}) = \frac{\Phi(\|\mathbf{x}\|^2 - r^2)}{\Phi(s^2 - \|\mathbf{x}\|^2) + \Phi(\|\mathbf{x}\|^2 - r^2)}, \; \mathbf{x} \in \mathbb{R}^m.$$

It follows by the chain rule that $\Theta_{r,s}$ is C^∞ (the square of the Euclidean norm is C^∞). We have $\operatorname{supp}(\Theta_{r,s}) = \overline{D}_s(\mathbf{0})$ and $\Theta_{r,s} \equiv 1$ on $\overline{D}_r(\mathbf{0})$.

Example 9.14.3 If $f : \mathbb{R}^m \to \mathbb{R}^n$ is C^∞, then $\Theta_{r,s}f \in C_c^\infty(\mathbb{R}^m, \mathbb{R}^n)$ and $\operatorname{supp}(\Theta_{r,s}f) \subset \overline{D}_s(\mathbf{0})$. Note that $\Theta_{r,s}f = f$ on $\overline{D}_r(\mathbf{0})$.   ♠

For maps defined on $\mathbb{R}^m$, it is useful to have a tabletop function with support a *hypercube* in $\mathbb{R}^m$—this is compatible with the coordinate structure on $\mathbb{R}^m$ and works well with multiple integrals. To this end, suppose $r_1, \cdots, r_m > 0$ and define

$$\eta_{r_1, \cdots, r_m}(\mathbf{x}) = \prod_{i=1}^m \Psi_{-r_i, r_i}(x_i), \; \mathbf{x} \in \mathbb{R}^m.$$

Since partial derivatives of $\eta_{r_1, \cdots, r_m}$ of all orders exist and are continuous, $\eta_{r_1, \cdots, r_m}$ is C^∞ and we have

$$\operatorname{supp}(\eta_{r_1, \cdots, r_m}) = \prod_{i=1}^m [-r_i, r_i].$$

If we take $r_i = r > 0$, $1 \le i \le m$, and set $\eta_{r_1, \dots, r_m} = \eta_r$, then $\operatorname{supp}(\eta_r)$ is the closed hypercube $\overline{C}(r) = [-r, r]^m$. If $f : \mathbb{R}^m \to \mathbb{R}^n$ is C^∞, then $\eta_r f$ is C^∞ and $\operatorname{supp}(\eta_r f) \subset \overline{C}(r)$.

Let K be a compact subset of $\mathbb{R}^m$. If $f \in C^p(\mathbb{R}^m, \mathbb{R}^n)$, define

$$\|f\|_p^K = \sum_{j=0}^p \sup_{\mathbf{x} \in K} \|D^j f_{\mathbf{x}}\|.$$

Lemma 9.14.4 *For all compact subsets K of $\mathbb{R}^m$, $\| \cdot \|_p^K$ defines a semi-norm on $C^p(\mathbb{R}^m, \mathbb{R}^n)$. That is,*

(1) $\|f\|_p^K \ge 0$, *for all* $f \in C^p(\mathbb{R}^m, \mathbb{R}^n)$.
(2) $\|f + g\|_p^K \le \|f\|_p^K + \|g\|_p^K$, *for all* $f, g \in C^p(\mathbb{R}^m, \mathbb{R}^n)$.
(3) $\|f\|_p^K = |\lambda| \|f\|_p^K$, *for all* $f \in C^p(\mathbb{R}^m, \mathbb{R}^n)$, $\lambda \in \mathbb{R}$.

Proof Routine and left to the exercises. □

Remark 9.14.5 Note that we can have $\|f\|_p^K = 0$ when $f \ne 0$. For example, if $K \cap \operatorname{supp}(f) = \emptyset$. ✳

If $f \in C_c^p(\mathbb{R}^m, \mathbb{R}^n)$, we may define

$$\|f\|_p = \|f\|_p^{\operatorname{supp}(f)} = \sum_{j=0}^p \sup_{\mathbf{x} \in \mathbb{R}^m} \|D^j f_{\mathbf{x}}\|.$$

It follows from Lemma 9.14.4 that, for $p \ge 0$, $\| \cdot \|_p$ defines a norm on $C_c^p(\mathbb{R}^m, \mathbb{R}^n)$, and we refer to $\| \cdot \|_p$ as the p-norm on $C_c^p(\mathbb{R}^m, \mathbb{R}^n)$.

The semi-norm $\| \cdot \|_p^K$ defines uniform convergence on K.

Lemma 9.14.6 *Let K be a compact subset of $\mathbb{R}^n$ and suppose that $(f_\ell) \subset C^0(\mathbb{R}^m, \mathbb{R}^n)$ is Cauchy with respect to $\| \cdot \|_0^K$. Then (f_ℓ) converges uniformly on K to a continuous function $f : K \subset \mathbb{R}^m \to \mathbb{R}^n$.*

Proof Set $g_\ell = f_\ell|K$, $\ell \in \mathbb{N}$, and apply Theorem 7.15.9 to (g_ℓ). □

Remark 9.14.7 Lemma 9.14.6 says nothing about the convergence of (f_ℓ) on $\mathbb{R}^m \smallsetminus K$. Indeed, it is easy to construct examples where (f_ℓ) converges on K but does not converge at any point of $\mathbb{R}^m \smallsetminus K$. ✳

We want to extend Lemma 9.14.4 to take account of differentiability. To keep matters simple, we restrict compact sets to the collection of closed hypercubes $\overline{C}(r) = [-r, r]^m$ and set $\| \cdot \|_p^{\overline{C}(r)} = \| \cdot \|_p^r$, $r > 0$. Let $C(r) = (-r, r)^m$ denote the open hypercube.

Lemma 9.14.8 *Let $(f_\ell) \subset C^1(\mathbb{R}^m, \mathbb{R}^n)$ and $r > 0$. If there exist maps $f, F_i : \overline{C}(r) \to \mathbb{R}^n$ satisfying*

$$\|f_\ell - f\|_0^r, \left\| \frac{\partial f_\ell}{\partial x_i} - F_i \right\|_0^r \to 0, \quad as\ \ell \to \infty,\ 1 \le i \le m,$$

then

(1) $f, F_i : \overline{C}(r) \to \mathbb{R}^n$ *are continuous,* $1 \le i \le m$.
(2) $f : C(r) \to \mathbb{R}^n$ *is* C^1 *with partial derivatives given by* $\frac{\partial f}{\partial x_i}(\mathbf{x}) = F_i(\mathbf{x})$, $1 \le i \le m$, $\mathbf{x} \in C(r)$.

Proof Since (f_ℓ), $(\frac{\partial f_\ell}{\partial x_i})$ converge uniformly on $\overline{C}(r)$, it follows that f, F_i are continuous on $\overline{C}(r)$, proving (1). Fix i, $1 \le i \le m$. For $(x_1, \cdots, x_m) \in \overline{C}(r)$, we have

$$f_\ell(x_1, \cdots, x_m) = \int_0^{x_i} \frac{\partial f_\ell}{\partial x_i}(x_1, \cdots, s, \cdots, x_m)\, ds + f_\ell(x_1, \cdots, 0, \cdots, x_m).$$

Since convergence is uniform, we may apply Proposition 4.7.1 for fixed x_j, $j \ne i$ (note Remark 4.7.3) and let $\ell \to \infty$ to obtain

$$f(x_1, \cdots, x_m) = \int_0^{x_i} F_i(x_1, \cdots, s, \cdots, x_m)\, ds + f(x_1, \cdots, \bar{x}, \cdots, x_m).$$

Hence f is continuously partially differentiable on $C(r)$, with $\frac{\partial f}{\partial x_i} = F_i$ on $C(r)$, and so f is C^1 on $C(r)$ by Theorem 9.10.11(2). $\square$

Theorem 9.14.9 *Let* $r > 0$ *and suppose* $(f_\ell) \subset C^p(\mathbb{R}^m, \mathbb{R}^n)$ *is Cauchy with respect to* $\| \cdot \|_p^r$. *Then there exists a continuous map* $f : \overline{C}(r) \to \mathbb{R}^n$ *such that*

(1) (f_ℓ) *converges to* f *uniformly on* $\overline{C}(r)$.
(2) $f : C(r) \to \mathbb{R}^m$ *is* C^p *and* $D^j f_\ell$ *converges uniformly to* $D^j f$ *on* $C(r)$ *for* $0 \le j \le p$.

Proof Statement (1) is Lemma 9.14.6. For (2) we have, again by Lemma 9.14.6, that $\partial^\alpha f_\ell$ converges uniformly to a continuous function $F_\alpha : \overline{C}(r) \to \mathbb{R}^n$ for all $\alpha \in \mathbb{Z}_+^m$, $|\alpha| \le p$. We use induction on $|\alpha|$ and Lemma 9.14.8 to show that $\partial^\alpha f = F_\alpha$ on $C(r)$, $|\alpha| \le p$. Hence, by Theorem 9.10.11, $f : C(r) \to \mathbb{R}^m$ is C^p. $\square$

Remark 9.14.10 If we replace $\| \cdot \|_p^r$ by $\| \cdot \|_p^K$, Theorem 9.14.9 continues to hold with $C(r)$ replaced by the interior of K.   ✠

Corollary 9.14.11 *Let* $(f_\ell) \subset C_c^p(\mathbb{R}^m, \mathbb{R}^n)$ *be Cauchy with respect to* $\| \cdot \|_p$. *Then* (f_ℓ) *converges to* $f \in C^p(\mathbb{R}^m, \mathbb{R}^n)$:

$$\lim_{\ell \to \infty} \| f_\ell - f \|_p = 0.$$

Proof Left to the exercises. In general, $f \notin C_c^p(\mathbb{R}^m, \mathbb{R}^n)$. $\square$

The semi-norm $\| \cdot \|_p^K$ is exactly what is needed to define *uniform approximation* of functions on a compact set K. We show that if $K \subset \mathbb{R}^m$ is compact and $f \in C^p(\mathbb{R}^m, \mathbb{R}^n)$, then for any $\varepsilon > 0$, we can find $\tilde{f} \in C^\infty(\mathbb{R}^m, \mathbb{R}^n)$ such that

$$\| f - \tilde{f} \|_p^K < \varepsilon.$$

This is uniform approximation of a function, and its first p derivatives, on K by a C^∞-function. The condition $\|f - \tilde{f}\|_p^K < \varepsilon$ implies that

$$\|D^j(f - \tilde{f})_\mathbf{x}\| < \varepsilon, \ \mathbf{x} \in K, j = 0, \cdots, p.$$

For our applications, it suffices to approximate by C^∞-functions rather than polynomials. Although the Weierstrass approximation theorem generalizes to $\mathbb{R}^m$, the Bernstein polynomial approach used in Chap. 5 only works well for the $\| \cdot \|_0$-norm. Our methods give uniform approximations of a C^p-function of compact support by a C^∞-function of compact support—this simplifies some of our proofs.

We define the C^∞ positive function $\gamma : \mathbb{R}^m \to \mathbb{R}$ by

$$\gamma(\mathbf{x}) = \eta_1(\mathbf{x}), \ \mathbf{x} \in \mathbb{R}^m,$$

where we recall that $\eta_1(\mathbf{x}) = \prod_{i=1}^m \Psi_{-1,1}(x_i)$. We have $\operatorname{supp}(\gamma) = \overline{C}(1)$ (the unit hypercube, centred at the origin) and

$$\int_{\mathbb{R}^m} \gamma = \int_\mathbb{R} \cdots \int_\mathbb{R} \Psi_{-1,1}(x_1) \cdots \Psi_{-1,1}(x_m) \, dx_1 \cdots dx_m$$

$$= \left(\int_\mathbb{R} \Psi_{-1,1}(s) \, ds \right)^m$$

$$= c,$$

where $c > 0$, since $\gamma > 0$ on $C(1)$. Replacing γ by $c^{-1}\gamma$ we may and shall assume that $\int_{\mathbb{R}^m} \gamma = 1$. For $\delta > 0$, define

$$\gamma_\delta(\mathbf{x}) = \frac{1}{\delta^n} \gamma \left(\frac{\mathbf{x}}{\delta} \right), \ \mathbf{x} \in \mathbb{R}^m.$$

Lemma 9.14.12 *For all $\delta > 0$,*

(1) $\int_{\mathbb{R}^m} \gamma_\delta = 1$.
(2) $\operatorname{supp}(\gamma_\delta) = \overline{C}(\delta)$.

Proof The first part is an elementary change of variables argument for $\int_\mathbb{R} \Psi_{-1,1}(s) \, ds$ (the non-trivial change of variables formula for multiple integrals is not needed). The second statement is obvious. □

Let $f \in C_c^p(\mathbb{R}^m, \mathbb{R}^n)$. For $\delta > 0$, define

$$f_\delta(\mathbf{x}) = \int_{\mathbb{R}^m} \gamma_\delta(\mathbf{x} - \mathbf{s}) f(\mathbf{s}) \, ds$$

$$= \int_\mathbb{R} \cdots \int_\mathbb{R} \gamma_\delta(x_1 - s_1, \cdots, x_m - s_m) f(s_1, \cdots, s_m) \, ds_1 \cdots ds_m.$$

Remarks 9.14.13

(1) The defining integral for f_δ can be evaluated over any hypercube $\overline{C}(r) \supset$ supp(f). Hence the integral lies within the elementary class of multiple integrals considered in this text (see Chap. 2, Exercises 2.8.10(10)).
(2) The integral defining f_δ is called the *convolution* of f and γ_δ.
(3) For the definition of f_δ, f was assumed to have compact support. However, the integral clearly converges for any continuous function $f : \mathbb{R}^m \to \mathbb{R}^n$ since, for fixed $\mathbf{x}$, $\gamma_\delta(\mathbf{x} - \mathbf{s})f(\mathbf{s})$ has compact support. ✠

Lemma 9.14.14 *If* $f \in C_c^p(\mathbb{R}^m, \mathbb{R}^n)$ *and* $\delta > 0$, *then*

$$f_\delta(\mathbf{x}) = \int_{\mathbb{R}^m} \gamma_\delta(\mathbf{x} - \mathbf{s})f(\mathbf{s}) \, ds = \int_{\mathbb{R}^m} \gamma_\delta(\mathbf{s})f(\mathbf{x} - \mathbf{s}) \, ds.$$

Proof Change variables from s_i to $s_i - x_i$, $1 \le i \le m$.   □

Remark 9.14.15 The second integral of Lemma 9.14.14 can be evaluated over the hypercube $\overline{C}(r + \delta)$, if $\overline{C}(r) \supset$ supp(f). ✠

Theorem 9.14.16 (Uniform Approximation by Smooth Functions) *Let* $p \ge 0$ *and* $f \in C_c^p(\mathbb{R}^m, \mathbb{R}^n)$. *Then*

(1) $f_\delta \in C_c^\infty(\mathbb{R}^m, \mathbb{R}^n)$, *for all* $\delta > 0$.
(2) $\lim_{\delta \to 0+} \|f - f_\delta\|_p = 0$.

Proof (1) Let $\boldsymbol{\alpha}$ be a multi-index, with $0 \le |\boldsymbol{\alpha}| \le p$. Since

$$f_\delta(x_1, \cdots, x_m)$$
$$= \int_{\mathbb{R}} \cdots \int_{\mathbb{R}} \gamma_\delta(x_1 - s_1, \cdots, x_m - s_m)f(s_1, \cdots, s_m) \, ds_1 \cdots ds_m,$$

and γ_δ is C^∞, we have (by the easy version of Lemma 6.1.6 for integrals of smooth functions over compact intervals) that all partial derivatives of f_ε exist, are continuous and are given by

$$\partial^{\boldsymbol{\alpha}} f_\delta(\mathbf{x}) = \int_{\mathbb{R}^m} \partial^{\boldsymbol{\alpha}} \gamma_\delta(\mathbf{x} - \mathbf{s})f(\mathbf{s}) \, ds.$$

Hence $f_\delta \in C_c^\infty(\mathbb{R}^m, \mathbb{R}^n)$, for all $\delta > 0$.
For (2), we use

$$f_\delta(\mathbf{x}) = \int_{\mathbb{R}^m} \gamma_\delta(\mathbf{s})f(\mathbf{x} - \mathbf{s}) \, ds.$$

We have (by the easy version of Lemma 6.1.6)

$$\partial^{\boldsymbol{\alpha}} f_\delta(\mathbf{x}) = \int_{\mathbb{R}^m} \gamma_\delta(\mathbf{s})\partial^{\boldsymbol{\alpha}} f(\mathbf{x} - \mathbf{s}) \, ds,$$

for all $\boldsymbol{\alpha}$, with $0 \le |\boldsymbol{\alpha}| \le p$. Since $\int_{\mathbb{R}^m} \gamma_\delta = 1$, it follows that

$$\partial^{\boldsymbol{\alpha}}(f_\delta - f)(\mathbf{x}) = \int_{\mathbb{R}^m} \gamma_\delta(\mathbf{s})(\partial^{\boldsymbol{\alpha}} f(\mathbf{x} - \mathbf{s}) - \partial^{\boldsymbol{\alpha}} f(\mathbf{x})) \, d\mathbf{s}.$$

Since supp(f) is compact, $\partial^{\boldsymbol{\alpha}} f$ is uniformly continuous on $\mathbb{R}^m$ for all $|\boldsymbol{\alpha}| \le p$. Hence, given $\varepsilon > 0$, we may choose $\delta > 0$ so that for all $|\boldsymbol{\alpha}| \le p$, $\|\partial^{\boldsymbol{\alpha}} f(\mathbf{x}-\mathbf{s}) - \partial^{\boldsymbol{\alpha}} f(\mathbf{x}))\| \le \varepsilon$ if $\|\mathbf{s}\|_\infty \le \bar{\delta}$. Since supp$(\gamma_\delta) = \overline{C}(\delta)$, we have

$$\|\partial^{\boldsymbol{\alpha}} f_\delta - \partial^{\boldsymbol{\alpha}} f\|_0 \le \varepsilon, \ |\boldsymbol{\alpha}| \le p, \ 0 < \delta \le \bar{\delta}.$$

It follows that $\lim_{\delta \to 0+} \|f - f_\delta\|_p = 0$. $\qquad\square$

Remark 9.14.17 Theorem 9.14.16 suffices for our main application in the next section. We indicate some extensions in the exercises. ✱

EXERCISES 9.14.18

(1) Let $p : \mathbb{R}^m \to \mathbb{R}^n$ be a polynomial of degree d. Show that supp(p) is compact iff $p \equiv 0$. (Hint: Look at $p|\mathbb{R}^m \smallsetminus D_r(0)$ for large r.)
(2) Provide the details of the proof of Lemma 9.14.4.
(3) Provide the proof for Corollary 9.14.11 and give examples to show that $(C_c^\infty(\mathbb{R}^m, \mathbb{R}^n), \| \cdot \|_p)$ is not complete for any $p \ge 0$.
(4) Show that if K is a compact subset of $\mathbb{R}^m$ and $(f_\ell) \subset C^p(\mathbb{R}^m, \mathbb{R}^n)$ is Cauchy with respect to $\| \cdot \|_p^K$, then there exists a C^p-function $f : \overset{\circ}{K} \to \mathbb{R}^n$ such that $D^j f_\ell$ converges uniformly to $D^j f$ on $\overset{\circ}{K}$, $1 \le j \le p$. (Hint: Use tabletop functions and Theorem 9.14.9.)
(5) Suppose that $f \in C^p(\mathbb{R}^m, \mathbb{R}^n)$ and K is a compact subset of $\mathbb{R}^m$. Show that for $\varepsilon > 0$, there exists an $\tilde{f} \in C_c^\infty(\mathbb{R}^m, \mathbb{R}^n)$ such that $\|f - \tilde{f}\|_p^K < \varepsilon$. (Hint: multiply f by a tabletop function which is identically one on a hypercube containing K and use Theorem 9.14.16.)
(6) Let U be an open subset of $\mathbb{R}^m$ and K be a compact subset of U. In this exercise we indicate how a C^p-function on U can be uniformly approximated on K (in $\| \cdot \|_p^K$) by smooth functions.

(a) Let $a = \inf\{d(\mathbf{x}, K) \mid \mathbf{x} \in \mathbb{R}^m \smallsetminus U\}$. Verify that $0 < a \le \infty$.
(b) For $r \in (0, a)$, define $K_r = \{\mathbf{x} \in \mathbb{R}^m \mid d(\mathbf{x}, K) \le r\}$. Verify that K_r is a compact subset of U.
(c) For $\delta > 0, r \in (0, a)$, define

$$\rho_{\delta,r}(\mathbf{x}) = \int_{\mathbb{R}^m} \gamma_\delta(\mathbf{x} - \mathbf{s}) d(\mathbf{s}, K_r) \, d\mathbf{s}.$$

Verify that $\rho_{\delta,r} \in C^\infty(\mathbb{R}^m, \mathbb{R})$ and that for sufficiently small $\delta > 0$, $\rho_{\delta,r} = 0$ on K and $\rho_{\delta,r} > 0$ on $\mathbb{R}^m \smallsetminus K_r$ ($r \in (0, a)$ is fixed).

(d) Let $2r \in (0, a)$. Using the compactness of K_r, choose a finite set $\psi_1, \cdots, \psi_N$ of bump functions such that $\sum_{j=1}^N \psi_j > 0$ on K_r and $\sum_{j=1}^N \psi_j = 0$ outside of K_{2r}.

(e) Show that, for sufficiently small $\delta > 0$, the function

$$\theta = \left(\sum_{j=1}^N \psi_j \right) \bigg/ \left(\sum_{j=1}^N \psi_j + \rho_{\delta,r} \right)$$

is C^∞ on $\mathbb{R}^m$, equal to 1 on K and equal to 0 outside K_{2r}.

(f) Let $f \in C^p(U, \mathbb{R}^n)$ and $\varepsilon > 0$. Using the function θ and Theorem 9.14.16, show that there exists an $\tilde{f} \in C_c^\infty(\mathbb{R}^m, \mathbb{R}^n)$ such that $\|f - \tilde{f}\|_p^K < \varepsilon$.

(7) In this exercise we indicate the steps in proving a strong version of the Weierstrass approximation theorem for C^p-functions on $\mathbb{R}^m$. The aim is to show that if K is a compact subset of $\mathbb{R}^m$ then any C^p-function $f : \mathbb{R}^m \to \mathbb{R}$ can be approximated by polynomials in the semi-norm $\| \cdot \|_p^K$.

(a) Show that it suffices to prove the result for f which have compact support in $(0, 1)^m$.

(b) For $p \geq 0$, define

$$\rho_p(x) = \begin{cases} c_p(1 - x^2)^p, & |x| \leq 1, \\ 0, & |x| > 1, \end{cases}$$

where $c_p > 0$ is chosen so that $\int_{-\infty}^{\infty} \rho_p = 1$. Show that for all $\varepsilon, \delta > 0$, there exists a $P = P(\varepsilon, \delta)$ such that if $p \geq P$ then $|\rho_p(x)| < \varepsilon$ for all $|x| > \delta$.

(c) If $f : \mathbb{R} \to \mathbb{R}$ is C^p and $\text{supp}(f) \subset (0, 1)$, show that $f_p(x) = \int_{-\infty}^{\infty} \rho_p(x - s) f(s) \, ds$ is a polynomial of degree at most $2p$.

(d) Show that $\|f - f_p\|_p^{[0,1]} \to 0$ as $p \to \infty$.

(e) Extend (c,d) to C^p-functions $f : \mathbb{R}^m \to \mathbb{R}$ with $\text{supp}(f) \subset (0, 1)^m$.

9.15 The Local C^r Existence Theorem for ODEs

Let $f : \mathbb{R}^m \to \mathbb{R}^m$ be a C^r vector field on $\mathbb{R}^m$, $r \geq 1$. We recall that a local flow is a continuous map $\phi : U \times (-\delta, \delta) \to \mathbb{R}^m$ defined on the open subset $U \times (-\delta, \delta)$ of $\mathbb{R}^m \times \mathbb{R}$ such that for each $x \in U$, $\phi_x : (-\delta, \delta) \to \mathbb{R}^m$ is the solution to $x' = f(x)$ with initial condition x. A local flow ϕ is C^r if ϕ is C^r (in (x, t)).

In this section we show that an ordinary differential equation $x' = f(x)$ has C^r local flows if the vector field f is of class C^r. It turns out that it is straightforward to prove the existence of C^r local flows if f is of class C^{r+1}. The improvement to C^r local flows if f is C^r requires a more sophisticated argument.

Lemma 9.15.1 *Suppose that f is a C^2 vector field on $\mathbb{R}^m$. Then $\mathbf{x}' = f(\mathbf{x})$ has a C^1 local flow on a neighbourhood of every point in $\mathbb{R}^m$.*

Proof If $\mathbf{x}_0 \in \mathbb{R}^m$, it follows by Theorem 9.7.2 that there is an open neighbourhood U of $\mathbf{x}_0$ and $\delta > 0$ such that the local flow $\phi : U \times (-\delta, \delta) \to \mathbb{R}^m$ is defined and C^0. We prove that ϕ is C^1—initially for a possibly smaller neighbourhood V of $\mathbf{x}_0$ and smaller $\delta > 0$.

Since ϕ is a local flow we have

$$\phi'(\mathbf{x}, t) = f(\phi(\mathbf{x}, t)), \quad (\mathbf{x}, t) \in U \times (-\delta, \delta). \tag{9.19}$$

Start by assuming ϕ is differentiable in $\mathbf{x}$. Differentiating (9.19) in the $\mathbb{R}^m$ variable, we obtain

$$D_1\phi'(\mathbf{x}, t) = Df_{\phi(\mathbf{x},t)} \circ D_1\phi(\mathbf{x}, t),$$

where $D_1\phi(\mathbf{x}, t)$ denotes the derivative of ϕ in the $\mathbf{x}$ variable at $(\mathbf{x}, t)$. Consequently, if ϕ is C^1, then $D_1\phi$ satisfies a linear differential equation in $L(\mathbb{R}^m, \mathbb{R}^m)$

$$\mathbf{U}'(\mathbf{x}, t) = Df_{\phi(\mathbf{x},t)} \circ \mathbf{U}(\mathbf{x}, t).$$

If we view $\mathbf{x}$ as a parameter in this equation and set $Df_{\phi(\mathbf{x},t)} = G(t)$, where $G : (-\delta, \delta) \to L(\mathbb{R}^m, \mathbb{R}^m)$, then $D_1\phi$ satisfies the linear differential equation

$$\mathbf{U}'(t) = G(t) \circ \mathbf{U}(t). \tag{9.20}$$

This equation is solvable if G is just continuous—Proposition 9.7.5. Since $\phi(\mathbf{x}, 0) = \mathbf{x}$, the initial condition we require for (9.20) is $\mathbf{U}(0) = I_m \in L(\mathbb{R}^m, \mathbb{R}^m)$. These observations suggest that we should solve the system on $\mathbb{R}^m \times L(\mathbb{R}^m, \mathbb{R}^m)$ defined by

$$\phi'(\mathbf{x}, t) = f(\phi(\mathbf{x}, t)), \tag{9.21}$$

$$\mathbf{U}'(\mathbf{x}, t) = Df_{\phi(\mathbf{x},t)} \circ \mathbf{U}(\mathbf{x}, t), \tag{9.22}$$

subject to the initial conditions $\phi(\mathbf{x}, 0) = \mathbf{x}$, $\mathbf{U}(\mathbf{x}, 0) = I$.

Since we are assuming f is C^2, (9.21), (9.22) satisfies the conditions of Theorem 9.7.2 and so there is a local flow $\Phi : V \times (-\delta', \delta') \to \mathbb{R}^m \times L(\mathbb{R}^m, \mathbb{R}^m)$, where V is an open neighbourhood of $\mathbf{x}_0$, with $\overline{V} \subset U$, and $0 < \delta' \leq \delta$. By uniqueness of solutions to (9.21), $\Phi(\mathbf{x}, t) = (\phi(\mathbf{x}, t), \mathbf{U}(\mathbf{x}, t))$, $(\mathbf{x}, t) \in V \times (-\delta', \delta')$.

It remains to show that ϕ is C^1 and $D_1\phi$ is equal to $\mathbf{U}(\mathbf{x}, t)$. This is not easy to infer directly from (9.21), (9.22). However, we can use part of the contraction mapping lemma together with an argument based on uniform convergence.

Specifically, let $X = C^0([-\delta', \delta'] \times \overline{V}, \mathbb{R}^m \times L(\mathbb{R}^m, \mathbb{R}^m))$ and define $T : X \to X$ in the usual way by

$$T(\psi, \mathbf{A})(\mathbf{x}, t) = (\mathbf{x}, I) + \left(\int_0^t f(\psi(\mathbf{x}, s) \, ds, \int_0^t Df_{\psi(\mathbf{x}, s)} \circ \mathbf{A}(\mathbf{x}, s) \, ds \right),$$

where $(\psi, \mathbf{A}) \in X$. Using the assumption that f is C^2, we may choose δ', V so that T is a contraction mapping. Define the sequence $(\phi_0, \mathbf{U}_n)$ inductively by $\phi_0 \equiv \mathbf{x}$, $\mathbf{U}_0 \equiv I$ and

$$(\phi_{n+1}, \mathbf{U}_{n+1}) = T(\phi_n, \mathbf{U}_n), \quad n \geq 0.$$

Obviously, $D_1 \phi_0 = \mathbf{U}_0$ and an easy induction shows that $D_1 \phi_n = \mathbf{U}_n$ for all $n \geq 0$. Since $(\phi_n, \mathbf{U}_n)$ converges *uniformly* to $(\phi, \mathbf{U})$, it follows by Theorem 9.14.9 that ϕ is C^1 and $D_1 \phi = \lim_{n \to \infty} \mathbf{U}_n = \mathbf{U}$.   □

Remarks 9.15.2

(1) The differential Eq. (9.22) is called the *equation of variations*.
(2) Even though it is relatively easy to solve the linear equation of variations, we still have to be careful to show that the solution does give the derivative $D_1 \phi$. If f is only C^1, we have to work harder.   ✱

Proposition 9.15.3 *Suppose that f is a C^{r+1} vector field on $\mathbb{R}^m$. Then $\mathbf{x}' = f(\mathbf{x})$ has a C^r local flow on some neighbourhood of every point $\mathbf{x}_0 \in \mathbb{R}^m$.*

Proof The result follows by induction on r using Lemma 9.15.1. We leave the details to the exercises.   □

It is conceivable that as r increases, the domain of the local flow given by Proposition 9.15.3 shrinks to $\{(\mathbf{x}_0, 0)\}$ and so, without further work, we cannot deduce that if f is C^∞, then there is a C^∞ local flow.

Suppose $\mathbf{x}' = f(\mathbf{x})$, where $f : \mathbb{R}^m \to \mathbb{R}^m$ is C^{r+1}. Let $\mathbf{x}_0 \in \mathbb{R}^m$. It follows from Theorem 9.7.2 that there is an open neighbourhood U of $\mathbf{x}_0$ and $\delta > 0$ such that the local flow $\phi : U \times (-\delta, \delta) \to \mathbb{R}^m$ is defined and C^0. We show that ϕ is C^r (same U, same δ).

Lemma 9.15.4 (1-Parameter Group Property for Local Flows) *Let $\phi : U \times (-\delta, \delta) \to \mathbb{R}^m$ be a local flow for $\mathbf{x}' = f(\mathbf{x})$. Then for all $\mathbf{x} \in U$, and $s, t \in \mathbb{R}$ such that $s, s + t \in (-\delta, \delta)$, we have*

$$\phi(\phi(\mathbf{x}, s), t) = \phi(\mathbf{x}, s + t).$$

Proof Fix $s \in (-\delta, \delta)$. Since $\phi'(\mathbf{x}, s + t) = f(\phi(\mathbf{x}, s + t))$, $\psi(t) = \phi(\mathbf{x}, s + t)$ is the solution to $\mathbf{x}' = f(\mathbf{x})$ with initial condition $\phi(\mathbf{x}, s)$. Note that ψ is defined and unique on $(-\delta - s, \delta - s)$. Differentiating with respect to t, we see that

$$\frac{d}{dt}(\phi(\phi(\mathbf{x}, s), t)) = \phi'(\phi(\mathbf{x}, s), t) = f(\phi(\phi(\mathbf{x}, s), t)), \ t \in (-\delta - s, \delta - s).$$

Since $\phi(\phi(\mathbf{x}, s), 0) = \phi(\mathbf{x}, s)$, the result follows by uniqueness of solutions. □

Lemma 9.15.5 *Let f be a C^{r+1} vector field on $\mathbb{R}^m$. If $\phi : U \times (-\delta, \delta) \to \mathbb{R}^m$ is the C^0 local flow given by Theorem 9.7.2, then ϕ is C^r.*

Proof By Proposition 9.15.3, we have a C^r local flow defined on a neighbourhood of every point $\phi(\mathbf{x}, s)$, $\mathbf{x} \in U$, $s \in (-\delta, \delta)$. Applying Lemma 9.15.4, we see that ϕ is C^r on a neighbourhood of every point in $U \times (-\delta, \delta)$ and so $\phi : U \times (-\delta, \delta) \to \mathbb{R}^m$ is C^r. □

Theorem 9.15.6 (Local Flows for Smooth Vector Fields) *Suppose that f is a C^∞ vector field on $\mathbb{R}^m$. Then $\mathbf{x}' = f(\mathbf{x})$ has a C^∞ local flow on a neighbourhood of every point in $\mathbb{R}^m$.*

Proof By Lemma 9.15.5 and Proposition 9.15.3, the C^0 local flows $\phi : U \times (-\delta, \delta) \to \mathbb{R}^m$ given by Theorem 9.7.2 are C^r for all $r \geq 0$. □

Theorem 9.15.7 (Existence of C^r Local Flows) *Let $1 \leq r \leq \infty$. If f is a C^r vector field on $\mathbb{R}^m$, then $\mathbf{x}' = f(\mathbf{x})$ has a C^r local flow on a neighbourhood of every point in $\mathbb{R}^m$.*

Proof We give the details for $r = 1$. The general case follows by induction on r. Fix $\mathbf{x}_0 \in \mathbb{R}^m$ and let $\overline{D}_1 = \overline{D}_1(\mathbf{x}_0)$.

We may assume that f has compact support (multiply f by a tabletop function which is 1 on the neighbourhood $\overline{D}_1$. By Theorem 9.14.16, we may choose a sequence $(f_n) \subset C_c^\infty(\mathbb{R}^m, \mathbb{R}^m)$ such that (f_n) converges to f in $\| \cdot \|_1$. By $\| \cdot \|_1$ convergence of (f_n) to f, there exist $M_1, M_2 > 0$ such that $\sup_{\mathbf{x} \in \mathbb{R}^m} \|g(\mathbf{x})\| \leq M_1$, $\sup_{\mathbf{x} \in \mathbb{R}^m} \|Dg_{\mathbf{x}}\| \leq M_2$, for all $g \in \{f\} \cup \{f_n \mid n \in \mathbb{N}\}$. It follows from the proof of Theorem 9.7.2 that we may choose open neighbourhoods $V \subset W$ of $\mathbf{x}_0$, with $\overline{V} \subset W$, and $\delta > 0$, so that every $g \in \{f\} \cup \{f_n \mid n \in \mathbb{N}\}$ has a local flow $\psi_g : W \times [-\delta, \delta] \to \mathbb{R}^m$. If $g = f$, set $\psi_g = \phi$, and if $g = f_n$, set $\psi_g = \phi_n$. Shrinking V, W and δ if necessary, we may assume that $\overline{\phi(W \times [-\delta, \delta])} \subset D_1(\mathbf{x}_0)$ (so that ϕ coincides with the local flow of the unmodified vector field f).

It follows from Lemma 9.15.5, that ϕ_n is C^1 on $W \times [-\delta, \delta]$, all $n \in \mathbb{N}$. Set $K = \overline{V} \times [-\delta, \delta]$. Using continuous dependence on parameters, or direct computation, (ϕ^n) converges uniformly to ϕ on K.

Next we consider the equation of variations. For $n \geq 1$, set $\mathbf{U}_n(\mathbf{x}, t) = D_1\phi^n(\mathbf{x}, t)$, $(\mathbf{x}, t) \in K$. We have

$$\mathbf{U}_n'(\mathbf{x}, t) = G_n(t) \circ \mathbf{U}_n(\mathbf{x}, t), \ n \in \mathbb{N}, \tag{9.23}$$

$$\mathbf{U}'(\mathbf{x}, t) = G(t) \circ \mathbf{U}(\mathbf{x}, t), \tag{9.24}$$

where $G_n(t) = D(f_n)_{\phi^n(\mathbf{x},t)}$ and $G(t) = Df_{\phi(\mathbf{x},t)}$. Now ϕ^n converges uniformly to ϕ on K and Df_n converges uniformly to Df on $\mathbb{R}^m$. We have

$$\|D(f_n)_{\phi^n} - Df_\phi\|_0^K \leq \|D(f_n)_{\phi^n} - D(f_n)_\phi\|_0^K + \|D(f_n)_\phi - Df_\phi\|_0^K.$$

The second term on the right-hand side converges to zero by the uniform convergence of Df_n. For the first term, observe that, since (Df_n) is an equicontinuous family on $\mathbb{R}^m$ (Lemma 7.16.9), given $\varepsilon > 0$, there exists a $\delta > 0$ such that $\|D(f_n)_X - D(f_n)_Y\| < \varepsilon$ whenever $X, Y \in K$ satisfy $\|X - Y\| < \delta$. Since (ϕ_n) converges uniformly to ϕ on K, given $\delta > 0$, there exists an $N \in \mathbb{N}$ such that $\|\phi_n - \phi\|_0^K < \delta$, for all $n \geq N$. Consequently, $\|D(f_n)_{\phi^n} - D(f_n)_\phi\|_0^K < \varepsilon$ for $n \geq N$. Hence the first term converges to zero as $n \to \infty$. Therefore, $\|D(f_n)_{\phi^n} - Df_\phi\|_0^K \to 0$ as $n \to \infty$. It now follows from our earlier result for linear differential equations (Proposition 9.7.5) that $\mathbf{U}_n = D_1\phi^n$ converges uniformly to the solution $\mathbf{U}$ of (9.24)—the equation of variations for $\mathbf{x}' = f(\mathbf{x})$. Hence ϕ is C^1 and $D_1\phi = \mathbf{U}$. $\square$

Example 9.15.8 We can use Theorem 9.15.6 to give an alternative proof of the inverse function theorem. We sketch the basic idea. Suppose that $f : \mathbb{R}^m \to \mathbb{R}^m$ is C^1, $f(\mathbf{0}) = \mathbf{0}$, and Df_0 is a linear isomorphism. We consider the problem of solving $f(\mathbf{x}) = \mathbf{y}$, for $\mathbf{y}$ close to the origin. Fixing $\mathbf{y}$, let us try to solve $f(\mathbf{x}) = t\mathbf{y}$, $t \in (-a, a)$, where $a > 1$. If we could solve the equation, we would get a family of solutions $\mathbf{x}(t)$ such that

$$f(\mathbf{x}(t)) = t\mathbf{y}, \ t \in (-a, a).$$

Differentiating we get $Df_{\mathbf{x}(t)}(\mathbf{x}'(t)) = \mathbf{y}$ and so $\mathbf{x}(t)$ satisfies the ODE

$$\mathbf{x}'(t) = (Df_{\mathbf{x}(t)})^{-1}(\mathbf{y}), \ \mathbf{x}(0) = \mathbf{0}.$$

We obtain a C^1 solution $\phi(t, \mathbf{y})$—in this case the initial condition is always $\phi(0, \mathbf{y}) = \mathbf{0}$, and the solution depends C^1 on the parameter $\mathbf{y}$ (a slight extension of the previous result to allow dependence on parameters). It is not hard to show that we can choose $r > 0$ such that $\phi(t, \mathbf{y})$ is defined for $|t| \leq 1$, provided that $\|\mathbf{y}\| \leq r$. We define the inverse map by $f^{-1}(\mathbf{y}) = \phi(1, \mathbf{y})$, $\|\mathbf{y}\| < r$. Since ϕ is C^1, it is immediate that f^{-1} is C^1.   ♠

EXERCISES 9.15.9

(1) Complete the proof of Proposition 9.15.3.
(2) Let U be an open subset of $\mathbb{R}^m$ and $f : U \to \mathbb{R}^m$ be C^1. Show that if $\mathbf{x}_0 \in U$, the ODE $\mathbf{x}' = f(\mathbf{x})$ has a C^0 local flow defined on a neighbourhood of $\mathbf{x}_0$. Extend to the case of C^r local flows, where f is C^r. (Hint: start by multiplying f by a smooth tabletop function Ψ which is equal to one on a neighbourhood W of $\mathbf{x}_0$ and equal to zero on a neighbourhood of $\mathbb{R}^m \smallsetminus U$. Apply Theorem 9.7.2 to $\mathbf{x}' = \Psi(\mathbf{x})f(\mathbf{x})$ to obtain a local flow $\phi : V \times [-a, a] \to \mathbb{R}^m$, where $\overline{V} \subset W$. Show that we can choose $b \in (0, a]$ so that $\phi : V \times [-b, b] \to U$ defines a local flow for $\mathbf{x}' = f(\mathbf{x})$.)

9.16 Diffeomorphisms and Flows

In Sect. 9.6, we gave the formal definition of a diffeomorphism between open subsets of $\mathbb{R}^m$. Yet we have been coy about giving specific examples. For example, what can one say about the group $\mathrm{Diff}^r(\mathbb{R}^m)$ of C^r diffeomorphisms of $\mathbb{R}^m$? The case $m = 1$ is easy—a C^r diffeomorphism of $\mathbb{R}$ is given by a strictly monotone C^r surjection of $\mathbb{R}$. The case $m > 1$ is not so transparent. Obviously, any linear map $A \in \mathrm{GL}(\mathbb{R}, m)$ defines a C^∞ diffeomorphism of $\mathbb{R}^m$—but this is a trivial example that needs no differential calculus or analysis for its elucidation. Using the method of proof of the inverse function theorem, it is straightforward to show that if $A \in \mathrm{GL}(\mathbb{R}, m)$, then $F(\mathbf{x}) = A\mathbf{x} + \rho(\mathbf{x})$ will be a C^∞ diffeomorphism of $\mathbb{R}^m$ for $\rho \in C_c^\infty(\mathbb{R}^m, \mathbb{R}^m)$ and $\|\rho\|_1$ sufficiently small. But this seems likely to give a small and unrepresentative class of diffeomorphisms of $\mathbb{R}^m$. If we look at polynomials $p \in P^d(\mathbb{R}^m, \mathbb{R}^m)$, $d > 1$, surprisingly little is known. We recall the *Jacobian Conjecture*:

> Suppose $p \in P^d(\mathbb{R}^m, \mathbb{R}^m)$, $d > 1$, and the Jacobian $\det(Dp)$ is constant on $\mathbb{R}^m$. Then p is a diffeomorphism of $\mathbb{R}^m$.

The conjecture was first made in 1939 by Ott-Heinrich Keller. At this time (2017), the conjecture is neither proved or disproved, even if $m = 2$. Many erroneous proofs have been proposed. Note that when p *is* a diffeomorphism, the inverse is a polynomial map.

For the remainder of the section we show how we can use the theory of ODEs, in particular the existence of local flows (Theorem 9.15.7), to construct many non-trivial examples of smooth diffeomorphisms of $\mathbb{R}^m$.

9.16.1 Smooth Flows

Definition 9.16.1 A map $\Phi : \mathbb{R}^m \times \mathbb{R} \to \mathbb{R}^m$ is a smooth or C^∞ *flow* on $\mathbb{R}^m$ if Φ is C^∞ and

$$\Phi(\mathbf{x}, 0) = \mathbf{x}, \text{ for all } \mathbf{x} \in \mathbb{R}^m. \tag{9.25}$$

$$\Phi(\mathbf{x}, t + s) = \Phi(\Phi(\mathbf{x}, t), s), \text{ for all } \mathbf{x} \in \mathbb{R}^m, s, t \in \mathbb{R}. \tag{9.26}$$

Remark 9.16.2 The definition, and most of what we do below, generalizes straightforwardly to C^r flows. ✠

Suppose that $\Phi : \mathbb{R}^m \times \mathbb{R} \to \mathbb{R}^m$ is a smooth flow. For $t \in \mathbb{R}$, let $\Phi_t : \mathbb{R}^m \to \mathbb{R}^m$ be the C^∞ map defined by $\Phi_t(\mathbf{x}) = \Phi(\mathbf{x}, t)$, $\mathbf{x} \in \mathbb{R}^m$.

Proposition 9.16.3 *If Φ is a smooth flow, then*

(1) $\Phi_0 = I_m$.
(2) *For all* $t, s \in \mathbb{R}$, $\Phi_{t+s} = \Phi_t \Phi_s = \Phi_s \Phi_t$.
(3) *For all* $t \in \mathbb{R}$, $\Phi_t \in \mathrm{Diff}^\infty(\mathbb{R}^m)$ *and has inverse* Φ_{-t}.

Proof Statement (1) is immediate from (9.25). Next observe that $\Phi(\mathbf{x}, t + s) = \Phi(\Phi(\mathbf{x}, t), s)$ for all $\mathbf{x}, t, s$ iff $\Phi_{t+s} = \Phi_s \Phi_t$ for all t, s. Since $\Phi(\mathbf{x}, t + s) = \Phi(\mathbf{x}, s + t)$, this proves (2). Finally (3) is immediate from (1,2), since $\Phi_t \Phi_{-t} = \Phi_{-t} \Phi_t = \Phi_0 = I_m$. $\qquad\square$

Remark 9.16.4 (1,2) of Proposition 9.16.3 are referred to as the *one-parameter group property* of a flow: the map $\rho : \mathbb{R} \to \mathrm{Diff}^\infty(\mathbb{R}^m)$ defined by $\rho(t) = \Phi_t$ is a group homomorphism. ✱

We continue to assume $\Phi : \mathbb{R}^m \times \mathbb{R} \to \mathbb{R}^m$ is a smooth flow. For $\mathbf{x} \in \mathbb{R}^m$, define $\Phi_\mathbf{x} : \mathbb{R} \to \mathbb{R}^m$ by $\Phi_\mathbf{x}(t) = \Phi(\mathbf{x}, t)$. We also define the C^∞ vector field $f = f_\Phi$ on $\mathbb{R}^m$ by

$$f(\mathbf{x}) = \frac{\partial \Phi}{\partial t}(\mathbf{x}, 0), \ \mathbf{x} \in \mathbb{R}^m. \tag{9.27}$$

Remark 9.16.5 If Φ is a C^r flow, then f is only C^{r-1}—this generates some complications and is the main reason why we restrict to smooth flows. ✱

Proposition 9.16.6 (Notation and Assumptions as Above) *For all* $\mathbf{x} \in \mathbb{R}^m$, $\Phi_\mathbf{x} : \mathbb{R} \to \mathbb{R}^m$ *is the unique solution to* $\mathbf{x}' = f(\mathbf{x})$ *with initial condition* $\mathbf{x}$.

Proof Fix $\mathbf{x} \in \mathbb{R}^m$. Differentiating the identity $\Phi(\Phi(\mathbf{x}, t), s) = \Phi(\mathbf{x}, t + s)$ with respect to s and setting $s = 0$, we get

$$f(\Phi_\mathbf{x}(t)) = \frac{d}{ds} \Phi_\mathbf{x}(t + s)|_{s=0} = \Phi_\mathbf{x}'(t), \ t \in \mathbb{R}.$$

Since $\Phi_\mathbf{x}(0) = \mathbf{x}$, it follows that $\Phi_\mathbf{x}$ is a solution to $\mathbf{x}' = f(\mathbf{x})$ with initial condition $\mathbf{x}$. That the solution is unique follows from Lemma 9.7.4. $\qquad\square$

Example 9.16.7 Let $A \in L(\mathbb{R}^m, \mathbb{R}^m)$ and consider the linear ODE $\mathbf{x}' = A\mathbf{x}$. The solution with initial condition $\mathbf{x}$ is given by $\Phi_\mathbf{x}(t) = \exp(At)\mathbf{x}$, where $\exp(At) = \sum_{n=0}^\infty \frac{A^n}{n!} t^n$. Since $\exp(At) \exp(As) = \exp(A(t + s))$, it follows that $\Phi(\mathbf{x}, t) = \exp(At)\mathbf{x}$ is a smooth (linear) flow. ♠

Definition 9.16.8 Let f be a smooth vector field on $\mathbb{R}^m$, $\mathbf{x}_0 \in \mathbb{R}^m$, and $-\infty \le a < b \le +\infty$. A *solution curve* $\phi : (a, b) \to \mathbb{R}^m$ to $\mathbf{x}' = f(\mathbf{x})$ with initial condition $\mathbf{x}_0$ is *maximal* if given any solution curve $\psi : (c, d) \to \mathbb{R}^m$, with initial condition $\mathbf{x}_0$, $(c, d) \subset (a, b)$.

Lemma 9.16.9 (Notation and Assumptions as Above) *For each initial condition* $\mathbf{x}_0 \in \mathbb{R}^m$, *there is unique maximal solution curve to* $\mathbf{x}' = f(\mathbf{x})$ *with initial condition* $\mathbf{x}_0$.

Proof Let $\{\phi_\lambda : I_\lambda \to \mathbb{R}^m \mid \lambda \in \Lambda\}$ denote the set of all solution curves to $\mathbf{x}' = f(\mathbf{x})$ with initial condition $\mathbf{x}_0$. Set $I = \cup_{\lambda \in \Lambda} I_\lambda$ and define $\phi(t) = \phi_\lambda(t)$, for $t \in I_\lambda \subset I$. The map $\phi : I \to \mathbb{R}^m$ is well defined—by Lemma 9.7.4—and obviously maximal by construction. Uniqueness follows from Lemma 9.7.4. $\qquad\square$

Given a smooth vector field $f : \mathbb{R}^m \to \mathbb{R}^m$, let $\Phi_{\mathbf{x}} : I_{\mathbf{x}} \to \mathbb{R}^m$ denote the maximal solution curve for $\mathbf{x}' = f(\mathbf{x})$ with initial condition $\mathbf{x}$. Set $\mathcal{D} = \cup_{\mathbf{x} \in \mathbb{R}^m} \{\mathbf{x}\} \times I_{\mathbf{x}} \subset \mathbb{R}^m \times \mathbb{R}$. Define $\Phi(\mathbf{x}, t) = \Phi_{\mathbf{x}}(t)$, $(\mathbf{x}, t) \in \mathcal{D}$.

Proposition 9.16.10 (Local Flows) *(Notations and assumptions as above.) We have*

(1) $\Phi(\mathbf{x}, 0) = \mathbf{x}$ *for all* $\mathbf{x} \in \mathbb{R}^m$.
(2) *If* $\mathbf{x} \in \mathbb{R}^m$, $s, t \in \mathbb{R}$, *and* $\Phi(\mathbf{x}, s), \Phi(\mathbf{x}, t + s) \in \mathcal{D}$, *then* $\Phi(\mathbf{x}, t + s) = \Phi(\Phi(\mathbf{x}, t), s)$.
(3) $\mathcal{D}$ *is an open subset of* $\mathbb{R}^m \times \mathbb{R}$ *containing* $\mathbb{R}^m \times \{0\}$.
(4) $\Phi : \mathcal{D} \to \mathbb{R}^m$ *is smooth.*
(5) *If* $\mathcal{D} = \mathbb{R}^m \times \mathbb{R}$, Φ *is a smooth flow.*

Proof (Sketch) Statement (1) follows by definition and (2) follows by uniqueness of maximal solution curves (if $I_{\mathbf{x}} = (a, b)$, then $I_{\Phi(\mathbf{x},t)} = (a - t, b - t)$). It follows from the previous section that at every point $\mathbf{y} = \Phi_{\mathbf{x}}(t) \in \mathcal{D}$ there is a local C^∞ flow. Together with uniqueness of maximal solutions, it follows that $\mathcal{D}$ is an open neighbourhood of $\mathbb{R}^m \times \{0\}$. Statements (4,5) use Lemma 9.15.5. □

Remark 9.16.11 Proposition 9.16.10 holds if 'smooth' is replaced everywhere by 'C^r', $r \geq 1$. Note that Φ will be C^r (in $(\mathbf{x}, t)$) but C^{r+1} in t ($\Phi_{\mathbf{x}}' = f\Phi_{\mathbf{x}}$ which is C^r in t by the chain rule). As a result the vector field defined by Φ in (9.27) will be C^r, not just C^{r-1}. ✱

Example 9.16.12 The ODE $x' = x^2$, $x \in \mathbb{R}$, gives an example where $\mathcal{D}$ is a proper subset. A straightforward computation gives

$$\Phi(x, t) = \begin{cases} 0, & \text{if } x = 0, \\ \frac{x}{1 - tx}, & \text{if } x \neq 0. \end{cases}$$

If $x < 0$, then $I_x = (x^{-1}, \infty)$, if $x > 0$, $I_x = (-\infty, x^{-1})$, and if $x = 0$, $I_x = \mathbb{R}$. ♠
Our final result on smooth flows yields many examples of smooth flows and diffeomorphisms of $\mathbb{R}^m$.

Theorem 9.16.13 *Let f be a smooth vector field on $\mathbb{R}^m$. Suppose one of the following conditions holds*

(1) $f \in C_c^\infty(\mathbb{R}^m, \mathbb{R}^m)$.
(2) *There exist constants $A, B \in \mathbb{R}_+$ such that for all $\mathbf{x} \in \mathbb{R}^m$*

$$\|f(\mathbf{x})\| \leq A + B\|\mathbf{x}\|.$$

Then f has a smooth flow.

Proof Let $\mathbf{x}_0 \in \mathbb{R}^m$ and set $r_0 = \|\mathbf{x}_0\|$ and $\phi(t) = \Phi_{\mathbf{x}_0}(t)$, $t \in I_{\mathbf{x}}$. Assume (2) holds ((1) is easier). It follows that $\|\phi(t)\|$ is bounded by $r(t)$ where $r' = A + Br$, $r(0) = r_0$. That is, $\|\phi(t)\|$ can grow at most exponentially in t. It follows that $\phi(t)$

cannot go to infinity in finite time. Specifically, if $(a, b) \subset I_{\mathbf{x}}$, $-\infty < a < b < +\infty$, then there exists an $R > 0$ such that $\overline{\phi(a, b)} \subset \overline{D}_R$. Hence $\phi(a, b)$ is compact and from this it follows easily that we can choose $a' < a$, $b' > b$ such that $(a', b') \subset I_{\mathbf{x}}$. Consequently, $I_{\mathbf{x}} = \mathbb{R}$ (else require $I_{\mathbf{x}}$, (a, b) to share an end-point). $\square$

EXERCISES 9.16.14

(1) Fill in the details for the proof of Proposition 9.16.10.
(2) Show that if $f \in C^\infty(\mathbb{R}^m, \mathbb{R}^m)$ there exists a C^∞ map $\chi : \mathbb{R} \to \mathbb{R}(> 0)$ such that χf satisfies (2) of Theorem 9.16.13. Verify that f, χf have the same trajectories—that is, $\Phi_{\mathbf{x}}(I_{\mathbf{x}}) = \Phi_{\mathbf{x}}^\chi(\mathbb{R})$, where $\Phi_{\mathbf{x}} : I_{\mathbf{x}} \to \mathbb{R}$ is the maximal solution curve for $\mathbf{x}' = f(\mathbf{x})$ with initial condition $\mathbf{x}$ and $\Phi_{\mathbf{x}}^\chi$ gives the maximal solution curve for χf. (Hint: For $n \in \mathbb{N}$, define $A_n = 1 + \sup_{n-1 \le \|x\| \le n} \|f(\mathbf{x})\|$ and define $\chi(\mathbf{x}) = \sum_{n=1}^\infty A_n^{-1} \Pi_n(\mathbf{x})$, where the Π_n are suitably chosen smooth functions with compact support.)

9.17 Concluding Comments

(a) The definition of derivative only used the existence of a norm on $\mathbb{R}^m$ and extends to general 'normed vector spaces', including infinite-dimensional spaces. For example, the space $C^0(I)$ of all continuous $\mathbb{R}$-valued functions on the closed interval $I = [0, 1]$ with norm defined by $\|f\| = \sup_{t \in [0,1]} |f(t)|$. If the normed space is complete (for example, $C^0(I)$) then the inverse and implicit function theorems apply [6].
(b) The contraction mapping lemma can be extended significantly. The version we gave involving parameters showed that under mild conditions the fixed point depends continuously on the parameter. This can be generalized to allow for the fixed point to depend differentiably on parameters. It is at this point that the generalizations sketched in (a) come into play and allow direct proofs of the existence and uniqueness theorem for ODEs as well as other foundational results in the theory of differential equations. For more details we refer the reader to the text *Smooth Dynamical Systems* by Irwin [16].
(c) The theory of smooth flows has far reaching generalizations to smooth compact manifolds—for example, the unit sphere in Euclidean space—and leads naturally into the subject of differentiable dynamical systems. From an extensive literature, we suggest John Milnor's monograph on differential topology [25] for a concise introduction to differential manifolds and the texts by Morris Hirsch et al. [14] and Stephen Strogatz [28] for introductions to differentiable dynamical systems.

9.18   Appendix: Finite-Dimensional Normed Vector Spaces

In Sect. 9.2, we showed how starting with the Euclidean vector spaces $\mathbb{R}^m$, $\mathbb{R}^n$ we arrived at a new normed vector space $L(\mathbb{R}^m, \mathbb{R}^n)$ and that the operator norm we defined on $L(\mathbb{R}^m, \mathbb{R}^n)$ was not generally the Euclidean norm obtained via the isomorphism $L(\mathbb{R}^m, \mathbb{R}^n) \approx \mathbb{R}^{mn}$, $[a_{ij}] \mapsto (a_{ij})$. The question arose as to whether we always get the same topology on $L(\mathbb{R}^m, \mathbb{R}^n)$—that is, might the open sets (and continuous functions) depend on the choice of norms on $\mathbb{R}^m$, $\mathbb{R}^n$? In this appendix, we resolve this issue and show that for finite-dimensional vector spaces, all norms define the same topology and hence the same continuous functions.

Theorem 9.18.1 *Any two norms on a finite-dimensional vector space V are equivalent. In particular,*

(1) *All norms define the same topology on V.*
(2) $(V, \| \cdot \|)$ *is complete with respect to any norm on V.*

The key step in the proof of Theorem 9.18.1 is given by the following lemma.

Lemma 9.18.2 *Let $(V, \| \cdot \|)$ be a normed vector space and suppose $L : V \to \mathbb{R}$ is linear. Then L is continuous iff $L^{-1}(0)$ is a closed subspace of V.*

Proof If L is continuous, then $L^{-1}(0)$ is a closed subspace of V by standard metric space theory. The proof of the converse is not so simple. We may assume L is not identically zero. Set $D = D_1(0) \subset V$. Observe that D is *balanced*: $tD \subset D$ for all $t \in [-1, 1]$. Since L is linear, $L(D) \subset \mathbb{R}$ is also balanced. It suffices to prove that if L is not continuous then $L^{-1}(0)$ is not a closed subspace of V. If L is not continuous at $\mathbf{x} = \mathbf{0}$, then $L(D)$ is not a bounded subset of $\mathbb{R}$ (continuity at $\mathbf{x} = \mathbf{0}$ implies there exists an $r > 0$ such that $L(D_r(\mathbf{0})) \subset [-1, 1]$. But then $L(D) \subset [-1/r, 1/r]$). Since $L(D)$ is balanced, we must therefore have $L(D) = \mathbb{R}$. Let $\mathbf{x} \in V$, $\varepsilon > 0$. Since $L(D) = \mathbb{R}$, there exists a $\mathbf{z} \in D$ such that $L(\mathbf{z}) = \varepsilon^{-1}L(\mathbf{x})$. That is $L(\mathbf{x} - \varepsilon\mathbf{z}) = 0$ and so $\mathbf{x} - \varepsilon\mathbf{z} \in L^{-1}(0)$ and $d(\mathbf{x}, L^{-1}(0)) < \varepsilon$. Since this is so for all $\varepsilon > 0$, $\mathbf{x} \in \overline{L^{-1}(0)}$. Our argument proves that $L^{-1}(0)$ is dense in V. Since $L \neq \mathbf{0}$, $L^{-1}(0) \neq V$ and so $L^{-1}(0)$ is not closed.   □

Proof of Theorem 9.18.1. Our proof is by a double induction. For $n \in \mathbb{N}$, let E_n be the statement that all norms on an n-dimensional vector space are equivalent and C_n be the statement that every n-dimensional normed vector space is complete (in the associated metric). The induction depends on showing that: E_1 is true, $E_n \implies C_n$, $C_n \implies E_{n+1}$. We leave the verification of E_1 to the exercises.

$E_n \implies C_n$. Let V be an n-dimensional vector space. We start by noting that if $\| \cdot \|_1$ is equivalent to $\| \cdot \|_2$ then V is complete in the metric defined by $\| \cdot \|_1$ iff V is complete in the metric defined by $\| \cdot \|_2$ (this is easy to check as both metrics have the same Cauchy sequences). Consequently, to verify that $E_n \implies C_n$ it is enough to find one norm on V relative to which V is complete. For this, choose a linear isomorphism $A : V \to \mathbb{R}^n$ and define $\|\mathbf{x}\| = \|A\mathbf{x}\|_2$, where $\| \cdot \|_2$ is the Euclidean norm on $\mathbb{R}^n$.

$C_n \implies E_{n+1}$. Suppose V is an $(n + 1)$-dimensional normed vector space. Fix a basis $\{v_1, \cdots, v_{n+1}\}$ for V and let $A : V \to \mathbb{R}^{n+1}$ be the linear isomorphism defined by $A(\sum_{i=1}^{n+1} x_i v_i) = (x_1, \cdots, x_{n+1})$. Define the norm $\| \cdot \|_\star$ on V by $\|x\|_\star = \|Ax\|_\infty$, where $\|(x_1, \cdots, x_{n+1})\|_\infty = \max_i |x_i|$. Suppose $\| \cdot \|$ is a norm on V. It suffices to prove that $\| \cdot \|$ and $\| \cdot \|_\star$ are equivalent. Denote the components of A by $a_i : V \to \mathbb{R}$, $i = 1, \cdots, n + 1$. Fix i and set $a = a_i$. Since A is a linear isomorphism, $a \neq 0$ and so $E = a^{-1}(0)$ is an n-dimensional linear subspace of V. Take the induced norm on V. By hypothesis C_n, $(E, \| \cdot \|)$ is complete. A subspace of a metric space which is complete in the induced metric contains all its limit points and hence is closed. Therefore, E is a proper closed subset of V and a is continuous by Lemma 9.18.2. Hence

$$\|x\|_\star = \|Ax\| = \|(a_1(x), \cdots, a_{n+1}(x))\|_\infty \leq K\|x\|,$$

where $K = \sup\{|a_i(x)| \mid 1 \leq i \leq n + 1, \|x\| \leq 1\} < \infty$. On the other hand, $A^{-1}(x_1, \cdots, x_{n+1}) = \sum_{i=1}^{n+1} x_i v_i$ and so

$$\|x\| = \| \sum_{i=1}^{n+1} x_i v_i \| \leq \max_i \{\|v_i\|\} \max_i |x_i| = \max_i \{\|v_i\|\} \|x\|_\star.$$

Hence the norms $\| \cdot \|$ and $\| \cdot \|_\star$ are equivalent.   □

Corollary 9.18.3 *If $(V, \| \cdot \|)$ and $(W, \| \cdot \|)$ are finite-dimensional normed vector spaces then every linear map $A : V \to W$ is continuous.*

Proof If we choose bases for V and W we can always assume by Theorem 9.1.4 that the norms on $V \cong \mathbb{R}^m$ and $W \cong \mathbb{R}^n$ are the Euclidean norms. Apply Proposition 9.2.6.   □

We end with a topological characterization of finite-dimensional normed vector spaces.

Theorem 9.18.4 (F. Riesz) *Let $(E, \| \cdot \|)$ be a normed vector space. Then the closed unit disk $\overline{D}_1(0)$ in E is compact iff E is finite-dimensional.*

Proof If E is of finite dimension n, then we may fix an isomorphism $E \cong \mathbb{R}^n$. Let $\| \cdot \|_2$ denote the induced Euclidean norm on E. Now all closed disks $\overline{B}_r(0)$ are compact in $(E, \| \cdot \|_2)$. Since $\| \cdot \|, \| \cdot \|_2$ are equivalent norms, we may choose $R > 0$ such that $\overline{D}_1(0) \subset \overline{B}_R(0)$. Hence $\overline{D}_1(0)$ is compact.

Conversely, suppose that $\overline{D}_1(0)$ is a compact subset of E. By compactness,[2] we may choose a finite subset $f_1, \cdots, f_k \in \overline{D}_1(0)$ such that $d(x, \{f_1, \cdots, f_k\}) < 1$ for all $x \in \overline{D}_1(0)$. Let F be the finite-dimensional subspace of E spanned by $\{f_1, \cdots, f_k\}$. By Theorem 9.1.4(2), $(F, \| \cdot \|)$ is a *closed* normed vector subspace of E. It suffices to show $F = E$. If not, we may choose $x \in E \smallsetminus F$ such that $d(x, F) > 0$. Every closed

---

[2]Either sequential compactness or the open cover definition.

disk with centre $\mathbf{0}$ in F is contained in $\lambda \overline{D}_1(\mathbf{0})$ for some $\lambda > 0$ and so all closed disks with centre $\mathbf{0}$ in F are compact. Since $d(\mathbf{x}, \mathbf{f}) = \|\mathbf{x} - \mathbf{f}\|$ is a continuous function of $\mathbf{f} \in F$, it follows that the lower bound $d(\mathbf{x}, F) = \inf_{\mathbf{f} \in F} \|\mathbf{x} - \mathbf{f}\|$ is attained at some point $\mathbf{y} \in F$. That is, $d(\mathbf{x}, F) = \|\mathbf{x} - \mathbf{y}\|$, where $\mathbf{y} \in F$. Set $\mathbf{z} = (\mathbf{x} - \mathbf{y})/\|\mathbf{x} - \mathbf{y}\| \in \overline{D}_1(\mathbf{0})$. We have

$$
\begin{aligned}
d(\mathbf{z}, F) &= d\left(\frac{\mathbf{x} - \mathbf{y}}{\|\mathbf{x} - \mathbf{y}\|}, F\right) \\
&= \frac{1}{\|\mathbf{x} - \mathbf{y}\|} d(\mathbf{x} - \mathbf{y}, F), \quad \text{scalar invariance of } d \\
&= \frac{1}{\|\mathbf{x} - \mathbf{y}\|} d(\mathbf{x}, F), \quad \text{translation invariance of } d \\
&= 1.
\end{aligned}
$$

But $d(\mathbf{z}, F) \le d(\mathbf{z}, \{\mathbf{f}_1, \cdots, \mathbf{f}_k\}) < 1$. Contradiction. Hence $E = F$ and E is finite-dimensional. $\qquad \square$

EXERCISES 9.18.5

(1) Show that if $\| \cdot \|_1$ is equivalent to $\| \cdot \|_2$ and $\| \cdot \|_2$ is equivalent to $\| \cdot \|_3$ then $\| \cdot \|_1$ is equivalent to $\| \cdot \|_3$.
(2) Prove that all norms on $\mathbb{R}$ are equivalent (statement E_1 of the proof of Theorem 9.1.4).
(3) Show directly that the norms $\| \cdot \|_2$, $\| \cdot \|_1$ and $\| \cdot \|_\infty$ on $\mathbb{R}^n$ are all equivalent. Specifically, prove that for all $x \in \mathbb{R}^n$, we have

$$
\|x\|_\infty \le \|x\|_1 \le n\|x\|_\infty, \qquad \|x\|_\infty \le \|x\|_2 \le \sqrt{n}\|x\|_\infty.
$$

Generalize to all p-norms $\| \cdot \|_p, p \ge 1$.
(4) Define the norm $\| \cdot \|_1$ on $C^0([0, 1])$ by

$$
\|f\|_1 = \int_0^1 |f(t)| \, dt, \ f \in C^0([0, 1]).
$$

(a) Verify that $\| \cdot \|_1$ does define a norm on $C^0([0, 1])$.
(b) By considering the sequence of functions $(f_n) \subset C^0([0, 1])$ defined by

$$
f_n(t) = \begin{cases} n^2 t^2, & 0 \le t \le 1/n, \\ n^3(t - 2/n)^2, & 1/n \le t \le 2/n, \\ 0, & t \ge 2/n, \end{cases}
$$

show that the norm $\| \cdot \|_1$ is not equivalent to the L^2-norm $| \cdot |_2$ on $C^0([0, 1])$ (see Sect. 5.6 for the definition of $| \cdot |_2$).

(c) By considering the sequence $(g_n) \subset C^0([0, 1])$ defined by

$$g_n(t) = \begin{cases} 1, & 0 \leq t \leq 1/2 - 1/n, \\ (n/2 + 1 - nt)/2, & 1/2 - 1/n \leq t \leq 1/2 + 1/n, \\ 0, & t \geq 1/2 + 1/n, \end{cases}$$

show that $(C^0([0, 1]), \| \cdot \|_1)$ is not complete.

References

1. R. Abraham, J.W. Robbin, *Transversal Mappings and Flows* (Benjamin, New York, 1967)
2. M. Barnsley, *Fractals Everywhere* (Academic, New York, 1988)
3. P. Błaszczyk, M.G. Katz, D. Sherry, Ten misconceptions from the history of analysis and their debunking. Found. Sci. **18**(1), 43–74 (2013)
4. A.V. Borovnik, *Mathematics under the Microscope. Notes on Cognitive Aspects of Mathematical Practice* (American Mathematical Society, Providence, RI, 2009)
5. T.J.I'A. Bromwich, *Theory of Infinite Series*, 2nd edn. (Macmillan and Co., London, 1959)
6. J. Dieudonné, *Foundations of Modern Analysis* (Academic, New York, 1960)
7. M. Faà di Bruno, Note sur une nouvelle formule de calcul differentiel. Q. J. Pure Appl. Math. **1**, 359–360 (1857)
8. K. Falconer, *Fractal Geometry: Mathematical Foundations and Applications*, 2nd edn. (Wiley, New York, 2003)
9. M. Field, *Differential Calculus and Its Applications* (Van Nostrand Reinhold, New York, 1976)
10. M.J. Field, Stratification of equivariant varieties. Bull. Aust. Math. Soc. **16**, 279–296 (1977)
11. M. Field, M. Golubitsky, *Symmetry in Chaos: A Search for Pattern in Mathematics, Art and Nature*, 2nd edn. (Society for Industrial and Applied Mathematics, Philadelphia, 2009)
12. A. Fraenkel, *Abstract Set Theory* (North Holland, Amsterdam, 1953)
13. A. Fraenkel, Y. Bar-Hillel, A. Levy, *Foundations of Set Theory* (North Holland, Amsterdam, 1958)
14. M.W. Hirsch, S. Smale, R. Devanney, *Differential Equations, Dynamical Systems, and an Introduction to Chaos*, 3rd edn. (Academic, New York, 2013)
15. J.E. Hutchinson, Fractals and self similarity. Indiana Univ. Math. J. **30**, 713–747 (1981)
16. M.C. Irwin, *Smooth Dynamical Systems*. Advanced Series in Nonlinear Dynamics, vol. 17 (World Scientific, Singapore, 2001). The original book, published by Academic press, appeared in 1980
17. W.P. Johnson, The curious history of Faà di Bruno's formula. Am. Math. Mon. **109**, 217–234 (2002)
18. J.L. Kelley, *General Topology*. Graduate Texts in Mathematics, vol. 27 (Springer, New York, 1975). Originally published 1955, Van Nostrand Reinhold
19. S.G. Krantz, *Real Analysis and Foundations*, 2nd edn. (Chapman and Hall/CRC, Boca Raton, 2004)
20. S.G. Krantz, H.R. Parks, *A Primer of Real Analytic Functions*. Basler Lehrbücher, vol. 4 (Birkhäuser, Basel, 1992)
21. L. Kuipers, H. Niederreiter, *Uniform Distribution of Sequences* (Dover, New York, 2006)
22. J.W. Lamperti, *Probability*, 2nd edn. (Wiley, New York, 1996)

© Springer International Publishing AG 2017

M. Field, *Essential Real Analysis*, Springer Undergraduate Mathematics Series,
https://doi.org/10.1007/978-3-319-67546-6

23. D. Liberzon, *Switching in Systems and Control*. Systems and Control: Foundations and Applications (Birkhäuser, Basel, 2003)
24. P. Mandelbrot, *The Fractal Geometry of Nature* (W.H. Freeman and Co., New York, 1982)
25. J.W. Milnor, *Topology from the Differentiable Viewpoint* (Princeton University Press, Princeton, NJ, 1965)
26. H.-O. Peitgen, P.H. Richter, *The Beauty of Fractals* (Springer, New York, 1988)
27. W. Rudin, *Principles of Mathematical Analysis*, 3rd edn. (McGraw-Hill, New York, 1976)
28. S.H. Strogatz, *Nonlinear Dynamics and Chaos (Studies in Nonlinearity)*, 2nd edn. (Westview Press, Boulder, 2015)
29. A.N. Whitehead, *Science and the Modern World*, Paperback edn. (Macmillan Company, New York, 1925; Cambridge University Press, Cambridge, 2011)
30. S. Willard, *General Topology* (Dover, New York, 2004). Originally Published by Addison-Wesley, Reading, MA, 1970
31. W.H. Young, On the distinction of right and left at points of discontinuity. Q. J. Math. **39**, 67–83 (1908)

Index

$\binom{\alpha}{n}$, 147
$B^0(X, \mathbb{R})$, bounded continuous $\mathbb{R}$-valued
    functions on X, 299
$B(I)$, bounded functions on I, 132
$B(X, \mathbb{R})$, bounded $\mathbb{R}$-valued functions on X,
    249, 299
$B([a, b])$, bounded functions on $[a, b]$, 247
$B_n(f)$, Bernstein polynomial, 170
$C^0(I)$, continuous functions on I, 133
$C^0(X, \mathbb{R})$, continuous $\mathbb{R}$-valued functions on X,
    299
C^1-function, 152
C^∞-function, 68, 161, 423
$C^\omega(\mathbb{R})$, 162
$C_c^p(\mathbb{R}^m, \mathbb{R}^n)$, C^p-functions with compact
    support, 424
C^r-function, 68
$\overline{C}(r)$, closed hypercube, 425
$\Delta(X)$, 257
$\eta_{r_1, \dots, r_m}, \eta_r$, 424
e^x, 83
$\exp(x)$, 83
Γ-function, 211
$\Gamma(x)$, Gamma function, 211
$\mathcal{H}(\mathbb{R}^n)$, space of compact subsets of $\mathbb{R}^n$, 330
L^2-distance, 206
L^2-metric, 247
L^2-norm, 205
$L(\mathbb{R}^m, \mathbb{R}^n)$, 355
$\ln x$, 82
$\log x$, 82
$L_s^p(V; W)$, 401
$\|\cdot\|_p$, 352
1:1 function, 7
$P(\mathbb{R})$, polynomial functions, 161

$P^d(V, W)$, 403
$P^{(d)}(V, W)$, 402
$\Psi_{a,b}$, 165
S^m, unit sphere in $\mathbb{R}^{m+1}$, 351
$\star$ symmetrization operator, 417
$\Theta_{a,b}$, 166
$X \smallsetminus A$, complement, 4

absolute convergence, 113
accumulation point, 268
affine linear map, 339, 358
analytic
    continuation, 183
        maximal, 183
    function, 162, 178
        real analytic, 162
    functions
        composite, 181
        product, 181
        zeros, 182
anti-derivative, 75
Archimedean property, 20, 88
arithmetic mean, 47
Arzelà–Ascoli theorem, 307
Axiom of Choice, 5

balanced, 439
ball, 252
basis (for topology), 262
Bernoulli
    numbers, 223
    polynomials, 223, 225
Bernstein polynomials, 170
beta function, 222

big O notation, 361
bijection, 7
bilinear, 397
binomial series, 147, 185
Bolzano–Weierstrass theorem, 48
Borel's theorem, 164, 169
boundary of set, 262
bounded, 37
   above, 37
   below, 37
   function, 132
   variation, 63
bump function, 165, 424

Cantor, 8
   set, 289, 329
Cantor–Bernstein theorem, 13
cardinality of set, 10
Cauchy sequence, 55, 297
   complex numbers, 66
   functions, 136
Cauchy's integral test, 96
Cauchy's test, 96
Cauchy–Schwarz inequality, 246
ceiling function, 62
chain rule, 365
   C^r-maps, 413
closed
   disk, 256
   subset, 255
closure, 258
cluster point, 269
collage theorem, 345
compact support, 423
comparison test, 93
complement of set, 4
complete metric space, 298
completion, 301
complex conjugation, 64
complex number, 63
   addition, 63
   modulus, 64
   multiplication, 64
composite
   mapping formula, 365
   of functions, 7
connected, 321
   components, 326
   subset, 321
constant sequence, 266
continuity
   at a point, 273
   metric spaces, 273

continuous
   family, 176
   function, 49
continuously differentiable, 68, 152
   C^p, 68, 411
contraction, 337
   constant, 312
   mapping, 312
   mapping lemma, 312
      with parameters, 314
convergence
   infinite series, 91
convex function, 215
convexity of log Γ, 215
convolution, 428
countable set, 10
countably infinite, 10

D'Alembert's test, 95
De Moivre's formula, 65
decimal expansion, 15
Dedekind numbers, 6
definite integral, 77
dense subset, 261
derivative, 67, 360
diagonal subspace, 257
diameter, 249
diffeomorphism, 376
   C^r, 413
differentiable function, 360
differentiation under integral sign, 213
directional derivative, 363
Dirichlet kernel, 193
disconnected, 321
discrete metric, 246
disk, 252
distance to a subset, 250
double series, 112
   absolute convergence, 113
dual implicit function theorem, 383, 414
duplication formula of Legendre, 223

embedding, 390
empty set, 2
equation of variations, 432
equicontinuous, 306
equivalence
   of sets, 8
   relation, 8
equivalent
   metrics, 249, 350
   norms, 350

Euclidean
    metric, 246
    norm, 351
Euler
    constant, 99, 220
    constant, computation, 238
    method for ODEs, 310
    product for zeta-function, 124
Euler's theorem, 370
Euler–Maclaurin formula, 231
    $r = 0$, 232
eventually
    increasing sequence, 41
    periodic decimal, 16
exponential function, 83

Faà di Bruno's formula
    history, 422
family
    of sets, 3
finite set, 9
first category, 303
fixed point, 52, 296, 312
floor function, 62
flow, 435
fnamily
    of functions, 176
formal power series, 421
Fourier
    coefficients, 188
    series, 188
        partial sum formula, 193
fractal, 294, 338
frontier of set, 262
function, 6
    analytic, 162, 178
    bijective, 7
    bounded
        continuous real-valued on X, 299
        on $[a, b]$, 247
        on I, 132
        real-valued on X, 249, 299
    bump, 165, 424
    C^1, 152
    C^∞, 68, 161, 423
    continuous, 49
        bounded real-valued on X, 299
        on I, 133
        real-valued on X, 299
    C^p with compact support, 424
    C^r, 68
    Gamma, 211
    injective, 7

    nowhere differentiable, 49, 156
        Weierstrass, 156
    1:1, 7
    1:1 onto, 7
    onto, 7
    polynomial, 161
    real analytic, 162
    smooth, 423
    surjective, 7
    tabletop, 166
fundamental theorem of calculus, 77

Gamma-function, 211
    properties, 215
general
    linear group, 376
    principle of uniform convergence, 136
generalized binomial coefficient, 147
geometric mean, 47
Gibbs phenomenon, 197
graph of function, 6
greatest lower bound, 38

Hausdorff metric, 330, 334
Heine–Borel theorem, 286
Hölder
    continuity, 306
    inequality, 353
homeomorphism, 276
homogeneous polynomial of degree d, 402
hypercube, 425

iff, 5
IFS, 338
image of function, 7
implicit function theorem, 380, 414
    dual version, 383
improper integral, 212
induced metric, 247
inf, 38
infimum, 38
infinite
    product, 116
        for $\sin x$, 121, 198
        general principle of convergence, 119
    series
        Cauchy's test, 96
        comparison test, 93
        D'Alembert's test, 95
        integral test, 96
        necessary condition for convergence, 92

positive terms, 92
ratio test, 94
set, 9
infinitely differentiable, 411
initial condition, 316, 391
injective function, 7
inner product, 351
of functions, 204
integral curve, 316
interior, 258
point, 259
intersection (of sets), 3
inverse
function theorem, 377, 414
on $\mathbb{R}$, 318
image (of set), 7
map, 7
invertible linear maps, 376
irrational number, 14, 18
isolated point, 257, 268
isometry, 275, 301
iterated function system, 338

Jacobian, 364
matrix, 364
jump discontinuity, 60, 187

least upper bound, 38
Legendre polynomials, 208
Leibniz rule, 417
lim inf, 59
limit
of sequence, 32
point, 268
of sequence, 269
lim sup, 59
linear map, 353
Lipschitz, 306, 312
inverse function theorem, 321
local
diffeomorphism, 377
flow, 392
locally compact, 296
logarithm
Napierian, 82
natural, 82
logistic map, 47
lower semi-continuous, 62

M-test, 141
map, 6

maximal solution curve, 436
mean square convergence, 204
mean value theorem, 67
for integrals, 80
metric, 246
space, 246
completion, 301
topology, 254
middle-thirds Cantor set, 289
Minkowski's inequality, 352
modulus and argument, 65
monomial, 404
of degree d, 404
multi-index notation, 404
multi-linear, 397

Napierian logarithm, 82
natural
isomorphism, 399
logarithm, 82
neighbourhood, 263
non-analytic smooth function, 162
norm, 350
of a linear map, 355
p-norm, 352
normed vector space, 350
nowhere dense, 265, 303
nowhere differentiable function, 49, 156
Weierstrass, 156

ODE, 391
one-parameter group property, 436
one-sided limits, 59
onto function, 7
open
cover, 285
disk, 252
neighbourhood, 263
subset, 252
operator norm, 355, 396
ordinary differential equation, 316, 391
integral curve, 316
solution curve, 316
orthogonal functions, 205

partial
derivative, 364
sum, 91
path-connected, 324
perfect set, 294
periodic function, 186

p-fold tensor product, 398
piecewise continuous function, 187
p-linear map, 397
p-norm, 352, 425
pointwise
    bounded set, 305
    convergence, 130
    limit, 130
polarization lemma, 405
polynomial
    Bernstein, 170
    function, 161
    of degree d, 403
power series, 142
    product, 144
    radius of convergence, 143
    reciprocal, 145
    sum, 144
power set, 4, 8
product
    metric, 248, 327
    norm, 397, 400
    of power series, 144
    of sets, 5
proper, 288
    map, 390
    subset, 3
Pythagoras' theorem, 206

radius of convergence, 143
random iteration, 345
range of function, 7
rank theorem, 385, 414
ratio test, 94
rational number, 13
real analytic function, 162
reciprocal of power series, 145
rectangle, 81
removable discontinuity, 60
repeated series, 114
Riemann integral, 77
Riesz's theorem, 440
Rolle's theorem, 67
Russell's paradox, 5, 8

scalar invariance of metric, 352
second
    category, 303
    countable, 262, 285
    derivative, 407
    Weierstrass approximation theorem, 187
self-similar, 294

self-similarity, 338
semi-continuity, 62
semi-norm, 425
separable, 261, 285
separated sets, 325
sequence, 32
    complex numbers, 66
    convergence, 32
    increasing, 41
    limit, 32
    metric space, 266
    metric space convergence, 267
sequences diverging to $\pm\infty$, 40
sequential continuity, 50, 279
sequentially compact, 281
Sierpiński triangle, 340, 341
small o notation, 359
smooth
    flow, 435
    function, 68, 161, 423
    non-analytic function, 162
solution of differential equation, 391
space of linear maps, 355
square wave, 197
squeezing lemma, 34
Stirling's
    formula, 232, 237
    series, 238
subcover, 285
submanifold, 390
subsequence, 35
subset, 2
subspace
    isolated point, 268
sum
    by columns, 114
    by rows, 114
    of power series, 144
sup, 38
superset, 3
support (of function), 423
supremum, 38
surjective function, 7
symmetric
    p-linear map, 401
    group, 401
symmetrization, 403

tabletop function, 166, 424
tangent
    line, 358
    plane, 358
Tannery's theorem, 101, 120

Taylor series, 69, 161
   analytic function, 178
Taylor's theorem, 68, 414
    Cauchy remainder, 70
    integral remainder, 68, 416
    Lagrange remainder, 70
    remainder estimate, 70
terminating decimal, 15
ternary expansion, 295
Tietze extension theorem, 278
topological space, 254
topology, 254
totally disconnected, 294, 322
trajectory of differential equation, 316
translation invariance of metric, 352
triangle inequality, 89, 133, 246
trigonometric
    polynomial, 186
      degree, 186
    series, 187
twice differentiable, 407

uncountable, 10
uniform
    approximation, 131, 170, 426
      C^r-functions, 423
    convergence, 134
      Abel test, 147

Dirichlet test, 147
    series, 139
    metric, 246, 249, 299, 300, 305, 392
    norm, 132
uniformly
    bounded set, 306
    distributed, 203
union (of sets), 3
unit sphere, 351
upper semi-continuous, 62
Urysohn's lemma, 277

Wallis' formula, 200, 203
Weierstrass
    approximation theorem, 170, 174, 261,
        423, 430
      continuous families, 177
      trigonometric polynomials, 187
    inequalities, 118, 123
    nowhere differentiable function, 156
Weyl criterion, 203

Young's theorem, 62, 276

Zariski topology, 258

Printed in the United States
By Bookmasters